Food Exploitation by Social Insects

Ecological, Behavioral, and Theoretical Approaches

CONTEMPORARY TOPICS in ENTOMOLOGY SERIES

THOMAS A. MILLER Editor

Insect Symbiosis
Edited by Kostas Bourtzis and Thomas A. Miller

Insect Sounds and Communication: Physiology, Behaviour, Ecology, and Evolution
Edited by Sakis Drosopoulos and Michael F. Claridge

Insect Symbiosis, Volume 2
Edited by Kostas Bourtzis and Thomas A. Miller

Insect Symbiosis, Volume 3
Edited by Kostas Bourtzis and Thomas A. Miller

Food Exploitation by Social Insects: Ecological, Behavioral, and Theoretical Approaches
Edited by Stefan Jarau and Michael Hrncir

CONTEMPORARY TOPICS in ENTOMOLOGY SERIES

THOMAS A. MILLER Editor

Food Exploitation by Social Insects

Ecological, Behavioral, and Theoretical Approaches

Edited by

Stefan Jarau • Michael Hrncir

CRC Press
Taylor & Francis Group
Boca Raton London New York

CRC Press is an imprint of the
Taylor & Francis Group, an **informa** business

CRC Press
Taylor & Francis Group
6000 Broken Sound Parkway NW, Suite 300
Boca Raton, FL 33487-2742

First issued in paperback 2019

ISBN-13: 978-1-4200-7560-1 (hbk)
ISBN-13: 978-0-367-38561-3 (pbk)

Library of Congress Cataloging-in-Publication Data

Food exploitation by social insects : ecological, behavioral, and theoretical approaches / editors: Stefan Jarau, Michael Hrncir.
    p. cm. -- (Contemporary topics in entomology series)
   Includes bibliographical references and index.
   ISBN 978-1-4200-7560-1 (hardcover : alk. paper)
   1. Insect societies. 2. Insects--Food. 3. Animal communication. I. Jarau, Stefan. II. Hrncir, Michael. III. Title. IV. Series.

  QL496.F66 2009
  595.7'153--dc22
                                                  2009009972

**Visit the Taylor & Francis Web site at
http://www.taylorandfrancis.com**

**and the CRC Press Web site at
http://www.crcpress.com**

Contents

Part III: Modeling Social Insect Foraging

Preface

Social insects are well known to most, if not all, people. This is for good reason. They are omnipresent in virtually all terrestrial ecosystems, only the extreme arctic regions being excluded. Probably everybody has had his or her experiences with one species or the other—nice ones, like listening to the comforting humming of a bumble bee passing by while one is lying in a spring meadow enjoying some warming sunbeams, or bad ones, like the sting of a honey bee or wasp. Furthermore, in addition to the undisputable ecological importance of social insects, many species came into interest for economical reasons, either as beneficial animals or as pests. The most prominent examples here may be the commercialization of products harvested from honey bee hives, the increase of tomato harvest through pollination by bumble bees, or the damage caused to crops by certain ant species and the impact of invasive species (e.g., red imported fire ants and Africanized honey bees) on native species. It is, therefore, not surprising that the study of social insects has attracted a vast number of researchers. As a consequence, a huge amount of information about the many aspects of their biology and ecology has accumulated over the last centuries. One of these aspects—the one we both got hooked on—is how social insects find food sources and how they efficiently exploit them for the benefit of the entire colony. A symposium on this topic that we had organized at the 3rd European Congress on Social Insects, held in St. Petersburg, Russia in August 2005, not only showed us how much brand new and exciting information is available but also made the lack of a book that combines this information quite obvious. It was the hour of birth of the idea for the volume in hand. We were aware from the beginning that writing such a book could hardly be done by two authors alone, and an edited volume was the choice. It was also clear to us that the book can hardly contain *all* information about, for example, which search strategies the various kinds of social insects apply while looking for food, which and how patterns of food exploitation emerge, how foragers of the different species recruit nestmates and communicate with each other, which information they use, or how models of social insect foraging help us to understand the way it works. The compilation of all this information would certainly result in an encyclopedia rather than a single book! Topics had to be chosen, therefore, and we omitted those that are well represented by other books and book chapters or by recent review articles (see Introduction for more details). Still other topics are not covered by the book, unfortunately, simply due to the lack of authors available at the time of its preparation.

The book contains sixteen chapters that cover a wide range of aspects related to social insect foraging, including the most recent scientific developments and studies that, in some cases, changed our view about outmoded, yet still taught, "wisdom." In particular, we asked the authors for a fresh look on the topics they had written. Our sincere thanks go to all the colleagues who accepted our invitation to contribute to this book and prepared their chapters—doubtless a time-consuming and sometimes nerve-racking task. We also thank John Sulzycki and Pat Roberson from CRC Press of Taylor and Francis Group for their assistance during the editing process as well as for their patience and understanding for repeatedly extended deadlines.

We hope that this volume will serve as a useful source of information to many people who are interested in—or even work with—social insects and that it will be an incentive to study the many puzzling questions that still remain in regard to food exploitation by social insects.

Stefan Jarau and Michael Hrncir
Heredia, Costa Rica and Ribeirão Preto, Brazil

The Editors

Stefan Jarau is currently a scientific employee at the University of Ulm, Germany. He obtained his PhD in 2003 from the University of Vienna, Austria, where he studied biology with a focus on zoology. Since 1996 his research has focused on the recruitment behavior and social nest organization in tropical stingless bees, with a particular emphasis on communication mediated by pheromones and other chemical volatiles. His field projects have repeatedly taken him to Brazil and Costa Rica, where he cooperates with colleagues from the University of São Paulo at Ribeirão, Preto and the Center for Tropical Bee Research at the National University of Costa Rica in Heredia, respectively. In 1996 he started to work with Michael Hrncir, a collaboration that continued for almost a decade and led to a warm and continuing friendship. In 2005 they organized a symposium on the topic of the book in hand at the 3rd European Congress on Social Insects in St. Petersburg, Russia, which was met with great interest by the participants and ultimately led to the idea to compile this volume.

Michael Hrncir started his academic career at the University of Vienna, Austria, studying film and theatre. After one utterly wasted year, he decided to switch to something more meaningful (and less boring), a decision that guided him—to make a long story short—into the world of bees (somewhere along the way there were kiwis—the birds, not the fruit—and spiders). In 1996 he set out for Brazil, together with Stefan Jarau, with the mission to unravel the communication of stingless bees—a sheer impossible task, as both soon found out. And even today, more than a decade later, this mission is still far from being accomplished. After a diploma thesis (1998) as well as a doctoral thesis (2003) on the potential message of the "sounds" produced by stingless bee foragers, Michael Hrncir strengthened his passion for these animals during a postdoctoral fellowship from the University of Vienna (2004–2006), during which he started to investigate the physiological background and sensory mechanisms underlying the communication of bees. Currently, he is a Young Scientist Research Fellow at the University of São Paulo in Ribeirão Preto, Brazil, where he studies "how bees find a flower." The present book has resulted from his ongoing and firm friendship with Stefan Jarau, which was borne out of a coincidentally shared scientific interest, and his fascination for the world of social insects.

Contributors

Takaya Arita
Graduate School of Information Science
Nagoya University
Nagoya, Japan
arita@nagoya-u.jp

Madeleine Beekman
Behaviour and Genetics
  of Social Insects Laboratory
School of Biological Sciences
University of Sydney
Sydney, Australia
mbeekman@bio.usyd.edu.au

Martin Bollazzi
Behavioral Physiology and Sociobiology,
  Biocenter
University of Würzburg
Würzburg, Germany
bollazzi@biozentrum.uni-wuerzburg.de

Allen Cheung
Queensland Brain Institute
and
School of Information Technology
  and Electrical Engineering
University of Queensland
Brisbane, Australia
a.cheung@uq.edu.au

Lars Chittka
Research Centre for Psychology
School of Biological and Chemical Sciences
Queen Mary, University of London
London, United Kingdom
l.chittka@qmul.ac.uk

Jean-Louis Deneubourg
Unit of Social Ecology (CP. 231)
Université Libre de Bruxelles
Brussels, Belgium
jldeneub@ulb.ac.be

Claire Detrain
Unit of Social Ecology (CP.231)
Université Libre de Bruxelles
Brussels, Belgium
cdetrain@ulb.ac.be

Audrey Dussutour
School of Biological Sciences
University of Sydney
Sydney, Australia
adussutour@usyd.edu.au

Walter M. Farina
Grupo de Estudio de Insectos Sociales
Departamento de Biodiversidad y
  Biología Experimental
Faculdad de Ciencias Exactas y Naturales
Universidad de Buenos Aires
Buenos Aires, Argentina
walter@fbmc.fcen.uba.ar

Dave Goulson
School of Biological and
  Environmental Sciences
University of Stirling
Stirling, United Kingdom
dave.goulson@stir.ac.uk

Christoph Grüter
Division of Behavioral Ecology
Ethologische Station Hasli
University of Bern
Hinterkappelen, Switzerland
christoph.grueter@esh.unibe.ch

Current address:
Department of Biology and
  Environmental Science
University of Sussex
Brighton, United Kingdom
cg213@sussex.ac.uk

Michael Hrncir
Department of Biology, FFCLRP
University of São Paulo
Ribeirão Preto-SP, Brazil
michael.hrncir@gmx.at

William O. H. Hughes
Institute of Integrative
  and Comparative Biology
Faculty of Biological Sciences
University of Leeds
Leeds, United Kingdom
w.o.h.hughes@leeds.ac.uk

Stefan Jarau
Institute for Experimental Ecology
University of Ulm
Ulm, Germany
stefan.jarau@uni-ulm.de

Robert L. Jeanne
Department of Entomology
University of Wisconsin
Madison, Wisconsin
jeanne@entomology.wisc.edu

Ellouise Leadbeater
Research Centre for Psychology
School of Biological and Chemical Sciences
Queen Mary, University of London
London, United Kingdom

Current address:
Department of Biology and
  Environmental Science
University of Sussex
Brighton, United Kingdom
e.leadbeater@sussex.ac.uk

Flávia N. S. Medeiros
Departamento de Zoologia
Instituto de Biologia
Universidade Estadual de Campinas
Campinas-SP, Brazil
fnatercia@yahoo.com

Yoshiyuki Nakamichi
Information Technology Center
Numazu National College of Technology
Shizuoka, Japan
nakamiti@emergence.jp

Paulo S. Oliveira
Departamento de Zoologia
Instituto de Biologia
Universidade Estadual de Campinas
Campinas-SP, Brazil
pso@unicamp.br

Juliet L. Osborne
Department of Plant and Invertebrate Ecology
Rothamsted Research
Harpenden, Hertfordshire, United Kingdom
juliet.osborne@bbsrc.ac.uk

Nigel E. Raine
Research Centre for Psychology
School of Biological and Chemical Sciences
Queen Mary, University of London
London, United Kingdom
n.e.raine@qmul.ac.uk

Judith Reinhard
Queensland Brain Institute
University of Queensland
Brisbane, Australia
j.reinhard@uq.edu.au

Flavio Roces
Behavioral Physiology
  and Sociobiology, Biocenter
University of Würzburg
Würzburg, Germany
roces@biozentrum.uni-wuerzburg.de

E. Judith Slaa
Research Office
Faculty of Medicine and Health
University of Leeds
Leeds, United Kingdom
e.j.slaa@leeds.ac.uk

Mandyam V. Srinivasan
Queensland Brain Institute
and
School of Information Technology
  and Electrical Engineering
University of Queensland
Brisbane, Australia
m.srinivasan@uq.edu.au

Benjamin J. Taylor
Department of Zoology
University of Wisconsin
Madison, Wisconsin
bjtaylor1@wisc.edu

Introduction

Michael Hrncir and Stefan Jarau

Probably everyone who has set out for a relaxing picnic on a sunny afternoon has gotten into the annoying situation that, all of a sudden, he or she had to compete for the favorite salami (or something similarly delicious) with several dozens of ants, which apparently turned up out of nowhere. Whereas many people, for sure, are merely irritated by this, others might marvel at the agility and speed with which these tiny animals find and collect food. In fact, the fascination and knowledge about the foraging efficiency of eusocial insects, particularly of ants, dates back several thousand years and has, for instance, found its way into Aesop's fables and the Old Testament: "Go to the ant, thou sluggard! Consider her ways, and be wise: which, having no guide, overseer, or ruler, provideth her meat in the summer, and gathereth her food in the harvest" (Book of Proverbs 6:6–8).

Nowadays, the industriousness of ants no longer serves as a mandatory example of how to behave in human society. However, the efficiency with which eusocial insects exploit food sources has by no means lost its fascination. Far from it! More and more researchers are intrigued by the highly structured organization of foraging processes and by the way a social insect society is able to solve a variety of rather complex tasks in spite of the "simplicity" of its members and of the rules they follow. The way by which these animals share and collectively process information, thereby enabling a colony to function as a single purposeful unit, is not solely of interest to biologists. During the last decade, the functional principles of social insect colonies have greatly inspired the design of intelligent artificial systems that find application in communication networks and robotics (e.g., Bonabeau et al. 1999; Dorigo and Stützle 2004).

This recent development called for a sound biological basis, which consequently resulted in a constant and rapid increase in information about task organization and information management in social insect societies. Here, the exploitation of food is probably one of the best-studied examples. Yet, in consequence of the steadily increasing knowledge about food exploitation by social insects, one can easily get lost in the constantly growing network of scientific literature. The present book aims at providing a comprehensive overview of the most recent insights into social insect foraging by combining ecological, behavioral, and theoretical studies. This offers the possibility to take a look at the fascinating aspects of social insect colony organization and coordination from different points of view. For obvious reasons, however, a single book cannot contain *all* information available about social insect foraging. We therefore excluded topics that we think have already extensively been covered by excellent books or review articles, such as, for example, the dance communication of the honey bee,* *Apis mellifera* (von Frisch 1965, 1967/1993; Seeley 1995; Dyer 2002). For readers interested in social insect foraging, we highly recommend the following publications:

---

* In our book, we will follow the "bee spelling rule" defined by Snodgrass (1956) as follows: "If the insect is what its name implies, write the two words separately; otherwise run them together. Thus we have such names as house fly, blow fly, and robber fly contrasted with dragonfly, caddicefly, and butterfly, because the latter are not flies, just as an aphislion is not a lion and a silverfish is not a fish. The honey bee is an insect and is pre-eminently a bee." Accordingly, we adopted the terms *honey bee* and *bumble bee* for our book instead of the frequently used *honeybee* and *bumblebee*.

For ants:

The Ants by Bert Hölldobler and Edward Wilson (1990)

Army Ants: The Biology of Social Predation by William Gotwald (1995)

Ants at Work: How an Insect Society Is Organized by Deborah Gordon (1999)

For bees:

The Social Behavior of the Bees: A Comparative Study by Charles Michener (1974)

Ecology and Natural History of Tropical Bees by David Roubik (1989)

The Wisdom of the Hive—The Social Physiology of Honey Bee Colonies by Thomas Seeley (1995)

"The Biology of the Dance Language" by Fred Dyer (2002)

Bumblebees—Their Behaviour and Ecology by Dave Goulson (2003)

Asian Honey Bees: Biology, Conservation, and Human Interactions by Benjamin Oldroyd and Siriwat Wongsiri (2006)

For wasps:

The Social Biology of Wasps edited by Kenneth Ross and Robert Matthews (1991)

"Social Wasp (Hymenoptera: Vespidae) Foraging Behavior" by Monica Raveret Richter (2000)

For termites:

Nourishment and Evolution in Insect Societies edited by James Hunt and Christine Nalepa (1994)

Termites: Evolution, Sociality, Symbioses, Ecology edited by Takuya Abe, David Bignell, and Masahiko Higashi (2000)

For self-organization processes in various taxa:

"Self-Organization in Social Insects" by Eric Bonabeau, Guy Theraulaz, Jean-Louis Deneubourg, Serge Aron, and Scott Camazine (1997)

Information Processing in Social Insects edited by Claire Detrain, Jean-Louis Deneubourg, and Jaques Pasteels (1997)

Self-Organization in Biological Systems by Scott Camazine, Jean-Louis Deneubourg, Nigel Franks, James Sneyd, Guy Theraulaz, and Eric Bonabeau (2003)

"The Principles of Collective Animal Behaviour" by David Sumpter (2006)

Apart from the eusocial insects covered by these books and articles, and also by the volume at hand, there are several species that do not have distinct female castes and a strict division of reproduction and labor but still exhibit some sort of social behavior—including the use of food sources. A marvelous and extensive summary of the biology of these insects is provided by a book written by James Costa (2006), *The Other Insect Societies*.

The chapters brought together in our book basically look at two different levels of social insect foraging, which are, however, inseparably interlaced with each other. The first is the macroscopic or colony level. Like a solitary animal roaming for food, the colony adapts its foraging decisions and strategies to the respective environmental situation and its own physiological condition in order to optimize its food intake. The second level is the microscopic or individual level. Foraging decisions by individual workers are largely based on information from other individuals and/or the local environment, but they are also influenced by innate preferences. At this level the processing of relevant information is crucial for optimizing the colony's food intake. Accordingly, the first part of the book (Chapters 1–5) focuses on *foraging decisions, patterns, and strategies* of social insect colonies, and the second part (Chapters 6–14) focuses on *information use and information transfer* by and between workers of social insects. Whereas these two sections contain information that is largely based on empirical studies, the third part of the book (Chapters 15 and 16) provides two examples of how this biological knowledge can be used as a basis for the construction of *mathematical and neural network models* that, in return, may help us to understand social insect foraging.

In Chapter 1, Nigel Raine and Lars Chittka address the question of how a particular foraging behavior contributes to the fitness of a social insect colony. Using a phylogenetic approach, the authors compare innate color preferences between different bumble bee species, and between different populations within a single species, respectively. If foraging preferences were adapted to the available food sources, local variations in flower traits (e.g., floral colors) should have driven selection of the bees' sensory bias toward particular colors. In addition, the authors evaluate how differences in learning performance among colonies of one bumble bee species affect their fitness. In order to examine to which extent a specific foraging trait is adapted to a given niche, the foraging success of an experimentally manipulated honey bee behavioral phenotype (directional information component of the dance language removed) is compared to that of the normal foraging phenotype.

In Chapter 2, Claire Detrain and Jean-Louis Deneubourg investigate in which way social cues influence the foraging patterns of colonies of trail-laying ants. Since in these species the efficiency of recruitment does not only depend on the behavior of the recruiting foragers, but also on the amount of potential recruits located within the trail's active space, the foraging dynamics should be particularly sensitive to the density of nestmates. Foraging individuals, therefore, should adjust their recruiting behavior and communication according to the social environment, which they can access via direct (contact) as well as indirect (area marking) cues from nestmates, but also from foreign ants.

In Chapter 3, Robert Jeanne and Benjamin Taylor explore how social foraging patterns can emerge in eusocial wasps, which apparently have not evolved specific signals to inform nestmates about the presence of valuable food sources. The foragers have to be highly sensitive to social cues, both in the environment and at the nest, as well as to the actual food availability within their foraging area in order to optimize the colony's food intake.

In Chapter 4, Flávia Medeiros and Paulo Oliveira illustrate the effects of seasonal changes of the environment on the foraging pattern of the Neotropical forest-dwelling ant *Pachycondyla striata*. This ant species shows a flexible foraging behavior that comprises a wide array of prey hunting techniques as well as the collection of plant material. It is, therefore, a suitable study object for improving our understanding of how ecological factors influence the foraging patterns and strategies of social insects.

In Chapter 5, Dave Goulson and Juliet Osborne evaluate different approaches to estimate the foraging ranges of bumble bees, comparing marking and homing experiments, theoretical calculations of possible flight distances, radar tracking, analyses of pollen collected from incoming foragers, and studies using molecular methods. The authors further discuss possible reasons for the observed differences in foraging ranges between different bumble bee species.

In Chapter 6, Madeleine Beekman and Audrey Dussutour investigate the costs and benefits of different recruitment mechanisms, how they allow insect colonies to adapt to changing conditions, and how social insects overcome constraints imposed by their ways of communication. After giving an overview of the different ways in which social insects recruit nestmates to food sources, the chapter mainly focuses on a comparison of two extreme and well-studied examples: mass recruitment via pheromone trails, where individuals only interact indirectly through the trail, and the honey bees' dance communication, where information is transferred directly from individual to individual.

In Chapter 7, Ellouise Leadbeater and Lars Chittka evaluate under which circumstances—and how—social learning and social information transmission ("teaching") add to the successful exploitation of food sources by foraging insects. By reviewing studies dealing with a variety of species, they reveal that, in spite of the high effectiveness of such social information systems, the underlying mechanisms are surprisingly simple.

A further and important aspect of social information use is discussed by Judith Slaa and William Hughes in Chapter 8, in which they outline the use or even exploitation of passively provided cues and of actively emitted signals from nestmates or non-nestmates, respectively. Although the role of local enhancement and local inhibition (use of social cues) as well as of eavesdropping (use of another colony's or species' signals) is relatively understudied in comparison to active recruitment systems, it is quite clear that these mechanisms considerably contribute to the foraging ecology of social insects.

In Chapter 9, Judith Reinhard and Mandyam Srinivasan investigate the role of scents in food collection processes of honey bees. While these bees employ a highly complex referential communication system to recruit nestmates to food sources, the dance "language," scents are important factors for the foraging and recruitment success of honey bee colonies. After a brief description of the morphology and neuroanatomy of the honey bees' sense of smell—the necessary basis to understand how bees learn odors and how they use them during foraging and recruitment—the authors explore the honey bees' world of odors, which are either of floral origin or emitted by other workers.

Inside the hive, honey bee workers learn the food odors while receiving nectar samples from foragers. The role of these trophallactic interactions for the exchange of chemosensory information between members of a colony is the topic of the contribution by Walter Farina and Christoph Grüter. In Chapter 10, they discuss which kind of information, in addition to the food's scent and taste, is transferred from foragers to receiver bees, and from the receivers on to other colony members—thereby creating information networks that connect the different groups of workers.

Far less studied than the honey bees are their mainly tropical relatives, the stingless bees. In Chapter 11, Michael Hrncir explores the potential messages and meanings of mechanical signals (jostling runs and pulsed thoracic vibrations) generated by stingless bee foragers during food exploitation processes. Due to the high competition over resources in the tropics, the strategy of these bees should aim at a quick increase in collecting foragers at valuable food sources once these become available. And indeed, empirical evidence indicates that the nest-internal behaviors by collecting bees mobilize additional foragers rather than inform recruits about the location of a food patch, as is known from honey bees.

In Chapter 12, Stefan Jarau reviews how odors, which originate from a variety of sources, are used by stingless bees during food exploitation processes. In addition to food odors, olfactory cues left by nestmates at food sources influence the decision of searching foragers on where to collect. Furthermore, several species employ pheromone marks at and near the food source, or even short pheromone trails, to guide recruits toward a specific food patch, and foragers of some stingless bees emit allomones to disrupt an organized defense by workers of nests they raid.

In Chapter 13, Dave Goulson shows how nectar foraging bees make use of olfactory cues left behind by previous visitors to decide whether to probe or reject a flower. Mainly focusing on bumble bees, he looks at the chemical compounds of scent marks, potential mechanisms that underlie the bees' behavior, and at its biological significance.

In Chapter 14, Flavio Roces and Martin Bollazzi describe the trade-offs underlying the organization of collective foraging processes in grass- and leaf-cutting ants by discussing different hypotheses why foragers of these ants cut leaf fragments below the size that would maximize the individual's delivery rate. After exploring the criteria used by workers during the selection of fragment sizes, they look at the adaptive value of cooperative vegetation delivery by means of transport chains, which might either serve to maximize the colony's leaf intake or improve the information transfer between workers.

In Chapter 15, Yoshiyuki Nakamichi and Takaya Arita present a model for the evolution of pheromone communication in ants based on computer simulations, in which neural networks of the agents evolve in accordance with their foraging success. The authors address the question whether this form of evolved pheromone communication is adaptive, and look at how the number of pheromones used by the agents influences their foraging efficiency. They end their chapter with a comparison of the effectiveness of the evolved pheromone communication with a human-designed one.

A critical aspect of social insect foraging is navigation. In Chapter 16, Allen Cheung provides an overview of various mathematical and neural network models for medium-range navigation in foraging insects. He compares different strategies to search for a resource without prior knowledge of its location (a new food source) and the principles of searching for a unique, stable target (like the nest), including path integration, map-based navigation (piloting), and grid-based navigation (view-based homing). Since social insect foragers may obtain useful navigational information through interactions with other individuals, particular emphasis is given to the effects of such social interactions on the navigation behavior of food collectors.

In a final chapter, we present some concluding thoughts, which basically arose from working through the single contributions of the book at hand.

We are aware that this volume certainly cannot answer all questions on food exploitation processes in social insects. However, we could hope for nothing better than that the effort made by all authors who contributed to this work will stimulate others to continue the exploration of this fascinating aspect of our and the insects' world.

REFERENCES

Abe T, Bignell DE, Higashi M (eds.). (2000). *Termites: Evolution, Sociality, Symbioses, Ecology*. Dordrecht, The Netherlands: Kluwer Academic Publishers.

Bonabeau E, Dorigo M, Theraulaz G. (1999). *Swarm Intelligence: From Natural to Artificial Systems*. Oxford: Oxford University Press.

Bonabeau E, Theraulaz G, Deneubourg JL, Aron S, Camazine S. (1997). Self-organization in social insects. *Trends Ecol Evol* 12:188–93.

Camazine S, Deneubourg JL, Franks NR, Sneyd J, Theraulaz G, Bonabeau E. (2003). *Self-Organization in Biological Systems*. Princeton, NJ: Princeton University Press.

Costa JT. (2006). *The Other Insect Societies*. Cambrige, MA and London: Belknap Press of Harvard University Press.

Detrain C, Deneubourg JL, Pasteels JM (eds.). (1997). *Information Processing in Social Insects*. Basel: Birkhäuser Verlag.

Dorigo M, Stützle T. (2004). *Ant Colony Optimization*. Cambridge, MA: MIT Press.

Dyer FC. (2002). The biology of the dance language. *Annu Rev Entomol* 47:917–49.

von Frisch K. (1965). *Tanzsprache und Orientierung der Bienen*. Berlin: Springer.

von Frisch K. (1967, second printing 1993). *The Dance Language and Orientation of Bees*. Cambridge, MA: Harvard University Press.

Gordon DM. (1999). *Ants at Work: How an Insect Society Is Organized*. New York: The Free Press.

Gotwald WH. (1995). *Army Ants: The Biology of Social Predation*. Ithaca, NY: Comstock Publishing Associates of Cornell University Press.

Goulson D. (2003). *Bumblebees—Their Behaviour and Ecology*. Oxford: Oxford University Press.

Hölldobler B, Wilson EO. (1990). *The Ants*. Cambridge, MA: Belknap Press of Harvard University Press.

Hunt JH, Nalepa CA (eds.). (1994). *Nourishment and Evolution in Insect Societies*. Boulder, CO: Westview Press.

Michener CD. (1974). *The Social Behavior of the Bees: A Comparative Study*. Cambridge, MA: Harvard University Press.

Oldroyd BP, Wongsiri S. (2006). *Asian Honey Bees: Biology, Conservation, and Human Interactions*. Cambridge, MA: Harvard University Press.

Raveret Richter M. (2000). Social wasp (Hymenoptera: Vespidae) foraging behavior. *Annu Rev Entomol* 45:121–50.

Ross KG, Matthews RW (eds.). (1991). *The Social Biology of Wasps*. Ithaca, NY: Cornell University Press.

Roubik DW. (1989). *Ecology and Natural History of Tropical Bees*. New York: Cambridge University Press.

Seeley TD. (1995). *The Wisdom of the Hive—The Social Physiology of Honey Bee Colonies*. Cambridge, MA: Harvard University Press.

Snodgrass RE. (1956). *Anatomy of the Honey Bee*. Ithaca, NY: Comstock Publishing Associates.

Sumpter DJT. (2006). The principles of collective animal behaviour. *Phil Trans R Soc B* 361:5–22.

Part I

Foraging Decisions, Patterns, and Strategies

1 Measuring the Adaptiveness of Social Insect Foraging Strategies
*An Empirical Approach**

Nigel E. Raine and Lars Chittka

CONTENTS

INTRODUCTION

Foraging has been one of the key research areas to define the development of behavioral ecology as a discipline (Krebs and Davies 1993; Owens 2006). However, despite the considerable research effort devoted to studying foraging behavior, a number of fundamental questions with respect to the adaptiveness of foraging strategies remain relatively unexplored. Generally, we still understand very little about how foraging strategies contribute to the fitness of animals in the wild. How well does a given foraging strategy perform relative to strategies used by other individuals, colonies, or species? Does variation in foraging-related traits actually translate into real differences in fitness?

The rich history of studying foraging has revealed a great deal about the economics of foraging decisions, such as which types of food an animal should choose to consume (Pyke et al. 1977; Waddington and Holden 1979; Stephens and Krebs 1986), when to abandon a foraging patch (Kacelnik and Krebs 1985; Cuthill et al. 1990), how variance in food supply affects forager choices (Real 1981; Shafir et al. 1999; Fülöp and Menzel 2000; Chittka 2002), and what currencies animals use when making decisions about food quality (Schmid-Hempel et al. 1985; McNamara et al. 1993). But can we conclude that the improvements in foraging success or efficiency brought about by these behavioral strategies actually lead to increases in fitness? In a field study of bumble bee foraging, Schmid-Hempel and Heeb repeatedly removed a large percentage of foragers (10–15% of all workers) at regular intervals during the colony cycle. Interestingly, they found no significant effects of this forager workforce decimation on colony growth, life history, or ultimate colony reproductive success in a natural environment (Schmid-Hempel and Heeb 1991). So how can the precise

---

subtleties of minute-to-minute foraging strategies of specific individuals matter if even the existence of those individuals does not appear to affect colony reproductive success? Schmid-Hempel and Heeb's findings are consistent with comparable worker removal studies in ants (Gentry 1974; Herbers 1980), and experiments in which food availability was restricted for bumble bee colonies (Sutcliffe and Plowright 1988) suggest that stress levels must be quite severe before colony development or reproductive output are adversely affected. Perhaps the real value of foraging strategies only becomes crucial under adverse conditions (Schmid-Hempel and Heeb 1991; Schmid-Hempel and Schmid-Hempel 1998), but this underlines the general point that we do not yet really understand how foraging strategies contribute to the fitness of animals under natural conditions.

The goal of this review is to determine whether particular behavioral traits represent actual adaptations in the context of foraging. To do this we revisit the central tenets of behavioral ecology, namely, the need to accurately quantify variation in behavior and then to use the subtle variation that exists in behavioral traits to test adaptive hypotheses (Krebs and Davies 1993; Owens 2006). We need to quantify variation in foraging-related traits, and then measure the fitness consequences associated with different competing behavioral strategies under the field conditions in which animals really operate. As assessing lifetime reproductive success is at best rather time-consuming and labor-intensive, and at worst impossible, it is perhaps unsurprising that there are very few studies of phenotypic selection on behavioral traits in the wild (Endler 1986; Kingsolver et al. 2001). The social bees provide a convenient and tractable biological system with which to examine the potential adaptiveness of a wide range of foraging traits under natural conditions. Bumble bees are particularly amenable to such studies, as their foraging behavior is easily studied in both the laboratory and field, and can be linked to colony reproductive success. Since bumble bee colonies produce males and new queens in proportion to the amount of food available to them (Schmid-Hempel and Schmid-Hempel 1998; Pelletier and McNeil 2003; Ings et al. 2006), we can use colony foraging performance as a robust proxy measure of colony fitness. Here, we focus on how empirical studies on social bees allow us to investigate the adaptiveness of foraging strategies.

First, we assess how the innate color preferences of bumble bees have been influenced by the particular floral markets in which they forage. Using a phylogenetic approach we compare the innate color biases of different bumble bee species, and distinct populations within a single bumble bee species, which are presumably adapted to forage in different conditions but diverged from a common ancestor. Second, we examine the hypothesis that local variation in flower traits drives selection for innate color biases. The performance of bees from nine colonies was compared in a laboratory color bias paradigm before assessing the foraging performance of the same colonies under the field conditions to which they should be locally adapted. Third, we examine the quantitative effects of variation in learning performance on fitness using a similar correlative approach. The learning performance of bumble bees from twelve colonies was evaluated in an ecologically relevant associative learning task under laboratory conditions, before testing the foraging performance of the same colonies under field conditions. Finally, we examine the ecological conditions in which the honey bee dance language is adaptive by comparing an experimentally manipulated foraging phenotype, in which the directional information component of the dance language is removed, to the normal behavioral phenotype.

THE ADAPTIVE SIGNIFICANCE OF SENSORY BIAS IN A FORAGING CONTEXT

Animals are constantly exposed to stimuli differing widely in their potential importance. Sensory systems and cognitive processes allow the animal to assess the relative salience of stimuli and select appropriate behavioral responses to the most important. One mechanism through which such adaptive behavioral outcomes are promoted is through sensory biases, within either the sensory system or subsequent cognitive processes, causing animals to respond more strongly to certain, more pertinent, stimuli (Basolo and Endler 1995; Endler and Basolo 1998). Although sensory biases have received attention in the context of animal signaling, predominantly relating to mate choice

(Dawkins and Guilford 1996; Collins 1999) and predator avoidance (Bruce et al. 2001), the potential adaptive role of such biases has only recently been studied in a foraging context where they could also be very influential (Smith et al. 2004). The flower choices of pollinators represent a good model system in which to study the adaptive role of sensory bias in the context of foraging. Flowers send out signals to attract the attention of potential pollinators in a competitive marketplace, and pollinators are attuned to particular traits, such as the color, morphology, scent, and temperature of the flowers they visit to find food (Heinrich 1979; Menzel 1985; Dyer et al. 2006; Raine et al. 2006a).

Naïve animals often initially use innate rules to find food. Pollinators, such as bees, use color as a cue to find flowers when first exploring the world (Lunau and Maier 1995; Chittka and Raine 2006). Therefore, sensory biases toward particular colors might help naïve bees find flowers, and perhaps even help them to locate the most profitable ones in the local area. Indeed, newly emerged bees, that have never seen flowers, show distinct sensory biases for certain colors (Giurfa et al. 1995; Lunau et al. 1996; Chittka et al. 2001). While these innate color preferences can be easily modified, or even reversed, as a result of individual experience (Menzel 1985; Raine et al. 2006b), there is evidence that in some situations (for example, when rewards are similar across flower species) bees will revert to their initial preferences (Heinrich et al. 1977; Banschbach 1994; Gumbert 2000). It therefore seems a reasonable hypothesis that these innate preferences reflect the traits of local flowers that are most profitable for bees.

PHYLOGENETIC APPROACHES TO UNDERSTANDING FLORAL COLOR PREFERENCES

Studying behavioral repertoires, such as those involved in foraging, can be complicated as animals typically proceed along multiple alternative evolutionary pathways to optimize foraging behavior, and constraints imposed by one foraging-related trait might easily be compensated for by modifications to another trait (Raine et al. 2006a). Therefore, attempting to identify evolutionary adaptations in foraging by focusing on a single species (Pyke 1978; Greggers and Menzel 1993), or sets of unrelated species (Dukas and Real 1991), can be problematic since optimality and correlation cannot be equated with adaptation (Maynard Smith 1978; Chittka 1996). However, interspecific comparisons can be more rewarding when applied to closely related species of known phylogeny. For example, we can infer that a trait is adapted for a particular task if it can be demonstrated that the ancestors of the animal in question, which did not share the same environment, also do not share the trait under scrutiny (Brooks and McLennan 1991; Losos and Miles 1994; Chittka and Briscoe 2001). The comparative phylogenetic method, which seeks to reconstruct the traits of ancestral species through comparing closely related extant species, is a powerful tool to study patterns of adaptation (Armbruster 1992; Ryan and Rand 1999; Phelps and Ryan 2000). Following earlier studies of adaptation in the foraging strategies in beetles (Betz 1998), birds (Barbosa and Moreno 1999), and primates (Clutton-Brock and Harvey 1977), this technique has more recently been applied to considering the adaptation in the innate floral color preferences of bees.

Giurfa et al. (1995) found a good correlation between the color preferences of naïve honey bees (*Apis mellifera* L.) and the nectar provided by different flowers in a nature reserve near Berlin. These honey bees preferred violet (bee UV-blue, i.e., flowers that stimulate the bees' UV and blue receptors most strongly) and blue (bee blue, i.e., stimulating predominantly the bees' blue receptors), which were also the colors most frequently associated with high nectar rewards. However, as correlation does not imply causality, a better test to examine whether color preferences actually evolved to match floral offerings would be to compare a set of closely related bee species that live in habitats in which the association of floral colors with reward is different.

The innate color preferences of eight bumble bee species were tested by presenting them with artificial flowers in a laboratory flight arena: four species from Europe (*Bombus terrestris* L., *Bombus lucorum* L., *Bombus pratorum* L., and *Bombus lapidarius* L.), three from temperate East Asia (*Bombus diversus* Smith, *Bombus ignitus* Smith, and *Bombus hypocrita* Pérez), and one from North America (*Bombus occidentalis* Greene) (Raine et al. 2006a). Colonies from all species were raised under identical temperature and humidity conditions in a dark laboratory, to minimize the chances that observed interspecific differences could be caused by nongenetic factors, and were never exposed to flower colors before experiments. The color preferences of each bee were tested individually by recording their choices in a flight arena containing eighteen unrewarding flowers of six different colors (i.e., three flowers of each color: violet (bee UV-blue), blue (bee blue), white (bee blue-green, i.e., producing a strong signal in the blue and green receptors of bees), yellow, orange, and red (all bee green, stimulating the bees' green receptors most strongly).

As all eight species showed a strong sensory bias for colors in the violet-blue range, this indicates a phylogenetically ancient preference (Figure 1.1). This preference is likely to be advantageous, since violet and blue flowers have been found to contain high nectar rewards in a variety of habitats (Menzel and Shmida 1993; Giurfa et al. 1995; Chittka et al. 2004; Raine and Chittka 2007c). However, there were also differences among species. *B. occidentalis* had a much stronger preference for red than any other species tested. This is intriguing because this species is frequently observed foraging, or robbing nectar, from red flowers whose morphology seems well adapted for pollination by hummingbirds (Chittka and Waser 1997; Irwin and Brody 1999). Comparative phylogenetic analysis strongly suggests that this preference is derived and is therefore likely to represent an adaptation to this unique foraging strategy of *B. occidentalis* (Chittka and Wells 2004; Raine et al. 2006a).

Comparing more recently diverged populations within the same species is also a useful approach to investigate patterns of adaptation among very closely related individuals operating under different ecological conditions. This can be especially true when comparisons include island populations that, by virtue of their small effective population size and risk of genetic bottlenecks, have increased potential for divergence from mainland populations either by chance processes or through more pronounced adaptation to local conditions due to limited gene flow with animals in different habitats (Ford 1955; Stanton and Galen 1997; Barton 1998; Chittka et al. 2004). Applying this phylogenetic approach to compare populations within a single species, Chittka and colleagues tested the innate color preferences of laboratory-raised colonies from eight *Bombus terrestris* populations—*B. t. terrestris* (L.) from Holland and Germany, *B. t. dalmatinus* (Dalla Torre) from Israel, Turkey, and Rhodes, *B. t. sassaricus* (Tournier) from Sardinia, *B. t. xanthopus* (Kriechbaumer) from Corsica, and *B. t. canariensis* (Pérez) from Tenerife—using the same paradigm as above (Chittka et al. 2001; Raine et al. 2006a). All populations preferred colors in the violet-blue range, but there were some differences in the relative preference for violet and blue (Figure 1.2). This pattern is similar to that seen across different bumble bee species (see above), and makes biological sense since these floral colors are frequently the most rewarding across a variety of habitats (Menzel and Shmida 1993; Giurfa et al. 1995; Chittka and Wells 2004; Raine and Chittka 2007c). Interestingly however, some island populations displayed a different pattern of color preference. *B. t. sassaricus* and *B. t. canariensis* exhibited a secondary red preference (Chittka et al. 2001; Raine et al. 2006a). Coupling this observation with the fact that both island populations are genetically differentiated, both from one another and the mainland population (Estoup et al. 1996; Widmer et al. 1998), and that such population differences are heritable (Chittka and Wells 2004), indicates evolutionary plasticity in flower color preference within *B. terrestris*.

The adaptive significance of the red preference shown by the Sardinian and Canary Island populations is not easy to explain. Red, UV-absorbing, pollen-rich flower species exist at the eastern end of the Mediterranean (e.g., Israel). However, the Israeli *B. terrestris* population shows no secondary red preference, and red flowers there are predominantly visited by beetles (Dafni et al. 1990; Menzel and Shmida 1993). In Sardinia, Chittka et al. (2004) found only a single red, UV-absorbing flower species (out of 140 measured). Conspicuously, almost all the red flower species (including those that reflect in the UV) in each habitat tested were pollen-only flowers, containing no nectar (Dafni et al. 1990; Chittka et al. 2004). Thus, a red preference would not be advantageous to the Sardinian population in terms of nectar foraging (the behavioral context tested by Chittka et al. 2004). Whether red flowers represent a particularly attractive pollen source remains to be tested,

a

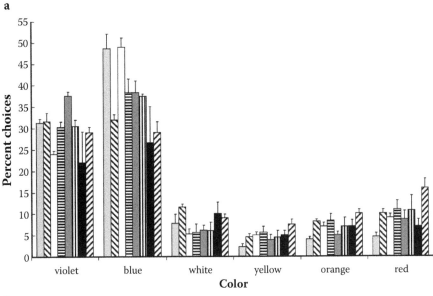

b

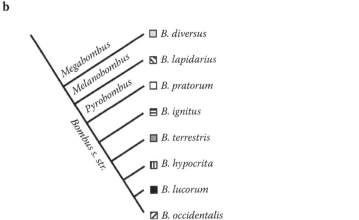

FIGURE 1.1 Color preferences of eight bumble bee species superimposed on their phylogeny (following Cameron et al. 2007). Each bee was color naïve at the start of the experiment, and only the first foraging bout was evaluated. At least three colonies were tested per species, and at least fifteen workers per colony. Bees were individually tested in a flight arena in which they were offered the colors violet (bee UV-blue), blue (bee blue), white (bee blue-green), yellow, orange, and red (the latter three are all bee green). Column height denotes the mean (+1 SE) of choice percentages. The sequence of species in the histogram (a), left to right, maps onto those from the phylogeny (b), top to bottom; hence, the leftmost column is *Bombus diversus*. (Data from Chittka et al. 2001, 2004, and Chittka and Wells 2004. Figure revised from Raine et al. 2006a. Copyright © 2006 Elsevier. Reproduced with permission.)

but there is no reason to expect the importance of such flowers species is higher in Sardinia than elsewhere. Several orange-red flower species are found on the Canary Islands (Vogel et al. 1984), which seem strongly adapted to bird pollination (Olesen 1985). Although birds visit several of these species (Valido et al. 2002), *B. t. canariensis* has not been observed to do so in the wild (Stelzer et al. 2007; Ollerton et al. 2009).

Thus, we are left with an interesting observation: flower color preferences are clearly variable among *B. terrestris* populations, and these differences are heritable (Chittka and Wells 2004).

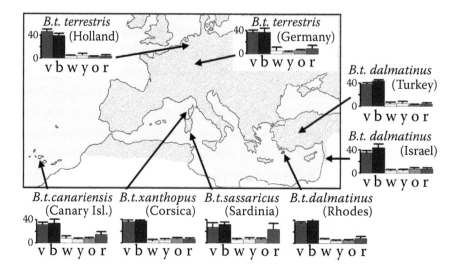

FIGURE 1.2 Biogeography of floral color preference in *Bombus terrestris* populations. Bees were individually offered the following colors: v, violet (bee UV-blue); b, blue (bee blue); w, white (bee blue-green); y, yellow; o, orange; and r, red (the latter three are bee green). Column height denotes the mean percentage (+1 SE) of colony choices. At least five colonies were tested per population. The shaded area shows the distribution of *B. terrestris* (this range was provided with the kind permission of Pierre Rasmont). (Data from Chittka et al. 2001, 2004. Figure after Raine et al. 2006a. Copyright © 2006 Elsevier. Reproduced with permission.)

However, we cannot easily correlate the color preferences in different habitats with differences in local floral colors. The possibility that genetic drift has produced the color preferences in some island populations certainly deserves consideration—especially since a secondary preference for red flowers does not appear to be selected against. However, it is also possible that the red preference of these bumble bee populations is a "behavioral fossil" that dates back to an age when red, bird-pollinated flowers were common in Europe (Raine et al. 2006a). If this is true, and if bumble bees exploited some of these flowers (as some species do in North America; Chittka and Waser 1997), then the red preference of some *B. terrestris* populations might be a result of historical adaptation rather than either recent adaptation or chance. Physiological data also suggest that Sardinian bees might be more sensitive to detecting red flowers than mainland conspecifics (Skorupski et al. 2007). However, differences between Sardinian and mainland populations in the peak wavelength sensitivity of their green photoreceptors (λ_{max}: *B. t. sassaricus* = 538 nm, *B. t. dalmatinus* = 533 nm) are insufficient in themselves to significantly improve wavelength discrimination in the red region of the spectrum. The clear behavioral differences between *Bombus terrestris* populations in color preference are probably not explained by any simple change at the photoreceptor level, but must instead be based on differences in postreceptor neuronal wiring.

SENSORY BIAS AND FORAGING PERFORMANCE

Innate preferences for violet and blue flowers appear to represent a useful sensory bias at the species and population level, but what about variation in sensory bias among colonies within a population? Can floral traits drive selection for local adaptation in bumble bee color preferences? In the flora near Würzburg, Germany, the average violet flower provides more than twice as much sugar reward as each blue flower (Raine and Chittka 2007a, 2007c), the next most productive flower color. If local floral traits do drive selection for local color biases, bees with a stronger sensory bias for violet (over blue) flowers should forage more effectively in this environment. As social insects, bumble bee reproduction is restricted to a subset of individuals within each colony. Hence, for bumble bees, intercolony (rather than interindividual) trait variation allows us to test the adaptive benefits of sensory bias variation when foraging in the local environment. Since bumble bee colonies produce males and new queens in proportion to the amount of food available to them (Schmid-Hempel and Schmid-Hempel 1998; Pelletier and McNeil 2003; Ings et al. 2006), we can use colony foraging performance as a robust measure of colony fitness. This approach explores intercolony variation of floral color bias within a natural population to measure the extent to which such sensory biases can be regarded as adaptive, i.e., improving colony foraging performance in their natural environment. This was achieved by comparing the performance of nine bumble bee (*B. t. terrestris*) colonies in color bias tests under laboratory conditions with the foraging performance of the same colonies under natural conditions. Using this approach allowed us to directly correlate trait variation in sensory bias with a proxy measure of colony fitness (foraging performance).

Colonies were raised in the laboratory from nest searching queens caught around Würzburg. Color preferences of color naïve workers (10–15 bees per colony, 101 in total) were tested in laboratory flight arenas using methods similar to those used in comparisons among species and populations— see above (Chittka et al. 2001; Raine et al. 2006a). The preference of each individual forager for violet (bee UV-blue) over blue (bee blue) was assessed in an arena containing sixteen unrewarded artificial flowers (eight of each color). After completion of laboratory tests, the nectar collection rate (mass of nectar collected per unit foraging time) of the same nine colonies was measured in the field (near Würzburg) in June–July 2002. The area is typical central European bumble bee habitat, giving colonies access to multiple flower species in bloom in dry grassland, deciduous forest, and farmland (Raine and Chittka 2007a).

Colonies with a stronger innate preference for violet (over blue) in the laboratory also harvested more nectar per unit time under natural conditions (Figure 1.3). Colonies with the strongest violet preference brought in almost 41% more nectar than the colony with the least strong bias when foraging from real flowers in the field. As violet flowers were on average twice as rewarding as blue flowers (the next most rewarding flower color) in the local area (Raine and Chittka 2007a, 2007c), these results support our hypothesis that colonies biased toward the more highly rewarding violet flowers collect more nectar per unit time. Comparing these findings with results from an earlier study (Raine and Chittka 2005) confirms that the pattern was very similar for consecutive years at the same location.

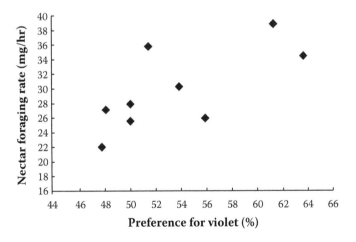

FIGURE 1.3 Correlation between strength of innate floral color preference for violet (over blue) in the laboratory and nectar foraging performance in the wild, measured for nine bumble bee colonies (*Bombus terrestris*) near Würzburg ($r_s = 0.678$, $n = 9$, $p = 0.045$). Each data point represents the colony median value for each of these traits. (After Raine and Chittka 2007a. With permission.)

It is easy to imagine how these innate color preferences help naïve bees to find flowers on their first foraging trip, as very few objects in natural landscapes are blue or violet except flowers. Presumably, color preferences lead bees to investigate violet or blue objects (flowers) in preference to leaves, rocks, etc. Following the same logic, if violet flowers are consistently more rewarding, it would make adaptive sense to prefer them (over blue). As bees gain foraging experience, by visiting hundreds or thousands of flowers each day (Heinrich 1979; Raine and Chittka 2007b), they establish an increasingly detailed picture of which flower species (or colors) are most profitable and when. Bees are easily able to learn to associate multiple floral traits, including color (Giurfa et al. 1995; Gumbert 2000), morphology (Laverty 1994; Raine and Chittka 2007d), and scent (Menzel 1985), with reward levels, and such learned associations allow individual foragers to modify, or even overwrite, their inbuilt sensory biases within a short time (Menzel 1985; Gegear and Laverty 2004; Raine et al. 2006b; Raine and Chittka 2008). Since overall flower visitation rates are largely dominated by the informed choices of experienced bees, rather than the result of the sensory biases of naïve foragers, this presumably prevents selection for flowers to exploit innate color preferences by reducing the amount of reward they provide while still maintaining similar pollination success (Raine and Chittka 2007a).

Earlier studies correlating color bias variation among bumble bee species (Chittka and Wells 2004; Raine et al. 2006a), or among populations within a single bumble bee species (Briscoe and Chittka 2001; Chittka et al. 2004), with differences in their respective foraging environments have provided valuable insights into patterns of bee color bias evolution within a phylogenetic framework. Changing the emphasis and focusing on the potential adaptive significance of color preference at the colony level adds the missing link, i.e., how variation in color biases actually affects foraging performance. Quantifying the level of local intercolony variation in a foraging-related trait (violet-blue bias) and assessing its potential effect on foraging performance (a robust proxy measure of reproductive success) using the same set of colonies provides a more direct test of the potential adaptive value of this sensory bias. This approach, linking demonstrations of trait variation in the laboratory with its effect on animals operating in their natural environment, represents a valuable tool that has the potential to be usefully applied to studying the adaptive value of many other foraging-related traits.

LEARNING AND FORAGING PERFORMANCE: BUMBLE BEES GAIN FITNESS THROUGH LEARNING

Learning, or the adaptive modification of behavior based on experience, affects virtually every aspect of animal behavior. However, despite the abundance of research on the mechanisms of learning in a wide variety of animal taxa, we still know very little about how learning performance is actually adapted to real ecological conditions (Shettleworth 1998; Dukas 2004). As different individuals or species vary widely in their learning capacities, it is commonly assumed that these differences reflect adaptations to the natural conditions under which such animals operate (Gallistel 1990; Dukas 1998; Shettleworth 1998). While it is intuitively appealing to assume that such variation in learning performance is adaptive (Johnston 1982; Dukas 1998), few studies have yet been conducted to specifically examine this link under natural conditions.

Laboratory studies, using grasshoppers (Dukas and Bernays 2000) and parasitoid wasps (Dukas and Duan 2000), suggest that animals able to form associations between cues (such as color, odor, or location) and rewards perform better than animals prevented from learning. Other laboratory studies, applying artificial selection to the learning ability of fruit flies, provide evidence for potential fitness costs associated with enhanced performance in associative learning (Mery and Kawecki 2003, 2004) or long-term memory (Mery and Kawecki 2005) tasks. While these results suggest that the ability to learn is useful (compared to being unable to learn) in highly controlled laboratory situations, and that enhanced learning appears to incur higher costs, they do not yet inform us directly about the potential fitness payoffs for animals with different learning abilities under natural conditions. Circumstantial evidence for the adaptive value of learning comes from comparisons between species (Dukas and Real 1991; Sherry and Healy 1998; Healy et al. 2005); for example, vole species with larger home ranges typically have better spatial memory, and their hippocampi (brain areas that store spatial memories) are typically larger (Sherry and Healy 1998). While such studies suggest that learning performance and ecologically important measures (such as home range size) are correlated, the species compared also vary in numerous other ecological requirements. Therefore, to make further progress in understanding the evolutionary and ecological relevance of learning abilities, the next step is to quantify how and to what extent learning differences within species affect animal fitness in nature (Papaj and Prokopy 1989; Dukas and Duan 2000).

We set out to address this question using the same successful approach applied to investigate the adaptive value of local color preferences (see "Sensory Bias and Foraging Performance" section): we directly correlated variation in learning performance with field foraging performance (a robust measure of fitness) for twelve bumble bee (*B. terrestris dalmatinus*) colonies (Raine and Chittka 2008). In laboratory learning trials, the bees' task was to overcome their innate preference for blue (Lunau et al. 1996; Chittka et al. 2004; Raine et al. 2006a) and learn to associate yellow as a predictor of floral reward. This is a simple associative task that bumble bees are able to learn, but individuals and colonies vary in their speed and accuracy (Chittka et al. 2004; Raine et al. 2006b). The task is ecologically relevant because foraging bees use a variety of cues, including floral color, pattern, and scent, to recognize, discriminate, and learn the flowers from which they collect nectar and pollen (Menzel 1985; Chittka and Raine 2006). We measured the nectar collection rate of the same twelve colonies in the field (Queen Mary College, London) in July–August 2005. Once outside the nest, bees could forage freely in an area containing abundant floral resources growing in numerous private gardens, several large parks, and other areas of open land (e.g., canal or railway embankments).

We found that colonies with higher learning speeds (in the laboratory tests) also harvested more nectar per unit time under field conditions (Figure 1.4). This positive correlation between learning and foraging performance indicates the slowest-learning colonies brought in 40% less nectar than the fastest-learning colonies. As foraging performance represents a robust proxy measure of fitness, these results indicate higher learning speed is closely associated with increased bumble bee colony fitness under natural conditions (Raine and Chittka 2008).

As bees forage in a complex and dynamic flower market in which rewards differ strongly among plant species and vary over time (Heinrich 1979; Willmer and Stone 2004), individual foragers must assess such differences and respond accordingly (Chittka 1998; Menzel 2001; Raine and Chittka 2007d, 2008). Rapid learning of salient floral cues, such as color, presumably assists bees to track

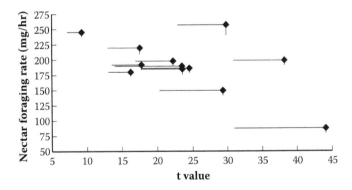

FIGURE 1.4 Correlation between learning speed and nectar foraging performance for twelve bumble bee (*Bombus terrestris dalmatinus*) colonies. High *t* values correspond to slow learning, while low values are generated by fast learners. Data presented are colony mean values (–1 SE) for both *t* value and foraging performance. On average, colonies with higher learning speeds (lower *t* values) brought in more nectar per unit time (Pearson's correlation: $r = -0.588$, $n = 12$, $p = 0.044$). (After Raine and Chittka 2008. With permission.)

changes in the floral rewards on offer, thereby improving bee foraging efficiency by allowing them to preferentially visit the current most profitable flower type (Raine et al. 2006a, 2006b). It would be interesting to examine whether colonies that learn more quickly in visual tasks (e.g., learning color associations) also show better learning performance in other modalities (e.g., odor or tactile cue learning). Laboratory studies (using the proboscis extension response paradigm) suggest that individual honey bees that are more sensitive to sucrose stimuli show improved learning in both odor and tactile conditioning experiments (Scheiner et al. 2001a, 2001b). These results suggest that performance levels in an associative learning task using one modality might indicate the relative performance for other modalities.

To date, discussion of the potential adaptive value of learning has concentrated on the environmental conditions under which learning is favored. Learning is predicted to be favored in most environments, except those that are either so changeable that prior experience has no predictive value, or so consistent (across generations) that genetically preprogrammed innate behaviors alone are sufficient (Johnston 1982; Shettleworth 1998; Dukas 2004). However, these studies do not allow assessment of how the subtle variation that exists between individuals in natural populations translates into fitness benefits. The correlative approach used here represents a first step toward examining potential fitness effects attributable to variation in learning performance under the real conditions to which animals are adapted (Raine and Chittka 2008). However, although variation in learning performance among bumble bee colonies appears the most likely explanation for observed differences in their foraging performance, further evidence is needed to establish a causal link. Ultimately, developing a more general understanding of the adaptive value of learning would require a direct examination of the fitness effects of variation in learning performance across a range of animal species and the environments to which they and their cognitive abilities are adapted.

MANIPULATION OF FORAGING PHENOTYPE: THE HONEY BEE DANCE

The honey bee dance language is regarded by many as one of the most intriguing communication systems in nonhuman animals (von Frisch 1955, 1967; Chittka 2004). A successful scout bee returns and advertises the location of a newly discovered food source to nestmates using a repetitive figure-eight-shaped sequence of movements. Shortly after these "waggle dances" commence, scores of newly recruited foragers will arrive at the food source being advertised (von Frisch 1955, 1967; Seeley 1995). But what were the ecological conditions under which such a dance language evolved, and what are its benefits to colony foraging performance?

Dornhaus and Chittka (2004b) investigated the potential adaptive significance of the dance language by measuring the foraging performance of honey bee (*A. mellifera*) colonies under natural conditions and comparing this to conditions under which the information flow between dancers and recruits was disrupted. Information transfer from dancer to recruit is disrupted with a simple trick. Under normal conditions in the darkness of the hive, the angle of the forager's waggle run relative to the direction of gravity on the vertical comb indicates the direction of the food source relative to the azimuth of the sun (von Frisch 1955, 1967). However, rotating the combs into a horizontal position eliminates the bees' ability to use gravity as a reference. Therefore, bees perform dances in random directions, effectively removing the directional information component. Having "interpreted" these nondirectional dances, recruits leave the hive in random directions (von Frisch 1967; Dornhaus and Chittka 2004a, 2004b). However, if bees can see the sun (or polarized light), then returning foragers can perform correctly oriented waggle dances on a horizontal surface using the sun rather than gravity for reference (von Frisch 1955, 1967). Using hives with horizontally arranged combs, Dornhaus and Chittka (2004b) controlled the ability of returning foragers to perform oriented dances by either covering or uncovering a window on the top of the hive (see Figure 7.1, p. 137).

Initially, the foraging success of colonies that were able, or unable, to communicate the direction of profitable food sources was compared in two temperate habitats: a typical Mediterranean habitat in the Sierra Espadán Nature Reserve, Spain, and a site near Würzburg, where agricultural land is mixed with natural meadows. A pair of hives, each containing about five thousand workers, was placed at each site. Each colony was switched from oriented to disoriented dancing every 2 days, by uncovering or covering the window on top of the hive. Foraging success was assessed using the daily weight gain of hives, which predominantly reflects nectar intake (Seeley 1995). Surprisingly, no differences in weight gain were recorded between days in which colonies could follow oriented or disoriented waggle dances from returning foragers at either site (Dornhaus and Chittka 2004b). These findings were confirmed by repeating the same experiment with two additional hives monitored in Würzburg from May to September (Dornhaus and Chittka 2004b).

Why do honey bees bother communicating the direction of profitable food sources if it makes no difference to foraging efficiency? It seems highly counterintuitive, especially when considering the enormous efficiency of the dance language to recruit bees to single points in space (Gould 1975; Towne and Gould 1988; Dyer 2002). Perhaps honey bee ecological history could help us understand their current behavioral patterns? It seems most likely that the evolutionary origins of the dance occurred in an open-nesting tropical ancestor of extant honey bees (Dyer and Seeley 1989), which foraged under conditions wholly different from those in which modern *A. mellifera* colonies operate. In tropical forests, an individual tree, or a local aggregation of trees, frequently offers many thousands of flowers at a very precise spatial location (Roubik 1992; Condit et al. 2000; Pitman et al. 2001), which might only persist or provide rewards for a short period of time (Stone et al. 1998, 2003; Raine et al. 2007). Under such conditions, it seems likely that an ability to accurately communicate the distance and direction of such foraging bonanzas to other foragers in the hive would be highly advantageous. This is in marked contrast to most temperate habitats, in which widely distributed herbs and shrubs form a significant component of a bee's diet (Heinrich 1979; Seeley 1995) and are easily located by individual foragers.

To test if the dance language is more important to efficient foraging in tropical than in temperate habitats, Dornhaus and Chittka repeated their experiment in an Indian tropical dry deciduous forest (Bandipur Biosphere Reserve). In this environment removing the direction component of the honey bee dance information reduced the number of successful foraging days by 85% (Figure 1.5).

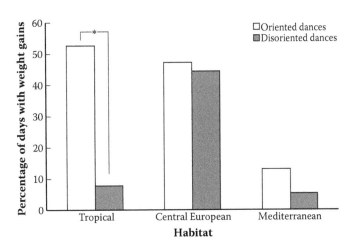

FIGURE 1.5 Foraging performance of honey bee *(Apis mellifera)* colonies with (oriented dances) and without (disoriented dances) the ability to communicate directional information about the location of food sources to nestmates through their dance language. Columns indicate the percentage of days on which each colony increased in weight. A significant effect of disrupting information between dancers and recruits was found in the tropical (indicated by the asterisk) but not in the two temperate habitats, where bees foraged equally well with and without directional communication about location of food sources. (Reprinted from Dornhaus and Chittka 2004a. Copyright © 2004 EDP Science. With permission.)

The median hive weight gain on days with oriented dances was 5 g, compared with a loss of 65 g on days when direction information was disrupted (Dornhaus and Chittka 2004b).

Maps of the flower patches visited by a hive can be made by decoding the dances of returning foragers (Visscher and Seeley 1982). Using this technique confirmed foraging sites were very patchily distributed within the Indian tropical forest (Dornhaus and Chittka 2004b), compared to temperate habitats for which comparable forage maps are available: temperate forest (Visscher and Seeley 1982), disturbed suburban habitat (Waddington et al. 1994), and disturbed habitat mixed with more natural open moors (Beekman and Ratnieks 2000).

These findings suggest that the honey bee dance language is an adaptation to the tropical conditions under which the genus *Apis* diversified, and may no longer be essential for efficient foraging in temperate habitats. Here, it might be maintained simply because it confers no selective disadvantage. In support of the argument that the dance language is more crucial under tropical conditions, Towne and Gould (1988) showed the precision of direction communication is higher in tropical than in temperate species. When food is less patchily distributed than in tropical forest, foraging by individual initiative, or communication through floral scent, may be as efficient as dance communication (Chittka 2004). Alternatively, stabilizing selection might have occurred through nonforaging dance applications, such as indicating the location of nesting sites (Weidenmüller and Seeley 1999; Beekman et al. 2008).

GENETIC BASIS OF FORAGING BEHAVIOR

Current questions being addressed in behavioral ecology require increasingly detailed understanding of the genetic and physiological mechanisms that underlie behavioral traits (Owens 2006). If we uncovered the genetic basis of foraging behavior it would clearly improve our understanding of the evolvability of traits that influence foraging, and the extent to which foraging behavior is adapted to a given niche (Ben-Shahar et al. 2002; Whitfield et al. 2003). Honey bee foraging is one of the best-understood behaviors at the genetic level (Fitzpatrick et al. 2005; Hunt et al. 2007; Oldroyd and Thompson 2007).

Although in some instances a single candidate gene can have profound behavioral effects—for example, a single genetic element (Gp-9) is responsible for determining variation in fire ant colony social organization (Ross and Keller 1998)—it seems likely that most behavioral traits are polygenic and linked through pleiotropies, i.e., correlated characters (Amdam et al. 2004; Owens 2006), which suggests that selection on any individual gene might have complex effects (Rueppell et al. 2004a, 2004b; Hunt et al. 2007). Studies investigating the genetic architecture and the physiological and molecular basis of a variety of foraging-related traits in honey bee behavior appear to confirm this idea (Robinson et al. 1989; Page and Robinson 1991; Page and Fondrk 1995; Pankiw et al. 2002; Rueppell et al. 2004a; Hunt et al. 2007). Page and colleagues started by selecting two strains of honey bee colonies for a single characteristic: the amount of pollen collected and stored (Page and Fondrk 1995; Page et al. 1995). Within a few generations, they had selectively bred two lines that differed strongly in the relative effort they devoted to nectar and pollen foraging. The resulting bee strains differed in multiple aspects of foraging behavior that could be linked, either directly or indirectly (via pleiotropic effects), to pollen foraging. The "high strain" colonies (those that hoarded more pollen) not only had more pollen foragers, and collected larger pollen loads, but also initiated foraging at a younger age, and collected smaller and less concentrated nectar loads (Pankiw and Page 2001). However, because foraging loads were not measured as a function of foraging flight duration, these data are not indicators of differential foraging performance. On the sensory level, proboscis extension reflex experiments showed that high-strain bees were more sensitive to low sucrose concentrations (Page et al. 1998), perhaps explaining their higher relative acceptance level for poor-quality nectar (Pankiw and Page 2000). Indeed, high-strain bees have higher sensitivity to other chemosensory stimuli, e.g., brood pheromones (Pankiw and Page 2001). This suggests a potential for overall improvement of sensory function in these bees, which could explain their superior performance in both olfactory and tactile learning paradigms (Scheiner et al. 2001b). However, there may also be changes at the level of the central nervous system: Humphries et al. (2003) found higher levels of protein kinases A and C in high-strain bee brains. Both enzymes play a role in memory consolidation and avoidance conditioning (Shobe 2002). It has also been proposed that these differences are pleiotropically linked to reproductive behavior (Amdam et al. 2004, 2006a), perhaps suggesting that division of labor in honey bee foraging evolved from an ancestral reproductive regulatory network (Hunt et al. 2007).

What are the implications of these findings for future studies of the adaptiveness of foraging behavior? The good news is that researchers are homing in on the genetic architecture underlying foraging behavior, that several foraging-related traits are heritable, and therefore that the raw material for selection exists (under both natural and experimental conditions). How quickly the genetic basis of behavioral trait variation will be uncovered will depend on the number of genetic elements involved and the strength of potential pleiotropic effects in each case (Oldroyd and Thompson 2007). Once we have this quantitative genetic information, it opens up the possibility of studying the adaptive benefits of these traits in the wild. However, the interpretation of the potential differences in fitness will be difficult because selection on any one trait is likely to drag along a host of other traits, which may all operate under a variety of environmental constraints, and might therefore affect fitness in different ways.

CONCLUDING REMARKS

In this review, we have illustrated the value of several empirical approaches to study the adaptive nature of foraging behavior under natural conditions. Comparing variation in a foraging trait among species, or geographically isolated subspecies, with a known phylogeny can allow us to distinguish adaptation from the effects of chance and history on the behavior in question (Clutton-Brock and Harvey 1977). Despite the potential power of the phylogenetic method, studies on the innate color preferences of bumble bees presented here (Chittka et al. 2001; Chittka and Wells 2004; Raine et al. 2006a) are among the few that use this approach to study behavioral traits related to foraging (e.g., Betz 1998). This could be because the method relies on well-supported phylogenies, which are not universally available, or that levels of trait variation among species are unpredictable. If traits are conserved across all species tested, for example, a strong preference for blue and violet flowers across all bumble bee species tested, this does not inform us about the adaptive value of this trait without testing additional outgroups with different ecological requirements. However, where trait variation exists among tested species, and can be linked to strong differences in ecology, this approach provides relatively strong, indirect evidence for trait adaptation, for example, the secondary preference for red in *Bombus occidentalis*, which robs nectar from bird-pollinated flower species.

Using a correlative approach, combining laboratory and field studies on the same animals (or colonies) represents a novel approach to studying whether variation in foraging-related traits actually translates into real differences in fitness. Positive correlations between two foraging-related traits in the laboratory and field foraging performance suggest the potential for strong directional selection for both a more extreme sensory bias for violet (over blue; Raine and Chittka 2007a) and higher learning speed (Raine and Chittka 2008) in bumble bees. This approach is well suited to studies of social insects, as they are central place foragers, and a discrete subset of workers from the same colony can be used for laboratory and field experiments. Furthermore, the foraging behavior of social insects (particularly bees and ants) is comparatively well studied, and numerous proxy measures or correlates of colony fitness are known. Although bumble bee foraging performance is strongly correlated with colony fitness (Schmid-Hempel and Schmid-Hempel 1998; Pelletier and McNeil 2003; Ings et al. 2006), it would be even better to determine a more direct measure of fitness, such as the number or biomass of sexual offspring (males and queens) produced. While this has been done in some field studies of parasitism in bumble bees (Müller and Schmid-Hempel 1992; Baer and Schmid-Hempel 1999), colony manipulations in these studies were minimal and could be conducted at (or before) the earliest stages of colony development, allowing colonies to grow and mature under field conditions. In contrast, the length of time needed to accurately quantify behavioral trait variation in the laboratory necessarily restricts the period of time during which colonies can develop in the natural environment. Under such conditions, it would be impossible to attribute any differences in colony reproductive output solely to differences in field foraging performance. While it is becoming potentially easier to monitor the foraging activity of social insect colonies, using radio frequency identification (RFID) technology to automatically monitor when workers enter and leave the nest (Streit et al. 2003; Sumner et al. 2007; Molet et al. 2008), there is no obvious solution to reducing the time required to quantify variation in foraging-related traits in the laboratory. Hence, given these constraints, foraging performance appears to be the best measure of colony fitness that can be achieved using this approach.

Manipulating the behavioral phenotype can allow a direct test of the potential adaptive value of a trait, particularly when the manipulation is completely reversible, allowing the opportunity to compare the manipulated and unmanipulated phenotype of the same individual (or colony). With their manipulation of the honey bee dance language, Dornhaus and Chittka (2004b) were able to elucidate that the dance language is highly advantageous when floral resources are highly clustered in time and space. The drawback of this approach is that there are very few, if any, other behavioral traits that would be amenable to such simple but effective, reversible manipulation. While this does not reduce the potential power of this approach, it means it is unlikely to be generally applicable.

Perhaps a more promising approach for future research would be manipulation of the foraging phenotype via the insight gained from studies of variation in behavioral traits at the genetic level. Researchers are closing in on isolating the genes that encode particular behavioral traits (Fitzpatrick et al. 2005; Hunt et al. 2007; Oldroyd and Thompson 2007), so it might soon be possible to selectively modify behavioral phenotypes by knocking out their expression using double-stranded RNA interference (dsRNAi; Fire et al. 1998), or perhaps by creating more traditional knockout mutants (Wolfer and Lipp 2000; Lipp 2002). While it is occasionally possible to create behavioral phenotypes for traits without genetic techniques, such as removing the ability to encode directional information in the honey bee waggle dance (see "Manipulation of Foraging Phenotype: The Honey Bee Dance" section), the use of dsRNAi could extend the potential of this powerful approach (i.e., modification of natural behavioral phenotypes) for many other traits of interest. This dsRNAi approach is already being used to study functional mechanisms by knocking out gene function. Blocking the octopaminergic pathway renders honey bees unable to learn an odor paired with sucrose reward, as octopamine mediates the unconditioned stimulus (the reward) in this associative learning task (Farooqui et al. 2003, 2004). Downregulation of vitellogenin activity using dsRNAi increases the gustatory responsiveness of workers, suggesting that vitellogenin is an important regulator of long-term changes in honey bee behavior (Amdam et al. 2006b). Continued advances in the search for other behaviorally important genes and refinements in dsRNAi techniques could herald the beginning of a very powerful future tool for the study of adaptation in behavioral ecology. The choice of study organism for such an approach is a trade-off between availability of techniques and its tractability for fitness studies under natural conditions. Social bees have long been a model system in the study of behavior. Now with recent advances in understanding social bee genetics, particularly since sequencing the honey bee genome (Honey Bee Genome Sequencing Consortium 2006), the combination of availability of techniques like dsRNAi and their tractability for fitness studies suggests social bees have a bright future for tests of the adaptiveness of foraging behavior under natural conditions.

ACKNOWLEDGMENTS

We thank Zainab Afzal, Denise Barrow, Samiya Batul, Adrienne Gerber-Kurz, Rosa Hardt, Amanda Hill, Sibel Ihsan, Anna Lo, Julia Mäschig, Nicole Milligan, Oscar Ramos Rodríguez, Claudia Rennemaier, Juliette Schikora, Miriam Sharp Pierson, Ralph Stelzer, and Matthew Wallace for their help with the experiments, and Alice Raine for helpful comments on an earlier version of this manuscript. This work was supported by the NERC (NER/A/S/2003/00469) and a combined grant from the Wellcome Trust, BBSRC, and EPSRC (BB/F52765X/1).

REFERENCES

Amdam GV, Csondes A, Fondrk MK, Page RE. (2006a). Complex social behaviour derived from maternal reproductive traits. *Nature* 439:76–78.

Amdam GV, Norberg K, Fondrk MK, Page RE. (2004). Reproductive ground plan may mediate colony-level selection effects on individual foraging behavior in honey bees. *Proc Natl Acad Sci USA* 101:11350–355.

Amdam GV, Norberg K, Page RE, Erber J, Scheiner R. (2006b). Downregulation of vitellogenin gene activity increases the gustatory responsiveness of honey bee workers (*Apis mellifera*). *Behav Brain Res* 169:201–5.

Armbruster WS. (1992). Phylogeny and the evolution of plant-animal interactions. *Bioscience* 42:12–20.

Baer B, Schmid-Hempel P. (1999). Experimental variation in polyandry affects parasite loads and fitness in a bumble-bee. *Nature* 397:151–54.

Banschbach VS. (1994). Colour association influences honey bee choice between sucrose concentrations. *J Comp Physiol A* 175:107–14.

Barbosa A, Moreno E. (1999). Evolution of foraging strategies in shorebirds: An ecomorphological approach. *Auk* 116:712–25.

Barton NH. (1998). Natural selection and random genetic drift as causes of evolution on islands. In Grant PR (ed.), *Evolution on Islands*. Oxford: Oxford University Press, pp. 102–23.

Basolo AL, Endler JA. (1995). Sensory biases and the evolution of sensory systems. *Trends Ecol Evol* 10:489.

Beekman M, Gloag RS, Even N, Wattanachaiyingcharoen W, Oldroyd BP. (2008). Dance precision of *Apis florea*—Clues to the evolution of the honeybee dance language? *Behav Ecol Sociobiol* 62:1259–65.

Beekman M, Ratnieks FLW. (2000). Long-range foraging by the honey-bee, *Apis mellifera* L. *Funct Ecol* 14:490–96.

Ben-Shahar Y, Robichon A, Sokolowski MB, Robinson GE. (2002). Influence of gene action across different time scales on behavior. *Science* 296:741–44.

Betz O. (1998). Comparative studies on the predatory behaviour of *Stenus* spp. (Coleoptera: Staphylinidae): The significance of its specialized labial apparatus. *J Zool* 244:527–44.

Briscoe AD, Chittka L. (2001). The evolution of color vision in insects. *Annu Rev Entomol* 46:471–510.

Brooks DR, McLennan DA. (1991). *Phylogeny, Ecology, and Behavior*. Chicago: University of Chicago Press.

Bruce MJ, Herberstein ME, Elgar MA. (2001). Signalling conflict between prey and predator attraction. *J Evol Biol* 14:786–94.

Cameron SA, Hines HM, Williams PH. (2007). A comprehensive phylogeny of the bumble bees (*Bombus*). *Biol J Linn Soc* 91:161–88.

Chittka L. (1996). Does bee color vision predate the evolution of flower color? *Naturwissenschaften* 83:136–38.

Chittka L. (1998). Sensorimotor learning in bumblebees: Long-term retention and reversal training. *J Exp Biol* 201:515–24.

Chittka L. (2002). Influence of intermittent rewards in learning to handle flowers in bumblebees (Hymenoptera: Apidae: *Bombus impatiens*). *Entomol Gener* 26:85–91.

Chittka L. (2004). Dances as windows into insect perception. *PLoS Biol* 2:898–900.

Chittka L, Briscoe A. (2001). Why sensory ecology needs to become more evolutionary—Insect color vision as a case in point. In Barth FG, Schmid A (eds.), *Ecology of Sensing*. Berlin: Springer Verlag, pp. 19–37.

Chittka L, Ings TC, Raine NE. (2004). Chance and adaptation in the evolution of island bumblebee behaviour. *Popul Ecol* 46:243–51.

Chittka L, Raine NE. (2006). Recognition of flowers by pollinators. *Curr Opin Plant Biol* 9:428–35.

Chittka L, Spaethe J, Schmidt A, Hickelsberger A. (2001). Adaptation, constraint, and chance in the evolution of flower color and pollinator color vision. In Chittka L, Thomson JD (eds.), *Cognitive Ecology of Pollination*. Cambridge, UK: Cambridge University Press, pp. 106–26.

Chittka L, Waser NM. (1997). Why red flowers are not invisible to bees. *Isr J Plant Sci* 45:169–83.

Chittka L, Wells H. (2004). Color vision in bees: Mechanisms, ecology and evolution. In Prete FR (ed.), *Complex Worlds from Simpler Nervous Systems*. Cambridge, MA: MIT Press, pp. 165–91.

Clutton-Brock TH, Harvey PH. (1977). Species differences in feeding and ranging behaviour in primates. In Clutton-Brock TH (Ed.), *Primate Ecology: Studies of Feeding and Ranging Behaviour in Lemurs, Monkeys and Apes*. London: Academic Press, pp. 557–84.

Collins SA. (1999). Is female preference for male repertoires due to sensory bias? *Proc R Soc Lond B* 266:2309–14.

Condit R, Ashton PS, Baker P, Bunyavejchewin S, Gunatilleke S, Gunatilleke N, Hubbell SP, Foster RB, Itoh A, LaFrankie JV, Lee HS, Losos E, Manokaran N, Sukumar R, Yamakura T. (2000). Spatial patterns in the distribution of tropical tree species. *Science* 288:1414–18.

Cuthill IC, Kacelnik A, Krebs JR, Haccou P, Iwasa Y. (1990). Starlings exploiting patches: The effect of recent experience on foraging decisions. *Anim Behav* 40:625–40.

Dafni A, Bernhardt P, Shmida A, Ivri Y, Greenbaum S, O'Toole C, Losito L. (1990). Red bowl-shaped flowers: Convergence for beetle pollination in the Mediterranean region. *Isr J Bot* 39:81–92.

Dawkins MS, Guilford T. (1996). Sensory bias and the adaptiveness of female choice. *Am Nat* 148:937–42.

Dornhaus A, Chittka L. (2004a). Information flow and regulation of foraging activity in bumble bees (*Bombus* spp.). Apidologie 35:183–92.

Dornhaus A, Chittka L. (2004b). Why do honey bees dance? *Behav Ecol Sociobiol* 55:395–401.

Dukas R. (1998). *Cognitive Ecology: The Evolutionary Ecology of Information Processing and Decision Making*. Chicago: University of Chicago Press.

Dukas R. (2004). Evolutionary biology of animal cognition. *Annu Rev Ecol Evol Syst* 35:347–74.

Dukas R, Bernays EA. (2000). Learning improves growth rate in grasshoppers. *Proc Natl Acad Sci USA* 97:2637–40.

Dukas R, Duan JJ. (2000). Fitness consequences of associative learning in a parasitoid wasp. *Behav Ecol* 11:536–43.

Dukas R, Real LA. (1991). Learning foraging tasks by bees: A comparison between social and solitary species. *Anim Behav* 42:269–76.

Dyer AG, Whitney HM, Arnold SEJ, Glover BJ, Chittka L. (2006). Bees associate warmth with floral colour. *Nature* 442:525.

Dyer FC. (2002). The biology of the dance language. *Ann Rev Entomol* 47:917–49.

Dyer FC, Seeley TD. (1989). On the evolution of the dance language. *Am Nat* 133:580–90.

Endler JA. (1986). *Natural Selection in the Wild*. Princeton, NJ: Princeton University Press.

Endler JA, Basolo AL. (1998). Sensory ecology, receiver biases and sexual selection. *Trends Ecol Evol* 13:415–20.

Estoup A, Solignac M, Cornuet JM, Goudet J, Scholl A. (1996). Genetic differentiation of continental and island populations of *Bombus terrestris* (Hymenoptera: Apidae) in Europe. *Mol Ecol* 5:19–31.

Farooqui T, Robinson K, Vaessin H, Smith BH. (2003). Modulation of early olfactory processing by an octopaminergic reinforcement pathway in the honeybee. *J Neurosci* 23:5370–80.

Farooqui T, Vaessin H, Smith BH. (2004). Octopamine receptors in the honeybee (*Apis mellifera*) brain and their disruption by RNA-mediated interference. *J Insect Physiol* 50:701–13.

Fire A, Xu S, Montgomery MK, Kostas SA, Driver SE, Mello CC. (1998). Potent and specific genetic interference by double-stranded RNA in *Caenorhabditis elegans*. *Nature* 391:806–11.

Fitzpatrick MJ, Ben-Shahar Y, Smid HM, Vet LEM, Robinson GE, Sokolowski MB. (2005). Candidate genes for behavioural ecology. *Trends Ecol Evol* 20:96–104.

Ford EB. (1955). Rapid evolution and the conditions which make it possible. *Cold Spring Harb Symp Quant Biol* 20:230–38.

von Frisch K. (1955). *The Dancing Bees*. London: Methuen.

von Frisch K. (1967). *The Dance Language and Orientation of Bees*. Cambridge, MA: Belknap Press.

Fülöp A, Menzel R. (2000). Risk-indifferent foraging behaviour in honeybees. *Anim Behav* 60:657–66.

Gallistel CR. (1990). *The Organization of Learning*. Cambridge, MA: MIT Press.

Gegear RJ, Laverty TM. (2004). Effect of a colour dimorphism on the flower constancy of honey bees and bumble bees. *Can J Zool* 82:587–93.

Gentry JB. (1974). Responses to predation by colonies of the Florida harvester ant, *Pogonomyrmex badius*. *Ecology* 55:1328–38.

Giurfa M, Núñez J, Chittka L, Menzel R. (1995). Colour preferences of flower-naive honeybees. *J Comp Physiol A* 177:247–59.

Gould JL. (1975). Honey bee recruitment: The dance-language controversy. *Science* 189:685–93.

Greggers U, Menzel R. (1993). Memory dynamics and foraging strategies of honeybees. *Behav Ecol Sociobiol* 32:17–29.

Gumbert A. (2000). Color choices by bumble bees (*Bombus terrestris*): Innate preferences and generalization after learning. *Behav Ecol Sociobiol* 48:36–43.

Healy SD, de Kort SR, Clayton NS. (2005). The hippocampus, spatial memory and food hoarding: A puzzle revisited. *Trends Ecol Evol* 20:17–22.

Heinrich B. (1979). *Bumblebee Economics*. Cambridge, MA: Harvard University Press.

Heinrich B, Mudge PR, Deringis PG. (1977). Laboratory analysis of flower constancy in foraging bumble bees: *Bombus ternarius* and *B. terricola*. *Behav Ecol Sociobiol* 2:247–65.

Herbers JM. (1980). On caste ratios in ant colonies: Population responses to changing environments. *Evolution* 34:575–85.

Honey Bee Genome Sequencing Consortium. (2006). Insights into social insects from the genome of the honeybee *Apis mellifera*. *Nature* 443:931–49.

Humphries MA, Müller U, Fondrk MK, Page RE. (2003). PKA and PKC content in the honey bee central brain differs in genotypic strains with distinct foraging behavior. *J Comp Physiol A* 189:555–62.

Hunt GJ, Amdam GV, Schlipalius D, Emore C, Sardesai N, Williams CE, Rueppell O, Guzmán-Novoa E, Arechavaleta-Velasco M, Chandra S, Fondrk MK, Beye M, Page RE. (2007). Behavioral genomics of honeybee foraging and nest defense. *Naturwissenschaften* 94:247–67.

Ings TC, Ward NL, Chittka L. (2006). Can commercially imported bumble bees out-compete their native conspecifics? *J Appl Ecol* 43:940–48.

Irwin RE, Brody AK. (1999). Nectar-robbing bumble bees reduce the fitness of *Ipomopsis aggregata* (Polemoniaceae). *Ecology* 80:1703–12.

Johnston TD. (1982). Selective costs and benefits in the evolution of learning. *Adv Stud Behav* 12:65–106.

Kacelnik A, Krebs JR. (1985). Learning to exploit patchily distributed food. In Silby RM, Smith R (eds.), *Behavioural Ecology*. Oxford: Oxford University Press, pp. 189–205.

Kingsolver JG, Hoekstra HE, Hoekstra JM, Berrigan D, Vignieri SN, Hill CE, Hoang A, Gibert P, Beerli P. (2001). The strength of phenotypic selection in natural populations. *Am Nat* 157:245–61.

Krebs JR, Davies NB. (1993). *An Introduction to Behavioural Ecology*. 3rd ed. Oxford: Blackwell Science.

Laverty TM. (1994). Bumble bee learning and flower morphology. *Anim Behav* 47:531–45.

Lipp HP. (2002). The tortuous path from genotype to phenotype: Genes and cognition in mutant mice. *Eur J Hum Genet* 10:55.

Losos JB, Miles DB. (1994). Adaptation, constraint, and the comparative method: Phylogenetic issues and methods. In Wainwright PC, Reilly SM (eds.), *Ecological Morphology: Integrative Organismal Biology*. Chicago: University of Chicago Press, pp. 60–98.

Lunau K, Maier EJ. (1995). Innate color preferences of flower visitors. *J Comp Physiol A* 177:1–19.

Lunau K, Wacht S, Chittka L. (1996). Colour choices of naive bumble bees and their implications for colour perception. *J Comp Physiol A* 178:477–89.

Maynard Smith J. (1978). Optimization theory in evolution. *Annu Rev Ecol Syst* 9:31–56.

McNamara JM, Houston AI, Weisser WW. (1993). Combining prey choice and patch use—What does rate-maximizing predict? *J Theor Biol* 164:219–38.

Menzel R. (1985). Learning in honey bees in an ecological and behavioral context. In Hölldobler B, Lindauer M (eds.), *Experimental Behavioral Ecology*. Stuttgart: Gustav Fischer Verlag, pp. 55–74.

Menzel R. (2001). Behavioral and neural mechanisms of learning and memory as determinants of flower constancy. In Chittka L, Thomson JD (eds.), *Cognitive Ecology of Pollination*. Cambridge, MA: Cambridge University Press, pp. 21–40.

Menzel R, Shmida A. (1993). The ecology of flower colours and the natural colour vision of insect pollinators: The Israeli flora as a case study. *Biol Rev* 68:81–120.

Mery F, Kawecki TJ. (2003). A fitness cost of learning ability in *Drosophila melanogaster*. *Proc R Soc Lond B* 270:2465–69.

Mery F, Kawecki TJ. (2004). An operating cost of learning in *Drosophila melanogaster*. *Anim Behav* 68:589–98.

Mery F, Kawecki TJ. (2005). A cost of long-term memory in *Drosophila*. *Science* 308:1148.

Molet M, Chittka L, Stelzer RJ, Streit S, Raine NE. (2008). Colony nutritional status modulates worker responses to foraging recruitment pheromone in the bumblebee. *Bombus terrestris*. *Behav Ecol Sociobiol* 62:1919–26.

Müller CB, Schmid-Hempel P. (1992). Correlates of reproductive success among field colonies of *Bombus lucorum*: The importance of growth and parasites. *Ecol Entomol* 17:343–53.

Oldroyd BP, Thompson GJ. (2007). Behavioural genetics of the honey bee *Apis mellifera*. *Adv Insect Physiol* 33:1–49.

Olesen JM. (1985). The Macronesian bird-flower elements and its relation to bird and bee opportunists. *Bot J Linn Soc* 91:395–414.

Ollerton J, Cranmer L, Stelzer RJ, Sullivan S, Chittka L. (2009). Bird pollination of Canary Island endemic plants. *Naturwissenschaften* 96:221–32.

Owens IPF. (2006). Where is behavioural ecology going? *Trends Ecol Evol* 21:356–61.

Page RE, Erber J, Fondrk MK. (1998). The effect of genotype on response thresholds to sucrose and foraging behavior of honey bees (*Apis mellifera* L.). *J Comp Physiol A* 182:489–500.

Page RE, Fondrk MK. (1995). The effects of colony-level selection on the social organization of honey bee (*Apis mellifera* L.) colonies: Colony level components of pollen hoarding. *Behav Ecol Sociobiol* 36:135–44.

Page RE, Robinson GE. (1991). The genetics of division of labor in honey bee colonies. *Adv Insect Physiol* 23:117–69.

Page RE, Waddington KD, Hunt GJ, Fondrk MK. (1995). Genetic determinants of honey bee foraging behaviour. *Anim Behav* 50:1617–25.

Pankiw T, Page RE. (2000). Response thresholds to sucrose predict foraging division of labor in honeybees. *Behav Ecol Sociobiol* 47:265–67.

Pankiw T, Page RE. (2001). Genotype and colony environment affect honeybee (*Apis mellifera* L.) development and foraging behavior. *Behav Ecol Sociobiol* 51:87–94.

Pankiw T, Tarpy DR, Page RE. (2002). Genotype and rearing environment affect honeybee perception and foraging behaviour. *Anim Behav* 64:663–72.

Papaj DR, Prokopy RJ. (1989). Ecological and evolutionary aspects of learning in phytophagous insects. *Ann Rev Entomol* 34:315–50.

Pelletier L, McNeil JN. (2003). The effect of food supplementation on reproductive success in bumblebee field colonies. *Oikos* 103:688–94.

Phelps SM, Ryan MJ. (2000). History influences signal recognition: Neural network models of túngara frogs. *Proc R Soc Lond B* 267:1633–39.

Pitman NCA, Terborgh JW, Silman MR, Núñez P, Neill DA, Cerón CE, Palacios WA, Aulestia M. (2001). Dominance and distribution of tree species in upper Amazonian terra firme forests. *Ecology* 82:2101–17.

Pyke GH. (1978). Optimal foraging: Movement patterns of bumblebees between inflorescences. *Theor Popul Biol* 13:72–98.

Pyke GH, Pulliam HR, Charnov EL. (1977). Optimal foraging: A selective review of theory and tests. *Q Rev Biol* 52:137–54.

Raine NE, Chittka L. (2005). Colour preferences in relation to the foraging performance and fitness of the bumblebee *Bombus terrestris*. *Uludag Bee J* 5:145–50.

Raine NE, Chittka L. (2007a). The adaptive significance of sensory bias in a foraging context: Floral colour preferences in the bumblebee *Bombus terrestris*. *PLoS One* 2: e556. doi:10.1371/journal.pone.0000556.

Raine NE, Chittka L. (2007b). Flower constancy and memory dynamics in bumblebees (Hymenoptera: Apidae: *Bombus*). *Entomol Gener* 29:179–99.

Raine NE, Chittka L. (2007c). Nectar production rates of 75 bumblebee-visited flower species in a German flora (Hymenoptera: Apidae: *Bombus terrestris*). *Entomol Gener* 30:191–92.

Raine NE, Chittka L. (2007d). Pollen foraging: Learning a complex motor skill by bumblebees (*Bombus terrestris*). *Naturwissenschaften* 94:459–64.

Raine NE, Chittka L. (2008). The correlation of learning speed and natural foraging success in bumble-bees. *Proc R Soc B* 275:803–8.

Raine NE, Ings TC, Dornhaus A, Saleh N, Chittka L. (2006a). Adaptation, genetic drift, pleiotropy, and history in the evolution of bee foraging behavior. *Adv Stud Behav* 36:305–54.

Raine NE, Ings TC, Ramos-Rodríguez O, Chittka L. (2006b). Intercolony variation in learning performance of a wild British bumblebee population (Hymenoptera: Apidae: *Bombus terrestris audax*). *Entomol Gener* 28:241–56.

Raine NE, Pierson AS, Stone GN. (2007). Plant-pollinator interactions in a Mexican *Acacia* community. *Arthropod-Plant Interact* 1:101–17.

Real LA. (1981). Uncertainty and pollinator-plant interactions: The foraging behavior of bees and wasps on artificial flowers. *Ecology* 62:20–26.

Robinson GE, Page RE, Strambi C, Strambi A. (1989). Hormonal and genetic control of behavioral integration in honey bee colonies. *Science* 246:109–12.

Ross KG, Keller L. (1998). Genetic control of social organization in an ant. *Proc Natl Acad Sci USA* 95:14232–37.

Roubik DW. (1992). Loose niches in tropical communities: Why are there so few bees and so many trees? In Hunter MD, Ohgushi T, Price PW (eds.), *Effects of Resource Distribution on Animal-Plant Interactions*. San Diego: Academic Press, pp. 327–54.

Rueppell O, Pankiw T, Nielsen DI, Fondrk MK, Beye M, Page RE. (2004a). The genetic architecture of the behavioral ontogeny of foraging in honeybee workers. *Genetics* 167:1767–79.

Rueppell O, Pankiw T, Page RE. (2004b). Pleiotropy, epistasis and new QTL: The genetic architecture of honey bee foraging behavior. *J Hered* 95:481–91.

Ryan MJ, Rand AS. (1999). Phylogenetic influence on mating call preferences in female túngara frogs, *Physalaemus pustulosus*. *Anim Behav* 57:945–56.

Scheiner R, Page RE, Erber J. (2001a). The effects of genotype, foraging role, and sucrose responsiveness on the tactile learning performance of honey bees (*Apis mellifera* L.). *Neurobiol Learn Mem* 76:138–50.

Scheiner R, Page RE, Erber J. (2001b). Responsiveness to sucrose affects tactile and olfactory learning in pre-foraging honey bees of two genetic strains. *Behav Brain Res* 120:67–73.

Schmid-Hempel P, Heeb D. (1991). Worker mortality and colony development in bumblebees, *Bombus lucorum* (L.) (Hymenoptera, Apidae). *Mitt Schweiz Entomol Ges* 64:93–108.

Schmid-Hempel P, Kacelnik A, Houston AI. (1985). Honeybees maximize efficiency by not filling their crop. *Behav Ecol Sociobiol* 17:61–66.

Schmid-Hempel R, Schmid-Hempel P. (1998). Colony performance and immunocompetence of a social insect, *Bombus terrestris*, in poor and variable environments. *Funct Ecol* 12:22–30.

Seeley TD. (1995). *The Wisdom of the Hive: The Social Physiology of Honey Bee Colonies*. Cambridge, MA: Harvard University Press.

Shafir S, Wiegmann DD, Smith BH, Real LA. (1999). Risk-sensitive foraging: Choice behaviour of honeybees in response to variability in volume of reward. *Anim Behav* 57:1055–61.

Sherry DF, Healy S. (1998). Neural mechanisms of spatial representation. In Healy S (ed.), *Spatial Representations in Animals*. Oxford: Oxford University Press, pp. 133–58.

Shettleworth SJ. (1998). *Cognition, Evolution, and Behavior*. Oxford: Oxford University Press.

Shobe J. (2002). The role of PKA, CaMKII and PKC in avoidance conditioning: Permissive or instructive? *Neurobiol Learn Mem* 77:291–312.

Skorupski P, Döring TF, Chittka L. (2007). Photoreceptor spectral sensitivity in island and mainland populations of the bumblebee, *Bombus terrestris*. *J Comp Physiol A* 193:485–94.

Smith C, Barber I, Wootton RJ, Chittka L. (2004). A receiver bias in the origin of three-spined stickleback mate choice. *Proc R Soc Lond B* 271:949–55.

Stanton ML, Galen C. (1997). Life on the edge: Adaptation versus environmentally mediated gene flow in the snow buttercup, *Ranunculus adoneus*. *Am Nat* 150:143–78.

Stelzer RJ, Ollerton J, Chittka L. (2007). Keine Nachweis für Hummelbesuch der Kanarischen Vogelblumen (Hymenoptera: Apidae). *Entomol Gener* 30:153–54.

Stephens DW, Krebs JR. (1986). *Foraging Theory*. Princeton, NJ: Princeton University Press.

Stone GN, Raine NE, Prescott M, Willmer PG. (2003). Pollination ecology of acacias (Fabaceae, Mimosoideae). *Aust Syst Bot* 16:103–18.

Stone GN, Willmer PG, Rowe JA. (1998). Partitioning of pollinators during flowering in an African *Acacia* community. *Ecology* 79:2808–27.

Streit S, Bock F, Pirk CWW, Tautz J. (2003). Automatic life-long monitoring of individual insect behaviour now possible. *Zoology* 106:169–71.

Sumner S, Lucas E, Barker J, Isaac N. (2007). Radio-tagging technology reveals extreme nest-drifting behavior in a eusocial insect. *Curr Biol* 17:140–45.

Sutcliffe GH, Plowright RC. (1988). The effects of food supply on adult size in the bumble bee *Bombus terricola* Kirby (Hymenoptera: Apidae). *Can Entomol* 120:1051–58.

Towne WF, Gould JL. (1988). The spatial precision of the honey bees' dance communication. *J Insect Behav* 1:129–55.

Valido A, Dupont YL, Hansen DM. (2002). Native birds and insects, and introduced honey bees visiting *Echium wildpretii* (Boraginaceae) in the Canary Islands. *Acta Oecol* 23:413–19.

Visscher PK, Seeley TD. (1982). Foraging strategy of honeybee colonies in a temperate deciduous forest. *Ecology* 63:1790–801.

Vogel S, Westerkamp C, Thiel B, Gessner K. (1984). Ornithophilie auf den Canarischen Inseln. *Plant Syst Evol* 146:225–48.

Waddington KD, Holden LR. (1979). Optimal foraging: On flower selection by bees. *Am Nat* 114:179–96.

Waddington KD, Visscher PK, Herbert TJ, Richter MR. (1994). Comparisons of forager distributions from matched honey bee colonies in suburban environments. *Behav Ecol Sociobiol* 35:423–29.

Weidenmüller A, Seeley TD. (1999). Imprecision in waggle dances of the honeybee (*Apis mellifera*) for nearby food sources: Error or adaptation? *Behav Ecol Sociobiol* 46:190–99.

Whitfield CW, Cziko AM, Robinson GE. (2003). Gene expression profiles in the brain predict behavior in individual honey bees. *Science* 302:296–99.

Widmer A, Schmid-Hempel P, Estoup A, Scholl A. (1998). Population genetic structure and colonization history of *Bombus terrestris* s.l. (Hymenoptera: Apidae) from the Canary Islands and Madeira. *Heredity* 81: 563–72.

Willmer PG, Stone GN. (2004). Behavioral, ecological, and physiological determinants of the activity patterns of bees. *Adv Stud Behav* 34:347–466.

Wolfer DP, Lipp HP. (2000). Dissecting the behaviour of transgenic mice: Is it the mutation, the genetic background, or the environment? *Exp Physiol* 85:627–34.

2 Social Cues and Adaptive Foraging Strategies in Ants

Claire Detrain and Jean-Louis Deneubourg

CONTENTS

INTRODUCTION

Social insects have evolved different ways of sharing information about many aspects of their everyday life. In contrast to solitary insects, each decision taken by a colony member—for example, about where to forage or which food site to exploit—is no longer solely based on the individual's own experience, but involves the integration of information from other nestmates. The sharing of information between workers is the essence of sociality, because it is a prerequisite for the coordination of groups, whatever their size—from a few individuals up to several thousands of workers, as is the case in mature colonies of army ants.

Recruitment—the local increase of workers cooperating at a particular place—is the best-studied example of information sharing in insect societies. Recruitment is involved in many situations, such as food exploitation (for a review, see Hölldobler and Wilson 1990; Detrain et al. 1999), nest moving (e.g., Franks et al. 2002), territorial defense (e.g., Hölldobler 1981; Czechowski 1984; Detrain and Pasteels 1992), and nest building (e.g., Buhl et al. 2005). The mechanisms and signals used by social insects to recruit nestmates greatly vary among the species and can depend on their physiology, cognitive abilities, or phylogenetic background.

When discovering the meaning of the honey bees' dance, Karl von Frisch revealed a striking example on how complex and sophisticated the communication about food sources can be in social insects (von Frisch 1967). In regard to ants, many species use a probably less impressive but equally efficient method to recruit nestmates, which is by laying a pheromone trail. The basic scheme of interactions is similar for many ant species: As soon as an ant has discovered a food source, it goes back to the nest, depositing a chemical trail. The trail pheromone—sometimes in synergy with tactile and chemical "invitations" from the recruiter—triggers the exit of additional foragers and guides them, like an Ariadne's thread, to the food source. After feeding, each recruited ant can, in turn, reinforce the chemical trail and, consequently, stimulate other nestmates to forage. Trail recruitment, therefore, is a typical amplifying process, which means that the emission of a signal (i.e., the trail pheromone) increases the likeliness of its expression by additional nestmates (Wilson 1962; Sudd 1957).

Apart from the different levels of complexity, trail recruitment in ants also differs from the honey bee's dance in the quantity of potential recruits at which the recruitment signal is aimed. In fact, one honey bee recruits only a limited number of nestmates, regardless of how many bees are waiting in the hive in order to be recruited. This is due to the physical limit to the number of bees that can follow a dancer (Gould 1974, 1975; Seeley and Towne 1992; Seeley 1995; Dornhaus et al. 2006). By contrast, in trail recruitment, a single ant can potentially recruit large numbers of nestmates, depending on how many ants can perceive and follow the trail before the pheromone has completely evaporated (for a detailed discussion of these two contrasting recruitment strategies, see Chapter 6). Here, the limit to the number of potential recruits is set physically by the active space of the trail pheromone as well as physiologically by the perception threshold of the individual ants (Bossert and Wilson 1963; Pasteels et al. 1986). Hence, the intensity of trail recruitment in ants is determined not only by the behavior of a recruiting individual, but also by the number of potential recruits that are located within the trail's active space. In ants, therefore, the foraging dynamics (e.g., the rate of nest exits) should be particularly sensitive to colony size or, more precisely, the local density of nestmates.

FORAGING PATTERNS IN ANTS: A SOCIAL OUTCOME

A prime effect of an increasing colony size—and hence of a larger number of cooperating ants—is the higher potential for amplification processes. These positive feedback mechanisms, such as trail reinforcement by foragers, lead to a nonlinear increase in the number of recruits at food resources. In large ant colonies, the huge number of interacting individuals strengthens the impact of such amplifying processes and influences the benefits of certain recruitment strategies (Beckers et al. 1989; Anderson and McShea 2001; Beekman et al. 2001; Jun et al. 2003; Mailleux et al. 2003b; see also Chapter 6). In the case of foraging, this may speed up the mobilization of nestmates toward a specific food patch and, consequently, facilitate the monopolization of this resource by colonies with larger worker populations. Furthermore, the size of a colony strongly influences the pattern of the trail network that connects the nest with the food patches. Since each trail is the product of successive reinforcements by foragers, the final shape of the trail network strongly depends on the number of ants that have interacted with each other. When multiple food sources are available at the same time, large colonies tend to exploit only a subset of resources and travel along a few highly frequented trails (Deneubourg and Goss 1989; Camazine et al. 2001; Detrain and Deneubourg 2008). Conversely, workers of small colonies are scattered and forage at several food sources (Detrain et al. 1991; Nicolis and Deneubourg 1999; Beekman et al. 2001; Detrain and Deneubourg 2006).

All these examples clearly demonstrate that the social context of foraging, i.e., the number/density of interacting nestmates, plays a key role in the development of adaptive foraging strategies. Therefore, an ant colony should be able to adjust the density of individuals at any time and at any location in order to optimize its foraging strategies according to the resource opportunities.

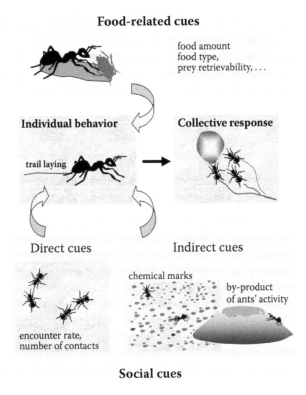

Food-related cues

food amount
food type,
prey retrievability, . . .

Individual behavior   **Collective response**

trail laying

Direct cues   Indirect cues

chemical marks

by-product
of ants' activity

encounter rate,
number of contacts

Social cues

FIGURE 2.1 Influence of food and social context in ant foraging. The foraging and trail-laying behavior of ant individuals is influenced by features of the food (honeydew quality, prey size, etc.) as well as by the social context of foraging. This social context can be assessed either by direct social cues, such as contact rates with nestmates, or by indirect social cues, such as by-products of the ants' activity (e.g., height of the nest crater) or the amount of area marking (see text for further explanations). See color insert following page 142.

The adjustment of the number of cooperating foragers is achieved mainly by the tuning of recruitment signals, e.g., the recruitment trail, according to the food characteristics and the current social context of foraging (Figure 2.1). First, the information collected directly at the food sources by ant scouts shapes their recruiting behavior. Depending on the estimated profitability of a food source, scouts decide whether or not to lay a trail, tune the amount of the released pheromone, and thus adjust the number of workers recruited in such a way that they preferentially exploit the resources with the highest rewards (e.g., Hölldobler and Wilson 1990; Detrain et al. 1999; Deneubourg et al. 2005; Detrain and Deneubourg 2008). To assess the value of a discovered food, ant scouts often rely on rather crude but functional decision criteria instead of making sophisticated measurements of food features (Detrain and Deneubourg 1997, 2002, 2006; Detrain et al. 1999; Mailleux et al. 2000, 2003a, 2005, 2006). These simple recruitment rules, which are directly related to food features, most likely were shaped during the course of evolution in a way that the ant density at a food site best fits the cooperation level needed for an efficient exploitation of the respective resource.

In addition to food features, the social context of foraging also strongly influences the tuning of recruiting signals by ants, as well as their behavior on foraged areas (Figure 2.1). In the present chapter, we will review in detail the current knowledge about the role of such social cues—in particular, those related to the density and activity of individual foragers—and will discuss their influence on the ants' foraging behavior. Special emphasis will be put on the costs and benefits that arise for the ants when they use either direct or indirect cues to assess their social context.

HOW MANY FORAGERS? ASSESSING THE SOCIAL CONTEXT OF FORAGING

Today, many publications claim that information from con- and heterospecifics plays a dominant role in many fitness-enhancing decisions (Danchin et al. 2004). The use of socially acquired information has been extensively investigated in animals with sophisticated cognitive abilities, such as vertebrates. In these cases, an individual can improve its behavioral strategy either by acquiring experience on its own or by observing how conspecifics solve problems and succeed in accessing resources, avoiding predators, etc. (see Chapter 7 for a discussion of the social transmission of information in insects).

Trying to understand how ants make foraging decisions would be useless if one focused only on how ants assess food features. In fact, there is also an extremely rich body of information provided to foraging ants by their nestmates, by conspecifics, which may even be competitors, and by the by-products arising from the activity of other individuals. As emphasized by Seeley (1998) when describing the organization of the honey bee hive: "the process of colony integration is largely a matter of information flow from colony to individual, so that each individual can adjust its behavior in accordance with the activities of the other colony members" (Seeley 1998, p. 73). Likewise, in ants, acquiring information about the current social context of foraging can be important for scouts before deciding whether to lay a recruitment trail or not. For instance, nestmate density at a food patch may indicate how many workers are already potentially available for cooperation, and whether additional foragers should be recruited. The ability of a scout to adjust its recruiting behavior according to the current social context, i.e., to the number or the density of nestmate workers, substantially improves the efficiency of food exploitation. Depending on its perception of ant density, a forager may decide to reinforce the trail in order to speed up the mobilization of additional workers and, consequently, to increase cooperation or, conversely, to stop recruitment in order to slow down the number of newcomers arriving at overcrowded food sources.

Group-level indicators, which may be used by individuals to assess the social context of foraging, most often are cues. Unlike signals, which have been molded by natural selection for the purpose of information transfer, cues are not specifically designed for communication and convey information only incidentally (Seeley 1998). Social cues can be obtained from any behavior of fellow nestmates or from any by-product of their activities. The challenge is to identify the social cues that are actually used by ants to fine-tune their foraging behavior, and that lead to an adaptive response at the colony level (Figure 2.1).

In the present chapter, we approach this question by reviewing the current knowledge about how social cues influence the foraging strategies of ants (Figure 2.1). There are two possibilities for ants to obtain information about the presence or activity of conspecifics. First, social information can be acquired by direct sampling, i.e., by contacts between individuals. The second major channel of social information transfer is the shared environment, where an individual perceives modifications of the environment by other group members. This form of information transfer is related to the "stigmergy" concept, which was originally proposed by Grassé (1959) to explain termite nest-building behavior. Stigmergy accounts for the fact that changes in the behavior of an individual can be induced by the perception of the work that has already been accomplished by its nestmates. Modifying the local environment, therefore, represents an indirect way of information exchange between individuals (e.g., during joint comb building; Darchen 1959; Hepburn 1986; Skarka et al. 1990; Camazine and Sneyd 1991). Unlike direct interactions between nestmates, indirect social cues easily allow an asynchronous transfer of information between the sender and the receiver. In this chapter, we will describe in detail how chemical cues of conspecifics can be used as indirect social cues that allow individuals to assess their social environment, i.e., the nestmate density at a certain location, even in the absence of the emitters of these cues.

DIRECT SOCIAL CUES: GETTING INFORMATION THROUGH CONTACTS

In honey bees, it is a well-established fact that direct cues play an essential role in work organization, especially for the management of the food supply/food demand within the hive. Foragers directly obtain social information about the colony's food demand through interactions with bees working inside the hive. Shortly after a nectar-collecting honey bee forager returns to her colony, she approaches her nestmates and tries to unload her crop to a receiver bee, or to distribute the collected nectar among several receivers in multiple bouts of unloadings (Seeley 1989; Hart and Ratnieks 2001; Huang and Seeley 2003; see also Chapter 10). Here, the delay a forager experiences prior to contacting a nectar receiver serves as a cue that informs her about the colony's capacity to process the incoming food (Seeley 1989, 1995; Seeley and Tovey 1994; Ratnieks and Anderson 1999; Chapter 10). A short delay indicates that the work capacity of receivers exceeds that of nectar-collecting bees, and the active forager usually performs a waggle dance to recruit additional foragers to the flower patch. A long delay, on the other hand, usually causes the forager to perform another type of dance—the tremble dance—that recruits additional receivers to unload the incoming nectar foragers (Seeley 1992; Kirchner and Lindauer 1994). These contact delays, i.e., the time spent in finding a nectar receiver, are a direct social cue that influences an active forager's decision of whether to start, pursue, or stop foraging or to recruit nestmates. This well-studied example shows that patterns of direct encounters between nestmates are crucial for the organization of work in the bee hive. In line with the findings in honey bees, direct interactions among ant workers play a key role in improving the foraging efficiency of a colony, too. A growing body of evidences indicates that encounters between nestmates are a major channel for information transfer in ant societies. Many studies have investigated task regulation or information exchange associated with trophallactic interactions inside the nest (e.g., Cassill and Tschinkel 1995, 1999; Lenoir et al. 1999; McCabe et al. 2006). In the present chapter, however, we will mainly focus on direct contacts *without* food exchange.

ALLOCATING ANTS TO FORAGING TASKS

Ant colonies have to respond to changing conditions by adjusting the number of workers engaged in various tasks, such as foraging or work required within the nest. Since the process of task allocation operates without a central control, it should result from the responses of workers to local social cues. In this context, the pattern of antennal contacts with other workers can inform an ant about the number of individuals that are currently engaged in a particular activity. When two ants meet and touch each other with their antennae, they perceive cuticular chemical compounds that inform them of not only whether the other individual is a nestmate, but also which task it has recently performed. For example, *Pogonomyrmex barbatus* foragers, which are more exposed to light and dry conditions when walking outside the nest, have higher portions of n-alkanes than n-alkenes and branched alkanes in their cuticular hydrocarbon profiles than ants that work inside the nest (Wagner et al. 2001). Based on the contacts with other ants, an individual can thus evaluate to which extent a specific task has recently been performed. Harvester ants are a well-studied example in this context and have been demonstrated to use direct social cues for the regulation of task allocation. For instance, a *Pogonomyrmex* worker more likely changes its task to do midden work when its rate of encounters with midden workers is high (Gordon and Mehdiabadi 1999). Similarly, in the context of foraging, the safe return of patrollers stimulates the foragers to leave the nest. Here, the crucial cue triggering the onset of foraging is the rate at which patrollers return, or the time interval between interactions of patrollers and prospective foragers, respectively (Greene and Gordon 2003, 2007). Both examples indicate that an ant's decision to switch between tasks is not based on single encounters with other ants but on multiple interactions with her nestmates (Gordon and Mehdiabadhi 1999). Thus, each individual may improve its "knowledge" about the current task allocation within the entire colony by collecting information via the frequency of contacts with nestmates performing the same or different tasks.

IMPROVING FORAGING EFFICIENCY

Direct encounters between inbound and outbound workers traveling along trails or walking around within the foraging areas may substantially improve the foraging efficiency of the colony. During these head-on encounters, each individual perceives not only the partner's identity but also the type of resource the other ant currently exploits, or even the profitability of a particular foraging patch, through the frequency with which it encounters laden workers. It is well documented that in honey bees (Seeley 1995) and bumble bees (Dornhaus and Chittka 2004) floral odors clinging to the body of a scout are detected by other workers, which make use of this olfactory information during their search for food outside the hive (see also Chapters 9 and 10). Similarly, in *Lasius niger*, residues of sucrose on the mandibles or antennae of scouts are perceived by recruited nestmates, and convey information about the nature of the discovered food. Likewise, in leaf-cutting ants, outbound and inbound workers laden with food exchange information about the food characteristics during encounters on the foraging trails (Roces 1990; Roces and Núñez 1993; Howard et al. 1996; see also Chapter 14). That information actually is transferred by such encounters is indicated by the fact that recruited workers display the same food-correlated search tactics as the scouts after reaching the site of food discovery (Le Breton and Fourcassié 2004). Hence, the mere contact with food items or food residues on a recruiter's body informs the workers about the characteristics of the resource prior to their arrival at the feeding site, and triggers a specific locomotor program that increases the probability of finding the respective kind of food within the patch.

Encounters with nestmates may also increase the foraging efficiency along the ants' trails. At first sight, a high rate of interactions on the trails seems detrimental to foraging, because frequent head-on encounters on overcrowded trails may lead to a reduction in the overall progression of the ants. This effect, however, is reduced by the formation of distinct traffic lanes (Couzin and Franks 2003) or by regulating the recruitment intensity along a trail according to the ants' density on it (Jaffe and Howse 1979). U-turns, which are induced when an ant encounters other ants traveling in the opposite direction, may also regulate the traffic by acting as a dispersive force that counteracts trail congestion (Dussutour et al. 2004, 2005, 2006). In addition, several studies have convincingly demonstrated that a high rate of contacts between workers improves the food retrieval efficiency of the colony instead of diminishing it. For instance, in the ant *Dorymyrmex doetschi*, the colony's foraging success, i.e., the amount of captured prey, increases significantly with the number of brief antennal contacts experienced by the foragers on their return to the nest (Torres-Contreras and Vásquez 2007). Likewise, in *Atta* leaf-cutting ants, a crowding of the foraging trails increases the rate of leaf retrieval (Burd and Aranwela 2003; Dussutour et al. 2007). Outbound workers encountering inbound laden ants are thus stimulated to engage in cutting leaf fragments and transport them to the nest. In leaf- and grass-cutting ants, the transfer of information about the type or quality of the harvested plant is enforced by dropping the cut fragments on the trail and by the formation of transport chains, whereby several individuals consecutively carry the fragments part of the way to the nest (see Chapter 14 for details).

Direct interactions that disseminate food-related cues may generally speed up the information transfer and may influence the outbound foragers' decisions concerning food exploitation and food retrieval.

DEFENDING FORAGING AREAS

There is much evidence that ants can respond to changes in the number of non-nestmate conspecifics in the surroundings of their nests, and that they evaluate their rival's strength during competition for foraging grounds (Hölldobler 1981; Adams 1990; Sakata and Katayama 2001). This ability of individuals to assess the relative force of opponents often relies on direct contacts with competing ants. Indeed, in several species (e.g., *Myrmecocystus* honeypot ants, Hölldobler 1981; *Lasius niger*, Czechowski 1984; *Polyrhachis laboriosa*, Mercier et al. 1997) between-colony contests take the form of tournaments with fighting displays. For instance, when two honeypot ant workers from different nests meet, they intensively drum on the opponent with their antennae and kick it with their legs (Hölldobler 1981). After a while, one individual usually gives up, and the winner can either meet another opponent or return to the nest and recruit dozens of workers to the fighting area. The recruitment decision of the ants is determined by their evaluation of the opponent's strength during the tournaments. Since an individual is unable to perform a "head counting" of all fighters, the enemy's strength could be assessed by the rate at which opponents are encountered, as well as by the queuing delay between two subsequent encounters with opponents. A long queuing delay indicates a small rival density, whereas a short waiting time is an indicator of a large, dominant colony (Lumsden and Hölldobler 1983). Honeypot ants, therefore, use a direct social cue—the rate at which they encounter non-nestmates—before deciding whether to engage in escalating fights or to flee from the battleground.

Several studies have confirmed that the frequency of encounters with rival ants actually provides a social cue for decision making concerning defensive behavior (Gordon et al. 1993; Heinze et al. 1996; Knaden and Wehner 2003; Thomas et al. 2007). In the harvester ant *Messor andrei*, for instance, encounters between workers from neighboring colonies at a food site increase the probability that the colonies return to forage at this particular site. By doing so, even though putting themselves in danger by foraging at contested food sources, the ants may gain important information about changes in the local number of competing ants (Brown and Gordon 2000). Contact rates are thus ideal means for ants to perceive the presence of competitors and to decide about an appropriate response in order to defend (or leave) a food site.

Do Ants Tune Their Contact Rates?

As shown above, antennal contacts are of great importance for the social organization of an ant colony. The frequency of such brief contacts can provide a direct cue about local worker density (nestmates and non-nestmates). In a group of freely moving ants that collide at random, the number of pairwise encounters that each individual experiences per unit of time increases as a function of ant density (ideal gas model, Hutchinson and Waser 2007). As a corollary, following the laws of Brownian motion, the total number of antennal contacts should increase quadratically with ant density. However, under undisturbed conditions, the contact rates between individuals are not a simple consequence of random collisions. Instead, ants appear to regulate their contact rates by walking less randomly than assumed from theoretical calculations for different worker densities (Gordon et al. 1993). At low densities, ants tend to aggregate (e.g., along the edge of flat arenas), which increases the overall contact rates. Conversely, when densities are high, ants seem to level off their contact rates. This regulation is thought to stabilize both the flow of information between nestmates and the decision making at the colony level (Adler and Gordon 1992; Gordon et al. 1993). Yet, the conclusion that ants actively avoid encounters with nearby ants at high worker densities could be hasty. The fact that contact rates are lower than expected from theoretical calculations may simply be the consequence of the kinetics of the collisions. Indeed, because each encounter lasts a finite time (τ), the maximum number of interactions during a period T is limited to the value T/τ and can never increase beyond this limit, even for high ant densities (Nicolis et al. 2005). This fact, however, does not strictly rule out the possibility of an active regulation of direct interactions below the limit set by the duration of the interactions. A promising research topic certainly is to investigate whether a tuning of contacts actually could help ants to maintain the flow of information at a level that is sufficiently high to make efficient decisions.

Sampling Effort and Accuracy of Direct Contact Cues

An ant must achieve multiple contacts if it has to figure out the average ant density at a given moment and location. Assuming that ants move randomly, the number of contacts experienced

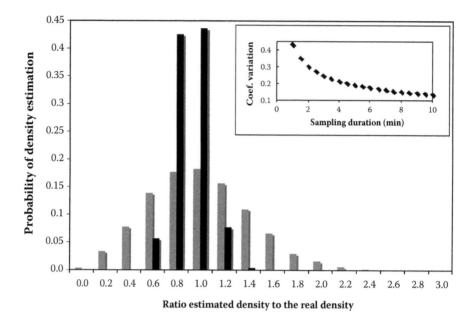

FIGURE 2.2 Distribution of the density values estimated through the number of contacts experienced by an ant in relation to the actual worker density. A ratio of 1 indicates that the ants have correctly estimated the nestmate density. For the simulations, the following assumptions were made: probability of encountering nestmates = 0.08 s^{-1}, which corresponds to a walking speed of 1.5 cm/s; perception range = 1 cm; actual ant density = 0.05 ants/cm^2. Light gray bars: sampling duration of 1 min; dark gray bars: sampling duration of 10 min. Inset: Coefficient of variation of the estimated density (ratio of the standard deviation to the mean value) as a function of sampling duration. The decrease of the coefficient of variation indicates that the probability for an individual to wrongly estimate nestmate density decreases as the sampling duration increases.

during a sampling period provides each ant with an estimate of her nestmates' density. The accuracy of this estimate depends on the sampling effort, i.e., the time spent encountering nestmates (Figure 2.2). For a short sampling duration (1 min), the distribution of the estimated density values—i.e., the number of contacts experienced by the ant—is characterized by a high variability (Figure 2.2). Consequently, many individuals would either underestimate or overestimate the actual ant density. Not surprisingly, for longer sampling periods (e.g., 10 min), the estimated density is less variable. When calculating the coefficient of variation for each density estimate distribution, it becomes clear that the coefficients of variation actually decrease as the sampling time increases (see inset in Figure 2.2). This illustrates well the classical speed versus accuracy trade-off that ants have to face when estimating ant density through their contacts with nestmates: in order to achieve an accurate estimate of nestmate density, ants have to engage in a lengthy sampling process. It should be mentioned here that in the case of area marking, i.e., ants adding one unit of chemical mark per unit time while walking, the amount of volatiles at a location can also serve as an indicator of ant density (see below). However, this indicator is less sensitive to sampling time: even for short sampling periods of 60 s, the distribution of nestmate densities estimated via the perception of scent marks is narrower and closer to the actual density value (Detrain and Deneubourg, in preparation). Hence, when the sampling effort or the ants' density is high, both contact rates and perception of area marking are equally efficient to assess the density of individuals. In the case that the sampling effort or the ants' density is low, however, direct cues based on contacts provide less reliable estimates for the individuals than indirect cues perceived via area markings.

As shown above, ants use encounters with nestmates as direct social cues that allow them to assess worker density and to accordingly regulate task allocation, defensive strategies, and traffic over trails, or to enhance the efficiency of food retrieval. Direct encounters play a key role in various decision-making processes. For instance, in *Temnothorax,* direct interactions with nestmates at a new nest site are a decisive social cue in the context of nest moving. As soon as the interaction rate reaches a threshold value (quorum size), scouts switch to a more rapid recruitment, which allows the colony to select the most occupied and suitable nest site (Franks et al. 2002; Pratt 2005). Furthermore, several studies on gregarious insects have confirmed that contact rates are ubiquitous social cues. Indeed, high encounter rates due to crowding cause profound changes in the physiology, behavior, and morphology of these animals. A well-known example is the density-dependent phase polyphenism of locusts, where crowding stimulates individuals to change from the shy, cryptically colored, solitary phase into the conspicuously colored, swarm-forming, gregarious phase. It has been demonstrated that direct physical contacts—the touching of another individual's hind legs— are the primary cause of this switch (Simpson et al. 2001). Likewise, the soldier production in social aphid species is elicited by mechanosensory inputs when, at high population densities, the rate of direct contacts exceeds a certain threshold value (Shibao et al. 2004). These examples (see also Amé et al. 2006; Costa 2006) make clear that quite different arthropods make use of direct social cues in order to assess the density of conspecifics.

INDIRECT SOCIAL CUES: GETTING INFORMATION THROUGH AREA MARKING

An ant may assess nestmate density without physically encountering or directly contacting other workers. In fact, an additional and efficient way to acquire information about the social context is through the perception of indirect cues. The term *indirect* expresses that some time may elapse between the production of such cues by one ant and their perception by another ant. The number of such indirect cues is potentially unlimited since they can be any by-product of the ants' activities, or excreted compounds related to the ants' physiology. It is the amount or the intensity of such indirect cues that allows an ant to assess the local density/activity of nestmates or of con- or heterospecifics. Unlike direct contacts, indirect cues immediately provide the receiver with an estimate for the average worker density, or for their activity level, without necessarily requiring much sampling effort. Moreover, since indirect cues can be perceived at any moment, there are fewer time constraints upon the information acquisition by a receiver ant. In other words, whereas the sampling of worker densities through direct contacts is costly in terms of energy and time, the use of indirect cues minimizes the costs for acquiring this kind of social information.

A wide array of social cues consists of by-products of the ants' activity (Figure 2.1). Any environmental alterations in the vicinity of the nest may indicate the occupation level of the area or the size of the working force of a colony. For instance, the height of the soil crater surrounding the nest entrance of some ground-dwelling ant species reflects the size of a colony and its demography. In harvester ants, the size of the chaff pile near the nest may also be a social cue that contains information about the foraging potential of the colony as well as about its efficiency in seed retrieval and seed discarding (Detrain and Tasse 2000). Likewise, trails with heavy traffic that have been cleared from cumbersome items by foragers (Howard 2001; Detrain et al. 2000) reflect the foraging investment of a colony at a specific food patch. The width of such physical trails could indicate the current—or the past—traffic flow of foragers in the nest surroundings. Many other indirect cues, such as the length of nest galleries, the size of the colony midden, or the amount of food reserves (e.g., the amount of seeds in granaries), are potential indicators of colony size and, more specifically, social context of foraging.

Another wide array of indirect social cues comprises physiological waste products, whose amount or concentration is directly related to the ants' density. For instance, the amount of carbon dioxide within a colony directly results from the ants' respiratory activity, and therefore may provide a good estimate for the local worker density as well as for the level of nestmate activity. Indeed, the carbon dioxide concentration influences the aggregation behavior of ants, affects their spatial distribution within the nest (Cox and Blanchard 2000), and its gradient, which increases with the depth below soil surface, and can release specific building activities at different parts of the nest (Franks and Deneubourg 1997; Camazine et al. 2001; Tschinkel 2004). Actually, any chemical trace of the ants' presence potentially indicates the frequentation level of an area by a colony (Figure 2.1). For instance, workers may passively leave "footprints" on the substrate while walking. Since such chemical cues accumulate with each individual that passes by, the local concentration of footprint marks directly reflects the frequentation level of this area (Devigne and Detrain 2002).

Although the number of available indirect cues is potentially unlimited, there are only a few for which the use by ants has been clearly demonstrated. In the following, we will review the current knowledge about area markings as an indirect social cue, and investigate how passively deposited chemical compounds may provide useful information for ants to tune their foraging behavior.

Several studies have emphasized the key role of chemical cues, such as area markings of the nest entrance and its vicinity, of the foraging area, or of the food source, for decision-making processes in eusocial insects. In ants (e.g., *Manica* spp., Cammaerts and Cammaerts 1987; *Tetramorium* spp., Cammaerts and Cammaerts 2001), but also in other eusocial insects (e.g., *Apis mellifera*, Butler et al. 1969; *Bombus occidentalis*, Foster and Gamboa 1989; *Melipona seminigra*, Hrncir et al. 2004, Jarau et al. 2004; *Vespula germanica*, Jandt et al. 2005), area marking is well known to facilitate the orientation of individuals. For instance, harvester ants (*Messor capitatus*) mark the vicinity of their nests by depositing large quantities of material from the hindgut. This behavior has an informative value in the context of nest area identification. Anal fluids deposited in the nest's surroundings contain colony-specific chemical cues, which the ants use for distinguishing their own nest areas from foreign ones, even in the absence of nestmates or foreign workers (Grasso et al. 2000, 2005). Likewise, the foraging areas, or even the trails, can be chemically marked by the accumulation of hydrocarbons deposited from the ants' legs while walking. In *Lasius japonicus*, for example, foragers make use of the hydrocarbon profiles of such footprints to distinguish between trails from different colonies. This is due to the fact that workers from the same colony share the same footprint hydrocarbon profiles, whereas the respective profiles differ among colonies. As a result, the trail-following behavior in *L. japonicus* is colony specific, while the effectiveness of the actual trail pheromone to induce following behavior is not (Akino and Yamaoka 2005; see also Akino et al. 2005). In *Lasius niger*, area markings, passively left by walking ants, also consist of footprint hydrocarbons (Yamaoka and Akino 1994). In contrast to *L. japonicus*, however, there seems to be no colony specificity since workers of *L. niger* behave similarly on areas marked by their own nestmates or by ants from another colony (Devigne and Detrain 2002).

While walking or standing at a location, molecules from an ant's body surface can passively impregnate the substrate. Hence, an ant could use the amount of chemical marks on the substrate as an indirect social cue to assess the number of individuals that have already walked over this area, where they have settled, or even how long they have remained at a certain location. The following sections will demonstrate that ants indeed use these area markings as a reliable cue to assess nestmate density and to accordingly adjust several aspects of their foraging behavior.

KEEPING FORAGERS ALL TOGETHER

In insect societies, the nestmates' density at a certain area could be an important factor for estimating the quality or the safety of this location. The amount of chemical marks left passively by nestmates allows ants to assess the occupancy level of an area and, in case area marking is colony specific, may even define the territory or the home range of one colony. In fewer than 20% of the ant species reviewed by Levings and Traniello (1981), area marking is indeed used to mark a colony's territory.

In these cases, colony-specific compounds identify the home terrain that the colony defends against intra- and interspecific intruders (e.g., *Oecophylla longinoda*, Hölldobler and Wilson 1977; *Atta laevigata*, Whitehouse and Jaffe 1996; *Cataglyphis niger*, Wenseleers et al. 2002). In most cases, however, area markings simply indicate a colony's home range, in which ants are likely to stay together: it labels those areas that are currently—or have previously been—explored by a colony (Levings and Traniello 1981; Mayade et al. 1993; Devigne and Detrain 2002).

Area markings have been demonstrated to influence the behavior of ant foragers by modulating their foraging activity (Aron et al. 1986), improving their orientation (Fourcassié 1986; Cammaerts and Cammaerts 1987; Fourcassié and Beugnon 1988), and increasing their walking speed (Cammaerts and Cammaerts 1996). By affecting the movement pattern of ants and by facilitating the workers' motility, area markings increase the efficiency of food searching at the individual level and, consequently, strengthen the competitive ability of the resident colony against others in the frequent case of competition over a specific resource. Moreover, area markings facilitate the aggregation of nestmates (Depickère et al. 2004; Sempo et al. 2006) and, consequently, prevent the random scattering of ants all over a nest's surroundings (Nonacs 1991; Fourcassié and Deneubourg 1994). Thus, worker densities are kept at values that are high enough for the emergence of cooperative behaviors, such as the collective retrieval of large prey or the synergetic defense of food sources, which are beneficial for the entire colony.

Area marking is also the vector of a shared information strategy. Ants use the information arising from cues left by individuals with similar diet and food preferences, which allows them to assess incidentally the quality and productivity of alternative foraging areas. By using such a social cue, ants may build some external "memory" of the relative profitability of areas surrounding their nest and many accordingly adjust their motility and aggregation behavior.

PROMOTING FORAGING TOWARD KNOWN AREAS

Beside the effects of area markings on the ants' movement patterns and aggregation behavior, this kind of social cue influences the communication between nestmates during food exploitation. In the aphid-tending ant *Lasius niger*, for instance, area markings indicate the colony's occupation level of a foraging area to the workers, and considerably influence the foragers' behavior from the time of food discovery until their return to the nest (Devigne et al. 2004). First, the area markings significantly increase the motivation of the workers to forage and exploit the discovered food sources. Ants invest time to explore unmarked areas in order to get familiar with the surroundings of a resource. In already marked areas, by contrast, the foragers devote their time almost exclusively to food collection. Second, area markings keep the trail-laying behavior of foragers at high levels. On their way back to the nest, the intensity of trail laying decreases less when the home-bound ants walk on a marked substrate than when walking across an unexplored area. Hence, the amount of trail pheromones still deposited by ants when they reach the nest entrance depends on the social context of foraging and is higher in areas that have been previously visited by nestmate foragers.

Using a density cue, i.e., the amount of area marking, for tuning the colony's foraging dynamics is especially adaptive for mass-recruiting ant species. Indeed, the foraging strategies of these species strongly rely on amplification processes and are very sensitive to changes in local ant densities (Deneubourg and Goss 1989; Detrain et al. 1991; Beekman et al. 2001; Sumpter and Pratt 2003; Detrain and Deneubourg 2006). Instead of dispersing foragers between various different food patches, area marking seems to act as a catalyst for recruitment to particular food sites by promoting trail laying at locations that are safe and regularly frequented by nestmates. Thus, ants exhibit a conservative behavior that, on the one hand, hampers the extension of the colony's home range but, on the other hand, maintains local nestmate densities at values favorable for the cooperative exploitation and monopolization of food resources (Devigne et al. 2004).

GETTING INFORMATION FOR FREE: ASSESSING FOOD DISTANCE

In many ant species, the intensity of the trail that reaches the nest entrance differs according to the distance of the resource: closer food sources will elicit higher rates of nestmate recruitment and, consequently, will be more often selected than remote resources. As ants are central place foragers, both the density of scouts and the amount of area marking decrease with distance from the nest. Therefore, the amount of area marking may be a reliable cue for ants to assess their distance from the colony and to tune their recruitment behavior. Alternatively, scouts may rely on a direct measure of distance per se using internal odometric parameters such as the time or energy spent on a foraging trip, step counting (Wittlinger et al. 2006), or path integration (Collett et al. 2006; Graham and Collet 2006; Ronacher et al. 2006). Experiments carried out with *Lasius niger* to test the different possibilities (Devigne and Detrain 2006) demonstrated that when ants foraged on an unexplored area (no area marking present), the distance experienced by a scout on its way to the food source did not modify its trail-laying behavior when homing back to the nest. This fact clearly demonstrates that *L. niger* ants do not use an internal odometer for tuning their recruiting behavior according to the food's distance from the nest. For a modulation of communication with respect to distance, the substrate leading to food sources (bridges in the experiments by Devigne and Detrain 2006) has to be previously explored and area marked. The perception of area marking stimulates the foragers to maintain their trail-laying behavior on their way to the nest, and ultimately leads to a more intense recruitment toward nearby food sources. Hence, the amount of home-range marking not only is a nestmate density cue, but also provides useful information about the distance of a feeding site from the areas regularly frequented by the colony, such as the nest itself or the major trunk trails.

SMELLING THE PAST

Indirect cues such as area markings have the advantage that they can be perceived at any time and even in the absence of their emitters. In return, however, it may be difficult for individuals to know whether indirect cues, like chemical marks, have been laid a few seconds, several minutes, or even hours before. Thus, the question arises: How are ants able to figure out whether area markings reflect the current occupation level of an area, or whether they provide merely out-of-date information about past colony activities? In the case that area markings only consist of a single compound, ants will not be able to deduce the time elapsed since the deposition of the marks, because a low amount of area marking means either (1) a recent low density/activity of nestmates or (2) a high density/activity that already ended some time ago.

The following model demonstrates that at least two chemical compounds with different volatilities should be present to enable the ants to distinguish between a recent and a past marking. When area markings are passively laid by walking ants, the amount (Φ) of compound i deposited at a location at time t ($\Phi_i(t)$) is proportional to the current ant density at this location ($A(t)$) and to the quantity of this compound deposited per time unit and per individual (α_i):

$$\Phi_i(t) = \alpha_i A(t) \qquad i = 1,2 \tag{2.1}$$

The rate of decay of compound i ($\lambda_i C_i$) (for two compounds, $i = 1$ or 2) is proportional to its concentration (C_i) and to a constant (λ_i), which is the inverse of its mean lifetime (Beckers et al. 1993; Detrain et al. 2001; Jeanson et al. 2003). The evolution of the concentration of compound i (C_i) is

$$\frac{dC_i}{dt} = \Phi_i(t) - \lambda_i C_i \qquad i = 1,2 \tag{2.2}$$

For the purpose of simplification, we assume that (1) the area marking comprises only two compounds deposited during a very short period of time, so that the deposition of the respective blend at time $t = 0$ can be considered as punctual, and (2) the initial amount of area marking is the same for both compounds ($\Phi_1 = \Phi_2 = \Phi$). Consequently, the temporal decay of each compound obeys the law (solution of Equation 2.2)

$$C_i = \Phi e^{-\lambda_i t} \qquad i = 1, 2 \tag{2.3a}$$

while the ratio R between both concentrations is

$$R = \frac{C_1}{C_2} = e^{-(\lambda_1 - \lambda_2)t} \tag{2.3b}$$

Initially, at time $t = 0$, the amount of both compounds is identical ($C_1 = C_2 = \Phi$) under the assumption made above, and consequently, their ratio is $R = 1$. In the case that the two area marking components differ in their decay rates ($\lambda_1 \neq \lambda_2$), their ratio will change over time (Figure 2.3). Thus, for each moment following the deposition of the area marking ($t = 0$), there is a specific ratio between the two compounds. Thus, the age of the marking (t) can be simply deduced from the relative concentrations of the blend's single compounds:

$$t = -\frac{\ln R}{\lambda_1 - \lambda_2} = -\frac{\ln C_1 - \ln C_2}{\lambda_1 - \lambda_2} \tag{2.4a}$$

Likewise, the initial level of marking is

$$\ln \Phi = -\frac{\lambda_2}{\lambda_1 - \lambda_2} \ln C_1 + \frac{\lambda_1}{\lambda_1 - \lambda_2} \ln C_2 \tag{2.4b}$$

In other words, by perceiving the concentration of each of the two compounds, ants likely are able to estimate the age of the marking perceived at time t as well as the initial level of marking at time $t = 0$. This computation can be done in an analogous way by simple neural networks (Copelli et al. 2002).

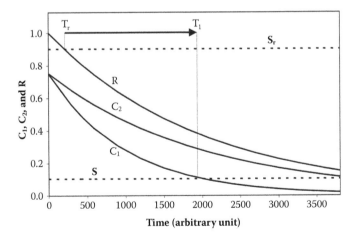

FIGURE 2.3 Changes of area markings over time. The concentration of compounds 1 and 2 (C_1, C_2) as well as their ratio ($R = C_1/C_2$) is given as a function of time. At time $t = 0$ the compounds' concentrations are $C_1 = C_2 = 0.75$. The constants (the inverse of mean lifetime; see text) of compounds 1 and 2 are $\lambda_1 = 0.001$ and $\lambda_2 = 0.0005$. S_r and S are the perception thresholds for the ratio R and for the concentration C_1 and C_2, respectively. $T_1 - T_r$ is the time during which ants are able to assess both the age and the initial amount of the area marking. See text for details.

The equations given above, however, neglect an important biological trait of animals, which is the existence of perception thresholds. The perception of ants can be characterized by two types of thresholds:

1. The perception threshold (S_i) for a single compound; if $C_i < S_i$ $(i = 1, 2)$, the insect does not detect the component i.
2. The perception threshold S_r for the compounds' ratio R; if $R > S_r$, the insect is unable to perceive the ratio R as different from its initial value $(R = 1)$.

If we assume that $S_1 = S_2 = S$, the system is characterized by two physicochemical parameters (λ_1, λ_2) and two neurophysiogical parameters (S, S_r).

Due to the existence of a perception threshold of the compounds' ratio (S_r), there is a time period, T_r (Equation 2.5a), during which the ants are unable to estimate the age of the marking because $R > S_r$, and thus is too close to the initial ratio $(R = 1)$. Furthermore, there is a time T_i (Equation 2.5b) at which C_i will be equal to the threshold S. Hence, for any time $t > T_i$, compound i can no longer be detected by the ants, since its concentration C_i is lower than its detection threshold (S).

T_r and T_i are described by the equations:

$$T_r = -\frac{\ln S_r}{\lambda_1 - \lambda_2} \tag{2.5a}$$

$$T_i = \frac{\ln \Phi - \ln S}{\lambda_i} \tag{2.5b}$$

In the case that compound 1 has the higher rate of decay, ants can perceive the compound ratio—and hence are able to estimate the initial amount and the age of area marking—only during the time interval $T_1 - T_r$ (Equation 2.5c).

$$T_1 - T_r = \frac{\ln \Phi - \ln S}{\lambda_1} + \frac{\ln S_r}{\lambda_1 - \lambda_2} \tag{2.5c}$$

There are two further time intervals that are important for the information processing by the ants: $0 \rightarrow T_r$ and $T_1 \rightarrow \infty$, respectively. During the time interval $0 \rightarrow T_r$, the ants are unable to precisely assess the age of the marking because the compounds' ratio R is too close to its initial value. Nevertheless, ants may extract information because they know that they are faced with a recently laid area marking. By contrast, for times later than T_1 (i.e., for the interval $T_1 \rightarrow \infty$), the ants cannot extract any reliable information at all, because they then detect only one compound (Figure 2.3).

The time interval $T_1 - T_r$, during which ants are able to confidently assess the history of area marking, can be extended through a higher initial ant density/activity at time $t = 0$ (high initial Φ values), or by a higher perception sensitivity of the individuals (high S_r or low S thresholds). In addition, concerning the chemical compounds of the marking, for a given λ_2 value there is a λ_1 value $(\lambda_{1\max}$; Equation 2.6a) that maximizes the time interval $T_1 - T_r$ $[(T_1 - T_r)_{\max}$; Equation 2.6b] (Figure 2.4):

$$\lambda_{1\max} = \frac{\lambda_2(1 + (-u)^{0.5})}{u+1}; \qquad u = \frac{\ln S_r}{\ln \Phi - \ln S} \tag{2.6a}$$

$$(T_1 - T_r)_{\max} = \frac{(u+1)}{\lambda_2}\left(\frac{\ln \Phi - \ln S}{1 + (-u)^{0.5}} + \frac{\ln S_r}{(-u)^{0.5} - u}\right) \tag{2.6b}$$

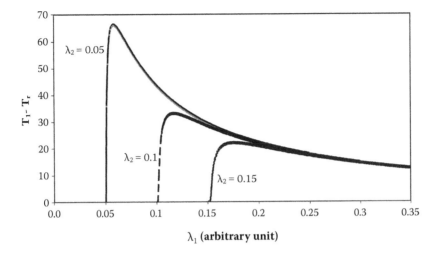

FIGURE 2.4 The time during which an area marking can be used as a social cue $(T_1 - T_r)$ as a function of the constant (inverse of the mean lifetime) of compound 1 (λ_1), for $\lambda_2 = 0.05$, 0.10, and 0.15, respectively. For the simulations, the following assumptions were made: $S_r = 0.9$, $S = 0.01$, $\Phi = 1$. See text for details.

The differences in the decay rates of the compounds are due to different physicochemical processes, such as their degradation, evaporation, or adsorption. The different λ_i values, which result from the compounds' properties, are essential for the optimization of the time interval $T_1 - T_r$. An analysis shows that the time interval $T_1 - T_r$ reaches a maximum when the difference between the kinetic constants λ_1 and λ_2 is small (Figure 2.4), which suggests that both compounds should have similar physicochemical properties. One should also stress that the interval $T_1 - T_r$ must be optimized with respect to the temporal characteristics of the ants' activities, and not simply be maximized per se. For example, if a task is carried out within a time that is shorter than the decay rate of the area marking, the process described above would not work, since at the time a new marking is added, the previous one is not completely exhausted. This theoretical approach raises questions about the relationships between the number of components of a marking blend, the information that ants can extract from it, and the robustness of the whole decision-making procedure. In our simplified example of punctual area marking, the ants should be able to extract two major types of information from an indirect cue: (1) the initial density/activity of nestmates, and (2) the time that has elapsed since the activity stopped. This potentially allows ants to tune their behavioral response according to both the recent and the past activity of their colony. In regard to area markings, which are deposited repeatedly by several individuals, these marks can still provide an estimate of nestmate density, but it may become less accurate due to the mixture of marks with different ages (Detrain and Deneubourg, in preparation). At first sight, passive area marking might seem of little interest compared with any form of active chemical communication (e.g., recruitment trails). Yet, it offers the basis for various modulation processes. It was frequently postulated that the complexity of the marking blend has evolved for the purpose of nestmate recognition and territorial marking. Without denying such functional values, the blend complexity could as well be a source of information about the history of colony activities and a touchstone of foraging regulations. As long as experimental studies that validate this hypothesis are lacking, theoretical considerations, like the one presented above, contribute to an understanding of the relation between a blend's complexity and the amount of information that ants can extract from it under different social situations.

DIRECT VS. INDIRECT CUES: WHICH ONE IS THE BEST?

This last section of our chapter is an attempt to summarize the advantages as well as the disadvantages of direct social cues (i.e., contact rates) and indirect social cues (i.e., the amount of area marking) (Figure 2.5).

The main strength of direct social cues is their capability to transfer a potentially unlimited amount of chemical information. Through its antennal chemosensillae, the contacting ant perceives information about colony origin (alien or nestmate), the physiological state, or the recent activities of encountered ants. The variety of individual chemical profiles makes every encounter a unique event and provides each ant with quite an accurate picture of the social condition of the individual it is currently contacting.

Area marking, on the other hand, is more limited in terms of the amount of information conveyed since only a fraction of the ants' body odors are left on the substrate. These indirect cues provide information about how many ants have foraged at a location, and whether they were belonging to the same colony or came from a competing nest. It is unlikely, however, that area markings convey

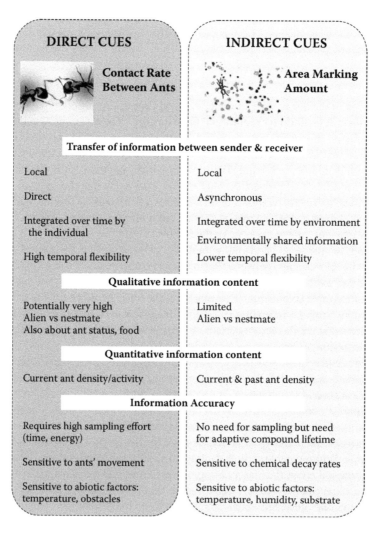

FIGURE 2.5 Comparison of the features of direct and indirect social cues.

any other type of information, such as the tasks performed or the retrieval efficiency of the ants that previously collected food in the foraging area.

Since only direct contacts allow ants to perceive each other's condition, it may be assumed that direct cues are the ideal means for nestmates to assess short-term changes in the number of active ants in each task group, in the number of successfully loaded foragers, or in the number of alerted nest defenders. It remains to be investigated, however, whether direct contact cues actually prevail over indirect cues in situations in which ants need to get a precise—albeit punctual—picture of the social context, such as in task allocation.

Since every contact is a kind of personal experience and, therefore, a potentially unique event, an ant should achieve multiple contacts in order to assess the average social context of foraging at a given moment and location. Indeed, multiple contacts improve information to foragers: they gain a more accurate estimate of the average occupation level of an area, the labor allocation, or the safety of the foraging site. This, however, may be a lengthy process, and the benefits of information improvement must compensate for the energy and time costs related to the increased sampling effort (Franks et al. 2003). Yet, in some cases, information can be acquired at low costs—without requiring an active sampling—simply from the encounters that workers experience automatically as they perform a task (e.g., when traveling on foraging trails). Further research is needed to investigate whether ants actually use an active sampling strategy in order to optimize the speed versus accuracy trade-off during decision-making processes. Another question that needs to be answered in this context is whether the sampling effort of ants varies with the nestmates' density at a location or with the need for a more accurate adjustment of the number of cooperating workers.

In contrast to direct social cues, area markings (indirect social cues) appear to be an easy, yet efficient, way of assessing the presence/activity of nestmates in a shared environment without relying on casual encounters with other individuals. The use of a shared environment for the transfer of information has an important advantage over direct contacts: it immediately provides the receivers with an estimate of the average occupancy level of a foraging area without necessarily requiring a time-consuming sampling. In fact, the amount of area marking is an integrative cue because it reflects the summation of tracks passively laid by ants over time. It is, therefore, an accurate indicator of the average ant density in the nest surroundings. Furthermore, although this remains to be thoroughly investigated, area marking putatively presents a way for ants to "smell the past": the perception of compound concentrations and ratios could allow individuals to assess how intensively, and for how long, an area has been occupied or foraged by nestmates. Future research should investigate how perceiving the age of area marking may influence the foragers' decisions. For instance, a recent marking may discourage ants from foraging at already overexploited locations. Conversely, an older area marking may trigger the ants' foraging activity, since it indicates an area that has previously been rewarding, and the resources may have already had time to be renewed. The adaptive value of assessing information about the time that has elapsed since ants foraged on a particular area should be studied especially in those ant species that exploit regularly renewing resources, such as seed patches or aphid colonies.

The contact rate is a means for ants to assess the density of conspecifics without counting them directly. We do not know yet whether ants perceive the contact rate rather as a frequency, i.e., as the number of contacts per unit time, or as the time interval between encounters. Whatever it may be, the contact rate experienced by an individual closely depends on the spatial configuration (e.g., the path sinuosity) as well as on the speed of the ants' movements. In addition, for the same actual ant density, an individual might experience quite different contact rates depending on abiotic conditions. For instance, direct cues are assumed to be sensitive to temperature fluctuations. Indeed, an increase of ground-level temperature, which is known to act upon an ant's speed, may result in a higher contact frequency between nestmates (Shapley 1920; Barnes and Kohn 1932). Likewise, indirect chemical cues are subject to weather conditions. For instance, high ground temperatures accelerate the decay of area markings and, consequently, reduce the time during which they can be perceived by ants. This dependency of the accuracy of social cues on weather conditions is worthwhile to be studied in more detail. For instance, an important question is whether cue-based estimates of nestmate density are less reliable outside the nest than inside, where changes in temperature and relative humidity are less extreme. Another interesting research topic would be to investigate to which extent direct contact rates are sensitive to the presence of obstacles (e.g., gravels, stones, twigs) and, consequently, may be less practicable to assess ant density outside of the nest than indirect chemical cues.

CONCLUDING REMARKS

The ability of worker ants to adapt their foraging behavior according to their social environment is imperative for the colony's growth and survival. Each ant should not behave, or make decisions, solely based on its own experience, but it should also integrate information provided directly or indirectly by its nestmates. Cues are the keystones of this socially acquired information: ant workers use them as statistical tools to assess the occupancy level of a location and to accordingly adjust their behavior and communication. Unfortunately, the use of social cues as feedback to regulate behavioral traits is still understudied for most social insects. Therefore, one promising line for future research is to link these social cues with the features of colony organization, and to see how they act on the various processes involved. This also raises questions about which kind of adaptive strategies could have emerged from either the sequential or the concurrent use of direct and indirect cues. To put this in a wider perspective, future research is needed to determine to what extent the use of social cues has opened the way for group living and for more elaborated forms of sociality among the insects.

REFERENCES

Adams ES. (1990). Boundary disputes in the territorial ant *Azteca trigona:* Effects of asymmetries in colony size. *Anim Behav* 39:321–28.

Adler FR, Gordon DM. (1992). Information collection and spread in networks of patrolling ants. *Am Nat* 140:373–400.

Akino T, Yamaoka R. (2005). Trail discrimination signal of *Lasius japonicus* (Hymenoptera: Formicidae). *Chemoecology* 15:21–30.

Akino T, Morimoto M, Yamaoka R. (2005). The chemical basis for trail recognition in *Lasius nipponensis* (Hymenoptera: Formicidae). *Chemoecology* 15:13–20.

Amé JM, Halloy J, Rivault C, Detrain C, Deneubourg JL. (2006). Collegial decision making based on social amplification leads to optimal group formation. *Proc Natl Acad Sci USA* 103:5835–840.

Anderson C, McShea DW. (2001). Individual versus social complexity, with particular reference to ant colonies. *Biol Rev* 76:211–37.

Aron S, Pasteels JM, Deneubourg JL, Boeve JL. (1986). Foraging recruitment in *Leptothorax unifasciatus*— The influence of foraging area familiarity and the age of the nest site. *Insect Soc* 33:338–51.

Barnes TC, Kohn HI. (1932). The effect of temperature on the leg posture and speed of creeping in the ant *Lasius*. *Biol Bull* 62:306–12.

Beckers R, Deneubourg JL, Goss S. (1993). Modulation of trail laying in the ant *Lasius niger* (Hymenoptera, Formicidae) and its role in the collective selection of a food source. *J Insect Behav* 6:751–59.

Beckers R, Goss S, Deneubourg JL, Pasteels JM. (1989). Colony size, communication and ant foraging strategy. *Psyche* 96:239–56.

Beekman M, Sumpter DJT, Ratnieks FLW. (2001). A phase transition between disordered and ordered foraging in Pharaoh's ants. *Proc Natl Acad Sci USA* 98:9703–6.

Bossert WH, Wilson EO. (1963). The analysis of olfactory communication among animals. *J Theor Biol* 5:443–69.

Brown MJF, Gordon DM. (2000). How resources and encounters affect the distribution of foraging activity in a seed-harvesting ant. *Behav Ecol Sociobiol* 47:195–203.

Buhl J, Deneubourg JL, Grimal A, Theraulaz G. (2005). Self-organized digging activity in ant colonies. *Behav Ecol Sociobiol* 58:9–17.

Burd M, Aranwela N. (2003). Head-on encounter rates and walking speed of foragers in leaf-cutting ant traffic. *Insect Soc* 50:3–8.

Butler CG, Fletcher DJC, Watler D. (1969). Nest entrance marking with pheromone by honeybee, *Apis mellifera* L., and by a wasp, *Vespula vulgaris* L. *Anim Behav* 17:142–47.

Camazine S, Sneyd J. (1991). A model of collective nectar source selection by honey bees: Self-organization through simple rules. *J Theor Biol* 149:547–71.

Camazine S, Deneubourg JL, Franks NR, Sneyd J, Theraulaz G, Bonabeau E. (2001). *Self-Organization in Biological Systems*. Princeton, NJ: Princeton University Press.

Cammaerts MC, Cammaerts R. (1996). Area marking in the ant *Pheidole pallidula* (Myrmicinae). *Behav Proc* 37:21–30.

Cammaerts MC, Cammaerts R. (2001). Marking of nest entrances and vicinity in two related *Tetramorium* ant species (Hymenoptera: Formicidae). *J Insect Behav* 14:247–69.

Cammaerts R, Cammaerts MC. (1987). Nest topology, nestmate recognition, territorial marking and homing in the ant *Manica rubida* (Hymenoptera, Formicidae). *Biol Behav* 12:65–81.

Cassill DL, Tschinkel WR. (1995). Allocation of liquid food to larvae via trophallaxis in colonies of the fire ant, *Solenopsis invicta*. *Anim Behav* 50:801–13.

Cassill DL, Tschinkel WR. (1999). Regulation of diet in the fire ant, *Solenopsis invicta*. *J Insect Behav* 12:307–28.

Collett M, Collett TS, Srinivasan MV. (2006). Insect navigation: Measuring travel distance across ground and through air. *Curr Biol* 16:R887–90.

Copelli M, Roque AC, Oliveira RF, Kinouchi O. (2002). Physics of psychophysics: Stevens and Weber-Fechner laws are transfer functions of excitable media. *Phys Rev E* 65: article 060901 (R).

Costa JT. (2006). *The Other Insect Societies*. Cambridge, MA: Belknap Press of Harvard University Press.

Couzin ID, Franks NR. (2003). Self-organized lane formation and optimized traffic flow in army ants. *Proc R Soc Lond B* 270:139–46.

Cox MD, Blanchard GB. (2000). Gaseous templates in ant nests. *J Theor Biol* 204:223–38.

Czechowski W. (1984). Tournaments and raids in *Lasius niger* (L.) (Hymenoptera, Formicidae). *Ann Zool* 38: 81–91.

Danchin E, Giraldeau LA, Valone TJ, Wagner RH. (2004). Public information: From nosy neighbours to cultural evolution. *Science* 305:487–91.

Darchen R. (1959). Les techniques de construction chez *Apis mellifica*. *Ann Sci Nat Zool* 12:113–209.

Deneubourg JL, Goss S. (1989). Collective patterns and decision-making. *Ethol Ecol Evol* 1:295–311.

Deneubourg JL, Nicolis S, Detrain C. (2005). Optimality of communication in self-organized social behaviour. In Hemelrijk C (ed.), *Self-Organization and Evolution in Social Systems*. New York: Cambridge University Press, pp. 25–35.

Depickère S, Fresneau D, Detrain C, Deneubourg JL. (2004). Marking as a decision factor in the choice of a new resting site in *Lasius niger*. *Insect Soc* 51:243–46.

Detrain C, Deneubourg JL. (1997). Scavenging by *Pheidole pallidula*: A key for understanding decision-making systems in ants. *Anim Behav* 53:537–47.

Detrain C, Deneubourg JL. (2002). Complexity of environment and parsimony of decision rules in insect societies. *Biol Bull* 202:268–74.

Detrain C, Deneubourg JL. (2006). Self-organized structures in a superorganism: Do ants "behave" like molecules? *Phys Life Rev* 3:162–87.

Detrain C, Deneubourg JL. (2008). Collective decision and foraging patterns in ants and honeybees. *Adv Insect Physiol*, 53:123–73.

Detrain C, Pasteels JM. (1992). Caste polyethism and collective defense in the ant *Pheidole pallidula*: The outcome of quantitative differences in recruitment. *Behav Ecol Sociobiol* 29:405–12.

Detrain C, Tasse O. (2000). Seed drops and caches by the harvester ant *Messor barbarus*: Do they contribute to seed dispersal in Mediterranean grasslands? *Naturwissenschaften* 87:373–76.

Detrain C, Deneubourg JL, Goss S, Quinet Y. (1991). Dynamics of collective exploration in the ant *Pheidole pallidula*. *Psyche* 98:21–31.

Detrain C, Deneubourg JL, Pasteels JM. (1999). Decision-making in foraging by social insects. In Detrain C, Deneubourg JL, Pasteels JM (eds.), *Information Processing in Social Insects*. Basel: Birkhäuser Verlag, pp. 331–54.

Detrain C, Natan C, Deneubourg JL. (2001). The influence of the physical environment on the self-organised foraging patterns of ants. *Naturwissenschaften* 88:171–74.

Detrain C, Tasse O, Versaen M, Pasteels JM. (2000). A field assessment of optimal foraging in ants: Trail patterns and seed retrieval by the European harvester ant *Messor barbarus*. *Insect Soc* 47:56–62.

Devigne C, Detrain C. (2002). Collective exploration and area marking in the ant *Lasius niger*. *Insect Soc* 49:357–62.

Devigne C, Detrain C. (2006). How does food distance influence foraging in the ant *Lasius niger*: The importance of home-range marking. *Insect Soc* 53:46–55.

Devigne C, Renon AJ, Detrain C. (2004). Out of sight but not out of mind: Modulation of recruitment according to home range marking in ants. *Anim Behav* 67:1023–29.

Dornhaus A, Chittka L. (2004). Information flow and regulation of foraging activity in bumblebees (*Bombus* spp.). *Apidologie* 35:183–92.

Dornhaus A, Klügl F, Oechslein C, Puppe F, Chittka L. (2006). Benefits of recruitment in honey bees: The effects of ecology and colony size in an individual-based model. *Behav Ecol* 17:336–44.

Dussutour A, Fourcassié V, Helbing D, Deneubourg JL. (2004). Optimal traffic organization in ants under crowded conditions. *Nature* 428:70–73.

Dussutour A, Deneubourg JL, Fourcassié V. (2005). Temporal organization of bi-directional traffic in the ant *Lasius niger* (L.). *J Exp Biol* 208:2903–12.

Dussutour A, Nicolis SC, Deneubourg JL, Fourcassié V. (2006). Collective decisions in ants when foraging under crowded conditions. *Behav Ecol Sociobiol* 61:17–30.

Dussutour A, Beshers S, Deneubourg JL, Fourcassié V. (2007). Crowding increases foraging efficiency in the leaf-cutting ant *Atta colombica*. *Insect Soc* 54:158–65.

Foster RL, Gamboa GJ. (1989). Nest entrance marking with colony specific odors by the bumble bee *Bombus occidentalis* (Hymenoptera, Apidae). *Ethology* 81:273–78.

Fourcassié V. (1986). Retour au nid et mécanismes d'orientation chez les ouvrières de la fourmi rousse des bois sur l'aire d'affouragement d'une colonie polycalique. *Actes Coll Insect Soc* 3:243–59.

Fourcassié V, Beugnon G. (1988). How do red wood ants orient when foraging in a 3 dimensional system? I. Laboratory experiments. *Insect Soc* 35:92–105.

Fourcassié V, Deneubourg JL. (1994). The dynamics of collective exploration and trail-formation in *Monomorium pharaonis*: Experiments and model. *Physiol Entomol* 19:291–300.

Franks N, Deneubourg JL. (1997). Self-organizing nest construction in ants: Individual worker behaviour and the nests dynamics. *Anim Behav* 54:779–96.

Franks NR, Pratt SC, Mallon EB, Britton NF, Sumpter DJT. (2002). Information flow, opinion polling and collective intelligence in house-hunting social insects. *Phil Trans R Soc Lond B* 357:1567–83.

Franks NR, Dornhaus A, Fitzsimmons JP, Stevens M. (2003). Speed versus accuracy in collective decision-making. *Phil Trans R Soc Lond B* 270:2457–63.

von Frisch K. (1967). *The Dance Language and Orientation of Bees*. Cambridge, MA: Harvard University Press.

Gordon DM, Mehdiabadi N. (1999). Encounter rate and task allocation in harvester ants. *Behav Ecol Sociobiol* 45:370–77.

Gordon DM, Paul RE, Thorpe K. (1993). What is the function of encounter patterns in ant colonies? *Anim Behav* 45:1083–100.

Gould JL. (1974). Honey bee communication. *Nature* 252:300–1.

Gould JL. (1975). Communication of distance information by honey bees. *J Comp Physiol A* 104:161–73.

Graham P, Collett TS. (2006). Bi-directional route learning in wood ants. *J Exp Biol* 209:3677–84.

Grassé PP. (1959). La reconstruction du nid et les coordinations interindividuelles chez *Bellicositermes natalensis* et *Cubitermes* sp. la théorie de la stigmergie: Essai d'interprétation du comportement des termites constructeurs. *Insect Soc* 6:41–80.

Grasso DA, Mori A, Le Moli F. (2000). The use of anal spots for nest-area marking in *Messor* ants (Hymenoptera, Formicidae). *Ins Soc Life* 3:61–66.

Grasso DA, Sledge MF, Le Moli F, Mori A, Turillazzi S. (2005). Nest-area marking with faeces: A chemical signature that allows colony-level recognition in seed harvesting ants (Hymenoptera, Formicidae). *Insect Soc* 52:36–44.

Greene MJ, Gordon DM. (2003). Social insects: Cuticular hydrocarbons inform task decisions. *Nature* 423:32.

Greene MJ, Gordon DM. (2007). Interaction rate informs harvester ant task decisions. *Behav Ecol* 18: 451–55.

Hart AG, Ratnieks FLW. (2001). Why do honey-bee (*Apis mellifera*) foragers transfer nectar to several receivers? Information improvement through multiple sampling in a biological system. *Behav Ecol Sociobiol* 49:244–50.

Heinze J, Foitzik S, Hippert A, Hölldobler B. (1996). Apparent dear-enemy phenomenon and environment-based recognition cues in the ant *Leptothorax nylanderi*. *Ethology* 102:510–22.

Hepburn HR. (1986). *Honeybees and Wax*. Berlin: Springer Verlag.

Hölldobler B. (1981). Foraging and spatiotemporal territories in the honey ant *Myrmecocystus mimicus* wheeler (Hymenoptera: Formicidae). *Behav Ecol Sociobiol* 9:301–14.

Hölldobler B, Wilson EO. (1977). Colony-specific territorial pheromone in the African weaver ant *Oecophylla longinoda* (Latreille). *Proc Natl Acad Sci USA* 74:2072–75.

Hölldobler B, Wilson EO. (1990). *The Ants*. Cambridge, MA: Belknap Press of Harvard University Press.

Howard JJ. (2001). Costs of trail construction and maintenance in the leaf-cutting ant *Atta columbica*. *Behav Ecol Sociobiol* 49:348–56.

Howard JJ, Henneman ML, Cronin G, Fox JA, Hormiga G. (1996). Conditioning of scouts and recruits during foraging by a leafcutting ant, *Atta colombica*. *Anim Behav* 52:299–306.

Hrncir M, Jarau S, Zucchi R, Barth FG. (2004). On the origin and properties of scent marks deposited at the food source by a stingless bee, *Melipona seminigra*. *Apidologie* 35:3–13.

Huang MH, Seeley TD. (2003). Multiple unloadings by nectar foragers in honey bees: A matter of information improvement or crop fullness? *Insect Soc* 50:330–39.

Hutchinson JMC, Waser PM. (2007). Use, misuse and extensions of "ideal gas" models of animal encounter. *Biol Rev* 82:335–59.

Jaffe K, Howse PE. (1979). The mass recruitment system of the leaf cutting ant, *Atta cephalotes* (L.). *Anim Behav* 27:930–39.

Jandt JM, Curry C, Hemauer S, Jeanne RL. (2005). The accumulation of a chemical cue: Nest-entrance trail in the German yellowjacket, *Vespula germanica*. *Naturwissenschaften* 92:242–45.

Jarau S, Hrncir M, Ayasse M, Schulz C, Francke W, Zucchi R, Barth FG. (2004). A stingless bee (*Melipona seminigra*) marks food sources with a pheromone from its claw retractor tendons. *J Chem Ecol* 30:793–804.

Jeanson R, Ratnieks FLW, Deneubourg JL. (2003). Pheromone trail decay rates on different substrates in the Pharaoh's ant, *Monomorium pharaonis*. *Physiol Entomol* 28:192–98.

Jun J, Pepper JW, Savage VM, Gillooly JF, Brown JH. (2003). Allometric scaling of ant foraging networks. *Evol Ecol Res* 5:297–303.

Kirchner WF, Lindauer M. (1994). The causes of the tremble dance of the honeybee, *Apis mellifera*. *Behav Ecol Sociobiol* 35:303–8.

Knaden M, Wehner R. (2003). Nest defense and conspecific enemy recognition in the desert ant *Cataglyphis fortis*. *J Insect Behav* 16:717–30.

Le Breton J, Fourcassié V. (2004). Information transfer during recruitment in the ant *Lasius niger* L. (Hymenoptera: Formicidae). *Behav Ecol Sociobiol* 55:242–50.

Lenoir A, Fresneau D, Errard C, Hefetz A. (1999). Individuality and colonial identity in ants: The emergence of the social representation concept. In Detrain C, Deneubourg JL, Pasteels J (eds.), *Information Processing in Social Insects*. Basel: Birkhäuser Verlag, pp. 219–37.

Levings SC, Traniello JFA. (1981). Territoriality, nest dispersion, and community structure in ants. *Psyche* 88:265–319.

Lumsden CJ, Hölldobler B. (1983). Ritualized combat and intercolony communication in ants. *J Theor Biol* 100:81–98.

Mailleux AC, Deneubourg JL, Detrain C. (2000). How do ants assess food volume? *Anim Behav* 59: 1061–69.

Mailleux AC, Deneubourg JL, Detrain C. (2003a). Regulation of ants' foraging to resource productivity. *Proc R Soc Lond B* 270:1609–16.

Mailleux AC, Deneubourg JL, Detrain C. (2003b). How does colony growth influence communication in ants? *Insect Soc* 50:24–31.

Mailleux AC, Detrain C, Deneubourg JL. (2005). Triggering and persistence of trail-laying in foragers of the ant *Lasius niger*. *J Insect Physiol* 51:297–304.

Mailleux AC, Detrain C, Deneubourg JL. (2006). Starvation drives a threshold triggering communication. *J Exp Biol* 209:4224–29.

Mayade S, Cammaerts MC, Suzzoni JP. (1993). Home-range marking and territorial marking in *Cataglyphis cursor* (Hymenoptera, Formicidae). *Behav Proc* 30:131–42.

McCabe S, Farina WM, Josens RB. (2006). Antennation of nectar-receivers encodes colony needs and food-source profitability in the ant *Camponotus mus*. *Insect Soc* 53:356–61.

Mercier JL, Lenoir A, Dejean A. (1997). Ritualised versus aggressive behaviours displayed by *Polyrhachis laboriosa* (F. Smith) during intraspecific competition. *Behav Proc* 41:39–50.

Nicolis SC, Deneubourg JL. (1999). Emerging patterns and food recruitment in ants: An analytical study. *J Theor Biol* 198:575–92.

Nicolis SC, Theraulaz G, Deneubourg JL. (2005). The effect of aggregates on interaction rate in ant colonies. *Anim Behav* 69:535–40.

Nonacs P. (1991). Exploratory behavior of *Lasius pallitarsis* ants encountering novel areas. *Insect Soc* 38:345–49.

Pasteels JM, Deneubourg JL, Verhaeghe JC, Boevé JL, Quinet Y. (1986). Orientation along terrestrial trails by ants. In Payne TL, Birch MC, Kennedy CEJ (eds.), *Mechanisms in Insect Olfaction*. Oxford: Oxford University Press, pp. 131–38.

Pratt S. (2005). Quorum sensing by encounter rates in the ant *Temnothorax albipennis*. *Behav Ecol* 10:488–96.

Ratnieks FLW, Anderson C. (1999). Task partitioning in insect societies. II. Use of queuing delay information in social insects. *Am Nat* 154:536–48.

Roces F. (1990). Olfactory conditioning during the recruitment process in a leaf-cutting ant. *Oecologia* 83:261–62.

Roces F, Núñez JA. (1993). Information about food quality influences load-size selection in recruited leaf-cutting ants. *Anim Behav* 45:135–43.

Ronacher B, Westwig E, Wehner R. (2006). Integrating two-dimensional paths: Do desert ants process distance information in the absence of celestial compass cues? *J Exp Biol* 209:3301–8.

Sakata H, Katayama N. (2001). Ant defence system: A mechanism organizing individual responses into efficient collective behaviour. *Ecol Res* 16:395–403.

Seeley TD. (1989). Social foraging in honeybees: How nectar foragers assess their colony's nutritional status. *Behav Ecol Sociobiol* 24:181–99.

Seeley TD. (1992). The tremble dance of the honey bee: Message and meanings. *Behav Ecol Sociobiol* 31:375–83.

Seeley TD. (1995). *The Wisdom of the Hive: The Social Physiology of Honey Bee Colonies*. Cambridge, MA: Harvard University Press.

Seeley TD. (1998). Thoughts on information and integration in honey bee colonies. *Apidologie* 29:67–80.

Seeley TD, Tovey CA. (1994). Why search time to find a food-storer bee accurately indicates the relative rates of nectar collecting and nectar processing in honey bee colonies. *Anim Behav* 47:311–16.

Seeley TD, Towne WF. (1992). Tactics of dance choice in honey bees: Do foragers compare dances? *Behav Ecol Sociobiol* 30:59–69.

Sempo G, Depickère S, Detrain C. (2006). Spatial organization in a dimorphic ant: Caste specificity of clustering patterns and area marking. *Behav Ecol* 17:642–50.

Shapley H. (1920). Thermokinetics of *Liometopum apiculatum* Mayr. *Proc Natl Acad Sci USA* 6:204–11.

Shibao H, Kutsukake M, Fukatsu T. (2004). The proximate cue of density-dependent soldier production in a social aphid. *J Insect Physiol* 50:143–47.

Simpson SJ, Despland E, Hägele BF, Dodgson T. (2001). Gregarious behavior in desert locusts is evoked by touching their back legs. *Proc Natl Acad Sci USA* 98:3895–97.

Skarka V, Deneubourg JL, Belic MR. (1990). Mathematical model of building behaviour of *Apis mellifera*. *J Theor Biol* 147:1–16.

Sudd JH. (1957). Communication and recruitment in Pharaoh's ant *Monomorium pharaonis* (L.). *Brit J Anim Behav* 5:104–9.

Sumpter DJT, Pratt SC. (2003). A modelling framework for understanding social insect foraging. *Behav Ecol Sociobiol* 53:131–44.

Thomas ML, Payne-Makrisâ CM, Suarez AV, Tsutsui ND, Holway DA. (2007). Contact between supercolonies elevates aggression in Argentine ants. *Insect Soc* 54:225–33.

Torres-Contreras H, Vásquez RA. (2007). Spatial heterogeneity and nestmate encounters affect locomotion and foraging success in the ant *Dorymyrmex goetschi*. *Ethology* 113:76–86.

Tschinkel WR. (2004). The nest architecture of the Florida harvester ant, *Pogonomyrmex badius*. *J Insect Sci* 4.21:1–19 (available online: http://insectscience.org/4.21).

Wagner D, Tissot M, Gordon D. (2001). Task-related environment alters the cuticular hydrocarbon composition of harvester ants. *J Chem Ecol* 27:1805–19.

Wenseleers T, Billen J, Hefetz A. (2002). Territorial marking in the desert ant *Cataglyphis niger*: Does it pay to play bourgeois? *J Insect Behav* 15:85–93.

Wilson EO. (1962). Chemical communication among workers of the fire ant *Solenopsis saevissima* (Smith): The organization of mass foraging. *Anim Behav* 10:134–47.

Whitehouse MEA, Jaffe K. (1996). Ant wars: Combat strategies, territory and nest defence in the leaf-cutting ant *Atta laevigata*. *Anim Behav* 51:1207–17.

Wittlinger M, Wehner R, Wolf H. (2006). The ant odometer: Stepping on stilts and stumps. *Science* 312: 1965–67.

Yamaoka R, Akino T. (1994). Ecological importance of cuticular hydrocarbons secreted from the tarsus of ants. In *Proceedings of the 12th Congress of the International Union for the Study of Social Insects*, Paris, p. 222.

3 Individual and Social Foraging in Social Wasps

Robert L. Jeanne and Benjamin J. Taylor

CONTENTS

INTRODUCTION

The social wasps differ from most species in the other major eusocial insect groups—termites, social bees, and ants—in several ways that affect foraging. First, their nests are constructed virtually entirely from foraged materials, typically some kind of vegetable fiber, although a few species of *Polybia* use mud (Richards 1978). Second, the diet of the social wasps is the least derived from that of the common ancestor of ants, bees, and wasps. That is, the wasps are the most carnivorous of the social insects, obtaining virtually all of their protein requirements from animal sources. Like bees and ants, however, they also require carbohydrates, particularly sugars. Finally, the wasps are unusual in that they appear to lack any of the sophisticated mechanisms of recruitment to resources that characterize many of the termites, bees, and ants.

In this chapter we review the biology of foraging in the social wasps. Although the emphasis is on food foraging, aspects of it are informed by work on foraging for nesting materials. Earlier reviews include Monica Raveret Richter's excellent summary of many details of foraging in social wasps (Raveret Richter 2000). Older reviews of the biology of the Vespinae include Spradbery (1973) and Edwards (1980). The volume edited by Kenneth Ross and Robert Matthews (1991) focuses on social behavior of wasps, with foraging treated as a minor aspect.

THE SOCIAL WASPS

Eusociality has arisen independently in two families of wasps. In the Sphecidae, subfamily Pemphredoninae, the tropical genera *Microstigmus*, *Arpactophilus*, and *Spilomena* include some social species (Matthews 1991). The more familiar social wasps are in three subfamilies of Vespidae: Stenograstrinae, Polistinae, and Vespinae. Because the latter two groups are much more widespread and abundant than the social sphecids and stenograstrines, most of the work on foraging behavior has been done on them, and therefore this chapter will focus mainly on the polistine and vespine wasps.

In our treatment below we divide these wasps into two behavioral groups according to how they initiate new colonies (Jeanne 1980). The first group comprises the independent founders, so called because colonies are founded by inseminated queens, independently of any workers. The Stenogastrinae and, with few exceptions, the Vespinae (hornets and yellowjackets) fall into this category, as do most of the Polistinae. It is clear that the independent-founding mode characterized the common ancestor of all eusocial vespids (Carpenter 1991). The remainder are swarm founders, in which new colonies are initiated by groups consisting of one or more queens and a larger number of workers. Swarm founding has evolved in four clades: the little-known tropical vespine genus *Provespa,* the ropalidiine genera *Ropalidia* (some species only) and *Polybioides* in the Old World tropics and subtropics, and in the Neotropical polistine tribe Epiponini, comprising twenty genera (Ross and Matthews 1991; Wenzel and Carpenter 1994).

FOOD SOURCE DIVERSITY, EUSOCIALITY, AND COLONY SIZE

Wasps can be thought of as the hunters and gatherers of the social insect world. As far as is known, all social wasps are omnivores, feeding on both animal protein and sugar-rich carbohydrate sources. Foraged proteins—predominantly live arthropods—provide nutrients for larval growth, whereas carbohydrates serve primarily as a source of energy—"aviation fuel"—for the adults. This dietary specialization by life stage is not absolute, however. Adults have trypsin-like and chymotrypsin-like proteases in the midgut, making them capable of protein digestion (Grogan and Hunt 1977; Hunt et al. 1987). It has been shown for *Polybia occidentalis* and *Polistes metricus* that at least the juices of captured prey enter the adult's midgut (Hunt 1984; Hunt et al. 1987). Conversely, adult wasps feeding on liquid carbohydrates colored with vegetable dye can be seen passing this to larvae; colored liquid is later found in the larval gut (Grinfel'd 1977; Hunt et al. 1987).

Colonial life creates dietary demands for the social wasps that most of their solitary relatives do not face. Many solitary wasps are specialists on one or a few related species of host or prey (O'Neill 2001). Often these are available during a limited part of the favorable season, and in many cases each wasp species' nesting and hunting season is restricted to a period of a few weeks or less. In contrast, social species have colony cycles that last for several months or longer. Food must be provided to larvae and adults throughout this period, precluding specialization on one or a few species of seasonally restricted prey. Thus, as a rule, the social wasps are generalist predators.

The primary source of protein for most species is live terrestrial arthropods. Because most social wasp foragers malaxate (chew up) their captures in the field, prey is largely unrecognizable when returning foragers are intercepted at the nest, making it difficult to compile extensive prey records (Figure 3.1). Nevertheless, we do know that the diets of the better-studied social wasp species are dominated by medium-sized, soft-bodied insects such as Lepidoptera larvae and adult Diptera, with Dermaptera, Orthoptera, Odonata, Hemiptera, Coleoptera, Hymenoptera, and Araneae comprising minor components (Spradbery 1973; Belavadi and Govindan 1981; Machado et al. 1987). For example, lepidopteran larvae from at least eight families comprised 95% of the prey taken by *Polistes versicolor* in Brazil (Prezoto et al. 2006). With larger adult insect prey, the wing muscles are prized as concentrated sources of protein; the wings, legs, and often the head and abdomen are chewed off before the remainder is malaxated into a ball and returned to the nest. With larger caterpillars,

FIGURE 3.1 *Nectarinella championi* forager arriving at the nest with a malaxated prey load. Atenas, Alajuela, Costa Rica. (Photograph by Robert L. Jeanne; copyright © Robert L. Jeanne.)

the gut is often removed, especially if the prey has been feeding on plant tissues containing allelochemicals (Rayor et al. 2007).

Some specializations do occur. The eusocial sphecids *Microstigmus comes* and *Arpactophilus mimi* prey on collembolans and plant lice (Psyllidae), respectively (Matthews 1968; Matthews and Naumann 1988). Other species of *Microstigmus* specialize on Thysanoptera or Cicadellidae (Matthews 1991). Specialization in these wasps may be possible by virtue of small colony size combined with seasonally unrestricted prey. The hornet *Vespa tropica* specializes on other social wasps, including *Polistes, Parapolybia, Ropalidia, Stenogaster,* and *Parischnogaster* (Turillazzi and Pardi 1982; Matsuura 1991; Landi et al. 2002). *Vespa mandarinia* takes mainly Coleoptera, although they also prey on other large, slow-moving insects, such as mantids and hornworms (Matsuura 1984). *Vespa crabro* is a semispecialist, with various cicada species comprising 95% of observed prey taken in the field (Matsuura 1984). No doubt the range in size of prey taken is restricted by the

FIGURE 3.2 *Polybia occidentalis* dismembering a grasshopper on the nest. Cañas, Guanacaste, Costa Rica. See color insert following page 142. (Photograph by Teresa León; copyright © Teresa León.)

body size of the wasp. Across species, social wasps vary in length from <4 mm in *Microstigmus* (Matthews 1991) to 38 mm in *Vespa mandarinia* (Matsuura and Yamane 1984).

The Neotropical swarm founder *Polybia occidentalis* and possibly other members of the genus are exceptions to the rule that foragers malaxate prey in the field; foragers of this species bring prey to the nest virtually intact (Hunt et al. 1987), making it possible to collect prey records by intercepting foragers as they arrive at the nest (Figure 3.2). Studies on several species of *Polybia* indicate that lepidopteran larvae make up the majority of prey, with several other orders comprising the rest (Gobbi et al. 1984; Gobbi and Machado 1985, 1986; Hunt et al. 1987). For a given colony, the proportion of the diet made up of each of the orders varies significantly during the year, presumably reflecting seasonal variations in the availability of prey (Gobbi et al. 1984; Gobbi and Machado 1985, 1986).

Many social species expand their diets opportunistically, capitalizing on food sources that are locally or temporally abundant. There are a few records of vespines preying on small, living vertebrates such as tadpoles, frogs, and nestling birds (Spradbery 1973; Matsuura and Yamane 1984). *Dolichovespula maculata* and *Polybia ignobilis* have been reported to capture flies attracted to carcasses (Heinrich 1984; Gomes et al. 2007). *Vespa velutina* is known to hawk honey bees returning to the hive (Tan et al. 2007). The hornets *V. mandarinia* and *V. orientalis* and the yellowjackets *Vespula germanica* and *V. vulgaris* occasionally prey on honey bee colonies, becoming serious pests in some places (Ishay et al. 1967; Edwards 1980; Matsuura 1984). *Vespa mandarinia* conducts group raids on honey bee colonies, as well as on colonies of other eusocial wasps, eventually occupying the nest and removing the pupae and larvae to feed their own developing brood.

An extreme form of opportunistic diet expansion is scavenging on vertebrate carrion, which has evolved independently in a few species of vespines and swarm-founding polistines. The *Vespula vulgaris* species group, a few species of *Vespa*, and *Vespula squamosa* have long-period nesting cycles and large colony sizes (Greene 1991; Matsuura 1991). Foragers of these species can often be pests at dumpsters, butcher shops, fish-cleaning stations, picnics, and other places where meat is exposed. Some even bite livestock, nestling birds, and humans, breaking the skin and imbibing the blood (Greene 1991). In contrast, the short-period nesters, including *Dolichovespula*, members of the *Vespula rufa* group, and several species of *Vespa,* have smaller colony sizes and rarely practice necrophagy (Matsuura and Yamane 1984; Matsuura 1991).

Among the swarm-founding Polistinae, a number of species scavenge on vertebrate carrion (O'Donnell 1995). The most prominent are several species of *Agelaia, Angiopolybia,* and *Polybia*, but species of *Brachygastra, Parachartergus, Protonectarina,* and *Synoeca* have also been recorded feeding at carrion (O'Donnell 1995; Silveira et al. 2005). Unlike the pattern in the vespines, necrophagy does not appear to correlate with colony size in the epiponines. Although the colonies of some species of *Agelaia* and *Brachygastra* can achieve extremely large size (Zucchi et al. 1995), colonies of *Agelaia cajennensis* and *Angiopolybia pallens* typically contain only a few hundred workers (Richards 1978). Of the six species taken in carrion traps in the Lower Amazon region, *A. pallens* was by far the most common, comprising 43% of trapped social wasps (Silveira et al. 2005). Silveira et al. (2005) also found that the number of wasps trapped was higher in the dry season than in the wet. It is not known whether this is a reflection of the season when colonies are actively rearing brood, or the absence of alternative food resources, such as insects. Although it is clear that the necrophagous vespines are facultative carrion feeders, it is possible that some of the epiponines may be obligate necrophages.

Unlike bees, wasps do not rely on floral nectar as a major source of sugars. Although they do obtain nectar from flowers with shallow corollas (Spradbery 1973), the short tongues of wasps preclude them from accessing most floral nectaries. Instead, they rely heavily on other sources of sugars, including extrafloral nectaries, honeydew, tree saps, and ripe and rotting fruit (Figures 3.3 and 3.4), and in some species, honey from bees (Edwards 1980). Some hornets (*Vespa*) are even reported to use mushrooms as a carbohydrate source (Matsuura 1984).

Opportunistic exploitation of carbohydrate sources also occurs. Vespines in the temperate zone are frequently attracted to anthropogenic sources of sugar, such as juices and sodas (Figure 3.5), and can become serious pests at bakeries, wineries, farmers markets, and picnics, especially toward the end of the nesting season, when colony sizes reach a maximum (Edwards 1980).

HOW RESOURCES ARE FOUND

Descriptive accounts of how wasps search for food have been provided by several authors. *Vespula vulgaris* and *V. rufa* "hunt slowly and deliberately around stems, in bushes, and under leaves low down near the ground" (Arnold 1966, cited by Edwards 1980, p. 141). In contrast, *V. germanica* and *Dolichovespula sylvestris* hunt in the open for prey, flying swiftly among vegetation until prey is encountered, then rapidly pouncing on it (Edwards 1980). Many vespids walk or crawl when hunting in dense vegetation (Akre 1982). In contrast, the "hover wasps," true to their name, have never been observed to forage on foot. The stenogastrine genera *Parischnogaster* and *Eustenogaster* hover at spider webs and snatch prey from them (Turillazzi 1983). *Mischocyttarus drewseni* also steals prey from spiders, but does so by climbing freely on the orb webs without becoming entangled (Jeanne 1972). At least some *Mischocyttarus* take the spiders themselves from their orb webs (RLJ, personal observation) (Figure 3.6). When gathering carbohydrates, foragers move slowly among honeydew producers, or move from flower to flower as bees do (Brian and Brian 1952).

When hunting live, fast-moving prey, wasps appear to rely heavily on visual cues to locate potential food items (Spradbery 1973). *Dolichovespula maculata* foragers pounce from flight on potential prey items, then determine if the item is prey or nonprey (Heinrich 1984). For example, Duncan (1939)

FIGURE 3.3 *Synoeca septentrionalis* forager feeding on ripe guava fruit on the tree. Cañas, Guanacaste, Costa Rica. See color insert following page 142. (Photograph by Robert L. Jeanne; © Robert L. Jeanne.)

FIGURE 3.4 *Vespula maculifrons* feeding on honeydew at pine tortoise scale (*Toumeyella parvicornis*) on pine. Sauk City, Wisconsin. See color insert following page 142. (Photograph by Robert L. Jeanne; copyright © Robert L. Jeanne.)

FIGURE 3.5 *Vespula germanica* feeding on soda. Madison, Wisconsin. See color insert following page 142. (Photograph by Robert L. Jeanne; copyright © Robert L. Jeanne.)

FIGURE 3.6 *Mischocyttarus* sp. preying on a spider on its web. Escazu, San José, Costa Rica. (Photograph by Robert L. Jeanne; copyright © Robert L. Jeanne.)

observed this species pouncing on dark-colored nail heads on a shed wall, apparently mistaking them for flies. Similarly, *Mischocyttarus drewseni* foragers fly slowly among vegetation and pounce on swellings on grass stems and twigs and even the barbs on barbed-wire fences (Jeanne 1972).

For less active prey items, olfactory cues appear to play a role in detection. When returning to cached prey, *Polybia sericea* foragers used visual and olfactory cues, but olfactory cues elicited landing more often than did visual cues (Raveret Richter and Jeanne 1985). Biting the prey was elicited by a combination of tactile, chemotactile, and possibly visual cues. Not surprisingly, olfactory cues have been shown to be of primary importance in tests using baits of fish or processed meats (Rau 1944; Akre et al. 1980; Ross et al. 1984; Reid et al. 1995). However, the modalities used can be situation dependent. *V. germanica* foragers used olfactory cues to locate Mediterranean fruit flies (Diptera: Tephritidae) gathering en masse in reproductive leks, apparently attracted to them from a distance by the kairomone of calling males (Hendrichs et al. 1994). In dense citrus foliage, where visibility is reduced, foragers responded with a similar number of approaches and landings to containers with and without visual cues, but in the less dense foliage of fig and mulberry there were more landings on containers that offered visual cues in conjunction with olfactory cues (Hendrichs et al. 1994; for a detailed review of the parasitic use of alien cues and signals by social insects, see Chapter 8).

Odor cues are also utilized in finding carbohydrate sources. Both *Vespula maculifrons* and *V. germanica* foragers reportedly approach baits from downwind and, when presented with feeders containing either honey or the relatively odorless glucose, ignore the glucose (Gaul 1952; Overmyer and Jeanne 1998). Foragers are also strongly attracted to the smell of fruit emanating from orchards, as well as from bakeries and cafés (Akre et al. 1980; Edwards 1980; Day and Jeanne 2001).

If a *Vespa* or *Polistes* forager is successful at a particular location, she performs an orientation flight, apparently learning the nearby visual landmarks. She will return to that site on subsequent foraging trips, even for a period of several days after the food has been removed (Ishay et al. 1967; Takagi et al. 1980). The time to extinction of this behavior is positively correlated with the magnitude of the initial reward (Free 1970; Lozada and D'Adamo 2006).

Wasps are able to learn both colors and odors associated with food rewards, but odors are apparently more readily learned than colors (Beier 1983; Jander 1998; McPheron and Mills 2007). *Vespula vulgaris* foragers are capable not only of associating color with reward, but also of forming an expectation of the reward-color association based on previous experience, and will exhibit flower constancy much as bees do, albeit less pronounced (Real 1981).

Once a location, visual cue, or odor is learned, the cues used when relocating that food source also appear to be situation dependent (D'Adamo and Lozada 2003; Moreyra et al. 2006). *V. germanica* foragers trained to a food source were given a choice between the original training site and visual and odor stimuli, both displaced 50–60 cm away from the site. They were found to behave differently depending on the type of resource (D'Adamo and Lozada 2003; Moreyra et al. 2006). Carbohydrate foragers tended to orient at close range to the visual cue of the bait (feeder) itself, even if it had been moved a short distance from its learned position. In contrast, protein foragers more often oriented to the location of the original feeding site, using nearby visual landmarks as cues. As interpreted by the investigators, carbohydrate resources frequently occur in patches (flowers, fallen fruit, honeydew producers), which make the precise location of the visited resource in the patch less important to learn. Carrion, on the other hand, is typically at a single point source in space, so it is adaptive to learn its location in relation to nearby landmarks (D'Adamo and Lozada 2003; Moreyra et al. 2006).

The pattern that emerges from these studies suggests that foragers use visual, olfactory, and tactile cues—learned or not—in whatever combination provides the most reliable information about the respective resource, and that the relative importance of the different cue modes varies with the type of resource as well as with the context.

FOOD RECRUITMENT

Recruitment is communication that brings nestmates to a place where work is required (Wilson 1971). Whether recruitment in wasps occurs or not depends on one's definition of communication. The most useful definition restricts communication to the transfer of information by signals (Lloyd 1983; Dawkins 1986). A signal is an organism's structure or action that has been shaped by natural selection to convey information that benefits both the sender and the receiver (Lloyd 1983). Cues, in contrast, are structures or actions that have not been shaped by natural selection to convey information; that is, they convey information incidentally (Lloyd 1983; Seeley 1995). By this narrow definition, except for one species—*Vespa mandarinia*—wasps appear not to recruit to food. That is, there is no evidence as yet that a successful forager produces any kind of signal that informs a nestmate about the location of the resource it has found.

On the other hand, there is ample evidence that foragers take advantage of cues of various kinds that enhance their success in returning food resources to the colony. Cues can be available at the food source itself, or at the nest.

INFORMATION AT THE FOOD SOURCE

Social facilitation is behavior "initiated or increased in pace or frequency by the presence or actions of another animal" (Wilson 1975, p. 51). One variant of this is local enhancement, in which the presence of one or more foragers at a patch of food increases the attractiveness of the patch to foragers flying nearby (Thorpe 1963; see also Chapter 8). The most extensive work on local enhancement has been done on species in the *Vespula vulgaris* group. For *V. germanica* the attractiveness to foragers of a dish of honey solution increases with the number of wasp decoys (dead wasps) on it (Parrish and Fowler 1983; Raveret Richter and Tisch 1999; D'Adamo and Lozada 2005). In contrast, *V. maculifrons* exhibited local inhibition, preferentially feeding at dishes where fewer wasps were present (Parrish and Fowler 1983; Raveret Richter and Tisch 1999). Reid et al. (1995), on the other hand, found that *V. maculifrons* did exhibit local enhancement in response to live foragers feeding at protein. It is not clear whether the difference is due to the use of live wasps or of protein bait. *V. flavopilosa*, another member of the *V. vulgaris* group, showed no preference for feeding at sites occupied by conspecifics, whereas *V. vidua* and *V. consobrina*, both members of the *V. rufa* species group, preferred unoccupied over occupied feeders (Raveret Richter and Tisch 1999). *V. koreensis*, in the sister group to the *V. vulgaris* group (Carpenter 1991), was also found to choose unoccupied sugar water feeders over feeders with dead decoys (Kim et al. 2007).

Among the swarm-founding polistines, *Agelaia pallipes* foragers preferentially landed on protein feeders provided with decoys (dead wasps) over unoccupied feeders (Fowler 1992). On the other hand, there is no evidence for a similar preference in *A. multipicta* or *A. hamiltoni* (Jeanne et al. 1995) (Figure 3.7).

Foragers of *Polybia occidentalis* were more attracted to occupied feeding dishes, for both protein and carbohydrate, and regardless of whether dead or live wasps were used as cues (Raveret Richter 1990; Hrncir et al. 2007). Interestingly, *P. diguetana*, a close relative, avoids occupied dishes under

FIGURE 3.7 *Agelaia multipicta* foragers on hamburger bait in an experiment to test for nest-based recruitment to food. Cojedes, Venezuela. (Photograph by Robert L. Jeanne; copyright © Robert L. Jeanne.)

the same conditions (Raveret Richter 1990). The independent-founding *Polistes fuscatus* prefers to feed at occupied feeders (Raveret Richter and Tisch 1999).

D'Adamo et al. (2000) asked whether odor plays a role in local enhancement in *V. germanica* at protein baits. Feeders providing only visual cues of wasps were less attractive to foragers than feeders with the odor of foragers. Extending the work to combine the visual and odor cues of the bait with those of the wasps, D'Adamo et al. (2003) reinforced the conclusion that odor cues are more effective than visual cues, and found that a combination of olfactory and visual cues is more attractive than either alone. Taken together, these results suggest that wasps are most attracted to feeders with the greatest amount of information available.

The adaptive advantage of local enhancement has not been systematically investigated. The visual cue of wasps at a food resource may convey two kinds of information: that food is available and that predators are absent. Thus, it may increase the efficiency of resource acquisition by decreasing the average search time without increasing risk of mortality. But this does not explain why different wasp species respond differently to the information.

One hypothesis is that the colony benefits if most or all visitors to a food resource are nestmates. This would predict that species forming large colonies should tend to show local enhancement, since most of the foragers in an area would likely be nestmates. This prediction is not borne out, however: some *Vespula* species with large colonies do not show local enhancement, whereas *Polistes*, with small colonies, does.

An alternative hypothesis is that local enhancement is body size dependent, such that a positive response to others at a resource is present in large species and absent in small ones. Larger-bodied species could benefit by responding positively to the cue of feeding wasps of any species, which they could push aside and successfully dominate at the resource (O'Donnell 1995). In other words, it may be a mechanism for resource theft (Raveret Richter and Tisch 1999). Smaller species, in contrast, may not only be at a competitive disadvantage, but also at risk of predation by larger wasps when foraging in close quarters with them. This could explain Raveret Richter's (1990) result with *Polybia occidentalis* and *P. diguetana*. The two species are sympatric where she did her study. They are very similar in appearance, but *P. diguetana*, which avoids occupied feeders, is slightly smaller than *P. occidentalis*. Parrish and Fowler (1983), who found that *V. maculifrons* avoided occupied feeders, did their study in a site where *V. germanica*, a larger wasp, was also foraging. Likewise, Raveret Richter and Tisch (1999) studied *V. maculifrons* in a habitat where the larger *V. vidua* was common. In contrast, Reid et al. (1995), who found that *V. maculifrons* was attracted to occupied feeders, did their tests at a site where *V. germanica* was rare or absent. If this hypothesis has any merit, it raises the question of whether the presence or absence of the local enhancement response is a function of the local mix of species. It also suggests that local enhancement or inhibition might be learned, again depending on the local mix of species (for further discussion on potential causes underlying local enhancement/inhibition in social insects, such as aggressiveness and dominance of a certain species in a food patch, or the experience of the foragers with a certain resource, see Chapter 8).

Another way that foragers could enhance recruitment to a rich food resource is by chemically marking it. Honey bees leave long-lasting attractants (Free et al. 1982; Free and Williams 1983; Seeley 1995; see also Chapter 9); the source of the chemicals remains unknown (Barth et al. 2008). They also apply short-term, repellent odors from the mandibular glands (Giurfa 1993; Stout and Goulson 2001). Stout and Goulson (2001) also found evidence for longer-lasting scent marks that are repellent. Foragers of some stingless bee species make marks on attractive sources using the labial and claw-retractor-tendon glands (Jarau et al. 2004, 2006; Schorkopf et al. 2007; Barth et al. 2008; see also Chapter 12). Bumble bees leave secretions that may be attractive (Schmitt et al. 1991) or repellent (Stout et al. 1998; Goulson et al. 2000; see also Chapter 13). Recent evidence suggests that the response is context dependent (attraction at *ad libitum* resources and avoidance at nonreplenished resources) and that the appropriate response may be learned (Saleh and Chittka 2006; Saleh et al. 2006, 2007; Witjes and Eltz 2007). The scent marks appear to be merely footprints left passively on flowers from the act of walking on them, suggesting a cue rather than a signal (Witjes and Eltz 2007; Wilms and Eltz 2008).

There is some evidence for wasps that previously visited feeders are more attractive to foraging wasps than clean ones (Beier 1983; Overmyer 1995; D'Adamo and Lozada 2005), and that extracts from heads are attractive to foraging *Vespula germanica* (D'Adamo et al. 2001, 2004; D'Adamo and Lozada 2005). However, in a controlled study Jandt et al. (2005) tested for scent marking in *V. germanica* and found no evidence that foragers chose feeders visited fifty or one hundred times over unvisited control feeders. Cuticular hydrocarbons appear to facilitate movement through nest entrances (Butler et al. 1969, Steinmetz et al. 2002, 2003), but have yet to be implicated in marking food resources. In sum, the evidence that social wasps scent-mark food sources, either with attractants or repellents, is weak at best.

The one exception to the absence of food marking by social wasps is the substance applied by *Vespa mandarinia* foragers during their autumn raids on honey bee hives and colonies of other vespine wasps, described in detail by Matsuura (1984). Upon finding a hive, individual foragers kill several worker honey bees and take them back to the nest to feed the brood (hunting phase). After several successful visits the forager wipes her abdomen across the hive, marking it with secretions from her van der Vecht's gland. The pheromone attracts nestmates (recruitment phase) (Ono et al. 1995). Nestmates in the area soon gather at the marked site and begin hunting individually, but after three or more foragers have arrived, the hornets begin to attack en masse, slaughtering bees by the thousands (slaughter phase). When the defending bee population is sufficiently weakened, the hornets move into the hive and begin removing the brood to feed their own developing larvae (occupation phase).

In summary, except for *Vespa mandarinia*, there is as yet no evidence for wasps that foragers at food sources signal in any way to others in the area. Local enhancement appears to be a simple response to cues—the presence of others feeding—that are reliable indicators of the presence of food and perhaps the absence of predators.

Why does food-site-marking behavior apparently occur in the social bees but not in the social wasps? Here, the exceptional case of *Vespa mandarinia* may be instructive. Honey bee colonies, while a very rich resource for a predator, are very well defended. An individual-hunting strategy, even by so large a hornet as *V. mandarinia,* cannot break through these defenses. By attracting a number of nestmates to the hive via scent marks, the hornets can shift to a cooperative mode of attack and successfully break through to reach the brood. For other social wasps, so far as is known, the nature of the prey taken does not demand such cooperation, so the benefits of marking may not outweigh the costs. Large, ephemeral protein sources such as carrion can be located easily by the odor of the rotting flesh itself, so adding a scent mark may provide little or no additional advantage to the colony. The same could be true for sources of carbohydrate such as honeydew. On the other hand, there could be a disadvantage to marking, in that it could attract conspecifics from competing colonies, diluting the advantage to the marking colony (for a detailed discussion of social insects "eavesdropping" on other species' recruitment signals, see Chapter 8).

INFORMATION AT THE NEST

Honey bees have evolved mechanisms for communicating information about food resource location to nestmates in the nest (von Frisch 1967). In most ant species and possibly some stingless bees, foragers can lead nestmates to a food resource via chemical trails or by other means (Wilson 1971, Nieh 2004; Barth et al. 2008; see also Chapters 2, 6, and 12). Do wasps have any such ability? In all of the studies to date, no evidence has been found that wasps lay recruitment trails to food sources, nor has there been any evidence that foragers can behaviorally encode distance or direction information in behavior performed at the nest.

There is, however, evidence for *Vespula germanica* and *V. vulgaris* that naïve foragers in the nest learn the odor of a recently arrived rich food load and use that information to help locate the source of the odor in the field. Maschwitz et al. (1974) presented naïve foragers with a choice between two carbohydrate solutions; one was scented identically to that fed on by trained foragers feeding on a separate feeder, and the other was a differently scented control. The naïve nestmates chose the test scent more often than the control. Overmyer and Jeanne (1998) repeated the experiment on *V. germanica*, using a design with tighter controls for the effect of local enhancement, and confirmed this result. Behavioral stimulation of nestmates by the forager may be involved, but has not yet been shown to be necessary. Indeed, scented sugar water experimentally injected into the nest is enough to bring foragers out of the nest and to an external source bearing the same odor (Jandt and Jeanne 2005).

Hrncir et al. (2007) recently showed that nest-based information from successful *Polybia occidentalis* foragers plays a role in activating naïve individuals to search for food. Once a naïve nestmate is stimulated to search for food, its choice of a feeder is strongly influenced by local enhancement at the feeder. The initial stimulus that activates foragers to search for food is not known, but the simplest explanation is that olfactory and gustatory information from successful foragers stimulated other naïve individuals to search for food, as shown for *V. germanica* (Overmyer and Jeanne 1998; Jandt and Jeanne 2005; for the importance of food odor information for foraging decisions in social bees, see Chapters 9, 10, and 12).

For other swarm-founding polistines, however, there is little evidence for information transfer via signals at the nest. Naumann (1970) described a "departure dance" performed by returned food foragers of *Protopolybia acutiscutis* (= *pumila*). The forager runs erratically over the nest, often wagging the abdomen from side to side and stopping frequently to engage in trophallaxis before taking off on her next trip. Dancing individuals appear to be attractive to other wasps, which follow them for short distances and lick and antennate them. The behavior appears to stimulate movement and sometimes departure by wasps on the nest, but whether the dance communicates information or stimulates others to leave the nest in search of food remains to be determined with carefully controlled experiments. The fact that numbers of *P. acutiscutis* foragers at a rich resource do not increase over time suggests that the effect of any such stimulus is weak at best (Naumann 1970). Forsyth (1978) studied ten species in the genera *Agelaia*, *Brachygastra*, *Metapolybia*, *Parachartergus*, *Polybia*, and *Synoeca* and claimed no evidence of recruitment, but provided no data to back up the assertion. In an experimental test, Jeanne et al. (1995) found no evidence for nest-based recruitment in *Agelaia multipicta* or *A. hamiltoni*. On the other hand, there is Lindauer's tantalizing result with *Polybia scutellaris* (Lindauer 1961). After ten foragers were trained to a bait 150 m from the nest, five to seven new foragers arrived every 30 min. It was not determined whether this increase was caused by information transfer at the nest or by local enhancement at the resource.

In a colony-level approach to the question of information transfer at the nest, several investigators have studied temporal patterns of traffic in and out of the nest. Blackith (1957) studied forager traffic at nests of *Vespula* spp. and concluded that entrances and departures were temporally clumped, i.e., nonrandom. He suggested that the clumping was a result of social facilitation, where the presence of a worker had a releaser effect on others, prompting the passage of more workers through the nest entrance. This was later elaborated on by Wilson (1971) as a type of group effect, wherein the return of a forager elicits a rise in activity and stimulates others to forage. Plowright (1979) contested Blackith's conclusion of nonrandomness, arguing that it was based on a flawed statistical analysis. Subsequent observations on *Dolichovespula arenaria* showed that traffic at the nest entrance closely fit a Poisson distribution, which suggested that the temporal pattern of entrances and departures was random (Pallett and Plowright 1979). Still a different result was claimed for *V. vulgaris* by Ruxton et al. (2001). Their analysis of interevent intervals (time between successive arrivals or exits) led them to conclude that arrivals and exits are neither random nor clumped, but occur at regular intervals.

These widely varying conclusions indicate that the question of temporal patterns in forager traffic is still unresolved. The disparate results could reflect real differences among species, or they could be attributable to methodological differences, such as time of day, size of colony, or whether data were analyzed as events per unit time or interevent intervals. As Ruxton et al. (2001) point out, further work is warranted. If nonrandomness in temporal patterning of traffic can be substantiated, it would be circumstantial evidence for some form of nest-based social facilitation. The next question would be if and how the interactions are adaptive in the context of food foraging.

Do Social Wasps Lack Recruitment to Food, and If So, Why?

Unlike their social relatives, the bees and the ants, wasps appear to lack sophisticated communication systems that inform nestmates of when and where to forage. Rather, wasp foragers seem to act solely as individuals. The question then becomes why no social wasp is known to have evolved a nest-based food recruitment mechanism. Several hypotheses have been suggested (Jeanne et al. 1995). One is that wasps have lacked the genetic variability sufficient to give rise to the structures or behavior to inform nestmates about the location of resources. This seems unlikely. Members of the swarm-founding Polistinae have exhibited enough genetic variability to have evolved trail pheromones to guide swarms to new nest sites (Jeanne 1981, 1991a), a mechanism that should be capable of functioning in food recruitment as well.

Another possibility is that the lack of a signal-based recruitment system may be due to a social constraint. A critical colony size may be needed in order for the benefits of recruitment to outweigh the costs (Beckers et al. 1989), and it could be argued that worker numbers in most wasp species are too low to effectively defend food sources against competitors such as stingless bees and ants. The limited ability to store resources may also make recruitment unprofitable. Most social wasps seem to live hand-to-mouth, the colony consuming food resources as fast as foragers can collect it. However, as mentioned below, some wasps store nectar, and others stock up on ant and termite alates, and yet no recruitment mechanism has yet been found in these species.

A third possibility is that ecological factors may make recruitment an unproductive strategy in the social wasps. For recruitment to pay off, resources need to be clumped in space and persist for appreciable periods of time. Theoretical considerations and empirical evidence both suggest that when sources are scattered and ephemeral, or are locally available and easily found by individual searching, there is less benefit to recruitment (Waddington et al. 1994; Portha et al. 2002; Dornhaus and Chittka 2004; Dechaume-Moncharmont et al. 2005). It is possible that concentrated protein and carbohydrate sources utilized by wasps occur too rarely to have been a strong selective force favoring recruitment. However, the fact that some stingless bees (Roubik 1989) and many ant species recruit to carrion (O'Donnell 1995), and that ants also recruit to carbohydrate sources such as honeydew (Portha et al. 2002), weakens this argument.

There is also a cost to obtaining information through recruitment (Dechaume-Moncharmont et al. 2005). By waiting for information, wasps may fail to find new resources. Perhaps the best strategy for wasps is to distribute themselves throughout their environment and remain opportunistic. It has been suggested that such a foraging strategy allows foragers to encounter new resources first (Deneubourg et al. 1983), and environments that have many small resources and rare, large, variable resources would select for an opportunistic foraging strategy as opposed to recruiting (Johnson 1983; Raveret Richter 2000).

In addition to the variables of space and time, resources must also be defensible against non-nestmates. Indeed, both polistines and vespines exhibit aggressiveness at resources (Parrish 1984; Raveret Richter 1990). In many areas, especially in the tropics, strong competition exists for clumped resources such as carrion and honeydew from species of bees and ants that recruit. Such species may accumulate foragers more quickly at clumped resources, making them more capable of defending them (Forsyth 1978; Roubik 1989; Silveira et al. 2005). Many ant species may even pose a predation risk to the wasps. In the face of such competition, recruitment, especially by small and moderately sized wasp colonies, may be a less effective strategy than exploiting new resources early and quickly, and then moving on when the competition builds up (Jeanne et al. 1995; Raveret Richter 2000).

Finally, it may be premature to generalize and conclude that all wasps lack food recruitment. Only a handful of species have been studied carefully enough to rule it out. The unstudied species include several potential candidates among the swarm-founding polistines. One is *Agelaia vicina*, with colonies of more than a million workers (Zucchi et al. 1995). If colony size is a critical factor, surely this species exceeds the threshold. A second is the genus *Pseudopolybia*, whose trail pheromone gland is much larger than in other genera, suggesting that its secretion is not just a trail pheromone used in recruitment to a new nest site, but may function in food recruitment as well (Jeanne et al. 1983). Two others are *Brachygastra mellifica* and *B. lecheguana*, commonly called honey wasps. These species store honey that is used to sustain the adults through the harsh season at the margins of the tropics, in amounts large enough that locals collect the nests to harvest the honey (Sugden and McAllen 1994). Species able to forage efficiently enough to store such amounts may well have a food recruitment system.

ORGANIZATION OF WORK

WORKER POLYETHISM

Worker polyethism (division of labor) is the differential performance of tasks among workers (Wilson 1971). Morphological worker subcastes are absent in eusocial wasps, but age polyethism (Wilson 1971) does occur. It is rudimentary in the independent-founding Polistinae, somewhat more pronounced in the Vespinae, and very well developed in the swarm-founding Polistinae (tribe Epiponini) (Jeanne et al. 1988; Jeanne 2003). Response-threshold models predict that the degree of specialization among workers should increase with colony size, a pattern generally supported by empirical evidence (Gautrais et al. 2002; Merkle and Middendorf 2004; Jeanson et al. 2007). In wasps, however, there are some exceptions to the pattern.

In *Polistes*, the first workers to eclose in the colony begin foraging after just a few days and continue to do much of the foraging for the colony throughout their lives (West-Eberhard 1969; Dew and Michener 1981). Later-emerging workers tend to work longer on the nest before adding foraging to their repertories (Post et al. 1988). *Ropalidia marginata* has a somewhat stronger temporal polyethism, with young workers remaining on the nest feeding larvae, and later engaging in nest construction, while older workers forage first for pulp, then for food (Gadagkar 2001).

In the Vespinae the extent of age polyethism correlates with colony size across species. It is virtually absent in *Dolichovespula* and *Vespa*. Workers begin foraging at age 2–4 days, but continue performing intranidal tasks throughout their foraging careers (Brian and Brian 1952; Matsuura 1984). In *V. simillima* and *V. analis*, which form the largest colonies in these two genera, workers tend to increase their frequency of foraging as they age. However, it is not unusual for older *Vespa* workers to perform several tasks, both intranidal and extranidal, in 1 day (Matsuura 1974, 1984). In both genera, as in the independent-founding polistines, there is a weak but detectable tendency for pulp foraging to be done earlier in the foraging career than prey and nectar (Brian and Brian 1952; Jeanne 1991b). In Brian and Brian's (1952) study of *D. sylvestris*, for example, pulp made up 59% of foraged loads during the first 3 days of a worker's life, dropping to 15% by age 8–9 days, with fluid becoming the predominant foraged material.

Age polyethism is more pronounced in the larger colonies of the *Vespula vulgaris* group than in *Vespa* and *Dolichovespula*. Workers start out as nurses, then progress to foraging (Shida 1959, cited by Matsuura 1984; Potter 1964; Montagner 1966; Akre et al. 1976; Reed and Akre 1983; Hurd et al. 2007). In *Vespula germanica* the onset of foraging begins toward the end of the first week of age (Hurd et al. 2007). As a rule, workers start out as pulp foragers, switch to carbohydrate foraging, and finally add prey, with long-lived foragers often ending up concentrating on fluid (Potter 1964; Akre et al. 1976). However, even after they take up foraging, workers do not give up nest work. Thus, despite the large colony sizes (>10^3 workers) achieved by some members of this genus, age specialization among workers appears to be rudimentary, and may be constrained by the incomplete task partitioning shown in the vespines (see below).

Age polyethism is most fully developed in the swarm-founding Polistinae. In *Polybia occidentalis*, the most thoroughly studied species, colonies range from a few hundred to a few thousand adults (Bouwma et al. 2003). Workers pass through three discrete temporal castes, corresponding to the three spatial regions of the nest where work is done: the nest interior, the outer surface of the nest, and away from the nest (Jeanne et al. 1988; Jeanne 1991b; O'Donnell and Jeanne 1995). Individuals vary tremendously in the rate at which they pass through the sequence (Jeanne et al. 1988), but the average worker spends her first week inside the nest. Although the nest envelope prevents observation of activities inside the nest, it is likely that she is working at brood-related tasks. By the beginning of the second week she appears on the outside of the nest, where she engages in nest construction and nest cooling, defends against ants, and receives loads from incoming foragers. She switches to foraging at about 3 weeks of age, although this ranges from 1 to 5 weeks among individuals. This is much later than in the independent-founding polistines and vespines. Furthermore, the switch to foraging is abrupt: after a worker begins to forage, she gives up nest tasks within a day or so (Jeanne et al. 1988). Thus, older workers become specialists on foraging. As in the vespines, foragers tend to specialize on water and pulp first, then on food (Jeanne et al. 1988). There is evidence for a similarly pronounced age-polyethic schedule in *Agelaia pallipes*, *Protopolybia acutiscutis*, *P. exigua*, and *Metapolybia* spp., representing colony sizes ranging from 100 to 35,000 adults (Naumann 1970; Simões 1977; Forsyth 1978; West-Eberhard 1978, 1981; Jeanne 2003).

In addition to a pronounced age polyethism, individual foragers of *Polybia occidentalis* show a strong tendency to specialize on one or two of the four foraged materials in two functional groups: prey or nectar (food), or pulp or water (nest material). In three colonies observed for 8–13 consecutive days, the identity of marked foragers explained 66–76% of the variance in effort devoted to each of the four foraged materials they specialized on (O'Donnell and Jeanne 1990). During periods of 15–180 minutes, foragers specialized on a single material for 95% or more of their trips. When they did switch, they were nearly four times as likely to switch to the other material in the same functional group, i.e., between nectar and prey or between water and pulp, as to switch between food and nest material (O'Donnell and Jeanne 1990). There is evidence that a similar specialization by individuals on single materials also occurs in *Metapolybia* spp. and in *Protopolybia exigua*, both of which have small colonies of one to two hundred adults (Simões 1977; Forsyth 1978; Karsai and Wenzel 2000). The larger the colony, the more likely foragers are to be faithful to one kind of material, a pattern that is supported on theoretical grounds (Gautrais et al. 2002) and appears to hold across as well as within species (based on two species of *Metapolybia* and one of *Polybia*) (Jeanne 1986b; Karsai and Wenzel 1998). Individual specialization on one or two types of resource may facilitate the increases in foraging success rates with experience, as documented for *P. occidentalis* (O'Donnell and Jeanne 1992c).

In summary, age polyethism and individual specialization among workers are poorly developed in the independent-founding polistines, somewhat stronger in the vespines, and very strong in the swarm-founding epiponines even in species with colony sizes toward the lower end of the range for the group.

TASK PARTITIONING

The collection and utilization of materials may be broken up into two or more component subtasks performed by different individuals (Jeanne 1986a). In the typical case, an arriving forager transfers her load to a nest worker, who handles its utilization on the nest. Theoretical aspects of task partitioning have been explored by Anderson and coworkers (Anderson and Ratnieks 1999, 2000; Ratnieks and Anderson 1999a, 1999b; Anderson and Franks 2001; Anderson et al. 2001). Results of these analyses suggest that complete task partitioning enables more efficient workflow in larger colonies (Anderson and Ratnieks 1999).

Task partitioning is most pronounced in the swarm-founding polistines (Jeanne 1991b, 2003). In *Polybia occidentalis*, all four foraged materials—nectar, prey, water, and pulp—are fully partitioned into collection by foragers and utilization by nest workers. Thus, foragers hand off their loads

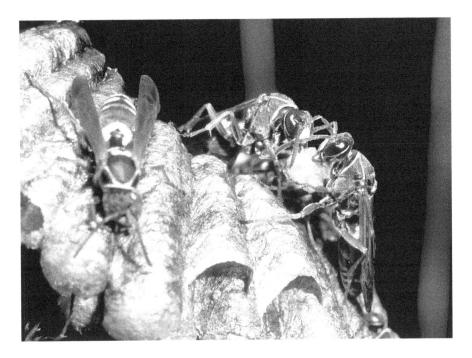

FIGURE 3.8 *Polistes snelleni* workers sharing a prey load. Sapporo, Japan. (Photograph by Robert L. Jeanne; copyright © Robert L. Jeanne.)

at the nest, then immediately leave on their next trip (Jeanne 1986b, 2003). Task partitioning is what enables the strong age polyethism shown by this wasp (Jeanne 1986a).

The independent-founding polistines show incomplete task partitioning (Jeanne 1991b, 2003). Pulp foragers always or nearly always utilize their own loads, rarely sharing them with nestmates. Prey foragers more frequently share their loads with one or more nestmates (Figure 3.8), but usually retain some to feed to larvae themselves.

When task partitioning does occur within the Vespinae, it seems to be best developed in the large-colony *Vespula* species. Of the four primary materials collected by vespines, only prey and carbohydrates are shared with, or completely given to, nestmates. Water and pulp, in contrast, are used at the nest by the foragers that collected them; that is, pulp foragers add their own loads to the nest, never sharing with nestmates (Greene 1991). The lack of partitioning of pulp collection and handling precludes the specialization of workers on nest work vs. foraging when it comes to nest construction; this in turn means that workers of all ages are doing both nest work and foraging. In *Vespula atropilosa* and *V. pensylvanica* prey is sometimes shared, but most often transferred entirely to a nest worker (Akre et al. 1976; Jeanne 1991b). *V. squamosa* shares prey with nestmates, as do *V. germanica* and *V. vulgaris*, although the latter two species may also at times give their whole loads to nestmates (Jeanne 1991b). When the load is carbohydrate food, it is always given, at least in part, to nestmates in both of these species. However, in *V. pensylvanica* and *V. atropilosa*, only in older colonies is carbohydrate handling completely partitioned (Akre et al. 1976).

POTENTIAL REASONS FOR DIFFERENCES IN WORK ORGANIZATION AMONG WASPS

The difference between the vespines and *Polybia occidentalis* in the degree of age polyethism and task partitioning is striking. Why, despite similar colony sizes, has the one evolved discrete age polyethism and complete task partitioning, while the other retains a much less pronounced system of specialization? One hypothesis is that the vespine system allows colonies the flexibility to respond to changing colony needs (Spradbery 1973; Matsuura 1984). But this begs the question of why the same advantage would not accrue to *Polybia*. Hurd et al. (2007) suggested that task generalist behavior may be an adaptation to details of life history in *V. germanica*. Workers of this species have much shorter life spans than those in species that have well-defined age polyethism, such as *A. mellifera* (Seeley 1995) and *P. occidentalis* (Jeanne et al. 1988). Species with shorter worker life spans may not be able to gain the same degree of benefit of task specialization and efficiency due to repeated task performance as species with longer life spans. A third possibility is that the initial-size bottleneck vespine colonies pass through—starting with a worker population of zero—has acted as a constraint on the evolution of a system that has efficiency payoffs only at large colony sizes. Still another is that it may be faster for a pulp forager to build with her own load than for her to find a receiver. *Ipso facto*, this forage-utilize sequence precludes the evolution of strong age polyethism (Jeanne 1986a).

These ideas are unlikely to exhaust the possible reasons behind the differences in organization of work among wasp groups. This pattern of distribution of social organization among the wasps remains unexplained and presents a potentially fertile area of research where additional hypotheses and tests are needed.

ALLOCATION OF FORAGING EFFORT

Success in the fitness wars is measured in terms of the number of surviving reproductives produced in generation F1 per reproductive in the parent generation relative to that of conspecific competitors. Reproductive success no doubt correlates closely with the colony's growth rate during the ergonomic phase: colonies that increase their worker numbers more quickly are likely to yield more reproductives at the end of the colony cycle. Colony growth rate, in turn, depends directly on how quickly and efficiently materials can be collected and brought to the nest. Natural selection should therefore favor colonies that allocate their foraging effort so as to maximize the appropriate objective function—perhaps net rate of energy gain in terms of some measure of foraging benefits discounted by some measure of foraging costs—given the choice variables and constraints the colony faces (Ydenberg et al. 2007). We assume that social wasps have evolved optimal solutions to at least three trade-offs: how much effort should be allocated to foraging vs. nonforaging activities, how the total foraging effort should be divided between foraging for food and foraging for nesting materials, and how food foragers should allocate their effort between prey hunting and carbohydrate gathering. For the most part, these questions have not been quantitatively addressed in wasps.

If foraging rates are increased or decreased from the optimum, net gain decreases. That is, either benefits decrease or costs rise, or both. Foraging is a costly activity for two reasons. First, flight is expensive compared to activities on the nest. Energetic cost of flight has not been measured in wasps, but in *Apis* and *Bombus* metabolic rate during flight exceeds that of resting bees by a factor of 20–50 (Hocking 1953; Kammer and Heinrich 1974). Thus, it takes a lot of energy to fuel foraging trips. Second, once a forager leaves the protection of the nest or swarm, she faces elevated risks of mortality due to predation, exposure, accident, and simply getting lost (Figure 3.9). After a *Polybia occidentalis* worker becomes a forager, her average remaining life expectancy is approximately a week, regardless of the age at which she makes the switch (O'Donnell and Jeanne 1992b). Increasing the foraging rate will therefore lead to an increase in worker mortality rate, reducing the number of workers, which in turn must lead to a decrease in colony growth rate. On the other hand, decreasing the foraging rate will reduce input of materials, which will also slow the colony's growth rate. How the optimal balance is struck and the behavioral mechanisms by which it is implemented have yet to be investigated.

The age-polyethic sequence of tasks seen in *Polybia occidentalis* and many other social insects may be another manifestation of the optimizing effects of selection. Putting off foraging, the riskiest task, until last in the sequence maximizes the colony's return on its investment in each worker

FIGURE 3.9 Mantid capturing foragers of *Polybia occidentalis* as they fly to and from an absconding swarm. Cañas, Guanacaste, Costa Rica. See color insert following page 142. (Photograph by Robert L. Jeanne; copyright © Robert L. Jeanne.)

(Jeanne 1986a). By the time the average worker begins to risk her life in the field, she has contributed three weeks of nest work to the colony (Jeanne et al. 1988).

Supply and demand also play a role. In a study of the effect of experimentally increasing the demand for nest material in colonies of *P. occidentalis*, O'Donnell and Jeanne (1992a) found that existing food foragers did not shift over to collecting nest materials, nor did existing nest material foragers increase their rate of work. Instead, nest workers switched to pulp and water foraging at an earlier age than in control colonies. When a subset of nectar foragers was removed from nests of *Polistes instabilis*, recovery of the overall rate took several days and occurred via replacement foragers recruited from among foragers of other materials or from nest workers, and not from an increase in rate of foraging of the remaining foragers (O'Donnell 1998). Recovery was accompanied by an increase in the frequency of dominance interactions, suggesting that dominance plays a role in regulating nectar foraging.

There is evidence for both independent- and swarm-founding polistines that colony-wide food foraging rates are closely correlated with the number of larvae in the nest, suggesting that this activity is demand driven (West-Eberhard 1969; Howard and Jeanne 2005). Supply also influences foraging rates. If the availability of food in the environment decreases, the amount of resource returned per trip also decreases, as must the benefit:cost ratio. Colonies may respond by increasing foraging effort, e.g., by fielding more foragers. This may increase the food input rate to the colony, but at the cost of increased energy expenditure and mortality risk. Thus, reduced supply will reduce the colony's growth rate, regardless of the colony's response.

There is also good evidence that colonies respond to the opposite situation. When food resources are abundant, cost per successful foraging trip, in terms of both energy and risk of mortality, is low, so the benefit:cost ratio rises. Colonies of at least some species take advantage by bringing in more food than is required to meet current demand. This is most clearly manifested in the storage of excess nutrients. Many social wasp species store excess nectar as sugary droplets on the walls of empty or egg-containing brood cells (Hunt et al. 1998). *Polistes annularis* colonies in parts of Texas even store enough honey to serve as a food supply for surviving adults during the winter months, after the colony has dissolved (Strassmann 1979). Even protein is stored in some cases. Several epiponine species stockpile hundreds or thousands of bodies of ant or termite reproductives following their nuptial flights (Richards 1978; RLJ, personal observation). These weak-flying insects are captured in mid-air or seized on the ground before they can take off (Forsyth 1978). Forsyth (1978) claimed that during these periods of glut, the number of *Polybia occidentalis* foragers increased from the normal 10–20% of the colony to three or four times that number. Thus, when resources are superabundant, wasps become opportunistic foragers.

Additional evidence that colonies allocate greater effort to foraging when costs are low is seen in the daily temporal patterning of wood-pulp foraging compared to food foraging. Foraging for wood pulp requires that water be collected first and carried to the source of dead wood, where it is used to moisten the fibers so they can be chewed loose (Jeanne 1986b). Wood sources are more likely to be moist, and free water (as dew) is more readily found, in the early morning hours. Not surprisingly, therefore, colonies of *Vespula vulgaris* (Potter 1964, Roland 1976) and *Polybia occidentalis* (RLJ, unpublished data) concentrate their pulp foraging and nest construction early in the day. In contrast, prey and carbohydrate foraging tend to be distributed throughout the daylight hours (Archer 2004). It is interesting that in *P. occidentalis* the rate of foraging for nest materials appears to be independent of food foraging rate. That is, the one is not engaged in at the expense of the other (O'Donnell and Jeanne 1990).

CONCLUDING REMARKS

We still know too little about the majority of species to allow us to fully characterize the range of foraging behavior in the social wasps. Although some broad patterns are evident, no doubt further study will reveal exceptions to currently recognized patterns.

As a rule, social wasps are generalist predators, taking soft-bodied terrestrial arthropods in a number of orders, but some are narrow specialists, notably the social sphecids and *Vespa tropica*. Although these are at opposite ends of the spectrum of body size among social wasps, too little is known about the dietary habits of the majority of species to speculate on whether body size influences dietary range. An unexplained peculiarity is the retrieval of largely intact prey by *Polybia occidentalis*. How widespread this is among the polistines and how it is adaptive remain to be investigated.

With the exception of *Vespa mandarinia* during its cooperative raids on colonies of other vespines and on honey bees, social wasps appear to forage as individual hunters and gatherers of protein and carbohydrate. Although foragers in several species have been shown to make use of a variety of cues, both at the resource and at the nest, the only signal so far identified in the context of food foraging is the use of scent marking by *V. mandarinia* during the recruitment phase of their raids on colonies of other social wasps and bees. Wasps appear to be unique among the social insects in lacking food recruitment via signals issued at the nest, a pattern that has yet to be adequately explained. On the other hand, only a fraction of vespid species has been studied carefully in this regard. Surprises may await us, and exceptions often provide insight into what is behind the rule. Nevertheless, Seeley's (1995) suggestion that cues are more important than signals in coordinating social behavior appears to be supported in the case of food foraging in the wasps.

Wasps are capable of learning to associate odor cues with food rewards and to use the information to help locate resources. We know just enough about the occurrence of local enhancement in a few species to begin to wonder whether the appropriate response to the cue of other feeding wasps is also learned in local populations, rather than being an innate, species-specific trait.

A challenging question about age polyethism is why it is so well developed in some of the swarm-founding polistines, yet so rudimentary in the vespines. In ants and bees, the degree of specialization appears to correlate positively with colony size, but the correlation breaks down across these two wasp groups: colonies of some vespines are one or two orders of magnitude larger than some of the swarm founders.

The phenomenon of tempo may have a bearing on this puzzle (Oster and Wilson 1978). Species differences in tempo, which influences rates of foraging, colony growth rate, and mortality rate, have been explored in honey bees (Dyer and Seeley 1991) and ants (Davidson 1997), but not yet in wasps. It strikes us that at least some vespines may be high-tempo species, while swarm founders such as *Polybia occidentalis* are low-tempo.

Finally, some of the most promising areas for future work in the social wasps have to do with the question of allocation of colony effort to foraging. Much remains to be worked out about how decisions at the individual level lead to the emergence of adaptive responses at the colony level.

ACKNOWLEDGMENTS

We gratefully acknowledge Teresa León for the photograph used in Figure 3.2. We thank Michael Hrncir and Stefan Jarau for their support and encouragement during the preparation of this review. James Hoey assisted with the figures. Research was supported by the College of Agricultural and Life Sciences, University of Wisconsin–Madison.

REFERENCES

Akre RD. (1982). Social wasps. In Hermann H (ed.), *Social Insects,* Vol. IV. New York: Academic Press, pp. 1–105.

Akre RD, Garnett WB, MacDonald JF, Greene A, Landolt PJ. (1976). Behavior and colony development of *Vespula pensylvanica* and *V. atropilosa* (Hymenoptera: Vespidae). *J Kansas Entomol Soc* 49:63–84.

Akre RD, Greene A, MacDonald JF, Landolt PJ, Davis HG. (1980). *The Yellowjackets of America North of Mexico.* Agriculture Handbook 552, U.S. Department of Agriculture.

Anderson C, Franks NR. (2001). Teams in animal societies. *Behav Ecol* 12:534–40.

Anderson C, Franks NR, McShea DW. (2001). The complexity and hierarchical structure of tasks in insect societies. *Anim Behav* 62:643–51.

Anderson C, Ratnieks FLW. (1999). Task partitioning in insect societies. I. Effect of colony size on queueing delay and colony ergonomic efficiency. *Am Nat* 154:521–35.

Anderson C, Ratnieks FLW. (2000). Task partitioning in insect societies: Novel situations. *Insect Soc* 47: 198–99.

Archer ME. (2004). All-day foraging characteristics of successful underground colonies of *Vespula vulgaris* (Hymenoptera, Vespidae) in England. *Insect Soc* 51:171–78.

Arnold TS. (1966). Biology of social wasps: Comparative ecology of the British species of social wasps belonging to the family Vespidae. Diploma thesis, University of London.

Barth FG, Hrncir M, Jarau S. (2008). Signals and cues in the recruitment behavior of stingless bees (Meliponini). *J Comp Physiol A* 194:313–27.

Beckers R, Goss S, Deneubourg JL, Pasteels JM. (1989). Colony size, communication and ant foraging strategy. *Psyche* 96:239–56.

Beier W. (1983). Observations and experimental approach to the foraging behavior of the German wasp *Paravespula germanica*. *Zool Bietr* 28:321–48.

Belavadi VV, Govindan R. (1981). Nesting habits and behaviour of *Ropalidia (Icariola) marginata* (Hymenoptera: Vespidae) in South India. *Colemania* 1:95–101.

Blackith RE. (1957). The analysis of social facilitation at the nest entrance of some Hymenoptera. *Physiol Comp Oecol* 4:388–402.

Bouwma AM, Bouwma PE, Nordheim EV, Jeanne RL. (2003). Founding swarms in a tropical social wasp: Adult mortality, emigration distance, and swarm size. *J Insect Behav* 16:439–52.

Brian MV, Brian AD. (1952). The wasp, *Vespula sylvestris* Scopoli: Feeding, foraging and colony development. *Trans R Entomol Soc Lond* 103:1–26.

Butler CG, Fletcher DJC, Watler D. (1969). Nest-entrance marking with pheromones by the honeybee, *Apis mellifera* L., and by a wasp, *Vespula vulgaris* L. *Anim Behav* 17:142–47.

Carpenter JM. (1991). Phylogenetic relationships and the origin of social behavior in the Vespidae. In Ross KG, Matthews RW (eds.), *The Social Biology of Wasps*. Ithaca, NY: Comstock Publishing Associates of Cornell University Press, pp. 7–32.

D'Adamo P, Lozada M. (2003). The importance of location and visual cues during foraging in the German wasp (*Vespula germanica* F.) (Hymenoptera: Vespidae). *NZJ Zool* 30:171–74.

D'Adamo P, Corley JC, Lozada M. (2001). Attraction of *Vespula germanica* (Hymenoptera: Vespidae) foragers by conspecific heads. *J Econ Entomol* 94:850–52.

D'Adamo P, Corley J, Sackmann P, Lozada M. (2000). Local enhancement in the wasp *Vespula germanica*: Are visual cues all that matter? *Insect Soc* 47:289–91.

D'Adamo P, Lozada M. (2005). Conspecific and food attraction in the wasp *Vespula germanica* (Hymenoptera: Vespidae), and their possible contributions to control. *Ann Entomol Soc Am* 98:236–40.

D'Adamo P, Lozada M, Corley J. (2003). Conspecifics enhance attraction of *Vespula germanica* (Hymenoptera: Vespidae) foragers to food baits. *Ann Entomol Soc Am* 96:685–88.

D'Adamo P, Lozada M, Corley JC. (2004). An attraction pheromone from heads of worker *Vespula germanica* wasps. *J Insect Behav* 17:809–21.

Davidson DW. (1997). The role of resource imbalances in the evolutionary ecology of tropical arboreal ants. *Biol J Linnean Soc* 61:153–81.

Dawkins MS. (1986). *Unravelling Animal Behaviour*. Essex: Longman Scientific.

Day SE, Jeanne RL. (2001). Food volatiles as attractants for yellowjackets (Hymenoptera: Vespidae). *Environ Entomol* 30:157–65.

Dechaume-Moncharmont FX, Dornhaus A, Houston AI, McNamara JM, Collins EJ, Franks NR. (2005). The hidden cost of information in collective foraging. *Proc R Soc B* 272:1689–95.

Deneubourg JL, Pasteels JM, Verhaeghe JC. (1983). Probabilistic behaviour in ants: A strategy of errors? *J Theor Biol* 105:259–71.

Dew HE, Michener CD. (1981). Division of labor among workers of *Polistes metricus* (Hymenoptera: Vespidae): Laboratory foraging activities. *Insect Soc* 28:87–101.

Dornhaus A, Chittka L. (2004). Why do honey bees dance? *Behav Ecol Sociobiol* 55:395–401.

Duncan C. (1939). A contribution to the biology of North American vespine wasps. *Stanford Univ Publ Biol Sci* 8:1–272.

Dyer FC, Seeley TD. (1991). Nesting behavior and the evolution of worker tempo in four honey bee species. *Ecology* 72:156–70.

Edwards R. (1980). *Social Wasps: Their Biology and Control*. East Grinstead, UK: Rentokil.

Forsyth AB. (1978). Studies on the Behavioral Ecology of Polygynous Social Wasps. Doctoral thesis, Harvard University, Cambridge, MA.

Fowler HG. (1992). Social facilitation during foraging in *Agelaia* (Hymenoptera: Vespidae). *Naturwissenschaften* 79:424.

Free JB. (1970). The behavior of wasps (*Vespula germanica* L. and *Vespula vulgaris* L.) when foraging. *Insect Soc* 17:11–20.

Free JB, Williams IH. (1983). Scent-marking of flowers by honeybees. *J Apicult Res* 18:128–35.

Free JB, Williams I, Pickett JA, Ferguson AW, Martin AP. (1982). Attractiveness of (Z)-11-eicosen-1-ol to foraging honeybees. *J Apicult Res* 21:151–56.

von Frisch K. (1967). *The Dance Language and Orientation of Bees*. Cambridge, MA: Harvard University Press.

Gadagkar R. (2001). *The Social Biology of Ropalidia marginata: Toward Understanding the Evolution of Eusociality*. Cambridge, MA: Harvard University Press.

Gaul AT. (1952). The awakening and diurnal flight activities of vespine wasps. *Proc R Entomol Soc Lond A* 27:33–38.

Gautrais J, Theraulaz G, Deneubourg JL, Anderson C. (2002). Emergent polyethism as a consequence of increased colony size in insect societies. *J Theor Biol* 215:363–73.

Giurfa M. (1993). The repellent scent-mark of the honeybee *Apis mellifera ligustica* and its role as communication cue during foraging. *Insect Soc* 40:59–67.

Gobbi N, Machado VLL. (1985). Material capturado e utilizado na alimentação de *Polybia (Myrapetra) paulista* Ihering, 1896 (Hymenoptera – Vespidae). *An Soc Entomol Bras* 14:189–95.

Gobbi N, Machado VLL. (1986). Material capturado e utilizado na alimentação de *Polybia (Trichothorax) ignobilis* (Haliday, 1836) (Hymenoptera, Vespidae). *An Soc Entomol Bras* 15:117–24.

Gobbi N, Machado VLL, Tavares Filho JA. (1984). Sazonalidade das presas utilizadas na alimentação de *Polybia occidentalis occidentalis* (Olivier, 1791) (Hym., Vespidae). *An Soc Entomol Bras* 13:63–69.

Gomes L, Gomes G, Oliveira HG, Morlin Junior JJ, Desuó IC, Silva IM, Shima SN, von Zuben CJ. (2007). Foraging by *Polybia (Trichothorax) ignobilis* (Hymenoptera, Vespidae) on flies at animal carcasses. *Rev Bras Entomol* 51:389–93.

Goulson D, Stout JC, Langley J, Hughes WOH. (2000). Identity and function of scent marks deposited by foraging bumblebees. *J Chem Ecol* 26:2897–911.

Greene A. (1991). *Dolichovespula and Vespula*. In Ross KG, Matthews RW (eds.), *The Social Biology of Wasps*. Ithaca, NY: Comstock Publishing Associates of Cornell University Press, pp. 263–305.

Grinfel'd EK. (1977). The feeding of the social wasp *Polistes gallicus* (Hymenoptera, Vespidae). *Entomol Rev* 56:24–29.

Grogan DE, Hunt JH. (1977). Digestive proteases of two species of wasps of the genus *Vespula*. *Insect Biochem* 7:191–96.

Heinrich B. (1984). Strategies of thermoregulation and foraging in two vespid wasps *Dolichovespula maculata* and *Vespula vulgaris*. *J Comp Physiol B* 154:175–80.

Hendrichs J, Katsoyannos BI, Wornoayporn V, Hendrichs MA. (1994). Odour-mediated foraging by yellow-jacket wasps (Hymenoptera: Vespidae): Predation on leks of pheromone-calling Mediterranean fruit fly males (Diptera: Tephritidae). *Oecologia* 99:88–94.

Hocking B. (1953). The intrinsic range and speed of flight of insects. *Trans R Entomol Soc Lond* 104:223–345.

Howard KJ, Jeanne RL. (2005). Shifting foraging strategies in colonies of the social wasp *Polybia occidentalis* (Hymenoptera, Vespidae). *Behav Ecol Sociobiol* 57:481–89.

Hrncir M, Mateus S, Nascimento FS. (2007). Exploitation of carbohydrate food sources in *Polybia occidentalis*: Social cues influence foraging decisions in swarm-founding wasps. *Behav Ecol Sociobiol* 61:975–83.

Hunt JH. (1984). Adult nourishment during larval provisioning in a primitively eusocial wasp, *Polistes metricus* Say. *Insect Soc* 31:452–60.

Hunt JH, Jeanne RL, Baker I, Grogan DE. (1987). Nutrient dynamics of a swarm-founding social wasp species, *Polybia occidentalis* (Hymenoptera: Vespidae). *Ethology* 75:291–305.

Hunt JH, Rossi AM, Holmberg NJ, Smith SR, Sherman WR. (1998). Nutrients in social wasp (Hymenoptera: Vespidae, Polistinae) honey. *Ann Entomol Soc Am* 91:466–72.

Hurd CR, Jeanne RL, Nordheim EV. (2007). Temporal polyethism and worker specialization in the wasp, *Vespula germanica*. *J Insect Sci* 7:1–13.

Ishay J, Bytinski-Salz H, Shulov A. (1967). Contributions to the bionomics of the Oriental hornet (*Vespa orientalis* Fabricius). *Isr J Entomol* 2:45–106.

Jander R. (1998). Olfactory learning of fruit odors in the eastern yellow jacket, *Vespula maculifrons* (Hymenoptera: Vespidae). *J Insect Behav* 11:879–88.

Jandt JM, Jeanne RL. (2005). German yellowjacket (*Vespula germanica*) foragers use odors inside the nest to find carbohydrate food sources. *Ethology* 111:641–51.

Jandt JM, Riel L, Crain B, Jeanne RL. (2005). *Vespula germanica* foragers do not scent-mark carbohydrate food sites. *J Insect Behav* 18:19–31.

Jarau S, Hrncir M, Ayasse M, Schulz C, Francke W, Zucchi R, Barth FG. (2004). A stingless bee (*Melipona seminigra*) marks food sources with a pheromone from its claw retractor tendons. *J Chem Ecol* 30: 793–804.

Jarau S, Schulz CM, Hrncir M, Francke W, Zucchi R, Barth FG, Ayasse M. (2006). Hexyl decanoate, the first trail pheromone compound identified in a stingless bee, *Trigona recursa*. *J Chem Ecol* 32:1555–64.

Jeanne RL. (1972). Social biology of the Neotropical wasp *Mischocyttarus drewseni*. *Bull Mus Comp Zool Harvard* 144:63–150.

Jeanne RL. (1980). Evolution of social behavior in the Vespidae. *Annu Rev Entomol* 25:371–96.

Jeanne RL. (1981). Chemical communication during swarm emigration in the social wasp *Polybia sericea* (Olivier). *Anim Behav* 29:102–13.

Jeanne RL. (1986a). The evolution of the organization of work in social insects. *Monitore Zoologico Italiano* 20:119–33.

Jeanne RL. (1986b). The organization of work in *Polybia occidentalis:* Costs and benefits of specialization in a social wasp. *Behav Ecol Sociobiol* 19:333–41.

Jeanne RL. (1991a). The swarm-founding Polistinae. In Ross KG, Matthews RW (eds.), *The Social Biology of Wasps*. Ithaca, NY: Comstock Publishing Associates of Cornell University Press, pp. 191–231.

Jeanne RL. (1991b). Polyethism. In Ross KG, Matthews RW (eds.), *The Social Biology of Wasps*. Ithaca, NY: Comstock Publishing Associates of Cornell University Press, pp. 389–425.

Jeanne RL. (2003). Social complexity in the Hymenoptera, with special attention to the wasps. In Kikuchi T, Azuma N, Higashi S (eds.), *Genes, Behaviors and Evolution of Social Insects*. Sapporo: Hokkaido University Press, pp. 81–130.

Jeanne RL, Downing HA, Post DC. (1983). Morphology and function of sternal glands in polistine wasps (Hymenoptera: Vespidae). *Zoomorphology* 103:149–64.

Jeanne RL, Downing HA, Post DC. (1988). Age polyethism and individual variation in *Polybia occidentalis*, an advanced eusocial wasp. In Jeanne RL (ed.), *Interindividual Behavioral Variability in Social Insects*. Boulder: Westview Press, pp. 323–57.

Jeanne RL, Hunt JH, Keeping MG. (1995). Foraging in social wasps: *Agelaia* lacks recruitment to food (Hymenoptera: Vespidae). *J Kansas Entomol Soc* 68:279–89.

Jeanson R, Fewell JH, Gorelick R, Bertram SM. (2007). Emergence of increased division of labor as a function of group size. *Behav Ecol Sociobiol* 62:289–98.

Johnson LK. (1983). Foraging strategies and the structure of stingless bee communities in Costa Rica. In Jaisson P (ed.), *Social Insects in the Tropics*. Paris: Université Paris-Nord, pp. 31–58.

Kammer AE, Heinrich B. (1974). Metabolic rates related to muscle activity in bumblebees. *J Exp Biol* 61: 219–27.

Karsai I, Wenzel JW. (1998). Productivity, individual-level and colony-level flexibility, and organization of work as consequences of colony size. *Proc Natl Acad Sci USA* 95:8665–69.

Karsai I, Wenzel JW. (2000). Organization and regulation of nest construction behavior in *Metapolybia* wasps. *J Insect Behav* 13:111–40.

Kim KW, Noh S, Choe JC. (2007). Lack of field-based recruitment to carbohydrate food in the Korean yellowjacket, *Vespula koreensis*. *Ecol Res* 22:825–30.

Landi M, Coster-Longman C, Turillazzi S. (2002). Are the selfish herd and dilution effects important in promoting nest clustering in the hover wasp *Parischnogaster alternata* (Stenogastrinae: Vespidae: Hymenoptera). *Ethol Ecol Evol* 14:297–305.

Lindauer M. (1961). *Communication among Social Bees*. Cambridge, MA: Harvard University Press.

Lloyd JE. (1983). Bioluminescence and communication in insects. *Annu Rev Entomol* 28:131–60.

Lozada M, D'Adamo P. (2006). How long do *Vespula germanica* wasps search for a food source that is no longer available? *J Insect Behav* 19:591–600.

Machado VLL, Gobbi N, Simões D. (1987). Material capturado e utilizado na alimentação de *Stelopolybia pallipes* (Olivier, 1791) (Hymenoptera-Vespidae). *An Soc Entomol Brasil* 16: 73–79.

Maschwitz U, Beier W, Dietrich I, Keidel W, Beier W, Menzel R. (1974). Futterverständigung bei Wespen der Gattung *Paravespula*. *Naturwissenschaften* 61:506.

Matsuura M. (1974). Intracolonial polyethism in *Vespa*. Part 1. Behavior and its change of the foundress of *Vespa crabro flavofasciata* in early nesting stage in relation to worker emergence. *Kontyû* 42:333–50.

Matsuura M. (1984). Comparative biology of the five Japanese species of the genus *Vespa* (Hymenoptera, Vespidae). *Bull Fac Agr Mie Univ* 69:1–131.

Matsuura M. (1991). *Vespa* and *Provespa*. In Ross KG, Matthews RW (eds.), *The Social Biology of Wasps*. Ithaca, NY: Comstock Publishing Associates of Cornell University Press, pp. 232–62.

Matsuura M, Yamane SK. (1984). *Comparative Ethology of the Vespine Wasps*. Sapporo: Hokkaido University Press.

Matthews RW. (1968). Nesting biology of the social wasp *Microstigmus comes* (Hymenoptera: Sphecidae, Pemphredoninae). *Psyche* 75:23–45.

Matthews RW. (1991). Evolution of social behavior in the sphecid wasps. In Ross KG, Matthews RW (eds.), *The Social Biology of Wasps*. Ithaca, NY: Comstock Publishing Associates of Cornell University Press, pp. 570–602.

Matthews RW, Naumann ID. (1988). Nesting biology and taxonomy of *Arpactophilus mimi*, a new species of social sphecid (Hymenoptera: Sphecidae) from northern Australia. *Aust J Zool* 36:585–97.

McPheron LJ, Mills NJ. (2007). Discrimination learning of color-odor compounds in a paper wasp (Hymenoptera: Vespidae: Pompilinae: *Mischocyttarus flavitarsis*). *Entomol Gen* 29:125–34.

Merkle D, Middendorf M. (2004). Dynamic polyethism and competition for tasks in threshold reinforcement models of social insects. *Adapt Behav* 12:251–62.

Montagner H. (1966). Le mécanisme et les consequences des comportements trophallactiques chez les guêpes du genre *Vespa*. Doctoral thesis, Faculté des Sciences de l'Université de Nancy, France.

Moreyra S, D'Adamo P, Lozada M. (2006). Odour and visual cues utilised by German yellowjackets (*Vespula germanica*) while relocating protein or carbohydrate resources. *Aust J Zool* 54:393–97.

Naumann MG. (1970). The nesting behavior of *Protopolybia pumila* in Panama (Hymenoptera, Vespidae). Doctoral thesis, University of Kansas, Lawrence.

Nieh JC. (2004). Recruitment communication in stingless bees (Hymenoptera, Apidae, Meliponini). *Apidologie* 35:159–82.

O'Donnell S. (1995). Necrophagy by Neotropical swarm-founding wasps (Hymenoptera: Vespidae, Epiponini). *Biotropica* 27:133–36.

O'Donnell S. (1998). Effects of experimental forager removals on division of labour in the primitively eusocial wasp *Polistes instabilis* (Hymenoptera: Vespidae). *Behaviour* 135:173–93.

O'Donnell S, Jeanne RL. (1990). Forager specialization and the control of nest repair in *Polybia occidentalis* Olivier (Hymenoptera: Vespidae). *Behav Ecol Sociobiol* 27:359–64.

O'Donnell S, Jeanne RL. (1992a). The effects of colony characteristics on life span and foraging behavior of individual wasps (*Polybia occidentalis*, Hymenoptera: Vespidae). *Insect Soc* 39:73–80.

O'Donnell S, Jeanne RL. (1992b). Lifelong patterns of forager behaviour in a tropical swarm-founding wasp: Effects of specialization and activity levels on longevity. *Anim Behav* 44:1021–27.

O'Donnell S, Jeanne RL. (1992c). Forager success increases with experience in *Polybia occidentalis* (Hymenoptera: Vespidae). *Insect Soc* 39:451–54.

O'Donnell S, Jeanne RL. (1995). Worker lipid stores decrease with outside-nest task performance in wasps: Implications for the evolution of age polyethism. *Experientia* 51:749–52.

O'Neill KM. (2001). *Solitary Wasps: Behavior and Natural History*. Ithaca, NY: Comstock Publishing Associates of Cornell University Press.

Ono M, Igarashi T, Ohno E, Sasaki M. (1995). Unusual thermal defence by a honeybee against mass attack by hornets. *Nature* 377:334–36.

Oster GF, Wilson EO. (1978). *Caste and Ecology in the Social Insects*. Princeton, NJ: Princeton University Press.

Overmyer SL. (1995). Scent-mediated recruitment and food-site marking in the German yellowjacket, *Vespula germanica* (Hymenoptera: Vespidae). Master's thesis, University of Wisconsin, Madison.

Overmyer SL, Jeanne RL. (1998). Recruitment to food by the German yellowjacket, *Vespula germanica*. *Behav Ecol Sociobiol* 42:17–21.

Pallett MJ, Plowright RC. (1979). Traffic through the nest entrance of a colony of *Vespula arenaria* (Hymenoptera: Vespidae). *Can Entomol* 111:385–90.

Parrish MD. (1984). Factors influencing aggression between foraging yellowjacket wasps, *Vespula* spp. (Hymenoptera: Vespidae). *Ann Entmol Soc Am* 77:306–11.

Parrish MD, Fowler HG. (1983). Contrasting foraging related behaviors in two sympatric wasps (*Vespula maculifrons* and *Vespula germanica*). *Ecol Entomol* 8:185–90.

Plowright RC. (1979). Social facilitation at the nest entrances of bumble bees and wasps. *Insect Soc* 26: 223–31.

Portha S, Deneubourg J, Detrain C. (2002). Self-organized asymmetries in ant foraging: A functional response to food type and colony needs. *Behav Ecol* 13:776–81.

Post DC, Jeanne RL, Erickson Jr EH. (1988). Variation in behavior among workers of the primitively social wasp *Polistes fuscatus variatus*. In Jeanne RL (ed.), *Interindividual Behavioral Variability in Social Insects*. Boulder: Westview Press, pp. 283–321.

Potter NB. (1964). A study of the biology of the common wasp, *Vespula vulgaris* L., with special reference to the foraging behaviour. Doctoral thesis, University of Bristol, England.

Prezoto F, Santos-Prezoto HH, Machado VLL, Zanuncio JC. (2006). Prey captured and used in *Polistes versicolor* (Olivier) (Hymenoptera: Vespidae). *Neotrop Entomol* 35:707–9.

Ratnieks FLW, Anderson C. (1999a). Task partitioning in insect societies. *Insect Soc* 46:95–108.

Ratnieks FLW, Anderson C. (1999b). Task partitioning in insect societies. II. Use of queueing delay information in recruitment. *Am Nat* 154:536–48.

Rau P. (1944). A few notes on the behavior of *Vespula maculifrons* Buyss. *Bull Brooklyn Entomol Soc* 39: 177–78.

Raveret Richter M. (1990). Hunting social wasp interactions: Influence of prey size, arrival order, and wasp species. *Ecology* 71:1018–30.

Raveret Richter M. (2000). Social wasp (Hymenoptera: Vespidae) foraging behavior. *Annu Rev Entomol* 45: 121–50.

Raveret Richter MA, Jeanne RL. (1985). Predatory behavior of *Polybia sericea* (Olivier), a tropical social wasp (Hymenoptera: Vespidae). *Behav Ecol Sociobiol* 16:165–70.

Raveret Richter M, Tisch VL. (1999). Resource choice of social wasps: Influence of presence, size and species of resident wasps. *Insect Soc* 46:131–36.

Rayor LS, Mooney LJ, Renwick JA. (2007). Predatory behavior of *Polistes dominulus* wasps in response to cardenolides and glucosinolates in *Pieris napi* caterpillars. *J Chem Ecol* 33:1177–85.

Real LA. (1981). Uncertainty and pollinator-plant interactions: The foraging behavior of bees and wasps on artificial flowers. *Ecology* 62:20–26.

Reed HC, Akre RD. (1983). Comparative colony behavior of the forest yellowjacket *Vespula acadica* (Sladen) (Hymenoptera: Vespidae). *J Kans Entomol Soc* 56:581–606.

Reid BL, MacDonald JF, Ross DR. (1995). Foraging and spatial dispersion in protein-scavenging workers of *Vespula germanica* and *V. maculifrons* (Hymenoptera: Vespidae). *J Insect Behav* 8:315–30.

Richards OW. (1978). *The Social Wasps of the Americas, Excluding the Vespinae*. London: British Museum (Natural History).

Roland C. (1976). Approche éco-éthologique et biologique des sociétés de *Paravespula vulgaris* et *germanica*. Doctoral thesis, University of Nancy, France.

Ross DR, Shukle RH, MacDonald JF. (1984). Meat extracts attractive to scavenger *Vespula* in eastern North America (Hymenoptera: Vespidae). *J Econ Entomol* 77:637–42.

Ross KG, Matthews RW (eds.). (1991). *The Social Biology of Wasps*. Ithaca, NY: Comstock Associates of Cornell University Press.

Roubik DW. (1989). *Ecology and Natural History of Tropical Bees*. Cambridge, UK: Cambridge University Press.

Ruxton GD, Lee J, Hansell MH. (2001). Wasps enter and leave their nest at regular intervals. *Insect Soc* 48: 363–65.

Saleh N, Chittka L. (2006). The importance of experience in the interpretation of conspecific chemical signals. *Behav Ecol Sociobiol* 61:215–20.

Saleh N, Ohashi K, Thomson JD, Chittka L. (2006). Facultative use of the repellent scent mark in foraging bumblebees: Complex versus simple flowers. *Anim Behav* 71:847–54.

Saleh N, Scott AG, Bryning GP, Chittka L. (2007). Distinguishing signals and cues: Bumblebees use general footprints to generate adaptive behaviour at flowers and nest. *Arthropod-Plant Interactions* 1:119–27.

Schmitt U, Lübke G, Franke W. (1991). Tarsal secretion marks food sources in bumblebees (Hymenoptera: Apidae). *Chemoecology* 2:35–40.

Schorkopf DLP, Jarau S, Francke W, Twele R, Zucchi R, Hrncir M, Schmidt VM, Ayasse M, Barth FG. (2007). Spitting out information: *Trigona* bees deposit saliva to signal resource locations. *Proc R Soc Lond B* 274:895–98.

Seeley TD. (1995). *The Wisdom of the Hive—The Social Physiology of Honey Bee Colonies*. Cambridge, MA: Harvard University Press.

Shida T. (1959). *Vespula lewisii* in Musashino. *Nihon-Kontyuki* 1:77–145 [in Japanese].

Silveira OT, Esposito MC, dos Santos Jr JN, Gemaque Jr FE. (2005). Social wasps and bees captured in carrion traps in a rainforest in Brazil. *Entomol Sci* 8:33–39.

Simões D. (1977). Etologia e diferenciação de casta em algumas vespas sociais (Hymenoptera, Vespidae). Doctoral thesis, Universidade de São Paulo, Ribeirão Prêto, Brazil.

Spradbery JP. (1973). *Wasps: An Account of the Biology and Natural History of Solitary and Social Wasps.* Seattle: University of Washington Press.

Steinmetz I, Schmolz E, Ruther J. (2003). Cuticular lipids as trail pheromone in a social wasp. *Proc R Soc Lond B* 270:385–91.

Steinmetz I, Sieben S, Schmolz E. (2002). Chemical trails used for orientation in nest cavities by two vespine wasps, *Vespa crabro* and *Vespula vulgaris. Insect Soc* 49:354–56.

Stout JC, Goulson D (2001). The use of conspecific and interspecific scent marks by foraging bumblebees and honeybees. *Anim Behav* 62:183–89.

Stout JC, Goulson D, Allen JA. (1998). Repellent scent-marking of flowers by a guild of foraging bumblebees (*Bombus* spp.). *Behav Ecol Sociobiol* 43:317–26.

Strassmann JE. (1979). Honey caches help female paper wasps (*Polistes annularis*) survive Texas winters. *Science* 204:207–9.

Sugden EA, McAllen RL. (1994). Observations on foraging, population and nest biology of the Mexican honey wasp, *Brachygastra mellifica* (Say) in Texas (Vespidae: Polybiinae). *J Kans Entomol Soc* 67:141–55.

Takagi M, Hirose Y, Yamasaki M. (1980). Prey-location learning in *Polistes jadwigae* Dalla Torre (Hymenoptera, Vespidae). *Kontyû* 48:53–58.

Tan K, Radloff SE, Li JJ, Hepburn HR, Yang MX, Zhang LJ, Neumann P. (2007). Bee-hawking by the wasp, *Vespa velutina*, on the honeybees *Apis cerana* and *A. mellifera. Naturwissenschaften* 94:469–72.

Thorpe WH. (1963). *Learning and Instinct in Animals.* London: Methuen.

Turillazzi S. (1983). Extranidal behavior of *Parischnogaster nigricans serrei* (DuBuysson) (Hymenoptera, Stenogastrinae). *Z Tierpsychol* 63:27–36.

Turillazzi S, Pardi L. (1982). Social behavior of *Parischnogaster nigricans serrei* (Hymenoptera: Vespoidea) in Java. *Ann Entomol Soc Am* 75:657–64.

Waddington KD, Visscher PK, Herbert TJ, Raveret Richter M. (1994). Comparisons of forager distributions from matched honey bee colonies in suburban environments. *Behav Ecol Sociobiol* 35:423–29.

Wenzel JW, Carpenter JM. (1994). Comparing methods: Adaptive traits and tests of adaptation. In Eggleton P, Vane-Wright RI (eds.), *Phylogenetics and Ecology.* London: Academic Press, pp. 79–101.

West-Eberhard MJ. (1969). The social biology of polistine wasps. *Misc Publ Mus Zool Univ Michigan* 140: 1–101.

West-Eberhard MJ. (1978). Temporary queens in *Metapolybia* wasps: Nonreproductive helpers without altruism? *Science* 200:441–43.

West-Eberhard MJ. (1981). Intragroup selection and the evolution of insect societies. In Alexander RD, Tinkle DW (eds.), *Natural Selection and Social Behavior: Recent Research and New Theory.* New York: Chiron, pp. 3–17.

Wilms J, Eltz T. (2008). Foraging scent marks of bumblebees: Footprint cues rather than pheromone signals. *Naturwissenschaften* 95:149–53.

Wilson EO. (1971). *The Insect Societies.* Cambridge, MA: Belknap Press of Harvard University Press.

Wilson EO. (1975). *Sociobiology: The New Synthesis.* Cambridge, MA: Belknap Press of Harvard University Press.

Witjes S, Eltz T. (2007). Influence of scent deposits on flower choice: Experiments in an artificial flower array with bumblebees. *Apidologie* 38:12–18.

Ydenberg RC, Brown JS, Stephens DW. (2007). Foraging: An overview. In Stephens DW, Brown JS, Ydenberg RC (eds.), *Foraging: Behavior and Ecology.* Chicago: University of Chicago Press, pp. 1–28.

Zucchi R, Sakagami SF, Noll FB, Mechi MR, Mateus S, Baio MV, Shima SN. (1995). *Agelaia vicina*, a swarm-founding polistine with the largest colony size among wasps and bees (Hymenoptera: Vespidae). *J NY Entomol Soc* 103:129–37.

4 Season-Dependent Foraging Patterns

Case Study of a Neotropical Forest-Dwelling Ant (Pachycondyla striata; *Ponerinae*)

Flávia N. S. Medeiros and Paulo S. Oliveira

CONTENTS

INTRODUCTION

Ants are considered useful organisms to test hypotheses about foraging strategies because their foragers usually leave from a fixed nest site to collect food (e.g., Detrain and Deneubourg 2002). The development of models and hypotheses about ant foraging ecology, however, is constrained by the small amount of quantitative data on the foraging behavior of different ant taxa. The lack of data on the basic ecological features of ants is particularly evident in the Neotropical region, where these insects are extraordinarily abundant and diversified (Brown 2000).

Ants in the subfamily Ponerinae have retained many morphological and behavioral ancestral characteristics such as small colonies, simple nests, and solitary foraging (Peeters and Ito 2001). Because all members of the Ponerinae are armed with a sting and most species possess powerful mandibles, they are usually considered predators. However, a variety of feeding habits and foraging strategies can be observed among species in this subfamily. For instance, ponerines may search for food both on the ground and on plant substrates, and they include species that scavenge for dead arthropods, gather plant and insect exudates, and collect fruits and seeds (e.g., Oliveira and Brandão 1991; Dejean and Lachaud 1994; Fewell et al. 1996; Rico-Gray and Oliveira 2007). Moreover, whereas many ponerine species feed opportunistically on an array of food items (Duncan and Crewe 1994; Ehmer and Hölldobler 1995; Fourcassié and Oliveira 2002), others are highly prey specific (Freitas 1995; Dejean and Evraerts 1997). The foraging modes are also highly variable among ponerine species, and may range from solitary hunting without any cooperation during search and food retrieval to different levels of cooperative foraging associated with varying degrees of recruitment behavior among colony members (Peeters and Crewe 1987).

The foraging activity of ant colonies can be affected by biotic (e.g., competition, natural enemies) and abiotic (e.g., temperature, humidity) factors, and may vary both on a daily and on a seasonal basis (e.g., Carroll and Janzen 1973; Bernstein 1975, 1979; Orivel and Dejean 2002; Philpott et al. 2004; Cogni and Oliveira 2004a). Ants in the ponerine genus *Pachycondyla* are found in warm temperate areas, but are more common in tropical and subtropical regions throughout the world (Brown 2000). The different species of *Pachycondyla* can nest both in soil and in vegetation, and most of them include a wide array of animal (dead and alive) and plant (liquid and solid) material in their diets (e.g., Fresneau 1985; Hölldobler 1985; Orivel et al. 2000). Some species, however, have taxonomically narrow diets, such as the specialized termite hunters (e.g., Leal and Oliveira 1995).

This chapter presents a detailed field account of the foraging ecology of the forest-dwelling ant *Pachycondyla striata*, aiming at an integration of individual and colony-level components of its foraging behavior (Traniello 1989). By hunting in a species-rich and unpredictable environment such as the leaf litter of tropical forests (Levings 1983; Ward 2000), the large *P. striata* foragers with their visible and easily identifiable prey seem ideal organisms to study foraging ecology on temporal and spatial scales as well as to illustrate patterns appearing in their foraging decisions. To study these aspects we provide qualitative and quantitative field data on diet and foraging modes, and analyze colony activity rhythms and home ranges in different seasonal contexts.

FIELD OBSERVATIONS AND GENERAL PROCEDURE

Fieldwork was carried out in the Santa Genebra Reserve at Campinas, Southeast Brazil (22°49'45"S, 47°06'33"W), where the climate is warm and wet, with a dry winter from April to October and a wet summer from November to March. The average annual rainfall is 1381.2 mm and the mean annual temperature is 21.6°C. Most of the reserve is covered by a semideciduous mesophytic forest (Raimundo et al. 2008).

Activity Patterns, Diet, and Foraging Behavior

A total of fifty nests of *Pachycondyla striata* were tagged in the study area by following loaded workers attracted to sardine baits. We monitored the activity rhythm of four of these colonies (>10 m apart from each other) by recording all workers exiting or entering each nest within 24 h. Samplings consisted of counting ants continuously during the first 20 min of every hour. Simultaneously, we recorded the air temperature and humidity. The activity of each of the four colonies was evaluated once per season (each colony on a different day) in the dry/cold (July) and rainy/wet (January) period.

The food items retrieved by *P. striata* were surveyed by removing them from the mandibles of returning foragers from any of the fifty nests tagged. This procedure allowed the compilation of a large number of food items (N = 132). Foraging modes employed by workers of *P. striata* were determined by direct observation in the field and categorized as follows:

1. *Solitary scavenging*: Collection of dead organisms by a single worker.
2. *Group scavenging*: Collection of dead organisms by groups of two to five workers.
3. *Solitary predation*: Capture of live prey by a single worker.
4. *Predation in association with other ant species*: Capture of live termites in association with the obligate termitophagous ant, *Pachycondyla marginata* (see Leal and Oliveira 1995).
5. *Interspecific food robbing*: Stealing food from other ant species.

The food items taken from workers of *P. striata* were preserved in 70% alcohol and brought to the laboratory for their exact identification. The items were then kept in an oven at 60°C for 24 h, and their dry weights measured with a Mettler H51Ar analytical balance. In six cases the collection of the food item was not possible, yet they could be identified and were included in the survey and the respective foraging modes were documented.

COLONY HOME RANGES

To determine the foraging ranges of *P. striata* colonies, we monitored marked foragers of the same four colonies previously used for the investigation on activity rhythms. Sardine baits placed in the immediate vicinity of tagged nests induced ant foraging activity and allowed us to mark large numbers of workers from each colony. All individuals encountered outside the nests were then individually marked with dots of enamel paint (Testors Co., Rockford, Illinois) on the thorax and gaster, using a distinct color code for each colony. The foraging ranges of the colonies were assessed by following marked ants as they exited the nest and by recording the maximal distance they had walked before returning to the nest. Consecutively numbered flags were placed along a forager's route at approximately 1-min intervals. The maximal distances from the nest achieved by different ant foragers were measured, their directions determined with a compass, and the respective data points recorded on a map of the study plot. Maps of colony home ranges were then constructed based on the cumulative data of maximal distances achieved by the workers of each tagged colony. The foraging areas of the colonies were estimated as convex polygons created by connecting the outermost points at which workers were seen at intervals of 10° around a circumference with the nest entrance in the center. Colonies were monitored on nonconsecutive days during the dry/cold and rainy/hot season, mainly at the peak hours of their activity (see below). In total, nearly 60 h of observation per colony and season were carried out. Intra- and interspecific interactions involving *P. striata* were also documented during all field observation sessions.

To obtain data on nest structure and demography, four additional nests of *P. striata* were excavated. Ant voucher specimens are deposited in the Museu de Zoologia da Universidade de São Paulo (MZUSP), São Paulo, Brazil.

RESULTS

NEST STRUCTURE AND COLONY DEMOGRAPHY

Pachycondyla striata nests on the ground at shady sites, usually in close proximity to live trees and shrubs, under which the nest may extend among wooden roots. Although nests may have two to eight entrances beneath the leaf litter (20 to 80 cm apart from each other), most of the ant traffic occurs through a single main entrance. Excavated nests (N = 4) had five to six interconnected chambers located 5 to 80 cm beneath the ground surface. Pupae, large larvae, and winged females were found

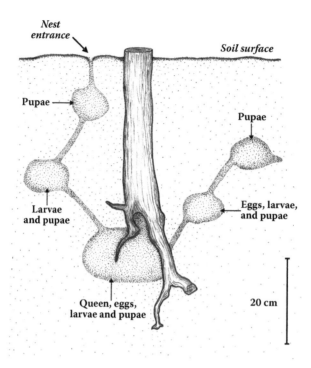

FIGURE 4.1 Vertical section through chambers and connecting galleries of a *Pachycondyla striata* nest next to the wooden roots of a live tree in a Brazilian forest. The profile is based on sketches from four excavated nests and shows their typical subterranean architecture during the rainy season when the brood is abundant.

in the more superficial chambers, whereas eggs, young larvae, and the queen were lodged in deeper chambers about 80 cm below the soil surface (Figure 4.1). The superficial chambers can be up to 1 m apart, suggesting a polydomous structure in which the colony is distributed in several spatial subunits.

The excavated *P. striata* colonies had one or no queen at all (so far, the occurrence of gamergates—mated, egg-laying workers—has not been documented in this species). The number of adult workers per colony ranged from 50 to 234 (145.5 ± 94.5, mean ± SD, N = 4). Immature stages (eggs, larvae, and pupae) were more abundant in the colonies excavated during the rainy/hot season (91 and 413, N = 2) than during the dry/cold season (15 and 20, N = 2).

ACTIVITY PATTERNS

In both seasons, the colony activity pattern was typically diurnal, with increased numbers of workers seen outside the nests at the warmest hours of the day (Figure 4.2). Worker activity was clearly lower in the dry/cold season than in the rainy/hot season. During winter, on average only up to two workers per sampling period (20 min of every hour) were seen during most of the day, with a slight increase in activity from 17.00 to 19.00 h, when air humidity increased and temperature was intermediate (Figure 4.2a). In contrast, during the summer, on average, at least four workers per sampling were seen outside the nest between 07.00 and 18.00 h, with a peak of activity from 11.00 to 17.00 h, when temperatures in the forest were usually above 25°C (Figure 4.2b).

DIET, BEHAVIOR, AND FORAGING MODES

Workers of *P. striata* forage for food mainly on the ground, above and beneath the leaf litter, and only occasionally climb on herbs to search for prey. Except for fast-moving ants and termites that are usually captured at once from behind (Figure 4.3a and b), potential prey organisms are

(a) Dry / cold season

(b) Rainy / hot season

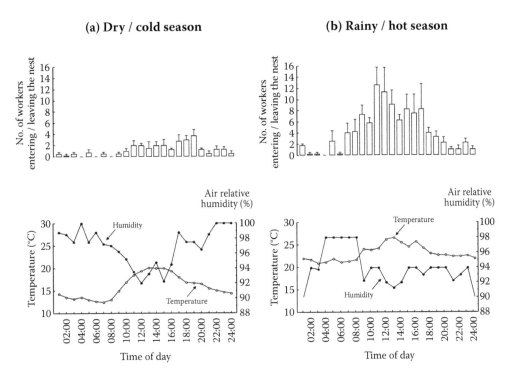

FIGURE 4.2 Daily and seasonal variation in the foraging activity of ground-dwelling *Pachycondyla striata* colonies in a Brazilian forest. The activity of four ant colonies was evaluated once during (a) the dry/cold season (July) and once during (b) the rainy/hot season (January). Bar graphs represent means (+1 SE; N = 4) of the colonies' foraging activity estimated as the number of workers entering or leaving the nest at a given time of the day. The temperature and humidity registered during the respective ant activity measurements are shown in the lower part of the figure.

FIGURE 4.3 Photographs illustrating the variety of food items retrieved by *Pachycondyla striata* foragers on the leaf litter of a Brazilian forest. Individual hunters attack from behind small- to medium-sized ground-dwelling insects, such as *Neocapritermes* termites (a) and *Odontomachus* ants (b), and transport them to the nest. Large prey items, such as centipedes (c), usually require that the forager returns to the nest and recruits nestmates through tandem runs. (d) The fleshy portion of large fruits and seeds, such as the red lipid-rich aril of the *Virola* seed shown here, is sequentially retrieved in pieces by individual foragers. See color insert following page 142. (Photographs by P. S. Oliveira (a, b, c) and M. A. Pizo (d).)

frequently inspected several times with the antennae before being attacked and retrieved by *P. striata* foragers. This latter behavior is commonly observed when the workers are confronted with large or well-armed prey items. From a total of 132 food items that were registered as part of the diet of *P. striata* in the study area, arthropods comprised the vast majority of the items retrieved by foraging ants (Figure 4.4a). Ants accounted for 32% of the prey items collected, including the following species: *Camponotus abdominalis*, *C. crassus*, *C. sericeiventris* (Formicinae), *Odontomachus chelifer*, *Pachycondyla marginata*, *P. striata*, *Pachycondyla* sp. (Ponerinae),

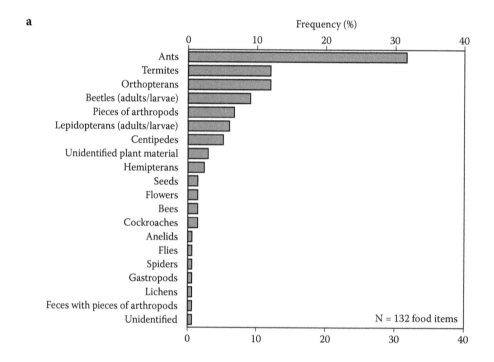

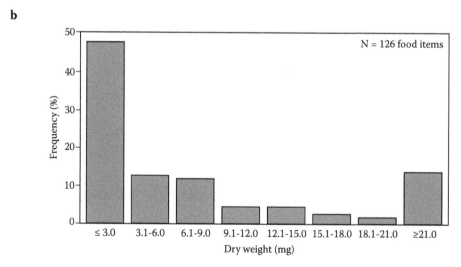

FIGURE 4.4 Relative abundance of the food items (a), and of their dry weights (b), that were retrieved by *Pachycondyla striata* foragers in a forest reserve in Brazil. Data are based on collections from returning ants of fifty observation nests.

Atta sexdens, *Pheidole* sp., and *Solenopsis* sp. (Myrmicinae). The vast majority (93%) of the ant prey consisted of dead workers, and only rarely did *P. striata* foragers kill other ants before retrieving them as food items (Figure 4.3b). Termites (workers, soldiers, and winged females) and orthopterans each accounted for 12% of the items captured by *P. striata*, followed by beetles (9%), parts of arthropods (7%), and lepidopterans (adults and larvae; 6%). Plant material, including fleshy fruits or seeds (Figure 4.3d) as well as flowers, and lichens accounted for only a small proportion of the food items sampled (Figure 4.4a). Nearly half of the food items retrieved by *P. striata* foragers consisted of insect parts or small insect prey such as ants and termites (dry weight ≤ 3.0 mg), whereas large arthropods such as katydids and centipedes (≥21.0 mg) accounted for ca. 14% of the ants' diet (Figures 4.3c and 4.4b).

Foragers of *P. striata* employ a variety of strategies to obtain food (Figure 4.5). The main feeding mode observed was scavenging for dead arthropods by solitary foragers, which accounted for 77% of the records in the field. In case the food item was too big to be carried by a single worker, solitary scouts usually returned to the nest in order to recruit other workers by means of tandem runs (observed in 5.5% of the cases). Predation on small- to medium-sized live arthropods by individual *P. striata* foragers comprised 10% of the feeding events seen in the field (Figure 4.5). Predation on termites was observed under two circumstances: First, individual *P. striata* foragers sometimes actively preyed on the winged sexuals during termite nuptial flights, as well as on workers and soldiers that aggregated in the vicinity of the nest entrances. In the second special case, *P. striata* joined raids on nests of *Neocapritermes opacus* (an abundant termite in the study area) performed by the specialized termitophagous ant *Pachycondyla marginata* (Leal and Oliveira 1995). In these cases, foragers of *P. striata* may take advantage of the intense excitement caused by the raid and enter the termite nest to capture workers, soldiers, or alates of *N. opacus*. Finally, interspecific food robbing was occasionally observed in *P. striata* foragers, which intercepted returning workers of other ant species such as *Odontomachus chelifer*, *Camponotus crassus*, and *Pheidole* sp., and robbed their insect prey, which included moths, beetle larvae, and winged ants (Figure 4.5).

COLONY HOME RANGES

The foraging home ranges of the four monitored *P. striata* colonies are shown in Figure 4.6. The maps are presented both separately for the dry/cold and rainy/hot seasons and overlapping in order to better illustrate seasonal variation in the colonies' foraging areas. The area used by the colonies to forage ranged from 1.5 m^2 (Colony I in the dry season) to 19.0 m^2 (Colony IV in the rainy season). Overall, home ranges tended to be larger in the rainy season than in the dry season (Mann-Whitney U test: $U = 15.5$; $P = 0.028$). The ants also considerably changed their foraging terrain during the different seasons. For instance, the foraging area of Colony III was not only larger in the rainy season but also rotated by nearly 180° compared to the home range observed in the dry period (Figure 4.6).

AGGRESSIVE INTERACTIONS

Foragers of *P. striata* were frequently engaged in intra- and interspecific combats with other ants at the border of their colony home ranges or in the vicinity of their nests. Such agonistic encounters sometimes lasted for more than 1 h, during which the ants usually bit and stung each other vigorously. In fights between non-nestmate *P. striata* foragers, the opponents frequently remained locked to one another for up to 30 min until one worker eventually killed or severely injured the rival, which was then carried to the nest as prey. *P. striata* ants with mutilated legs or antennae were commonly observed in the field, possibly as a result of such combats. Most of the fights seen in the field involved non-nestmate *P. striata* foragers, but dealated queens of *P. striata* that tried to intrude a foreign nest were also fiercely attacked by resident workers. Combats between *P. striata* and other

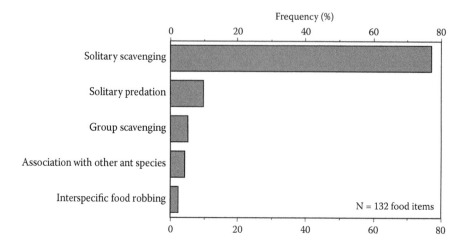

FIGURE 4.5 Frequency of occurrence of the various foraging modes employed by ground-dwelling *Pachycondyla striata* foragers in a Brazilian forest. Foraging techniques are defined as follows: *solitary scavenging*, retrieval of dead prey by a single forager; *group scavenging*, group retrieval of dead prey by two to five workers; *solitary predation*, capture of live prey by a single worker; *association with other ant species*, retrieval of live termites from their nests during raids by the termitophagous ant *Pachycondyla marginata*; and *interspecific food robbing*, theft of food from other ant species. Data are based on collections from returning foragers of fifty observation nests.

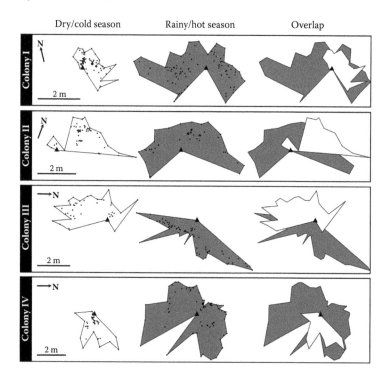

FIGURE 4.6 Seasonal variation of the home ranges of four *Pachycondyla striata* colonies in a forest reserve in Brazil. Maps are based on maximal distances achieved by individual foragers (represented by filled circles). The convex polygons for each colony were drawn by connecting the outermost points at which workers were seen, at intervals of 10° around the nest entrance (filled triangle). The home ranges of all four colonies were larger during the rainy/hot season (grey filled polygons) than during the cold/dry season (white polygons). Overlaps are shown to better illustrate the seasonal variation in the foraging areas of the colonies.

ant species involved *Pachycondyla marginata*, *Odontomachus chelifer*, and *Atta sexdens*. Pairwise interspecific aggression was usually observed near rich food resources (e.g., fallen fleshy fruits or large prey items), when the foraging paths of *P. striata* crossed that of other species, or during food robbing (see above). Foragers of *P. striata* and *O. chelifer* conspicuously avoided each other in the field, and frequently altered their routes in the imminence of an encounter. Finally, *P. striata* foragers themselves occasionally fell prey to other ant species that hunt cooperatively, as indicated by worker carcasses found on the refuse piles of *Pheidole* and *Solenopsis* ants.

DISCUSSION

Daily and Seasonal Activity Patterns

Abiotic conditions such as temperature and moisture can fluctuate widely in many natural habitats. Animals facing these environmental oscillations developed mechanisms to track them, and adjust their physiology and behavior in order to grow and reproduce efficiently; as animals switch between survival and reproductive strategies, their foraging behavior changes as well (Morse 1980; Stephens and Krebs 1986). Indeed, daily and seasonal activity shifts have been documented in a diversity of insect groups and are mediated mostly by temperature, relative humidity or moisture, and food availability (Heinrich 1993). Among the ants, every species operates within a certain temperature-humidity range. The tolerance of foraging individuals, however, is far more restricted than that of the entire colony, which can control microenvironmental conditions by moving between nest compartments or by clustering/dispersing colony members (Hölldobler and Wilson 1990).

Whereas in temperate climates the seasonal cycle clearly varies between warm months with high food availability and cold months with low food availability (Bernstein 1975, 1979), in the tropics moisture can mediate seasonal fluctuations, which are less extreme than the temperate seasons, but often are associated with fluctuating insect abundance and activity as well (Janzen and Schoener 1968; Wolda 1988). For instance, comparisons of the activity of litter-dwelling ants along moisture gradients in a seasonal Panamanian rainforest revealed a positive correlation between moisture and the workers' activity, which, in addition, increased rapidly upon experimental watering during the dry season (Levings 1983). Moist litter not only provides a suitable microhabitat for ant colonies, but sustains populations of microbes and microfauna that structure the litter food web (Levings and Windsor 1984). Increased activity in wet periods and habitats suggests that ant colonies are tracking decreased desiccation risks, increased food availability, or a combination of both (Kaspari and Weiser 2000).

Temperature is considered to be the primary control of colony activity and metabolism in ants (e.g., Porter and Tschinkel 1993). In addition, foraging periods have been found to correlate with soil surface temperature in many ant species (Bernstein, 1979). Moreover, ground temperature has recently been shown to mediate the preference for nesting sites and territorial behavior in ants (Sanada-Morimura et al. 2006). Temperature appears to be more important than humidity in determining daily activity rhythms in *Pachycondyla striata*, since increased activity matched well high daytime temperatures (Figure 4.2). In contrast, the activity schedule of the Amazonian *Dinoponera gigantea* is negatively associated with temperature and follows a bimodal pattern in which most activity is confined to early morning and late afternoon, with a marked decline around midday (Fourcassié and Oliveira 2002). Such an activity pattern is commonly seen in warm environments, including sand dunes (Oliveira et al. 1999), savannas (Lévieux 1979), and deserts (Hölldobler 1976). It is also exhibited by other ponerine species living in tropical forests (Lévieux 1977; Dejean and Lachaud 1994; Duncan and Crewe 1994). *Pachycondyla striata* nests in shady sites and in close proximity to trees. The closed forest canopy and the thick leaf litter likely allow its colonies to keep midday activity in summer by buffering high temperatures in the understory or by keeping moderate levels of soil moisture.

Foraging activity by *Pachycondyla striata* varies markedly along the year. In the rainy/hot season more ants leave the nest to forage, and foragers travel longer distances than in the dry/cold season. Seasonal variation in foraging activity has already been reported in other tropical, forest-dwelling ponerines (e.g., Dejean and Lachaud 1994; Raimundo et al. 2008). However, whereas *Pachycondyla striata* remained strictly diurnal year-round, other ant species may switch their daily activity peak across seasons in order to adjust it to more appropriate temperature ranges in the course of the year (Hölldobler and Wilson 1990). For instance, in the nomadic and termitophagous *Pachycondyla marginata*, both hunting and migratory activities were strongly affected by seasonal factors, with a shift in foraging toward the night in the hot season (Leal and Oliveira 1995). Similarly, in arid Australia the activity of *Odontomachus* sp. depends on the temperature, and colonies switch from crepuscular activity in spring toward nocturnal activity in summer (Briese and Macauley 1980).

The increased foraging activity by *P. striata* during the rainy/hot season corresponds to the period of greater quantity of brood in the colonies and to a threefold increase in the abundance of potential litter-dwelling prey, as compared with the dry/cold season (Raimundo et al. 2008). Similar results were obtained with the bromeliad-nesting *Gnamptogenys moelleri* (Ectatomminae) in a coastal Brazilian forest (Cogni and Oliveira 2004a). Because changes in abiotic conditions may influence food availability and the colonies' phenology (Kaspari et al. 2001), environmental fluctuations should also affect a colony's food intake (e.g., Weeks et al. 2004; Judd 2005). Although we have not evaluated food intake by *P. striata* colonies across seasons, evidence from different groups of social insects indicates that food preference is correlated with the types of individuals being reared in the colony (West-Eberhard 1969; Michener 1974; Oster and Wilson 1978; Roubik 1989; Raveret Richter 2000). Because larvae are normally chief consumers of protein, their presence should increase protein intake by the colony (Stradling 1987; Weeks et al. 2004). Indeed, Stein et al. (1990) demonstrated that the imported fire ant, *Solenopsis invicta*, foraged for protein in warmer months when the colonies were reproducing, and for carbohydrates in colder months when the colony was in its growth stage (i.e., the period with a strong increase in worker number; Oster and Wilson 1978). Similarly, Judd (2005) showed that the highest preference for protein in spring and summer by *Pheidole ceres* correlated with an increased number of larvae within the colonies during this period.

DIET AND FORAGING MODES

Pachycondyla striata showed an opportunistic foraging behavior, with a flexible diet that included a wide array of live and dead invertebrates as well as plant matter within a variable size range (Figures 4.3 and 4.4). The taxonomic diversity of the food items in *P. striata*'s diet is similar to that documented for other ponerine species living in tropical forests (Fresneau 1985; Dejean et al. 1993; Duncan and Crewe 1994; Ehmer and Hölldobler 1995). Although ponerines have generally been regarded as primarily carnivorous ants, several studies have demonstrated that species in the genera *Pachycondyla*, *Odontomachus*, *Dinoponera*, and *Rhytidoponera* also consume lipid- and protein-rich fleshy parts of seeds and fruits to complement their diets (Horvitz and Beattie 1980; Davidson and Morton 1981; Pizo and Oliveira 1998; Fourcassié and Oliveira 2002; Passos and Oliveira 2002, 2004; Araújo and Rodrigues 2006).

In addition to a diversified diet, *P. striata* also employed a variety of foraging modes to retrieve food (Figure 4.5). Behavioral flexibility in hunting techniques involving individual and group retrieving of prey is well documented for other ant species, including ponerines and ectatommines (Peeters and Crewe 1987; Hölldobler 1984; Breed et al. 1987; Dejean et al. 1993; Schatz et al. 1997; Cogni and Oliveira 2004a, 2004b). As a generalist species, *P. striata* workers are frequently confronted with variable circumstances while hunting, and they take advantage of an array of foraging strategies to collect a large diversity of prey items. Interference competition with other foragers by taking food directly away from their mandibles is well known in ants and may involve both solid and liquid food (Hölldobler 1986; Yamaguchi 1995; Richard et al. 2004). The technique employed by *P. striata* foragers consisted of stealing insect prey from returning workers of other ant species, which was also described for the African ponerine *Pachycondyla* (= *Hagensia*) *havilandi* (Duncan and Crewe 1994). Although the theft of prey by gaining entry into other ant colonies (e.g., Breed et al. 1990) was not observed in *P. striata*, this species did join the obligate termitophagous ant *P. marginata* to enter raided termite nests in order to retrieve prey. This foraging mode is likely to be relevant for *P. striata* in the study area, because the preferred termite prey (*Neocapritermes opacus*) is locally very abundant (1 nest/3 m), and *P. marginata* colonies hunt for termites every 2–3 weeks (Leal and Oliveira 1995).

COLONY HOME RANGES AND AGGRESSIVE INTERACTIONS

The home ranges of animals change over time, and the largest shifts generally occur on a seasonal basis (Wittenberger 1981). Although daily and seasonal foraging activities are well documented in ants, little is known about temporal changes in their foraging home ranges (but see Gordon 1995). In an environment such as the forest floor, where the distribution of resources can vary widely in space and time (Levings 1983), continual sampling of adjacent areas should result in more efficient foraging than concentrating entirely on the location of the most recent successful foraging site (Stephens and Krebs 1986).

Our results show that *P. striata* colonies markedly change their home ranges across the seasons, considerably expanding or switching them in the rainy/hot season (Figure 4.6), when brood is abundant and more prey is available in the forest environment (Raimundo et al. 2008). Indeed, there is evidence that other ground-dwelling ants can also make significant adjustments to the directional and temporal aspects of their home range orientation. For instance, Gordon (1995) demonstrated that the foraging ranges of harvester ant colonies (*Pogonomyrmex barbatus*) differed significantly between subsequent years: only about half of a colony's foraging range in one summer had been used by this colony in the previous summer (Gordon 1995). Similarly, although home ranges of the African *Pachycondyla* (= *Brachyponera*) *senaarensis* are of the same size in the rainy and dry seasons, there is a marked shift of the hunting terrains exploited by the colonies across the seasons (Dejean and Lachaud 1994). In the Brazilian Atlantic forest, workers of the arboreal *Gnamptogenys moelleri* colonies foraged only on their nest bromeliads during winter, but considerably expanded their home ranges to the ground and neighboring shrubs and trees in the summer. At this time of the year the amount of brood in the colonies is increased, arthropod prey is abundant, and *Gnamptogenys* foragers are nearly three times more efficient in retrieving food than in the winter (Cogni and Oliveira 2004a). The home range aboveground of the African stink ant, *Pachycondyla* (= *Paltothyreus*) *tarsata*, is generally not greater than 5 m (Hölldobler 1984). However, underground trunk routes with multiple exit holes enable a colony to considerably expand its territory and to have foragers collecting food up to 40 m away from the main nest, thereby covering a foraging area of 1,200 m^2 (Lévieux 1965; Dejean et al. 1993; Braun et al. 1994). This decentralization of the foraging territories through subterranean pathways reduces both predation on foragers and aggressive interactions with competitors, and in addition, minimizes the desiccation risk for workers (Hölldobler and Wilson 1990; Dejean et al. 1993).

Resource availability, costs of defense, and life history may all interact to influence territory size in animals and, more broadly, the way they use space (Wittenberger 1981; Gordon 1995). Colonies of *P. striata* have relatively small foraging areas (1.5 to 19.0 m^2), which is consistent with the home ranges of other forest-dwelling ponerines (Hölldobler 1984; Dejean et al. 1993; Duncan and Crewe 1994). The maintenance of a foraging area by *P. striata* colonies apparently is costly, since intra- and interspecific combats that can cause death or severe injury to foragers were frequently observed within their home ranges. Aggressive interactions with other ant species usually occurred near food sources and involved the theft of prey by *P. striata* and occasionally the death of the robbed ant. Similar observations were also reported for *Pachycondyla* (= *Hagensia*) *havilandi* in

Africa (Duncan and Crewe 1994), and fierce territorial fights have been described for several other ant species (see Hölldobler and Wilson 1990). In some species, however, ritualized contests rarely cause death, probably because bites are aimed mostly at the mandibles, thorax, or legs rather than at vital spots such as joints or the gaster (e.g., Sanada-Morimura et al. 2006). In the Amazon forest, Fourcassié and Oliveira (2002) found that *Dinoponera gigantea* foragers from neighboring colonies may engage in ritualized contests at the border of their foraging areas (ca. 10 m around the nest) that can last up to 30 min, during which the ants usually faced each other frontally and locked their mandibles together. As opposed to *P. striata*, however, intraspecific contests in *Dinoponera* caused no apparent injury to either of the ants involved (Fourcassié and Oliveira 2002).

CONCLUDING REMARKS

Forest-dwelling *Pachycondyla striata* ants exhibit a highly flexible foraging behavior that comprises an array of prey hunting techniques, including individual and group scavenging, solitary predation, predation in association with other species, and food robbing. Such a diversity of foraging modes allows *P. striata* workers to feed on live and dead invertebrates as well as on plant matter within a broad size range. Moreover, this ant species adjusts its activity patterns and the size and orientation of its home ranges according to seasonal oscillations in its environment. It is not clear to what extent (if at all) intra- and interspecific aggression with other ants mediates shifts in the home range boundaries of *P. striata* colonies. This question remains open for future investigations.

As a generalist species facing variable and complex ecological settings in both space and time, *P. striata* represents a suitable model to evaluate the integration of individual- and colony-level components of the foraging system of ant colonies. The field study presented in this chapter illustrates how the combination of natural history information with quantitative behavioral data can link ecological factors with the observed foraging patterns and strategies of a social insect.

ACKNOWLEDGMENTS

We are grateful to Stefan Jarau and Michael Hrncir for the invitation to contribute to this volume. A. V. Christianini, R. Cogni, A. V. L. Freitas, L. Kaminski, M. A. Pizo, P. Rodrigues, and M. Uehara-Prado provided helpful suggestions on the manuscript, and L. Kaminsky also assisted with the line drawings. S. Jarau and M. Hrncir considerably improved the final version of the manuscript. We thank the staff of the Santa Genebra Reserve for logistic support, and several colleagues for help in the field. Financial support to FNSM was provided by CAPES, and to PSO by the Brazilian Research Council (CNPq).

REFERENCES

Araújo A, Rodrigues Z. (2006). Foraging behavior of the queenless ant *Dinoponera quadriceps* Santschi (Hymenoptera: Formicidae). *Neotrop Entomol* 35:159–64.
Bernstein RA. (1975). Foraging strategies of ants in response to variable food density. *Ecology* 56:213–19.
Bernstein RA. (1979). Schedules of foraging activity in species of ants. *J Anim Ecol* 48:921–30.
Braun U, Peeters C, Hölldobler B. (1994). The giant nests of the African stink ant *Paltothyreus tarsatus* (Formicidae, Ponerinae). *Biotropica* 26:308–11.
Breed MD, Abel P, Bleuze TJ, Denton SE. (1990). Thievery, home ranges, and nestmate recognition in *Ectatomma ruidum*. *Oecologia* 84:117–21.
Breed MD, Fewell, JH, Moore AJ, Williams KR. (1987). Graded recruitment in a ponerine ant. *Behav Ecol Sociobiol* 20:407–11.
Briese DT, Macauley BJ. (1980). Temporal structure of an ant community in semi-arid Australia. *Austral Ecol* 5:121–34.

Brown WL. (2000). Diversity of ants. In Agosti D, Majer JD, Alonso LE, Schultz TR (eds.), *Ants: Standard Methods for Measuring and Monitoring Biodiversity*. Washington, DC: Smithsonian Institution Press, pp. 45–79.

Carroll CR, Janzen DH. (1973). Ecology of foraging by ants. *Annu Rev Ecol Syst* 4:231–57.

Cogni R, Oliveira PS. (2004a). Patterns in foraging and nesting ecology in the neotropical ant *Gnamptogenys moelleri* (Formicidae, Ponerinae). *Insect Soc* 51:123–30.

Cogni R, Oliveira PS. (2004b). Recruitment behavior during foraging in the neotropical ant *Gnamptogenys moelleri* (Formicidae: Ponerinae): Does the type of food matter? *J Ins Behav* 17:443–58.

Davidson DW, Morton SR. (1981). Myrmecochory in some plants (*F. chenopodiaceae*) of the Australian arid zone. *Oecologia* 50:357–66.

Dejean A, Beugnon G, Lachaud JP. (1993). Spatial components of foraging behavior in an African ponerine ant, *Paltothyreus tarsatus*. *J Ins Behav* 6:271–85.

Dejean A, Evraerts C. (1997). Predatory behavior in the genus *Leptogenys*: A comparative study. *J Ins Behav* 10:177–91.

Dejean A, Lachaud JP. (1994). Ecology and behavior of the seed-eating ponerine ant *Brachyponera senaarensis* (Mayr). *Insect Soc* 41:191–210.

Detrain C, Deneubourg JL. (2002). Complexity of environment and parsimony of decision rules in insect societies. *Biol Bull* 202:268–74.

Duncan FD, Crewe RM. (1994). Field study on the foraging characteristics of a ponerine ant, *Hagensia havilandi* Forel. *Insect Soc* 41:85–98.

Ehmer B, Hölldobler B. (1995). Foraging behavior of *Odontomachus bauri* on Barro Colorado Island, Panama. *Psyche* 102:215–24.

Fewell JH, Harrison JF, Lighton JRB, Breed MD. (1996). Foraging energetics of the ant, *Paraponera clavata*. *Oecologia* 105:419–27.

Fourcassié V, Oliveira PS. (2002). Foraging ecology of the giant Amazonian ant *Dinoponera gigantea* (Hymenoptera, Formicidae, Ponerinae): Activity schedule, diet, and spatial foraging patterns. *J Nat Hist* 36:2211–27.

Freitas AVL. (1995). Nest relocation and prey specialization in the ant *Leptogenys propefalcigera* Roger (Formicidae: Ponerinae) in an urban area in southeastern Brazil. *Insect Soc* 42:453–56.

Fresneau D. (1985). Individual foraging and path fidelity in a ponerine ant. *Insect Soc* 32:109–16.

Gordon DM. (1995). The development of an ant colony's foraging range. *Anim Behav* 49:649–59.

Heinrich B. (1993). *The Hot-Blooded Insects: Strategies and Mechanisms of Thermoregulation*. Cambridge, MA: Harvard University Press.

Hölldobler B. (1976). Recruitment behavior, home range orientation and territoriality in harvester ants, *Pogonomyrmex*. *Behav Ecol Sociobiol* 1:3–44.

Hölldobler B. (1984). Communication during foraging and nest-relocation in the African stink ant, *Paltothyreus tarsatus* Fabr. (Hymenoptera, Formicidae, Ponerinae). *Z Tierpsychol* 65:40–52.

Hölldobler B. (1985). Liquid food transmission and antennation signals in ponerine ants. *Israel J Entomol* 19: 89–99.

Hölldobler B. (1986). Food robbing in ants, a form of interference competition. *Oecologia* 69:12–15.

Hölldobler B, Wilson EO. (1990). *The Ants*. Cambridge, MA: Harvard University Press.

Horvitz CC, Beattie AJ. (1980). Ant dispersal of *Calathea* (Marantaceae) by carnivorous Ponerines (Formicidae) in a tropical rain forest. *Am J Bot* 67:321–26.

Janzen DH, Schoener TW. (1968). Differences in insect abundance and diversity between wetter and drier sites during a tropical dry season. *Ecology* 49:96–110.

Judd TM. (2005). The effects of water, season, and colony composition on foraging preferences of *Pheidole ceres* (Hymenoptera: Formicidae). *J Ins Behav* 18:781–803.

Kaspari M, Pickering J, Longino JT, Windsor D. (2001). The phenology of a Neotropical ant assemblage: Evidence for continuous and overlapping reproduction. *Behav Ecol Sociobiol* 50:382–90.

Kaspari M, Weiser MD. (2000). Ant activity along moisture gradients in a Neotropical forest. *Biotropica* 32: 703–11.

Leal IR, Oliveira PS. (1995). Behavioral ecology of the Neotropical termite-hunting ant *Pachycondyla* (=*Termitopone*) *marginata*: Colony founding, group raiding and migratory patterns. *Behav Ecol Sociobiol* 37:373–83.

Lévieux J. (1965). Description de quelques nids de fourmis de Côte d'Ivoire (Hym.). *Bull Soc Entomol Fr* 70: 259–66.

Lévieux J. (1977). La nutrition des fourmis tropicales. V. Éléments de synthèse. Les modes d'exploitation de la biocenose. *Insect Soc* 24:235–60.

Lévieux J. (1979). La nutrition des fourmis granivores IV. Cycle d'activité et régime alimentaire de *Messor galla* et de *Messor* (= *Cratomyrmex*) *regalis* en saison des pluies fluctuations annuelles. Discussion. *Insect Soc* 26:279–94.

Levings SC. (1983). Seasonal, annual, and among-site variation in the ground ant community of a deciduous tropical forest: Some causes of patchy species distributions. *Ecol Monogr* 53:435–55.

Levings SC, Windsor DM. (1984). Litter moisture content as a determinant of litter arthropod distribution and abundance during the dry season on Barro Colorado Island, Panama. *Biotropica* 16:125–31.

Michener CD. (1974). *The Social Behavior of the Bees: A Comparative Study*. Cambridge, MA: Harvard University Press.

Morse DH. (1980). *Behavioral Mechanisms in Ecology*. Cambridge, MA: Harvard University Press.

Oliveira PS, Brandão CRF. (1991). The ant community associated with extrafloral nectaries in the Brazilian cerrados. In Huxley CR, Cutler DF (eds.), *Ant-Plant Interactions*. Oxford: Oxford University Press, pp. 198–212.

Oliveira PS, Rico-Gray V, Díaz-Castelazo C, Castillo-Guevara C. (1999). Interaction between ants, extrafloral nectaries, and insect herbivores in Neotropical coastal sand dunes: Herbivore deterrence by visiting ants increases fruit set in *Opuntia stricta* (Cactaceae). *Funct Ecol* 13:623–31.

Orivel J, Dejean A. (2002). Ant activity rhythms in a pioneer vegetal formation of French Guiana (Hymenoptera: Formicidae). *Sociobiology* 39:65–76.

Orivel J, Souchal A, Cerdan P, Dejean A. (2000). Prey capture behavior of the arboreal ponerine ant *Pachycondyla goeldii* (Hymenoptera: Formicidae). *Sociobiology* 35:131–40.

Oster GF, Wilson EO. (1978). *Caste and Ecology in the Social Insects*. Princeton, NJ: Princeton University Press.

Passos L, Oliveira PS. (2002). Ants affect the distribution and performance of *Clusia criuva* seedlings, a primarily bird-dispersed rainforest tree. *J Ecol* 90:517–28.

Passos L, Oliveira PS. (2004). Interaction between ants and fruits of *Guapira opposita* (Nyctaginaceae) in a Brazilian sandy plain rainforest: Ant effects on seeds and seedlings. *Oecologia* 139:376–82.

Peeters C, Crewe R. (1987). Foraging and recruitment in ponerine ants: Solitary hunting in the queenless *Ophthalmopone berthoudi* (Hymenoptera: Formicidae). *Psyche* 94:201–14.

Peeters C, Ito F. (2001). Colony dispersal and the evolution of queen morphology in social Hymenoptera. *Annu Rev Entomol* 46:601–30.

Philpott SM, Maldonado J, Vandermeer J, Perfecto I. (2004). Taking trophic cascades up a level: Behaviorally-modified effects of phorid flies on ants and ant prey in coffee agroecosystems. *Oikos* 105:141–47.

Pizo MA, Oliveira PS. (1998). Interaction between ants and seeds of a nonmyrmecochorous neotropical tree, *Cabralea canjerana* (Meliaceae), in the Atlantic forest of Southeast Brazil. *Am J Bot* 85:669–74.

Porter SD, Tschinkel WR. (1993). Fire ant thermal preferences: Behavioral control of growth and metabolism. *Behav Ecol Sociobiol* 32:321–29.

Raimundo RLG, Freitas AVL, Oliveira PS. (2009). Activity schedule and diet of the ground-dwelling ant, *Odontomachus chelifer* (Formicidae: Ponerinae), in an Atlantic forest fragment. In preparation.

Raveret Richter M. (2000). Social wasp (Hymenoptera: Vespidae) foraging behavior. *Annu Rev Entomol* 45:121–50.

Richard FJ, Dejean A, Lachaud JP. (2004). Sugary food robbing in ants: A case of temporal cleptobiosis. *CR Biol* 327:509–17.

Rico-Gray V, Oliveira PS. (2007). *The Ecology and Evolution of Ant-Plant Interactions*. Chicago: University of Chicago Press.

Roubik DW. (1989). *Ecology and Natural History of Tropical Bees*. Cambridge, MA: Cambridge University Press.

Sanada-Morimura S, Satoh T, Obara Y. (2006). Territorial behavior and temperature preference for nesting sites in a pavement ant *Tetramorium tsushimae*. *Insect Soc* 53:141–48.

Schatz B, Lachaud JP, Beugnon G. (1997). Graded recruitment and hunting strategies linked to prey weight and size in the ponerine ant *Ectatomma ruidum*. *Behav Ecol Sociobiol* 40:337–49.

Stein MB, Thorvilson HG, Johnson JW. (1990). Seasonal-changes in bait preference by red imported fire ant, *Solenopsis invicta* (Hymenoptera, Formicidae). *Fla Entomol* 73:117–23.

Stephens DW, Krebs JR. (1986). *Foraging Theory*. Princeton, NJ: Princeton University Press.

Stradling DJ. (1987). Nutritional ecology of ants. In Slansky F, Rodriguez JG (eds.), *Nutional Ecology of Insects, Mites, Spiders and Related Invertebrates*. New York: Wiley, pp. 927–69.

Traniello JFA. (1989). Foraging strategies of ants. *Annu Rev Entomol* 34:191–210.

Ward PS. (2000). Broad-scale patterns of diversity in leaf litter ant communities. In Agosti D, Majer JD, Alonso LE, Schultz TR (eds.), *Ants: Standard Methods for Measuring and Monitoring Biodiversity*. Washington, DC: Smithsonian Institution Press, pp. 99–121.

Weeks RD, Wilson LT, Vinson SB, James WD. (2004). Flow of carbohydrates, lipids, and protein among colonies of polygyne red imported fire ants, *Solenopsis invicta* (Hymenoptera: Formicidae). *Ann Entomol Soc Am* 97:105–10.

West-Eberhard M. (1969). The social biology of polistine wasps. *Misc Publ Mus Zool Univ Mich* 140:1–101.

Wittenberger JF. (1981). *Animal Social Behavior*. Boston: Duxbury Press.

Wolda H. (1988). Insect seasonality: Why? *Annu Rev Ecol Syst* 19:1–18.

Yamaguchi T. (1995). Intraspecific competition through food robbing in the harvester ant, *Messor aciculatus* (Fr. Smith), and its consequences on colony survival. *Insect Soc* 42:89–101.

5 Foraging Range and the Spatial Distribution of Worker Bumble Bees

Dave Goulson and Juliet L. Osborne

CONTENTS

INTRODUCTION

All bee species, be they solitary or social, provision their broods by central place foraging, which means they gather pollen and nectar from flowers and use them to provision a nest. The foraging range of bees is thus a fundamental aspect of their ecology, for it determines the area of the habitat that an individual can exploit (Bronstein 1995; Westrich 1996). Understanding the spatial relationship between nest and food resources allows us to make predictions about colony survival and distribution (Nakamura and Toquenaga 2002; Williams and Kremen 2007). Also, if we are to study the effects of habitat management on bee reproduction (i.e., nest success), the appropriate scale of experimental plots depends on the area over which the occupants of each nest forage (whether it is a single bee or a large group of workers). Similarly, if we are to manage habitats to conserve particular bee species, their foraging range influences the distribution of forage that is considered optimal. For bees to thrive, presumably their nest sites and forage sites must be sufficiently close together that they can economically visit the forage. In the case of social species such as bumble bees with a nest that may persist for many months, the nest must be within foraging range of suitable flowers throughout its life. Bee foraging ranges are also important in the consideration of insect-mediated crop pollination. The distance a bee flies will determine which fields are visited within the vicinity of the nest, and also the distances over which pollen might be carried to effect gene flow within and between fields (Damgaard et al. 2008). This is particularly relevant when crops are grown for high seed purity, or when genetically modified crops are grown and the transfer of pollen between neighboring fields or farms is undesirable (Rieger et al. 2002; Cresswell et al. 2002; Damgaard and Kjellsson 2005; Weekes et al. 2005).

Flight ranges vary considerably among bee species. By far the most intensively studied species is the honey bee, *Apis mellifera*. Quantifying the foraging range of honey bees is relatively easy because the "waggle dance" of a returning forager describes the distance and direction of the food source (von Frisch 1967; Seeley 1985a). Honey bees have a foraging range of 1–6 km, very rarely up to 20 km when local resources are exceedingly scarce (Visscher and Seeley 1982; Seeley 1985b; Schneider and McNally 1992, 1993; Waddington et al. 1994; Schwarz and Hurst 1997). Honey bees are clearly not "doorstep foragers." Little is known of the foraging range of most other bee species, but those estimates that are available suggest that honey bees are not typical, and that most bees forage over much shorter distances. This is perhaps not surprising given their large size relative to most bee species, and the large number of workers per colony (which presumably necessitates exploitation of resources available over a large area). Other social bees that have been studied to date travel substantial but shorter distances: for example, the stingless bee *Melipona fasciata* travels up to 2.4 km (Roubik and Aluja 1983) and members of the Trigonini up to 1 km (Roubik et al. 1986). Solitary bee species are generally thought to travel only a few hundred meters (Gathmann and Tscharntke 2002). Greenleaf et al. (2007) recently reviewed the records of foraging range for sixty-two bee species, and correlated the evidence with body size, but data for the majority of bee species are lacking (Schwarz and Hurst 1997).

It has long been assumed that bumble bees (and, for that matter, other bee species) forage as close to their nest as possible, for, all else being equal, this would seem to be the most efficient strategy (Free and Butler 1959; Crosswhite and Crosswhite 1970; Heinrich 1976b; Teräs 1976; Bowers 1985; Free 1993). Indeed, optimality models predict (rather obviously) that choosing the closest foraging sites is the best strategy, all else being equal (Heinrich 1979). Of course, the same arguments ought to apply to honey bees, yet clearly they do sometimes travel great distances, perhaps driven by a paucity of forage near the nest. Remarkably, despite the wealth of literature on bumble bee foraging behavior, few studies have succeeded in providing measures of bumble bee foraging range (Bronstein 1995; Osborne et al. 1999; Cresswell et al. 2000). It is unexpectedly hard to quantify, and particularly to understand, its relationship to the quantity and quality of forage available.

MEASURING BUMBLE BEE FORAGING RANGE

CHALLENGES OF MARKING EXPERIMENTS

The simplest (or at least most obvious) approach to studying foraging range in bumble bees or any other central place forager is to mark the bees at the nest, and then search the flowers in the surrounding area to see where the marked bees are foraging. Unfortunately, bumble bee nests can be hard to find, so some researchers have resorted to marking the bees as they forage. Bowers (1985) marked *Bombus flavifrons* while foraging in discrete meadows surrounded by forest in Utah. Subsequent recaptures were all in the same meadows in which the bees had been marked, leading to the conclusion that bees did not move between meadows. However, the locations of the actual nests were not identified, so this tells us nothing about foraging range. One would expect individual bees to return to the same site repeatedly, having found it to be rewarding (see Thomson et al. 1997; Osborne and Williams 2001), but this does not mean that their nest was situated nearby.

It is far more informative to mark the bees at their nests if they can be located, and then search for them to discover where they go to forage. Unfortunately, this is also fraught with difficulties. Where forage is abundant and available close to the nest (for example, if the nest is in or adjacent to a flowering crop), a small proportion of marked bees has been observed foraging, and these bees tend to be close to the nest (Saville 1993; Schaffer and Wratten 1994). Where the rest of the marked bees are going is harder to discover. Studies of nests situated in more typical fragmented habitat with small, scattered sources of forage have found very few marked bees on flowers, even when many hundreds of bees were marked at the nest (Kwak et al. 1991; Dramstad 1996; Saville et al.

1997). For example, Dramstad (1996) found that few *B. terrestris/lucorum* workers were to be seen on patches of flowers located within 50 m of their nest, and searches at distances of up to 300 m located very few marked bees. Dramstad concluded that bumble bees do not forage close to their nests, but exactly where they do forage was not clear. Schaffer (1996) suggested that the leptokurtic distribution observed in earlier studies was an artifact of the experimental design. Observer effort was always biased toward searching areas close to the nest for the simple reason that the area to be searched increases as the square of the distance from the nest. To properly evaluate the distribution of marked bees within even a moderate distance of the nest such as 1 km would require 3.1 km^2 to be thoroughly searched—a formidable task. Suppose that there were fifty workers foraging at any one time (which would be quite a large and active nest); this would translate to an average of one marked bee per 6.2 ha. Wolf and Moritz (2008) overcame this problem by placing nests of *Bombus terrestris* in an intensively farmed agricultural "desert" landscape in Germany, where the only flowers within 3 km of the nests were along a single linear track that could be monitored relatively easily. They found that the majority of workers foraged near to their nest (mean distance 267 m, maximum 800 m). In more typical, patchy landscapes the mark–reobservation method is of little use without a huge team of observers to search for bees.

Homing Experiments

An alternative approach that has been used is to carry out homing experiments. In the late 1800s the famous entomologist Jean-Henri Fabre demonstrated that the solitary sphecid wasp *Cerceris tuberculata* and the gregarious bee *Chalicodoma muraria* could return to their nests when transported away from them several kilometers in darkened boxes (Fabre 1882). Similar experiments have since been performed on a range of solitary and social hymenopterans (reviewed by Wehner 1981; Southwick and Buchmann 1995; Chmurzynski et al. 1998; Capaldi and Dyer 1999; Greenleaf et al. 2007). Most of these studies have examined homing from distances ranging from 100 m to 3–4 km, and all have found that at least a proportion of the released insects return. The record holder among the Hymenoptera is the Euglossine bee *Euplusia surinamensis* (now *Eufriesea surinamensis*; Roubik and Hanson 2004), which has been found to successfully return home from 23 km (Janzen 1971). Homing experiments have been carried out on one bumble bee species, *B. terrestris* (Goulson and Stout 2001). The maximum distance from which a bee successfully returned to its nest was 9.8 km, with a clear decline in the probability of a bee returning as the distance over which it was displaced increased (Figure 5.1). Unexpectedly, no relationship was

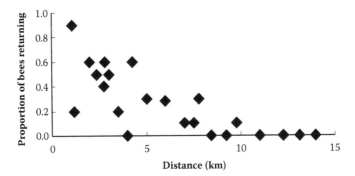

FIGURE 5.1 The proportion of marked *Bombus terrestris* workers returning to their nest after artificial displacement over distances up to 14 km. In total, 220 bees were displaced in batches of 10, so that each point marked is based on 10 bees. There was a clear negative relationship between the proportion of bees returning and the distance of the release site (linear regression, R^2 = 0.55, P < 0.001). (After Goulson and Stout 2001. With permission.)

apparent between the displacement distance and the time to return to the nest. Times from release to recapture varied from 6 h (from 2 km) to 9 days (from 3.5 km). Notably, one bee that returned to its nest after being displaced by 4.3 km was observed on a subsequent occasion gathering nectar at the release site, which contained a large patch of nectar-rich flowers.

What does this tell us about the foraging range of bumble bees? This is a matter for debate. Prolonged homing times have been found in studies of other species. For example, of 374 *Anthophora abrupta* females displaced up to 3.2 km from their nest, homing times varied from 20 min to 50 h (Rau 1929). This may provide a clue as to how Hymenoptera locate their nests. *B. terrestris* flies at a mean speed of 7.1 m/s when tagged with a harmonic radar transponder (Osborne et al. 1999; Riley et al. 1999), and so could theoretically return from up to 15 km within a little over 30 min. Pigeons are well known for their prodigious homing abilities, and they do so rapidly, flying more or less directly from the release site to their loft. Their homing abilities have been the subject of intensive studies, and it is thought that they possess some sort of coordinate system that enables them to determine their location relative to home, perhaps based on olfactory or magnetic stimuli (Wiltschko and Wiltschko 1988, Wallraff 1990, Able 1994). They may also be able to detect information as to the direction of transport during the outward journey, and simply reverse their path (Wallraff 1980, Wallraff et al. 1980). It seems highly unlikely that Hymenoptera possess either of these abilities. If they did, we would expect them to return swiftly to their nests. In reality, over moderate to long displacement distances they return very slowly (often taking days rather than hours), taking many times longer than one would expect if they flew directly home. Recent studies proposed a third mechanism: displaced insects are thought to engage in a search for familiar landmarks from which they can locate their nest (Collett and Collett 2002). Desert ants (*Cataglyphis* spp.) engage in systematic searches when displaced to unfamiliar terrain (Wehner 1996). Honey bees use visual landmarks to aid navigation between their nest and forage (Wehner 1981; Dyer 1996), and use a sun compass to relate the positions of landmarks and the nest (Wehner 1994). Honey bee homing is better when prominent horizon landmarks are present (Southwick and Buchmann 1995). Searching for familiar landmarks could lead to protracted homing times and so explain why, from more distant sites, many bees fail to return.

Evidence is growing, from directly tracking honey bee flight paths with radar, that when these bees are displaced to an unfamiliar location and begin searching for their colony—or when searching for a feeder in a known location—they adopt an optimal scale-free search strategy (Reynolds et al. 2007a, 2007b). The patterns of these flights can be described mathematically using a Levy distribution, which has been used to characterize search patterns of many animals (see also Chapter 16). Explained more simply, the bees tend to fly in a loop away from the release point and back, before taking another loop in a different direction, and gradually these loops may get larger over time (looping Levy flight patterns: Reynolds et al. 2007a, 2007b). The flights may result in their finding the goal they were searching for, or a familiar landmark allowing them to orientate on the goal. It seems probable that the homing mechanism used by Hymenoptera is a systematic search of the area surrounding the release site using an optimal, scale-free strategy of expanding loops, until a familiar landmark is recognized. If this is so, then we might expect all bees released within an area that they have previously explored to successfully return to the nest, and to do so rapidly, for they will soon recognize a familiar landmark. Of course, individual bees are likely to vary in their ability to home according to their foraging experience (they typically live for only a few weeks) and also according to the particular directions that they have previously explored, which will determine the number and distribution of familiar landmarks. Rau (1929) found that homing success in *Anthophora abrupta* was strongly related to age, with older (and presumably more experienced) bees being much more likely to return to the nest. Such variability may obscure relationships between displacement distance and success in homing.

It seems improbable that a bee released 9.8 km from its nest could find familiar landmarks unless its home range were several kilometers in radius (Goulson and Stout 2001). Since one marked bee that returned to its nest from 4.3 km was subsequently seen foraging at the site where it was released, bumble bees are clearly capable of remembering the location of forage at such distances and successfully navigating to and from these patches. Whether bumble bees locate forage at such distances from their nests naturally remains to be determined, but it seems likely that they do so if forced to by lack of forage near the nest. Since many bumble bees successfully returned from considerable distances (>5 km), it seems reasonable to conclude that *B. terrestris* naturally forage over several kilometers. However, we must be cautious when interpreting homing experiments, because we cannot directly observe the bees during most of their journey home. We do not know the size of the area that a bumble bee can search for familiar landmarks (if indeed that is what they do), and so we can only speculate as to the likely foraging area.

THEORETICAL APPROACHES

If we cannot easily directly measure the foraging range of bumble bees, perhaps we can calculate how far they ought to be able to fly. The economics of bumble bee foraging have been the subject of a substantial body of research, initiated by Bernd Heinrich in the 1970s (reviewed in Heinrich 1979). Detailed information is available on the energetic cost of bumble bee flight, on the amount of forage that they can carry, the time it takes them to handle flowers of different species, and the energetic rewards that they obtain per flower. Cresswell et al. (2000) combined these data to estimate the maximum distance that a bumble bee could travel to reach a patch of nectar-rich flowers and return with a net profit. The most accurate estimates of flight speeds are provided by recent studies using harmonic radar, which measured a mean airspeed of 7.1 m s^{-1} for *B. terrestris* (Riley et al. 1999). The energetic costs of flight have been estimated from oxygen consumption rates to be 1.2 kJ h^{-1} (Ellington et al. 1990). Estimates of the volume of the honey stomach (which determines the maximum volume of nectar that a bee can carry) vary from 60 to 100 µl (Allen et al. 1978; Heinrich 1979), and vary greatly according to the size of the worker (Goulson et al. 2002). Nectar concentrations vary greatly between flower species, and also with time of day, but they commonly fall within the range 40–60% sugars, (about 1–1.5 M sucrose or equivalent) (Cresswell et al. 2000). The metabolism of nectar sugars yields 16.7 kJ g^{-1} (Heinrich 1979). Let us assume that a worker leaves the nest with an empty honey stomach, and occasionally stops to feed at flowers to fuel the outward flight. If it returns with any nectar, it has thus made a profit. The limit to the distance from which the bee can return with a net profit is simply given by the time taken to burn up all of the sugars in the honey stomach on the return flight. For a 40% nectar solution and a honey stomach capacity of 80 µl, the maximum range is about 10 km (Cresswell et al. 2000). If the nectar is more concentrated, or if the bees concentrated the nectar within their honey stomach as they foraged, then the range could be larger.

Cresswell's model makes it clear that travel to flower patches constitutes a relatively small portion of a forager's time and energy budget. Bees should choose the forage patch closest to the nest, given a choice of equally rewarding patches, but because flight is swift relative to the time spent within patches of flowers, the more distant site does not have to be much more rewarding to be the better option. Thus, for example, if given the choice between a patch of flowers immediately adjacent to the nest, and a more distant patch in which average nectar rewards are twice as high, it may be worth flying up to 4 km farther to reach the more distant patch. If this model is approximately correct in its assumptions, then it may explain why bumble bees often do not seem to visit patches of apparently suitable forage close to their nests (Dramstad 1996). Even in the most fragmented and impoverished habitat there is likely to be a considerably better patch located within a radius of, say, 5 km (an area of 78 km^2).

The model is concerned only with nectar collection, and so tells us nothing about the economics of trips to gather pollen, or trips where both nectar and pollen are collected. Very little information is available on the rate at which bees are able to collect pollen, or on the relative values of pollen from different sources (Cook et al. 2003). Pollen is the only source of protein for bumblebees, and is vital for larval growth and development of eggs in the queen, so this represents a substantial knowledge gap. It has arisen simply because it is much easier to measure nectar volume and concentration than it is to quantify the amount and quality of pollen available per flower, so researchers tend to focus on the former. If pollen is in short supply near to the nest, it is conceivable that workers could engage in flights of more than 10 km to obtain it. Such flights would result in a net loss of energy, but could (theoretically) be fueled by visits to nearer nectar sources on both the outward and return journeys.

RADAR TRACKING

Perhaps the most innovative (and certainly the most expensive) approach yet deployed to measure forage range in bumble bees is the use of harmonic radar. This system for tracking insect movements was developed by the Natural Resources Institute in conjunction with the Institute of Arable Crop Research at Rothamsted, UK (Riley et al. 1996, 1998; Osborne et al. 1999). The technique involves attaching a 16 mm vertical aerial-like transponder to the thorax of the bumble bee (or other large insect). This captures the energy in radar emissions sent out from a base unit and re-emits the energy at a higher frequency, providing a harmonic of the original signal that can be tracked as the bee flies (Riley and Smith 2002). The transponder weighs 12 mg, and so is very light in comparison with the weight of a typical bumble bee forager (6–7% of the insects' body mass), and is much lighter than the normal foraging load, which can reach 90% of the body mass (Goulson et al. 2002, Peat and Goulson 2005). As far as can be ascertained, the transponder does not seem to interfere greatly with the behavior of the bumble bee, although bees with transponders do take longer to complete foraging trips than usual, perhaps because the transponder impedes their handling of flowers (Osborne et al. 1999). These studies have provided some fascinating insights into the flight paths of workers of the bumble bee *B. terrestris*. For example, they have revealed that bumble bees can compensate for crosswinds and manage to fly directly between the nest and patches of forage by flying at an angle to their intended course (Riley et al. 1999). Unfortunately, this technique has a major limitation with respect to determining foraging range, for bees can only be detected up to 1 km, and then only if they remain within a direct line of sight of the radar equipment. Osborne et al. (1999) found that very few foragers flew less than 200 m from their nest, even though there were patches of suitable forage just 50 to 100 m from the nest. In separate studies in June and August, Osborne et al. (1999) recorded mean maximum distances from the nest of 339 m (range 96–631 m) and 201 m (range 70–556 m), respectively (Figure 5.2). However, many of these bees flew beyond the range of the radar, being obscured from the radar's view behind hedges or other landscape features, so these do not represent the actual foraging ranges (13 of 35 tagged bees flew beyond radar range in June, and 14 of 30 in August). This study confirms that bumble bees do not necessarily forage on the closest available patches, but that a small majority of them do remain within 500–700 m of the nest. Therefore, although the harmonic radar approach is an elegant method to measure forage distances of bumble bees, it cannot tell us what the full distribution of foraging ranges is, or what the limit to foraging range might be.

MASS MARKING AND POLLEN ANALYSIS

More recently, the distribution of bumble bee foragers away from their colonies has been quantified using a combination of mass marking and pollen analysis in an agricultural landscape (Osborne et al. 2008). Unusually, this information on foraging range was also analyzed in the context of the spatial distribution of foraging habitats in the landscape, as determined from remote-sensed data. Experimental colonies were placed at 250 m intervals along a 1.5 km transect across UK farmland. A novel device was placed on the entrance to the colony (Martin et al. 2006), and this marked each exiting forager with powder dye (different colors for each position on the transect). A second device was used to capture returning foragers, so that their pollen loads could be sampled and analyzed (Martin et al. 2006). There was a borage field at one end of the transect, and the number of marked bees foraging on the borage, coming from colonies at each point on the transect, were recorded. The

FIGURE 5.2 Harmonic radar tracks of the bumble bee, *Bombus terrestris*, flying away from or returning to the nest. Rings are at 200 and 400 m from the radar. Shaded areas are patches of forage. Thick lines are hedges. Each symbol denotes an individual bee. Shown are thirty-five outward tracks by nine individuals. (Adapted from Osborne et al. 1999.)

proportion of pollen foragers returning to each colony with borage pollen was also recorded. Sixty three percent of pollen foragers at colonies next to the borage field collected borage pollen (and were therefore assumed to have foraged on the one field, as no other borage was growing in the area). The respective percentage declined, in an approximately linear fashion, to 17% of borage pollen foragers at 1.5 km (Osborne et al. 2008). Marked bees were also seen on the borage field from colonies 0 m to 1.5 km away. Once colony activity, the area surveyed, and the percentage contribution of borage to the forage landscape were taken into account, the relationship between the number of marked bees seen in the borage field and distance from their colony was a shallow curve with distance. While neither measurement can tell us the maximal foraging range, the distributions are comparable with predictions made by Dukas and Edelstein-Keshet (1998) for solitary foragers provisioning a nest (no communication of locality), based on maximization of energy intake.

ARE THERE FEWER BEES FORAGING CLOSE TO THE COLONY?

A recurring feature of studies of bumble bee foraging range is that a relatively small proportion of bees seem to forage close to the nest (but see Wolf and Moritz 2008). One suggestion as to why this may be so is that it minimizes intracolony competition (Dramstad 1996). Colonies of some species such as *B. terrestris* grow to contain up to four hundred workers, and a workforce of this size would quickly deplete resources close to the nest. Heinrich (1976a) found that bumble bees could remove

94% of the standing crop of nectar within an area. Since it is the cumulative foraging success of all foragers that determines colony success, one would predict that there is a trade-off between travel time and competition. Patches that are near the nest should always receive more visits than those that are farther away, but the difference need only be slight if travel is rapid (Dukas and Edelstein-Keshet 1998; Cresswell et al. 2000). Given that each individual bee may visit hundreds, sometimes even thousands, of flowers in a foraging bout (e.g., Creswell et al. 2002), one might expect only a very few bees to visit the nearest patches if the patches are small. This might explain the very low numbers of observed visits to patches close to the nest described by Dramstad (1996) and Osborne et al. (1999).

An alternative potential explanation for the apparent tendency of bumble bees to forage far from their nest relates to predation. It has been suggested that colonial food provisioners such as bumble bees may avoid foraging close to their nests so as to avoid attracting predators or parasites to the nest (Dramstad 1996). Bumble bee nests are attacked by a range of organisms, including queens of their own species and cuckoo bumble bees (subgenus *Psithyrus*) (reviewed in Alford 1975 and Goulson 2003). Dramstad (1996) suggested that a high concentration of workers close to the nest might attract *Psithyrus* queens and other enemies. It is not known how *Psithyrus* females locate the nests of their hosts. If foragers remained close to their nests, then the chemical cues deposited on flowers to aid foraging (Goulson et al. 1998; see also Chapter 13) could potentially provide a "scent magnet" to cuckoo bees (Dramstad 1996). Similarly, if foragers were all concentrated in the area close to the nest, then they might attract aggregations of predators. Gentry (1978) found that aggregations of pollinators on the flowers of tropical trees attracted large numbers of bee-eating birds. Little is known of the risk of predation faced by foraging bumble bees. Many researchers have considered predation on foraging bumble bees to be rare and of minimal ecological importance (Brian 1965; Pyke 1978; Zimmerman 1982; Hodges 1985), but others suggest predation rates are significant (Rodd et al. 1980; Goldblatt and Fell 1987). There are also likely to be big differences between regions; for example, the UK has very few bee-eating birds, but they are common in southern Europe (Alford 1975).

The difficulty with these hypotheses concerning the risk of attack by predators, parasites, or inquilines is that they are hard to test since the available evidence suggests that bumble bees do not forage close to their nests. It is not easy to demonstrate that this is a result of past predation pressure. A comparison of the foraging ranges of species with and without these enemies could in theory provide an insight, if such species exist, but our sketchy knowledge of both the foraging ranges and natural enemies of different bumble bee species prevents a meaningful comparison at present. It is probably fair to say that the most likely explanation for the low density of bees close to their nest is simply that flight is relatively cheap and intracolony competition would be high if many workers stayed close to their nests. Consequently, the most economic strategy for a colony is to distribute its foragers widely in space.

VARIATION BETWEEN BUMBLE BEE SPECIES IN FORAGING RANGE

Interestingly, there is now evidence that bumble bee species differ markedly in foraging range and that this correlates with colony size (Table 5.1). Experiments with bees marked at the nest (Kreyer et al. 2004) and anecdotal observations suggest that species such as *B. pascuorum*, *B. sylvarum*, *B. ruderarius*, and *B. muscorum* are "doorstep foragers," mostly remaining within 500 m of their nests, while *B. lapidarius* forage farther afield (mostly <1,500 m), and *B. terrestris* regularly forage over more than 2 km away from their nests (Walther-Hellwig and Frankl 2000). These experiments suffer from a reduced intensity of sampling at greater distances from the nest, and also from rather small sample sizes, which, however, cannot explain the differences that were found between species. These differences appear to correspond to the known nest sizes of the bee species: *B. terrestris* and *B. lapidarius* nests grow to a large size (100–400 workers), while those of *B. pascuorum*, *B. sylvarum*, *B. ruderarius*, and *B. muscorum* (all commonly known as carder bees, and belonging to the subgenus *Thoracobombus*) are generally small (20–100 workers) (Alford 1975).

TABLE 5.1
Summary of Bumble Bee (*Bombus sp.*) Foraging Range Data Available to Date

		B. terrestris	B. lapidarius	B. pascuorum	B. pratorum
Worker thorax width range (foragers)		3.9–6.8 mm	3.4–5.8 mm	3.3–5.2 mm	3.2–4.6 mm
Average colony size (no. workers)		Large (<400)	Large (<300)	Medium (<200)	Small (<100)
Method	**Study**	**Foraging Range**			
Pollen and marking	Osborne et al. 2008	>1,500 m	—	—	—
Genetics	Knight et al. 2005	758 m	450 m	449 m	674 m?[b]
Genetics	Darvill et al. 2004	>312 m	—	<312 m	—
Mark–recapture	Walther-Hellwig and Frankl 2000	1,750 m	1,500 m	500 m	—
Visit response to landscape[a]	Westphal et al. 2006	3,000 m	2,750 m	1,000 m	250 m
Mark–recapture	Wolf and Moritz 2008	800 m	—	—	—

Note: Thorax widths are from DG, unpublished data. Colony sizes are from Benton (2006).

[a] Westphal et al. 2006 used statistical estimation of how well flower visitation response correlated with landscape structure at different scales.

[b] Knight et al. 2005 suggest that their results for *B. pratorum* should be treated with caution due to anomalies in the genetic data.

Recent attempts to measure foraging range of bumble bees have adopted a very different approach, but largely confirm the findings of Walther-Hellwig and Frankl (2000). Darvill et al. (2004) and Knight et al. (2005) used molecular (microsatellite) markers to identify sister pairs along a transect of samples of foragers. This approach is aided by the haplodiploid genetics of bumble bees and the monogamous nature of queens, so that all workers from a nest should be 75% related to one another. The distribution of sisters along the transect can be used to estimate foraging range. This approach has advantages, in that the bees are from wild nests and are foraging naturally. The disadvantages, however, are that the actual sites of the nests are not known, and also that sister foragers at the outer limits of their foraging range are at very low density. Thus, the tail of the forager distribution is unlikely to be detected (of course, this criticism applies to other approaches, too). Nonetheless, this method provides a powerful test for differences between species in their foraging range. The maximum detected foraging range was greatest for *B. terrestris* (758 m), least for *B. pascuorum* (449 m) and *B. lapidarius* (450 m), and intermediate for *B. pratorum* (674 m). Since the area of forage available to nests increases as the square of foraging range, these differences correspond to a threefold variation in area exploited by bumble bee nests of different species.

Westphal et al. (2006) examined the spatial scales at which the density of mass-flowering crops in the surrounding area affects recruitment of the same four bumble bee species. Interpretation of this approach is difficult, but the authors found that abundance of different bee species was best described by the density of mass-flowering crops at different radii from a focal point: 3,000 m for *B. terrestris*, 2,750 m for *B. lapidarius*, 1,000 m for *B. pascuorum*, and 250 m for *B. pratorum*. The foraging range is best interpreted as half these figures, since, for example, in the case of *B. terrestris*, a nest up to 1,500 m from the focal point with a foraging radius of 1,500 m will be foraging at the focal point, and up to 1,500 m in the opposite direction, at 3,000 m from the focal point. These estimates compare well with those of Walther-Hellwig and Frankl (2000), and are broadly comparable to but generally larger than those obtained by Darvill et al. (2004) and Knight et al. (2005) using molecular methods.

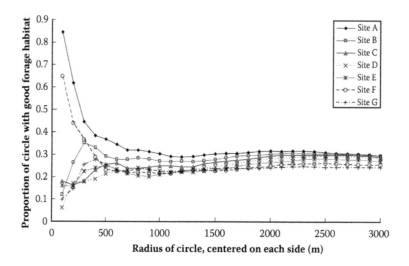

FIGURE 5.3 Forage landscape for colonies at each site A–G, estimated by calculating the proportion of the landscape containing good forage habitats at different radial distances from each site. (After Osborne et al. 2008. With permission.)

Three quite different approaches to measuring or inferring foraging range have reached broadly similar conclusions: there are large differences in foraging range between species, with *B. terrestris* having a long foraging range and *B. pascuorum* foraging much closer to the nest (Table 5.1). Such fundamental differences between the ecology of bumble bee species have clear implications for management of bumble bees for pollination or conservation purposes. In their foraging range experiment, Osborne et al. (2008) quantified the proportion of the landscape containing habitats with good nectar and pollen sources at different distances from each set of experimental colonies. In that landscape there was great variation in forage availability within a radial distance of 500 m of the colonies, but little variation beyond 1 km (Figure 5.3), suggesting that this is the scale at which the "forage landscape" operates. The authors hypothesized that the scale of *Bombus terrestris* foraging (>1.5 km) would be large enough to buffer effects of forage patch and flowering crop heterogeneity in this landscape, while bee species with shorter foraging ranges are likely to be more vulnerable to this spatial variation (Osborne et al. 2008). Thus, positioning of colonies in the landscape could have serious consequences for the survival and reproduction of these doorstep foragers.

COMMUNICATION BETWEEN FORAGERS AS TO FORAGE LOCATIONS

The waggle dance of the honey bee has been frequently described and much studied, and is one of the most complex systems in insect communication. In contrast, it has long been assumed that foraging in bumble bees is essentially a solitary endeavor—that workers do not communicate with each other about good sources of forage, so that each individual has to learn for itself which flowers provide reward. Indeed, it has been known for many years that bumble bees (of a range of species) are unable to recruit nestmates to particular places (Jacobs-Jessen 1959; Esch 1967; von Frisch 1967; Kerr 1969). However, it has become apparent that bumble bee foragers do communicate, and that recruitment does occur, but not to specific locations. In an elegantly simple experiment, Dornhaus and Chittka (1999, 2001) demonstrated that, on their return to the nest, successful foragers of *B. terrestris* stimulate other workers to forage, and communicate to them the scent of the food source that they have located. The returning forager runs around on the surface of the nest in an excited manner, frequently bumping into nestmates and buzzing her wings (very similar behavior occurs in some stingless bees; see Chapter 11). This stimulates workers to leave the nest and search for the source of the scent. This communication system is less complex than that of honey bees, for the new recruits do not appear to be given any positional information as to the location of the food source. It appears that there is a pheromone signal released by the returning forager: chemical compounds in the pheromone have now been identified (Granero et al. 2005), and they have been shown to stimulate foragers (Dornhaus et al. 2003; Granero et al. 2005). Granero et al. (2005) hypothesize that the evolutionary origins of the bumble bee alert system and the honey bee waggle dance are separate since honey bees use the Nasanov pheromone primarily to mark food sources, rather than alert foragers (for a discussion on pheromones during foraging and recruitment in honey bees see Chapter 9).

Why do bumble bees not communicate positional information? Dornhaus and Chittka (1999) argue that conveying the location of food sources may be less important to bumble bees than to honey bees; honey bees evolved in tropical ecosystems where they rely heavily on flowering trees, a highly clumped resource that may be several kilometers from the nest and so would be difficult to locate. In the temperate habitats in which bumble bees probably evolved, the herbaceous plants that are their main food source are generally more scattered. There is nothing to be gained in recruiting more workers to a specific small patch that one bee can adequately exploit single-handedly. However, communication as to the types of flowers that are providing rewards will allow the colony to rapidly recruit to feeding on a rewarding plant species when it comes in to flower, and so keep track of the changing seasonal availability of different species. A second possibility is that bumble bees forage over shorter distances than do honey bees—perhaps as a result of their smaller colony size—thus rendering communication as to the precise location of forage less important (if forage is near the nest, it will not take long for a forager to find it even without any clues as to its location). Alternatively, it has been hypothesized that the honey bee's dance language evolved not in the context of foraging but as a means to communicate the location of new nest sites. Since honey bees need to guide an entire swarm (including the valuable queen) to a new site, they need to accurately convey the location (Beekman and Bin Lew 2008). The annual life cycle of bumblebees renders this unnecessary.

There is a cost to conveying location information. Honey bee recruits can take over an hour to decide where to forage when presented with just two returned foragers advertising different locations. They also take a long time to find the food source that is being advertised, for the locational information is not precise (Wenner and Wells 1990). It may be that, in bumble bees, the costs of conveying this information outweigh any gains.

CONCLUDING REMARKS

Studies using a diversity of approaches, including direct observation, harmonic radar, theoretical modeling of energetics, and molecular markers, have provided a range of measures as to how bumble bee foragers distribute themselves around their nests. However, only a small number of species have been studied (predominantly *B. terrestris*), and it is apparent that there is considerable variation between species. The ecological/evolutionary explanation for these differences remains unclear. Greenleaf et al. (2007) show that, across a range of bees (mainly solitary species), size is a reliable predictor of foraging range, and the evidence that is available for bumble bees would appear to support this (see Table 5.1; the mean sizes of these species are *B. terrestris* > *B. lapidarius* > *B. pascuorum* > *B. pratorum*). However, further comparative studies are required to establish whether this is a general relationship, and if it is, this then begs the question of why size varies between bumble bee species. We also have little idea at present how flexible bumble bees are with regard to foraging range. Estimates for *B. terrestris* vary from >312 m (mixed farmland, southern UK) to 3,000 m (intensively farmed arable land, Germany) (Table 5.1), but whether this represents a response to differing floral resource density and distribution or differences between experimental approaches is unclear. The mass-marking approach of Martin et al. (2006) probably represents the most cost-effective technique for studying bumble bee foraging range, and comparative studies of bee species in a range of landscapes using this approach would be invaluable in resolving these issues.

REFERENCES

Able KP. (1994). Magnetic orientation and magnetoreception in birds. *Prog Neurobiol* 42:449–73.

Alford DV. (1975). *Bumblebees*. London: Davis-Poynter.

Allen T, Cameron S, McGinley R, Heinrich B. (1978). The role of workers and new queens in the ergonomics of a bumblebee colony (Hymenoptera: Apoidea). *J Kans Entomol Soc* 51:329–42.

Beekman, M, Lew JB. (2008). Foraging in honeybees—When does it pay to dance? *Behav Ecol* 19:255–62.

Benton T. (2006). *Bumblebees— The Natural History and Identification of the Species Found in Britain*. London: Collins New Naturalist Series.

Bowers MA. (1985). Bumblebee colonization, extinction, and reproduction in subalpine meadows in Northeastern Utah. *Ecology* 66:914–27.

Brian MV. (1965). *Social Insect Populations*. London: Academic Press.

Bronstein JL. (1995). The plant-pollinator landscape. In Hansson L, Fahrig L, Merriam G (eds.), *Mosaic Landscapes and Ecological Processes*. London: Chapman & Hall, pp. 256–88.

Capaldi EA, Dyer FC. (1999). The role of orientation flights on homing performance in honeybees. *J Exp Biol* 202:1655–66.

Chmurzynski JA, Kieruzel M, Krzysztofiak A, Krzysztofiak L. (1998). Long-distance homing ability in *Dasypoda altercator* (Hymenoptera, Melittidae). *Ethology* 104:421–29.

Collett TS, Collett M. (2002). Memory use in insect visual navigation. *Nat Rev Neurosci* 3:542–52.

Cook SM, Awmack CS, Murray DA, Williams IH. (2003). Are honey bees' foraging preferences affected by pollen amino acid composition? *Ecol Entomol* 28:622–27.

Cresswell JE, Osborne JL, Bell S. (2002). A model of pollinator-mediated gene flow between plant populations, with numerical solutions for bumble bees pollinating oilseed rape. *Oikos* 98:375–84.

Cresswell JE, Osborne JL, Goulson D. (2000). An economic model of the limits to foraging range in central place foragers with numerical solutions for bumblebees. *Ecol Entomol* 25:249–55.

Crosswhite FS, Crosswhite CD. (1970). Pollination of *Castilleja sessiflora* in southern Wisconsin. *Bull Torrey Bot Club* 97:100–5.

Damgaard C, Kjellsson G. (2005). Gene flow of oilseed rape (*Brassica napus*) according to isolation distance and buffer zone. *Agr Ecosyst Environ* 108:291–301.

Damgaard C, Simonsen V, Osborne JL. (2008). Prediction of pollen-mediated gene flow between fields of red clover (*Trifolium pratense*). *Environ Model Assess* 13:483–90.

Darvill B, Knight ME, Goulson D. (2004). Use of genetic markers to quantify bumblebee foraging range and nest density. *Oikos* 107:471–78.

Dornhaus A, Brockman A, Chittka L. (2003). Bumblebees alert to food with pheromone from tergal gland. *J Comp Physiol A* 189:47–51.

Dornhaus A, Chittka L. (1999). Evolutionary origins of bee dances. *Nature* 401:38.

Dornhaus A, Chittka L. (2001). Food alert in bumblebees (*Bombus terrestris*): Possible mechanisms and evolutionary implications. *Behav Ecol Sociobiol* 50:570–76.

Dramstad WE. (1996). Do bumblebees (Hymenoptera: Apidae) really forage close to their nests? *J Insect Behav* 9:163–82.

Dukas R, Edelstein-Keshet L. (1998). The spatial distribution of colonial food provisioners. *J Theor Biol* 190:121–34.

Dyer FC. (1996). Spatial memory and navigation by honeybees on the scale of the foraging range. *J Exp Biol* 199:147–54.

Ellington CP, Machin KE, Casey TM. (1990). Oxygen consumption of bumblebees in forward flight. *Nature* 347:472–73.

Esch H. (1967). Die Bedeutung der Lauterzeugung für die Verständigung der stachellosen Bienen. *Z Vergl Physiol* 56:199–220.

Fabre JH. (1882). *Souvenirs Entomologiques Études sur l'Instinct et les Moeurs des Insectes*. 2ᵉ Sér. Paris: Delagrave.

Free JB. (1993). *Insect Pollination of Crops*. 2nd ed. London: Academic Press.

Free JB, Butler CG. (1959). *Bumblebees*. London: Collins.

von Frisch K. (1967). *The Dance Language and Orientation of Bees*. Cambridge, UK: Harvard University Press.

Gathmann A, Tscharntke T. (2002). Foraging ranges of solitary bees. *J Anim Ecol* 71:757–64.

Gentry AH. (1978). Anti-pollinators for mass-flowering plants? *Biotropica* 10:68–69.

Goldblatt JW, Fell RD. (1987). Adult longevity of workers of the bumble bee *Bombus fervidus* (F.) and *Bombus pennsylvanicus* (De Geer) (Hymenoptera: Apidae). *Can J Zool* 65:2349–53.

Goulson D. (2003). *Bumblebees: Their Behaviour and Ecology.* Oxford: Oxford University Press.

Goulson D, Hawson SA, Stout JC. (1998). Foraging bumblebees avoid flowers already visited by conspecifics or by other bumblebee species. *Anim Behav* 55:199–206.

Goulson D, Peat J, Stout JC, Tucker J, Darvill B, Derwent LC, Hughes WHO. (2002). Can alloethism in workers of the bumblebee *Bombus terrestris* be explained in terms of foraging efficiency? *Anim Behav* 64:123–30.

Goulson D, Stout JC. (2001). Homing ability of the bumblebee *Bombus terrestris. Apidologie* 32:105–11.

Granero AM, Sanz JMG, Gonzalez FJE, Vidal JLM, Dornhaus A, Ghani J, Serrano AR, Chittka L. (2005). Chemical compounds of the foraging recruitment pheromone in bumblebees. *Naturwissenschaften* 92:371–74.

Greenleaf SS, Williams NM, Winfree R, Kremen C. (2007). Bee foraging ranges and their relationship to body size. *Oecologia* 153:589–96.

Heinrich B. (1976a). Resource partitioning among some eusocial insects: Bumblebees. *Ecology* 57:874–89.

Heinrich B. (1976b). The resource specializations of individual bumblebees. *Ecol Monogr* 46:105–28.

Heinrich B. (1979). *Bumblebee Economics.* Cambridge, MA: Harvard University Press.

Hodges CM. (1985). Bumble bees foraging: Energetic consequences of using a threshold departure rule. *Ecology* 66:188–97.

Jacobs-Jessen UF. (1959). Zur Orientierung der Hummeln und einiger anderer Hymenopteren. *Z Vergl Physiol* 41:597–641.

Janzen DH. (1971). Euglossine bees as long-distance pollinators of tropical plants. *Science* 171:203–5.

Kerr WE. (1969). Some aspects of the evolution of social bees (Apidae). *Evol Biol* 3:119–75.

Knight ME, Martin AP, Bishop S, Osborne JL, Hale RJ, Sanderson RA, Goulson D. (2005). An interspecific comparison of foraging range and nest density of four bumblebee (*Bombus*) species. *Mol Ecol* 14:1811–20.

Kreyer D, Oed A, Walther-Hellwig K, Frankl R. (2004). Are forests potential landscape barriers for foraging bumblebees? Landscape scale experiments with *Bombus terrestris* agg. and *Bombus pascuorum* (Hymenoptera, Apidae). *Biol Conserv* 116:111–18.

Kwak MM, Kremer P, Boerrichter E, van den Brand C. (1991). Pollination of the rare species *Phyteuma nigrum* (Campanulaceae): Flight distances of bumblebees. *Proc Exp Appl Entomol* 2:131–36.

Martin AP, Carreck NL, Swain JL, Goulson D, Knight ME, Hale RJ, Sanderson RA, Osborne JL. (2006). A modular system for trapping and mass-marking bumblebees: Applications for studying food choice and foraging range. *Apidologie* 37:341–50.

Nakamura H, Toquenaga Y. (2002). Estimating colony locations of bumble bees with moving average model. *Ecol Res* 17:39–48.

Osborne JL, Clark SJ, Morris RJ, Williams IH, Riley JR, Smith AD, Reynolds DR, Edwards AS. (1999). A landscape study of bumble bee foraging range and constancy, using harmonic radar. *J Appl Ecol* 36:519–33.

Osborne JL, Martin AP, Carreck NL, Swain JL, Knight ME, Goulson D, Hale RJ, Sanderson RA. (2008). Bumblebee flight distances in relation to the forage landscape. *J Anim Ecol* 77:406–15.

Osborne JL, Williams IH. (2001). Site constancy of bumble bees in an experimentally patchy habitat. *Agr Ecosyst Environ* 83:129–41.

Peat J, Goulson D. (2005). Effects of experience and weather on foraging efficiency and pollen versus nectar collection in the bumblebee, *Bombus terrestris. Behav Ecol Sociobiol* 58:152–56.

Pyke GH. (1978). Optimal body size in bumblebees. *Oecologia* 34:255–66.

Rau P. (1929). Experimental studies in the homing of carpenter and mining bees. *J Comp Psychol* 9:35–70.

Reynolds AM, Smith AD, Menzel R, Greggers U, Reynolds DR, Riley JR. (2007a). Displaced honey bees perform optimal scale-free search flights. *Ecology* 88:1955–61.

Reynolds AM, Smith AD, Reynolds DR, Carreck NL, Osborne JL. (2007b). Honeybees perform optimal scale-free searching flights when attempting to locate a food source. *J Exp Biol* 210:3763–70.

Rieger MA, Lamond M, Preston C, Powles SB, Roush RT. (2002). Pollen-mediated movement of herbicide resistance between commercial canola fields. *Science* 296:2386–88.

Riley JR, Reynolds DR, Smith AD, Edwards AS, Osborne JL, Williams IH, McCartney HA. (1999). Compensation for wind drift by bumblebees. *Nature* 400:126.

Riley JR, Smith AD. (2002). Design considerations for an harmonic radar to investigate the flight of insects at low altitude. *Comput Electron Agric* 35:151–69.

Riley JR, Smith AD, Reynolds DR, Edwards AS, Osborne JL, Williams IH, Carreck NL, Poppy GM. (1996). Tracking bees with harmonic radar. *Nature* 379:29–30.

Riley JR, Valeur P, Smith AD, Reynolds DR, Poppy GM, Löfstedt C. (1998). Harmonic radar as a means of tracking the pheromone-finding and pheromone-following flight in male moths. *J Insect Behav* 11:287–96.

Rodd FH, Plowright RC, Owen RE. (1980). Mortality rates of adult bumble bee workers (Hymenoptera: Apidae). *Can J Zool* 58:1718–21.

Roubik DW, Aluja M. (1983). Flight ranges of *Melipona* and *Trigona* in tropical forests. *J Kans Entomol Soc* 56:217–22.

Roubik DW, Hanson PE. (2004). *Abejas de Orquídeas de la América Tropical: Biología y Guía de Campo* [Orchid Bees of Tropical America: Biology and Field Guide]. Santo Domingo de Heredia, Costa Rica: Editorial INBio.

Roubik DW, Moreno JE, Vergara C, Wittman D. (1986). Sporadic food competition with the African honey bee: Projected impact on neotropical social species. *J Trop Ecol* 2:97–111.

Saville NM. (1993). Bumblebee ecology in woodlands and arable farmland. PhD thesis, University of Cambridge, England.

Saville NM, Dramstad WE, Fry GLA, Corbet SA. (1997). Bumblebee movement in a fragmented agricultural landscape. *Agr Ecosyst Environ* 61:145–54.

Schaffer MJ. (1996). Spatial aspects of bumble bee (*Bombus* spp: Apidae) foraging in farm landscapes. Master's thesis, University of Lincoln, New Zealand.

Schaffer M, Wratten SD. (1994). Bumblebee (*Bombus terrestris*) movement in an intensive farm landscape. In *Proceedings of the 47th New Zealand Plant Protection Conference*, Rotorua, New Zealand, pp. 253–56.

Schneider SS, McNally LC. (1992). Factors influencing seasonal absconding in colonies of the African honeybee, *Apis mellifera scutellata. Insect Soc* 39:403–23.

Schneider SS, McNally LC. (1993). Spatial foraging patterns and colony energy status in the African honeybee, *Apis mellifera scutellata. J Insect Behav* 6:195–210.

Schwarz MP, Hurst PS. (1997). Effects of introduced honey bees on Australia's native bee fauna. *Vic Natur* 114:7–12.

Seeley TD. (1985a). *Honeybee Ecology*. Princeton, NJ: Princeton University Press.

Seeley TD. (1985b). The information-center strategy of honey bee foraging. In Hölldobler B, Lindauer M (eds.), *Experimental Behavioral Ecology and Sociobiology: In Memoriam Karl von Frisch*. Sunderland: Sinauer Associates, pp. 75-90.

Southwick EE, Buchmann SL. (1995). Effects of horizon landmarks on homing success of honeybees. *Am Nat* 146:748–64.

Teräs I. (1976). Flower visits of bumblebees, *Bombus* Latr. (Hymenoptera: Apidae) during one summer. *Ann Zool Fenn* 13:200–32.

Thomson JD, Slatkin M, Thomson BA. (1997). Trapline foraging by bumble bees. II. Definition and detection from sequence data. *Behav Ecol* 8:199–210.

Visscher PK, Seeley TD. (1982). Foraging strategy of honeybee colonies in a temperate deciduous forest. *Ecology* 63:1790–801.

Waddington KD, Visscher PK, Herbert TJ, Raveret Richter M. (1994). Comparisons of forager distributions from matched honey bee colonies in suburban environments. *Behav Ecol Sociobiol* 35:423–29.

Wallraff HG. (1980). Does pigeon homing depend on stimuli perceived during displacement? I. Experiments in Germany. *J Comp Physiol A* 139:193–201.

Wallraff HG. (1990). Navigation by homing pigeons. *Ethol Ecol Evol* 2:81–115.

Wallraff HG, Foà A, Ioalé P. (1980). Does pigeon homing depend on stimuli perceived during displacement? II. Experiments in Italy. *J Comp Physiol A* 139:203–8.

Walther-Hellwig K, Frankl R. (2000). Foraging distances of *Bombus muscorum, Bombus lapidarius* and *Bombus terrestris* (Hymenoptera, Apidae). *J Insect Behav* 13:239–46.

Weekes R, Deppe C, Allnutt T, Boffey C, Morgan D, Morgan S, Bilton M, Daniels R, Henry C. (2005). Crop-to-crop gene flow using farm scale sites of oilseed rape (*Brassica napus*) in the UK. *Transgenic Res* 14:749–59.

Wehner R. (1981). Spatial vision in arthropods. In Autrum H, Jury R, Loewenstein WR, Mackay DM, Teuber HL (eds.), *Handbook of Sensory Physiology*. Vol VII/6C. Berlin: Springer Verlag, pp. 98–201.

Wehner R. (1994). The polarization-vision project: Championing organismic biology. In Schildberger K, Elsner N (eds.), *Neural Basis of Behavioural Adaptation*. Stuttgart: Gustav Fischer, pp. 103–43.

Wehner R. (1996). Middle scale navigation: The insect case. *J Exp Biol* 199:125–27.

Wenner AM, Wells PH. (1990). *Anatomy of a Controversy. The Question of a "Language" among Bees*. New York: Columbia University Press.

Westphal C, Steffan-Dewenter I, Tscharntke T. (2006). Bumblebees experience landscapes at different spatial scales: Possible implications for coexistence. *Oecologia* 149:289–300.

Westrich P. (1996). Habitat requirements of central European bees and the problems of partial habitats. In Matheson A, Buchmann SL, O'Toole C, Westrich P, Williams IH (eds.), *The Conservation of Bees*. London: Academic Press, pp. 2–16.

Williams NM, Kremen C. (2007). Resource distributions among habitats determine solitary bee offspring production in a mosaic landscape. *Ecol Appl* 17: 910-921

Wiltschko W, Wiltschko R. (1988). Magnetic orientation in birds. In Johnston RF (ed.), *Current Ornithology*. Vol 5. New York: Plenum Press, pp. 67–121.

Wolf S, Moritz RFA. (2008). Foraging distance in *Bombus terrestris* L. (Hymenoptera: Apidae). *Apidologie* DOI:10.1051/apido:2008020.

Zimmerman M. (1982). Optimal foraging: Random movement by pollen collecting bumblebees. *Oecologia* 53:394–98.

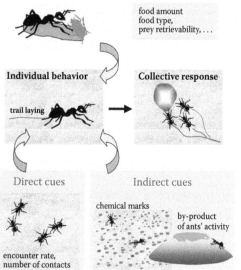

Food-related cues

food amount
food type,
prey retrievability, . . .

Individual behavior

trail laying

Collective response

Direct cues

encounter rate,
number of contacts

Indirect cues

chemical marks

by-product
of ants' activity

Social cues

COLOR FIGURE 2.1

COLOR FIGURE 3.4

COLOR FIGURE 3.5

COLOR FIGURE 3.2

COLOR FIGURE 3.3

COLOR FIGURE 3.9

COLOR FIGURE 4.3

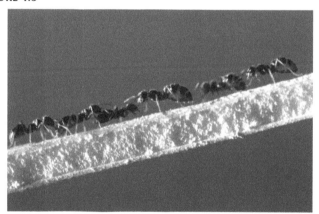

COLOR FIGURE 6.1

COLOR FIGURE 7.4

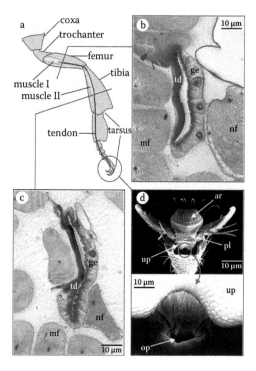

COLOR FIGURE 12.3

COLOR FIGURE 13.1

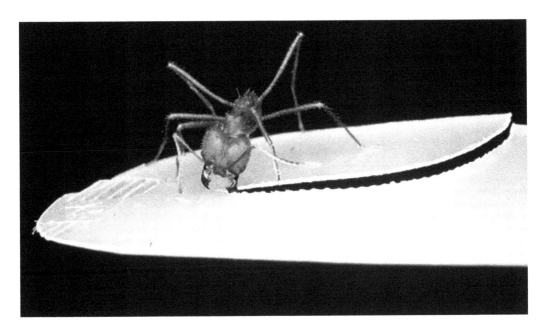

COLOR FIGURE 14.1

COLOR FIGURE 14.2

Part II

Information Use and Information Transfer

6 How to Tell Your Mates
Costs and Benefits of Different Recruitment Mechanisms

Madeleine Beekman and Audrey Dussutour

CONTENTS

INTRODUCTION

The evolution of sociality has created the need for information transfer among group members. Insects that live in societies cannot behave as if they were solitary; rather, actions by different group members need to be integrated to achieve desirable outcomes at the level of the group. The ways in which individuals exchange information depends largely on the size of the society. Small societies can rely on all individuals having knowledge about which tasks need to be done and where to go, but global knowledge about all tasks needed by a colony is much harder to achieve once group size increases (Deneubourg et al. 1987). Groups comprising a large number of individuals cannot rely on all individuals having perfect information about their colony's needs, but must instead rely on their interactions with their peers as well as information directly obtained from their immediate surroundings. The result is self-organized emergence of adaptive behavior at the group level (Camazine et al. 2001).

    Recruitment is an example of such self-organized behavior. Recruitment, a collective term for any behavior that results in an increase in the number of individuals at a particular place (Deneubourg et al. 1986), to food sources is probably the most widely studied, but recruitment also occurs in the context of building (e.g., Karsai and Penzes 1993; Bonabeau et al. 1998), excavation (e.g., Deneubourg and Franks 1995), defense (e.g., Millor et al. 1999), and nest site selection (Pratt et al. 2002; Visscher 2007).

Recruitment to food sources allows insect societies to forage efficiently in an environment in which food sources are patchily distributed or in which food sources are too large to be exploited by single individuals (Beckers et al. 1990; Beekman and Ratnieks 2000; Detrain and Deneubourg 2002). In addition, social insects that transfer information about the location of profitable food sources can exploit an area much larger than those that lack such sophisticated recruitment mechanisms. Honey bees are a prime example. Their sophisticated dance language (von Frisch 1967) allows them to forage at food sources as far as 10 km from the colony (Beekman and Ratnieks 2000).

The type of recruitment mechanism employed by social insects (see the next section) depends strongly on the size of the colony (Beckers et al. 1989). This is because colony size determines both the amount of food that is necessary for growth and maintenance and the number of potential recruits. In small colonies, individuals tend to explore the environment individually and do not recruit nestmates to food sources they find. For colonies of intermediate size, exploring individuals return to their colony once they find a profitable food source and recruit nestmates to that source. The larger the colony, the more recruitment relies on mass communication, mostly by means of a chemical trail, as in many species of ants and termites.

What are the costs and benefits of different kinds of recruitment mechanisms? This is the topic of our chapter. We will first give an overview of the different ways in which social insects recruit nestmates to food sources. In the rest of the chapter we will then mainly contrast the two extreme forms of recruitment mechanisms: mass recruitment via a pheromone trail, where individuals only interact indirectly via the trail, and the honey bees' dance language, where information is transferred directly from individual to individual. We will discuss how these two mechanisms allow the colony to adapt to changing conditions, the means insect colonies have evolved to overcome constraints imposed by their way of communication, and other implications of indirect and direct recruitment mechanisms. In the last section we will synthesize our findings.

THE TOWER OF BABEL: VARIETY IN INSECT LANGUAGES

Exact recruitment mechanisms vary greatly among the social insects but can be divided into two main classes: direct and indirect mechanisms. Mass recruitment via a chemical trail is an example of indirect recruitment. The recruiter and recruited are not physically in contact with each other; communication is instead via modulation of the environment: the trail. The recruiter deposits a pheromone on the way back from a profitable food source and recruits simply follow that trail (Figure 6.1). In a way, such a recruitment mechanism is comparable to radio broadcasting: information is disseminated without controlling who receives it. The other extreme is transferring information from individual to individual: direct recruitment. The best-known example of such recruitment mechanism is the honey bees' dance language. Successful foragers, the recruiters, perform a stylized dance that encodes information about the direction and distance of the food source found, and up to seven dance followers (Tautz and Rohrseitz 1998)—potential recruits—are able to extract this information, upon which they will leave the colony and try to locate the advertised food source. Recruitment trails and the honey bee dance language can be seen as the two extremes of a whole range of different mechanisms used by social insects to convey information about profitable food sources. Let us now consider the variety of insect "languages" and how the type of recruitment mechanism depends on colony size.

INDIRECT RECRUITMENT: TRAIL-BASED FORAGING

In many ants, foragers lay trails from food sources back to their nest. These trails allow nestmates to locate and exploit the source easily. As more individuals collect food they reinforce the trail. Because the pheromone is volatile, it can only be maintained if sufficient individuals are using it. Hence, the use of a volatile recruitment pheromone imposes constraints on minimum colony size because a stable trail is only possible if the trail is reinforced before the pheromone evaporates

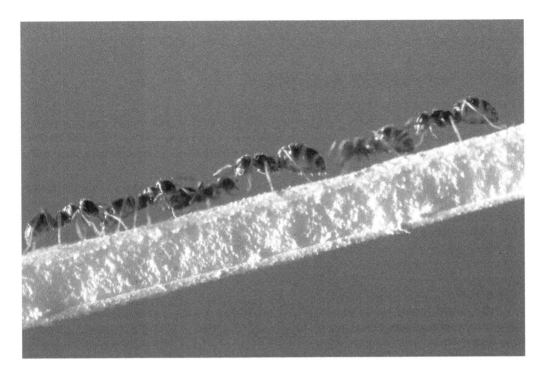

FIGURE 6.1 Ants on a trail. The ants coming from the right return from a food source, while the ants walking in the opposite direction come from the nest. Newly recruited ants are guided toward the food source by a chemical trail deposited by the ants that have been feeding at the source on their way back. See color insert following page 142. (Photograph by Perrin Emmanuel, taken from CNRS Phototheque. With permission.)

(Beekman et al. 2001). To overcome the constraint on minimum colony size set by the volatility of the trail pheromone, ants could presumably evolve less volatile pheromones. However, this is maladaptive for ants that feed on ephemeral food sources because the trail would outlast the food source.

Ants have evolved various mechanisms to overcome the constraint set by minimum colony size on pheromone trail foraging. For example, Pharaoh's ants, *Monomorium pharaonis* (Fourcassie and Deneubourg 1994), and other trail-recruiting species (Detrain et al. 1991) deposit an attracting pheromone when exploring new areas, even when no food has been found (see also Chapter 2). This mechanism may serve to increase the local ant density, thus increasing the probability of forming a trail when food is found. Some species with intermediate-size colonies (*Camponotus socius*, Hölldobler 1971; *Tetramorium caespitum*, Verhaeghe 1977; *T. impurum*, Verhaege 1982; *Ambliopone* sp., Ito 1993; *T. bicarinatum*, de Biseau et al. 1994; for reviews, see Beckers et al. 1989 and Hölldobler and Wilson 1990) use a combination of group recruitment, whereby a successful forager guides recruits directly to the food source (Beckers et al. 1989), and pheromone trails. Such a dual mechanism, combining both direct and indirect recruitment, with an initial period of group recruitment helps to establish a trail, thus overcoming the constraint set by colony size (Beekman et al. 2001).

Species that forage at food sources that are plentiful and stable, such as trees for collecting honeydew (*Lasius fuliginosus*, Quinet et al. 1997), leaves (*Atta* sp., Fowler and Robinson 1979; Shepherd 1982, 1985; Howard 2001), seeds (*Pogonomyrmex rugosus* and *P. barbatus*, Hölldobler 1976; *Pheidole militicida*, Hölldobler and Möglich 1980; *Messor barbarus*, López et al. 1994; Detrain et al. 2000; Azcárate and Peco 2003), or dead wood (termites, Grassé 1986), construct long-lasting trails (trunk trails) that connect the nest to foraging locations. In some species, *Formica* or *Pogonomyrmex*, for example, trails are more or less permanent due to the ants' actively changing the environment by removing vegetation (Rosengren and Sundström 1987; Fewell 1988).

INDIRECT RECRUITMENT: AIRBORNE PHEROMONE TRAILS

Even though odors play an important role in the location of food sources in many social insects, stingless bees are the only bees known to produce odor trails that can begin near the nest (but do not necessarily) and extend to the food source (Nieh 2004; for a critical review of pheromone trail use in stingless bees see Chapter 12). Such an odor trail is produced by foragers depositing odor droplets on vegetation along the vector from a food source to the nest (Lindauer and Kerr 1958). Unlike ant and termite trails, stingless bee odor trails are not continuous, and therefore need to be easy to find by unemployed foragers. It is probably for this reason that successful foragers leave their odor droplets on prominent vegetation (Nieh 2004). Judging from what we know of the physicochemical properties of stingless bee trail pheromone compounds that have been identified to date, it appears that their volatility makes them well suited to attract bees over longer distances (see Chapter 12).

DIRECT RECRUITMENT: DANCING BEES

Upon returning to her colony, a successful honey bee forager may perform a recruitment dance that informs her nestmates of the presence of the profitable food source. The dance encodes two main pieces of spatial information: the direction and the distance to the target. During a typical dance the dancer strides forward vigorously shaking her body from side to side (Tautz et al. 1996). This is known as the waggle phase of the dance. After the waggle phase the bee makes an abrupt turn to the left or right, circling back to start the waggle phase again. This is known as the return phase. At the end of the second waggle, the bee turns in the opposite direction so that with every second circuit of the dance she will have traced the famous figure-eight pattern of the waggle dance (von Frisch 1967).

The most information-rich phase of the dance is the waggle phase. During the waggle phase the bee aligns her body so that the angle of deflection from vertical is similar to the angle of the goal from the sun's current azimuth. Distance information is encoded in the duration of the waggle phase. Dances for nearby targets have short waggle phases, whereas dances for distant targets have protracted waggle phases.

Dance followers need to be in close contact with the dancer in order to be able to acquire the directional information (Rohrseitz and Tautz 1999), especially in cavity-nesting species such as *Apis mellifera* and *A. cerana*, where dancing is performed in the dark.

All species of *Apis* use the dance language to communicate directional information of food sources, although different species differ slightly in their dance details (Oldroyd and Wongsiri 2006). It is interesting that only honey bees have evolved a dance language; why not stingless bees? After all, many species of stingless bee have colony sizes comparable to those of *Apis* (Michener 2000), and both originated in the tropics, the environment in which the dance language is thought to provide the most benefits (Sherman and Visscher 2002; Dornhaus and Chittka 2004; see also Chapters 1 and 7). It could be that this fundamental difference in recruitment mechanism between honey bees and stingless bees is due to their difference in foraging range. *Apis* exploit a vast area (Beekman and Ratnieks 2000), whereas stingless bees forage at a much shorter range (Roubik and Aluja 1983; van Nieuwstadt and Ruano 1996). However, this is an obvious chicken or egg problem: Do honey bees forage over a large range because they communicate effectively, or did the evolution of the dance language allow them to forage farther afield?

Recently it was hypothesized that the honey bee's dance language evolved not in the context of foraging but as a means to communicate the location of new nest sites (Beekman et al. 2008; Beekman and Lew 2008). Honey bee swarms go through a period of homelessness when they leave the mother colony and form a temporary cluster. Scout bees search the environment for a suitable new home (reviewed in Winston 1987) and need to provide spatial information about the location of potential nest sites found to other bees in the swarm, which they do via recruitment dances (Lindauer 1955). In contrast, stingless bees move into a new nest gradually over several weeks or months (Michener 1974). Most likely pheromone marking is used to mark the chosen nest, in a way similar to the marking of food sources. Thus, stingless bees do not need a symbolic dance mechanism in order to communicate the location of new nest sites (Lindauer and Kerr 1958).

DIRECT RECRUITMENT: IN-NEST EXCITATION

The only reason why flowers smell is to attract pollinators. It is therefore not surprising that odor plays an important role in foraging in the social bees. In Chapter 9, Judith Reinhard and Mandyam V. Srinivasan describe the importance of floral odors in honey bee foraging where the scent of past-profitable sites triggers renewed interest by foragers in flowers that carry that scent (Reinhard et al. 2004; Beekman 2005). In honey bees floral odors serve as a means to reactivate foragers, but successful foragers themselves also produce odors that entice more foragers to leave the colony (Thom et al. 2007). In bumble bees the transfer of scent from a returning forager to unemployed bees is the only means of recruitment. Bumble bees returning from profitable forage sites perform excitatory runs upon their return to the colony and carry with them the odor of the plant species they have visited. Unemployed foragers receive information not only about the presence of profitable forage sites but also about the smell of the plants in flower. Even though such recruitment mechanism does not convey directional information, it is remarkably effective: when presented with artificial feeders with different odors, recruits mainly visited the feeder at which the initial foragers had been foraging (Dornhaus and Chittka 1999). The same mechanism is found in some species of stingless bee (Nieh 2004; see also Chapter 12), where, in addition, the activation of the forager force is particularly enhanced though mechanical signals (see Chapter 11).

In some species of ant, returning foragers perform excitatory displays while facing nestmates. This behavior motivates nestmates to follow the forager when it leaves the colony, and up to thirty ants will follow the experienced ant to the food source (Hölldobler 1971; Lenoir and Jaisson 1982). In some mass-recruiting species, only ants that have been stimulated by this in-nest excitation will follow a pheromone trail (*Formica fusca*, Möglich and Hölldobler 1975; *Myrmecocystus mimicus*, Hölldobler 1981). In other species, such excitation increases recruitment efficiency because it alerts a large number of individuals to leave the colony (*Monomorium venustum*, *M. subopacum*, Szlep and Jacobi 1967; *Camponotus pensylvanicus*, Traniello 1977; *Neivamyrmex* sp., Topoff and Mirenda 1978; *Pheidole pallidula*, Detrain and Pasteels 1991).

DIRECT RECRUITMENT: TANDEM AND GROUP RECRUITMENT

Because the use of pheromone trails is only possible in species with large colonies, ants with small colonies such as ponerines and *Temnothorax* often use tandem or group recruitment in which the returned forager directly guides one (tandem) or several (group) individuals toward the food source (Hölldobler and Wilson 1990). In this way, recruits are taught the route to the food source. Similarly, in some species of stingless bee, foragers were observed to actively guide recruits toward food sources, for example, by flying in a zigzag manner toward the food (Nieh 2004). These recruitment mechanisms are relatively short range and are most likely used in combination with scent marking of the food source and its surroundings (see Chapter 12).

STABLE VERSUS DYNAMICALLY CHANGING ENVIRONMENTS

The decentralized nature of most recruitment mechanisms allows an insect colony to choose the best forage site out of several alternatives. This is achieved without individual foragers directly comparing the quality of the sites on offer. Instead, positive feedback mechanisms ensure the collective choice of the best site (Bonabeau et al. 1997; Camazine et al. 2001).

Most studies on the allocation of foragers to food sources have used stable environments in which the feeders or forage sites were kept constant (e.g., Beckers et al. 1990; Seeley et al. 1991; Sumpter and Beekman 2003). However, natural conditions are rarely stable. Can we predict how a particular mechanism will work when conditions are not stable but change dynamically? How should recruitment mechanisms allow the optimal exploitation of food sources when conditions change?

When conditions are stable the optimal solution from the colony's point of view is to focus solely on the best food source (provided this food source is sufficiently large that it allows all recruited foragers to feed at it without crowding). As soon as conditions are unstable, however, it becomes important to have mechanisms that allow swapping to an alternative food source or food sources when they have become more profitable or when the initial food source has been depleted or has become unattractive because of competition with other species or other colonies. This means that in order to do well in a dynamically changing environment, insect colonies must allow storage of information about food patches that are currently exploited while at the same time allowing exploration for new sites, and retaining knowledge about previously profitable sites if they are likely to become profitable again.

At first sight it appears that direct and indirect recruitment mechanisms differ fundamentally in their ability to adapt to changing conditions. In indirect recruitment systems, information about forage sites is retained within the trail. Such systems are resilient to loss of individual foragers. In direct systems a colony can adapt more quickly to changes in the environment, but can render it more vulnerable to the loss of individuals. Consider a honey bee foraging site that is suddenly no longer profitable. Recruitment to that site ceases immediately, as returning foragers cease dancing for that site. Unemployed foragers are then recruited to alternative sites if these are available. The discovery of a new profitable site by just one forager is sufficient to initiate recruitment to that site. Pheromone trails, in contrast, are likely to outlast the food source even though trail pheromones are volatile. Consequently, there is a period in which the trail is still present when the food source has been depleted. Similarly, the formation of another trail to a newly discovered food source is likely to be slow until the old trail has sufficiently evaporated. On the other hand, the removal of a few individuals from an ant trail removes hardly any information, while the removal of a single honey bee returning from a forage site can destroy all information about that site.

The resilience of interaction networks can be visualized by presenting the interactions among individuals, e.g., ants on a trail or bees on the "dance floor" (Seeley 1995), in a network graph. We have done this in Figure 6.2 for mass-recruiting ants (Figure 6.2a) and dancing honey bees (Figure 6.2b). Because the ants mainly interact with each other via the pheromone trail, the network appears unstructured, with most links being between two and ten individuals. A network of honey bee interactions, on the other hand, shows many nodes: individuals that have many interactions with many other individuals. These nodes represent the dancing bees that interact with dance followers. From these interaction networks we can see how the network will react against perturbation. If we remove one node (a dancing bee) from the network, we will greatly affect the network as the removal of a node destroys a large number of interactions. In an unstructured network, as in the ants, removing an individual will hardly affect the structure of the network because an ant only interacts with a small number of ants.

Some forms of recruitment, for example, tandem recruitment, may seem very inefficient, as only one individual is guided toward the food source. However, tandem recruitment certainly is the most reliable of all recruitment mechanisms, as it often comes with a feedback mechanism that prevents information from being lost. As soon as the leading individual, the recruiter, realizes that the follower or followers have lost contact, she will stop and emit a pheromone that will attract the follower or followers (Deneubourg et al. 1986). When an individual of a mass-recruiting species loses the trail, she will not be able to find the food source unless the trail is rediscovered. Similarly, information about the location of food sources is stored in the memory of the individuals in most species of social insects apart from those that rely on mass recruitment only, as here all information

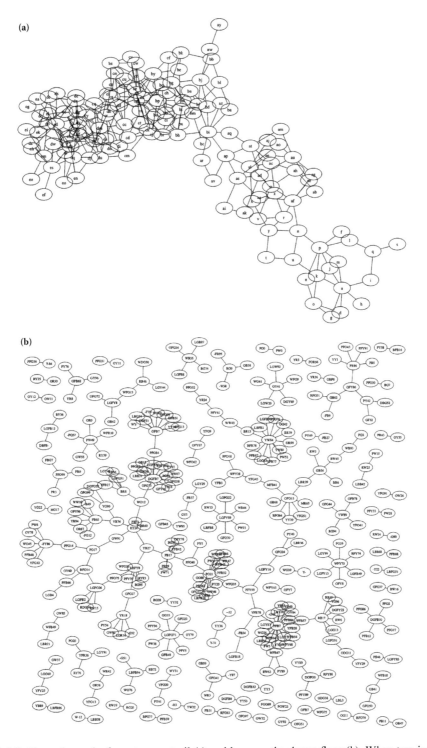

FIGURE 6.2 Network graphs for ants on a trail (a) and bees on the dance floor (b). When two individuals interact physically (ants) or when one individual follows a dance of another (bees), the two are connected by a line. Individuals are represented as ellipses, whereas the strings in the circles identify them. It is clear that most ants only interact with a small number of other ants, as the main means of interaction is via the trail. This contrasts with dancing bees, where some bees have many interactions as their dances are followed by a large number of other bees, which may dance in turn if they find the advertised food source.

is contained in the trail. Has the trail disappeared? Then so has the information of the food source it was leading to.

A comparative study by Beckers et al. (1990) indeed seems to suggest that the ability of ant colonies to adapt to changing foraging conditions depends on the type of recruitment employed. When a better-quality food source is introduced while a colony is already exploring a source of mediocre quality, *Tetramorium caespitum* (which uses a combination of group and mass recruitment) rapidly reallocates its foragers to the better food source. In contrast, *Lasius niger* (which depends solely on mass recruitment) is unable to shift its foraging efforts away from the mediocre food source. Likewise, workers of the stingless bee *Trigona recursa*, which uses scent trails for the recruitment of nestmates to food sources, failed to shift to a better food source when this became available after the bees were already foraging at a food source of lesser quality (Schmidt et al. 2006). The flexibility of *Tetramorium caespitum* colonies can be explained by two characteristics of its recruitment strategy. First, as soon as one individual has discovered the better food source, it can directly lead nestmates to that food source via group recruitment. This enables foragers to ignore the existing trail toward the less profitable food source. Second, individual workers modulate the amount of pheromone they deposit, depending on the quality of the food source found (measured in concentration of sugar) (Verhaeghe 1982). Hence, a pheromone trail initiated via group recruitment will rapidly build up to a better food source, resulting in a shift in the workers' collective foraging behavior without the system getting "trapped" into an existing pheromone trail toward a poorer food source.

A different way to achieve such flexibility is to use different chemical signals: a short-lasting trail pheromone and a long-lasting exploration pheromone. The latter acts as an external memory of the foraging environment (see also Chapter 2) and allows the colony to rapidly establish a new trail, whereas the former will quickly evaporate if it is no longer reenforced (*Pheidole megacephala*, Dussutour et al., unpublished data). In the Pharaoh's ants, *Monomorium pharaonis*, foragers can deposit a "no entry" signal when they return from an unrewarding food source, thereby allowing the colony to quickly reallocate its foragers (Robinson et al. 2005). These two examples indicate that ant colonies most likely have evolved multiple mechanisms for adapting to changing foraging conditions. The challenge now is to understand how different mechanisms have been shaped by natural selection to function optimally under particular ecological conditions.

STAYING INFORMED: SCOUTS VERSUS RECRUITS

The key to keeping track of changing conditions while gaining maximum advantage of existing food sources is the trade-off between exploitation and exploration: the use of existing information (exploitation) versus the collection of new information (exploration). How do different recruitment mechanisms allow for the discovery of new food sources? We will again discuss the two extreme recruitment mechanisms, trail-based foraging and the honey bee's dance language, and explore how both allow the discovery of new food sources.

The success of the pheromone trail mechanism is likely to be due, at least in part, to the nonlinear response of ants to pheromone trails where, for example, the distance that an ant follows a trail before leaving it is a saturating function of the concentration of the pheromone (Pasteels et al. 1986). In other words, the probability an ant will follow a trail is a function of trail strength (expressed as concentration of pheromone: the better the food source, the more pheromone the ants deposit; Sumpter and Beekman 2003), but ants never have a zero probability of losing a trail, irrespective of the strength of the trail. There will therefore always be a number of ants that become lost, and their number will be negatively correlated with the strength of the existing trail. Assuming that these lost ants are able to discover new food sources and thus serve as the colony's explorers or scouts, this "strategy of errors" (Deneubourg et al. 1983; Jaffe and Deneubourg 1992) allows the colony to fine-tune the number of scouts depending on the profitability of the food source that has already been exploited.

The regulation of scout numbers in honey bees similarly ensures a balance between the number of individuals allocated to exploration and exploitation. An unemployed forager (an individual that wants to forage but does not yet know where to forage) will first attempt to locate a dance to follow. If this fails because the number of dancers is low, she will leave the colony and search the surroundings, thereby becoming a scout (Beekman et al. 2007). As a result, the number of scouts is high when the colony has not discovered many profitable forage sites, as dancing will then be low, whereas the number of scouts will be low when forage is plentiful and the number of bees performing recruitment dances is high (Seeley 1983; Beekman et al. 2007).

In many social insects division of labor is thought to arise via individual differences in task thresholds—individuals with the lowest threshold for task X will be the first to engage in that task, therefore reducing the chance that other individuals, with a higher threshold for task X, will perform that task. The latter individuals are therefore more likely to perform another task, and division of labor results (Bonabeau et al. 1996; Beshers and Fewell 2001). Often this difference in individual thresholds has a genetic component; in species in which the queen mates multiply, different subfamilies (workers that share the same father) differ in their task thresholds (Robinson and Page 1989). Even though division of labor based on individual or subfamilial differences in thresholds has been shown for many tasks in many social insects (Oldroyd and Fewell 2007), such a mechanism does not seem to be flexible enough for the regulation of scouts and recruits (Beekman et al. 2007, but see Dreller 1998).

In the honey bee in particular, the "failed follower mechanism" provides the colony with the means to rapidly adjust its number of scouts depending on the amount of information available about profitable forage sites. Even when the colony is exploiting profitable patches, there may still be other, undiscovered, profitable sites. As soon as there is a reduction in the number of dances occurring in the colony, the probability that some unemployed foragers are unable to locate a dance increases, and the colony therefore sends out some scouts. Such fluctuations in the number of dances regularly occur in honey bee colonies, even when there is plenty of forage (Beekman et al. 2004).

Honey bees collect food over a vast area, often more than 100 km^2, changing their focus on a daily basis to adjust to the often-rapid changes in foraging conditions (Visscher and Seeley 1982; Schneider 1989; Waddington et al. 1994; Beekman and Ratnieks 2000; Beekman et al. 2004). This ability is the result of a sophisticated communication system in which foragers integrate a large amount of information about the conditions of the patch they themselves are exploiting, as well as information obtained both directly and indirectly from their nestmates. The failed follower mechanism is an elegant example of a regulatory feedback in this communication system (Seeley 1995; Fewell 2003; Sumpter 2005). Scouting is influenced by the amount of information, in the form of recruitment dances, being brought into the colony. When incoming information is low, the number of scouts is upregulated to gather more information. Potential scouts thus obtain a good estimate of the need to scout without leaving their colony.

THE INDIVIDUAL VERSUS THE COLLECTIVE

Bonabeau et al.'s (1997) influential paper on self-organization in social insects provided a clear thesis of the principles of self-organization that underlie many aspects of insect colony organization. Bonabeau et al.'s paper both summarized and stimulated the field of research as to how collective behavior is achieved through feedback mechanisms arising from the activities of individual insects (Camazine et al. 2001). Since then, self-organization principles have been used to study many other aspects than foraging, including nest construction (Camazine 1991; Karsai and Penzes 1993; Deneubourg and Franks 1995; Franks and Deneubourg 1997; Bonabeau et al. 1998), nest site selection (Camazine et al. 1999; Visscher and Camazine 1999; Mallon et al. 2001; Seeley and Visscher 2004; Pratt et al. 2005; Seeley et al. 2006; Visscher 2007; Janson et al. 2007), and coordinated movement (Couzin et al. 2005; Janson et al. 2005; Beekman et al. 2006). These studies have shown how, despite the simplicity of both the individuals and the rules they follow, social insects are capable of choosing the best nest site out of several possibilities, to build architecturally elaborate nests, and to coordinate the movement of largely ignorant individuals.

Despite the considerable progress in self-organization research, it has now become clear that self-organization principles are insufficient to explain many aspects of colony organization because self-organization explicitly ignores complexity and variability at the individual level. Consensus is now emerging that rather than considering individual insects as simple units that achieve complex collective behavior solely through feedback mechanisms, one should instead think of each insect as a relatively complex individual that directly exchanges information with other individuals (Mallon et al. 2001; Seeley and Buhrman 2001).

However, in ants at any rate, it appears that individual complexity decreases with increasing sophistication with respect to recruitment (Deneubourg et al. 1987). Solitary foragers cannot rely on a pheromone trail to find their way to and from the forage sites. Instead, they need to be able to navigate precisely, sometimes literally counting their steps (Wittlinger et al. 2006). Interestingly, in bees the trend seems to be reversed. Honey bees have the largest colonies and the most sophisticated recruitment mechanism, and the individuals are relatively complex, integrating vast amounts of information from different sources (Seeley 1995). This difference between bees and ants is most likely the result, at least in part, of the dimensionality of their foraging environment. Ants, as they do not fly, are faced with a two-dimensional environment that allows them to easily modify the environment, by forming trails, to recruit nestmates. Hence, the more sophisticated the recruitment mechanism, the more it relies on trails, and the easier it is for ants to simply follow the trail without even really knowing where they are going. It is therefore not surprising that ants with the largest colonies and the most elaborate trail networks, the army ants, are almost blind (Gotwald 1995).

Bees, on the other hand, fly to and fro, which makes it far more difficult to modify the environment. Odor-based recruitment in some species of stingless bees (see above) is only effective over relatively short distances (Nieh 2004; see also Chapter 12). Moreover, most bees forage at flowers (but note that there are some obligate necrophages among the stingless bees: Roubik 1982; Camargo and Roubik 1991), and many flowers only produce nectar at a certain time of day. Hence, bees need to not only be able to navigate, using both the sun's location and landmarks, but also know the time. The larger the colony, the further the bees forage and the more resources they need to collect. And when the distance to a particular food source needs to be communicated to nestmates, as in the honey bees' waggle dance, an individual not only needs to be able to measure this distance but also needs to code this into a waggle phase of the correct duration. A solitary foraging bee, on the other hand, even though it still needs to be able to navigate to and fro, does not need all the information processing skills of honey bees.

USING MORE THAN ONE LANGUAGE AND USE OF THE SAME LANGUAGE IN DIFFERENT CONTEXTS

We have already seen that individual ants can modulate their trail-laying behavior depending on the quality of the food source they have found: when the sugar concentration is high, ants will deposit more pheromone than when they return from a poor food source (*Solenopsis saevissima*, Wilson 1962a, 1962b, 1962c; *Monomorium* and *Tapinoma*, Szlep and Jacobi 1967; *Tetramorium impurum*, Verhaeghe 1982; *Myrmica sabuleti*, de Biseau et al. 1991; *Lasius niger*, Beckers et al. 1993; *Acromyrmex lundi*, Roces and Núñez 1993). Some species modulate their trail depending on the level of starvation (*Solenopsis geminata*, Hangartner 1969; *Oecophylla longinoda*, Hölldobler and Wilson 1978), the distance between the nest and the food source (*S. geminata*, Hangartner 1969), the deviation angle from the axis of the nest to the food source (*L. niger*, Beckers et al. 1992; Devigne and Detrain 2006), or the presence of territorial marking (*L. niger*, Devigne et al. 2004; see also Chapter 2). The ants' environment can indirectly affect pheromone trails. For example, trails can decay at different rates depending on the substrate (e.g., roots, log, or rocks), thus affecting the range over which the trail is effective (*S. saevissima*, Wilson 1962a, 1962b, 1962c; *Eciton* sp., Torgerson and Akre 1970; *L. niger*, Detrain et al. 2001; *M. pharaonis*, Jeanson et al. 2003). This can also apply to other information-carrying signals, such as vibrations produced by stridulating ants where the recruitment range might depend on the resonance properties of the substrate (Baroni-Urbani et al. 1988).

Although flexibility can potentially be achieved through the modulation of a single chemical signal, a process based on several signals seems intuitively more reliable. The modulation of a single signal needs to be precisely tuned to allow flexible foraging behavior, something that is difficult when the signal (i.e., the trail pheromone) is deposited on substrates that differ in their adsorption properties (Detrain et al. 2001). So far the use of different chemical signals has been poorly investigated (e.g., invitation behavior or trail pheromones that contain more than one component: Cammaerts-Tricot 1974; Hölldobler 1982a, 1982b; Vander Meer et al. 1988, 1990). However, some authors (Verhaeghe 1982; de Biseau et al. 1991; Dussutour et al., unpublished data) suggest that the use of two signals would allow the colony to adapt rapidly to changes in its environment.

Honey bees modulate their waggle dance depending on the profitability of the food source found. The more profitable the food source, the livelier and longer the dance (Seeley et al. 2000). As a result, bees dancing for highly profitable sites attract more dance followers than those that dance for mediocre sites.

Even though the honey bee's most famous dance is the waggle dance, other dances have been identified (von Frisch 1967). Of these, we only understand the meaning of the tremble dance, the "shaking signal," and the "buzz run." Tremble dances are performed by returning foragers when they are not able to find a receiver bee quickly to which to transfer their nectar. The result of the tremble dance is an increase in the number of receiver bees and a decrease in the number of foragers so as to balance the processing of incoming nectar with the number of available individuals (Seeley et al. 1996). The shaking signal is used by bees to encourage nestmates to greater activity (Seeley et al. 1998). One bee will climb on top of another bee and shake her for a few seconds. Buzz running is used in the context of nest site selection. Bees that have perceived that the decision-making process has come to an end will perform buzz runs on the swarm so that the swarm can prepare itself for the flight toward the new home (Seeley and Tautz 2001).

The same language can be used in different contexts. Ants use pheromone trails and honey bees use dances to recruit nestmates to new nest sites. Interestingly, bees modify their dance behavior when dancing for new nest sites. Whereas they will indicate the quality of a food source by dancing in a more excited way (Seeley et al. 2000), bees returning from a high-quality nest site will perform more waggle phases than bees that returned from a site of only mediocre quality (Seeley 2003). Each bee, irrespective of the quality of the nest site that she has discovered, will reduce the number of waggle phases performed after each return visit to the new nest site (Figure 6.3). As a result, individual bees dance for a limited period of time only (Seeley and Visscher 2004). This is in contrast to bees dancing for food, as a forager will dance for her site for as long as this site is profitable enough

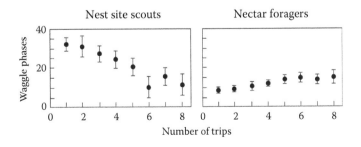

FIGURE 6.3 Number of waggle phases performed by bees that made successive trips to a potential nest site (left) and food (right). Bees visiting a potential nest site show a decline in the number of waggle phases performed, whereas foraging bees do not. (After Visscher 2007. With permission.)

(Seeley 1995) (Figure 6.3). It has been suggested that dance attrition is the crucial factor that allows a swarm to reach a decision about which nest site to move to (Britton et al. 2002; Myerscough 2003). If nest site scouts would not cease dancing, as we see in the context of foraging, all nest sites discovered would be advertised indefinitively and the swarm would not be able to select one only.

The ant *Temnothorax albipennis* uses a dual recruitment mechanism when recruiting to new nest sites. In the beginning of the process nestmates are recruited to potential new homes found by scouts via tandem recruitment. This relatively slow recruitment mechanism is abandoned as soon as the number of ants present in the new nest has reached a certain number (quorum). Once a quorum has been reached, individuals simply carry nestmates and brood to the chosen site, thereby significantly reducing the time needed to move into the new home (Mallon et al. 2001; Pratt et al. 2002, 2005). Such carrying does not take place in the context of foraging and is thus an adaptation to nest site selection.

For both the honey bee and the ant *T. albipennis* the context in which recruitment takes place, foraging or nest site selection, is clear. For honey bees it is clear because prior to selecting a nest site the bees comprising the swarm (the part of the colony that becomes homeless) leave the natal colony and form an interim cluster, while *T. albipennis* colonies tend to move home only when their old nest site has been destroyed. However, in many other social insects it may not be immediately obvious whether recruitment is for forage or nest site selection, especially in those species with colony fission, in which the move to a new nest site is a gradual process that can take days or even weeks. This raises the question of how recruited individuals know whether they are recruited toward a food source or a new nest site.

VARIATION IS THE SPICE OF LIFE

Insect colonies need to collect different resources, mainly carbohydrates (e.g., nectar) and protein (e.g., pollen or other insects). However, we know remarkably little about how a colony allocates its foragers over the two different types of food. Most, if not all, studies have focused on foraging for just one type of food (mostly carbohydrates) while offering the experimental colonies sufficient supplies of the other food source (protein). As protein is mainly used to feed the brood, the amount of brood must play an important role in determining how many individuals will devote their foraging effort to collecting protein (Sorensen et al. 1985). But how does the colony organize the simultaneous recruitment to carbohydrates and protein?

The division between nectar and pollen foragers in honey bees has a strong genetic basis (Calderone and Page 1988, 1991, 1992; Dreller et al. 1995; Page et al. 1995, 1998). Most likely, this genetic component determines an individual's threshold to forage for either pollen or nectar, without completely excluding foraging for both when the need arises (Fewell and Bertram 1999). Honey bees have a relatively simple mechanism that they use to determine if there is a need to forage for pollen. If the pollen stores are above a particular point, the number of pollen foragers is reduced; conversely, if the stores are below that point, their numbers are increased (Fewell and Winston 1992; Camazine 1993). The honey bees' direct recruitment mechanism makes it relatively easy to simultaneously recruit to pollen and nectar at the colony level. Both nectar and pollen foragers will dance on the dance floor, and unemployed foragers that are stimulated to forage for nectar or pollen can simply choose to follow either a nectar or pollen dance (Oldroyd et al. 1991). Hence, the only direct competition between nectar and pollen foraging is via the number of available foragers, but it seems unlikely that their numbers are ever limiting.

Specialization on foraging for either protein or carbohydrates appears to be common in the Formicinae ants (Higashi 1978; Wehner et al. 1983; Traniello et al. 1991; Sundström 1993; Quinet and Pasteels 1996), but for all other ants we do not know how they are able to balance their nutrient intake. Although adult ants need carbohydrates as a source of energy, brood rely mainly on proteins (Sorensen and Vinson 1981; Cassill and Tschinkel 1999). We thus expect the ants' foraging patterns to track colony demography and larval growth. Thus, when the amount of brood is high, colonies focus more on collecting protein than when the amount of brood is low (Dussutour and Simpson 2008).

TRAFFIC JAMS

Because ants move through two-dimensional space, they are vulnerable to traffic jams, for example, when some sections of the trail are too narrow to allow a high volume of traffic or debris obstructs traffic flow. Overcrowding can slow down the speed of the ants and thus lead to a decrease in overall traffic flow (Burd et al. 2002; Burd and Aranwela 2003; Dussutour et al. 2005, 2006, 2007). Ants have developed particular strategies to prevent traffic congestion on recruitment trails. On long-lasting recruitment trails (trunk trails) they can remove debris or obstacles from the trail (Howard 2001) or widen the trail to allow for a higher traffic flow (Berghoff et al. 2002; for termites, see also Bruinsma 1979). On more ephemeral trails congestion can be reduced without decreasing the overall traffic flow. If the trail is wide enough, ants can be spatially organized with a distinct central lane of returning laden foragers flanked by two lanes of outbound foragers (army ants, Couzin and Franks 2003; leaf-cutting ants, Dussutour 2004). When the trail is too narrow to allow such spatial segregation, either the flow can be temporarily organized with alternating clusters of inbound and outbound ants (Dussutour et al. 2005) or part of the traffic can be diverted onto another path (Dussutour et al. 2004).

Interestingly, in leaf-cutting ants the foraging efficiency of the colony increases when individual ants experience crowded conditions on the trail. The number of leaf fragments brought back to the colony per unit time was shown to increase when the ants were forced to forage on a narrow trail compared with when they used a wide trail (Dussutour et al. 2007). This may seem puzzling, as the width of the trail strongly affects the traffic flow, with more ants being able to use a wide trail than a narrow trail. It seems that the enforced close encounters between empty (outbound) ants and laden (inbound) ants inform the outbound ants about the presence of food and most likely increase their motivation to forage.

The honey bees' direct recruitment mechanism does not lead to traffic jams, irrespective of the number of dancing bees, because there is never a "bottleneck" in the system. However, as we discussed earlier, the flux of returning nectar foragers can become so high that there are insufficient numbers of receiver bees to handle the incoming nectar. If this is the case, returning foragers will no longer recruit more foragers, but receiver bees until the balance has been reestablished. Because pollen foragers deposit their pollen load in the cells directly, there is no such feedback mechanism for pollen collection. Instead, pollen foragers will cease collecting pollen when the amount of stored pollen has exceeded their individual threshold (Dreller et al. 1999).

CONCLUDING REMARKS

By dividing the different recruitment mechanisms into two broad categories, direct and indirect, we can sum up the benefits and costs of both (Table 6.1). Obviously, such categorization is a gross simplification, as within each category we find many different mechanisms. Moreover, within the same category, for example, mass-recruiting ants, we find idiosyncrasies that may reduce a specific cost under certain conditions. We have seen that although trail recruitment may result in ants getting "stuck" in a suboptimal solution because they are unable to establish a trail to a better food source, some species can circumvent this problem by using two different pheromones or by adjusting the amount of pheromone produced, depending on the quality of the food source. Honey bees not only rely on dance information but also remember the location and smell of forage sites visited previously. How many sites they can remember is unknown, and do not know how they balance private information (individual memory) with public information (dance communication) (Grüter et al. 2008).

It seems likely that the exact recruitment mechanism employed by a particular species is tuned to the environment in which it evolved. So far, however, the types of environments in which recruitment

TABLE 6.1
Broad Benefits and Costs of Indirect and Direct Recruitment Mechanisms

Recruitment Mechanism	Benefits	Costs
Indirect mechanisms	Robust to the removal of individuals (persistence of information)	Information (trail) can be altered by the environment
	Ability to focus on best food source under static conditions	Less flexible under dynamic conditions without extra mechanisms; trapped in suboptimal solution
	Rapid exploitation at food source (amplification phenomena)	Constraints on minimum colony size
	Reduction in individual complexity (ants)	
Direct mechanisms	Well-adapted to dynamic conditions without the need for extra mechanisms	Unable to focus on one source only under static conditions
	No constraints on colony size	Sensitive to the removal of individuals
		Slow buildup to food source

Note: Indirect mechanisms include foraging trails, while direct mechanisms include the honey bees' dance language as well as tandem and group recruitment. See text for more details.

mechanisms have been studied are rather limited, and experiments have mainly focused on static environments. We thus have inadequate knowledge about the capabilities of social insects to optimally adapt to changes in their environment. Thus, the time has come to study precisely the extent to which social insects with different recruitment mechanisms are capable of solving more realistic foraging problems. Only by offering species with distinct recruitment mechanisms similar foraging problems can we truly identify the costs and benefits of different communication mechanisms in the social insects.

REFERENCES

Azcárate FM, Peco B. (2003). Spatial patterns of seed predation by harvester ants (*Messor* Forel) in Mediterranean grassland and scrubland. *Insect Soc* 50:120–26.

Baroni-Urbani C, Buser MW, Schilliger E. (1988). Substrate vibration during recruitment in ant social organization. *Insect Soc* 35:241–50.

Beckers R, Goss S, Deneubourg JL, Pasteels JM. (1989). Colony size, communication and ant foraging strategy. *Psyche* 96:239–56.

Beckers R, Deneubourg JL, Goss S. (1992). Trails and U-turns in the selection of a path by the ant *Lasius niger*. *J Theor Biol* 159:397–415.

Beckers R, Deneubourg JL, Goss S. (1993). Modulation of trail laying in the ant *Lasius niger* (Hymenoptera: Formicidae) and its role in the collective selection of a food source. *J Insect Behav* 6:751–59.

Beckers R, Deneubourg JL, Gross S, Pasteels JM. (1990). Collective decision making through food recruitment. *Insect Soc* 37:258–67.

Beekman M. (2005). How long will honey bees (*Apis mellifera* L.) be stimulated by scent to revisit past-profitable forage sites? *J Comp Physiol A* 191:1115–20.

Beekman M, Fathke RL, Seeley TD. (2006). How does an informed minority of scouts guide a honeybee swarm as it flies to its new home? *Anim Behav* 71:161–71.

Beekman M, Gilchrist AL, Duncan M, Sumpter DJT. (2007). What makes a honeybee scout? *Behav Ecol Sociobiol* 61:985–95.

Beekman M, Gloag RS, Even N, Wattanachaiyingcharoen W, Oldroyd BP. (2008). Dance precision of *Apis florea*—Clues to the evolution of the honey bee dance language? *Behav Ecol Sociobiol* DOI:10.1007/s00265-008-0554-z.

Beekman M, Lew JB. (2008). Foraging in honeybees—When does it pay to dance? *Behav Ecol* 19:255–62.

Beekman M, Ratnieks FLW. (2000). Long range foraging by the honey-bee *Apis mellifera* L. *Funct Ecol* 14: 490–96.

Beekman M, Sumpter DJT, Ratnieks FLW. (2001). Phase transition between disordered and ordered foraging in Pharaoh's ants. *Proc Natl Acad Sci USA* 98:9703–6.

Beekman M, Sumpter DJT, Seraphides N, Ratnieks FLW. (2004). Comparing foraging behaviour of small and large honey-bee colonies by decoding waggle dances made by foragers. *Funct Ecol* 18: 829–35.

Berghoff SM, Weissflog A, Linsenmair KE, Hashim R, Maschwitz U. (2002). Foraging of a hypogaeic army ant: A long neglected majority. *Insect Soc* 49:133–41.

Beshers SN, Fewell JH. (2001). Models of division of labor in social insects. *Ann Rev Entomol* 46:413–40.

de Biseau JC, Deneubourg JL, Pasteels JM. (1991). Collective flexibility during mass recruitment in the ant *Myrmica sabuleti* (Hymenoptera: Formicidae). *Psyche* 98:323–36.

de Biseau JC, Schuiten M, Pasteels JM, Deneubourg JL. (1994). Respective contributions of leader and trail during recruitment to food in *Tetramorium bicarinatum* (Hymenoptera: Formicidae). *Insect Soc* 41: 241–54.

Bonabeau E, Theraulaz G, Deneubourg JL. (1996). Quantitative study of the fixed threshold model for the regulation of division of labour in insect societies. *Proc R Soc Lond B* 263:1565–69.

Bonabeau E, Theraulaz G, Deneubourg JL, Aron S, Camazine S. (1997). Self-organization in social insects. *Trends Ecol Evol* 12:188–93.

Bonabeau E, Theraulaz G, Deneubourg JL, Franks NR, Rafelsberger O, Joly JL, Blanco S. (1998). A model for the emergence of pillars, walls and royal chambers in termite nests. *Philos Trans R Soc Lond B* 353: 1561–76.

Britton NF, Franks NR, Pratt SC, Seeley TD. (2002). Deciding on a new home: How do honeybees agree? *Proc R Soc Lond B* 269:1383–88.

Bruinsma OH. (1979). An analysis of building behaviour of the termite *Macrotermes subhyalinus*. Doctoral thesis, Agricultural University of Wageningen, The Netherlands.

Burd M, Aranwela N. (2003). Head-on encounter rates and walking speed of foragers in leaf-cutting ant traffic. *Insect Soc* 50:3–8.

Burd M, Archer D, Aranwela N, Stradling DJ. (2002). Traffic dynamics of the leaf-cutting ant, *Atta cephalotes*. *Am Nat* 159:283–93.

Calderone NW, Page RE. (1988). Genotypic variability in age polyethism and task specialization in the honey bee, *Apis mellifera* (Hymenoptera: Apidae). *Behav Ecol Sociobiol* 22:17–25.

Calderone NW, Page RE. (1991). Evolutionary genetics of division of labor in colonies of the honey bee (*Apis mellifera*). *Am Nat* 138:69–92.

Calderone NW, Page RE. (1992). Effects of interactions among genotypically diverse nestmates on task specialization by foraging honey bees (*Apis mellifera*). *Behav Ecol Sociobiol* 30:219–26.

Camargo JMF, Roubik DW. (1991). Systematics and bionomics of the apoid obligate necrophages: The *Trigona hypogea* group (Hymenoptera: Apidae; Melipononae). *Biol J Linnean Soc* 44:13–40.

Camazine S. (1991). Self-organizing pattern formation on the combs of honey bee colonies. *Behav Ecol Sociobiol* 28:61–76.

Camazine S. (1993). The regulation of pollen foraging by honey bees: How foragers assess the colony's need for pollen. *Behav Ecol Sociobiol* 32:265–72.

Camazine S, Deneubourg JL, Franks NR, Sneyd J, Theraulaz G, Bonabeau E. (2001). *Self-Organization in Biological Systems*. Princeton, NJ: Princeton University Press.

Camazine S, Visscher PK, Finley J, Vetter RS. (1999). House-hunting by honey bee swarms: Collective decisions and individual behaviors. *Insect Soc* 46:348–62.

Cammaerts-Tricot MC. (1974). Piste et pheromone attractive chez la fourmi *Myrmica rubra*. *J Comp Physiol A* 88:373–82.

Cassill DL, Tschinkel WR. (1999). Regulation of diet in the fire ant, *Solenopsis invicta*. *J Insect Behav* 12: 307–28.

Couzin ID, Franks NR. (2003). Self-organized lane formation and optimized traffic flow in army ants. *Proc R Soc Lond B* 270:139–46.

Couzin ID, Krause J, Franks NR, Levin SA. (2005). Effective leadership and decision-making in animal groups on the move. *Nature* 433:513–16.

Deneubourg JL, Aron S, Goss S, Pasteels JM, Duerinck G. (1986). Random behaviour, amplification processes and number of participants: How they contribute to the foraging properties of ants. *Physica D* 2: 176–86.

Deneubourg JL, Franks NR. (1995). Collective control without explicit coding: The case of communal nest excavation. *J Insect Behav* 8:417–32.

Deneubourg JL, Goss S, Pasteels JM, Fresneau D, Lachaud JP. (1987). Self-organization mechanisms in ant societies (II): Learning in foraging and division of labor. In Pasteels JM, Deneubourg JL (eds.), *From Individual to Collective Behavior in Social Insects*. Basel: Birkhäuser, pp. 177–96.

Deneubourg JL, Pasteels JM, Verhaeghe JC. (1983). Probabilistic behaviour in ants: A strategy of errors? *J Theor Biol* 105:259–71.

Detrain C, Deneubourg JL. (2002). Complexity of environment and parsimony of decision rules in insect societies. *Biol Bull* 202:268–74.

Detrain C, Deneubourg JL, Goss S, Quinet Y. (1991). Dynamics of collective exploration in the ant *Pheidole pallidula. Psyche* 98:21–31.

Detrain C, Natan C, Deneubourg JL. (2001). The influence of the physical environment on the self-organised foraging patterns of ants. *Naturwissenschaften* 88:171–74.

Detrain C, Pasteels JM. (1991). Caste differences in behavioral thresholds as a basis for polyethism during food recruitment in the ant, *Pheidole pallidula* (Nyl.) (Hymenoptera: Myrmicinae). *J Insect Behav* 4: 157–76.

Detrain C, Tasse O, Versaen M, Pasteels JM. (2000). A field assessment of optimal foraging in ants: Trail patterns and seed retrieval by the European harvester ant *Messor barbarus. Insect Soc* 47:56–62.

Devigne C, Detrain C. (2006). How does food distance influence foraging in the ant *Lasius niger*: The importance of home-range marking. *Insect Soc* 53:46–55.

Devigne C, Renon AJ, Detrain C. (2004). Out of sight but not out of mind: Modulation of recruitment according to home range marking in ants. *Anim Behav* 67:1023–29.

Dornhaus A, Chittka L. (1999). Evolutionary origins of bee dances. *Nature* 401:38.

Dornhaus A, Chittka L. (2004). Why do honey bees dance? *Behav Ecol Sociobiol* 55:395–401.

Dreller C. (1998). Division of labor between scouts and recruits: Genetic influence and mechanisms. *Behav Ecol Sociobiol* 43:191–96.

Dreller C, Fondrk MK, Page RE. (1995). Genetic variability affects the behavior of foragers in a feral honeybee colony. *Naturwissenschaften* 82:243–45.

Dreller C, Page RE, Fondrk MK. (1999). Regulation of pollen foraging in honeybee colonies: Effects of young brood, stored pollen, and empty space. *Behav Ecol Sociobiol* 45:227–33.

Dussutour A. (2004). Organisation spatio-temporelle des déplacements collectifs chez les fourmis. Doctoral thesis, University of Toulouse III, France.

Dussutour A, Beshers S, Deneubourg JL, Fourcassié V. (2007). Crowding increases foraging efficiency in the leaf-cutting ant *Atta colombica. Insect Soc* 54:158–65.

Dussutour A, Deneubourg JL, Fourcassié V. (2005). Temporal organization of bi-directional traffic in the ant *Lasius niger* (L). *J Exp Biol* 208:2903–12.

Dussutour A, Fourcassié V, Helbing D, Deneubourg JL. (2004). Optimal traffic organization in ants under crowded conditions. *Nature* 428:70–73.

Dussutour A, Nicolis SC, Deneubourg JL, Fourcassié V. (2006). Collective decisions in ants when foraging under crowded conditions. *Behav Ecol Sociobiol* 61:17–30.

Dussutour A, Simpson SJ. (2008). Carbohydrate regulation in relation to colony growth in ants. *J Exp Biol.* 211:2224–32.

Fewell JH. (1988). Energetic and time costs of foraging in harvester ants, *Pogonomyrmex occidentalis. Behav Ecol Sociobiol* 22:401–8.

Fewell JH. (2003). Social insect networks. *Science* 301:1867–70.

Fewell JH, Bertram SM. (1999). Division of labor in a dynamic environment: Response by honeybees (*Apis mellifera*) to graded changes in colony pollen stores. *Behav Ecol Sociobiol* 46:171–79.

Fewell JH, Winston ML. (1992). Colony state and regulation of pollen foraging in the honey bee, *Apis mellifera* L. *Behav Ecol Sociobiol* 30:387–93.

Fourcassie V, Deneubourg JL. (1994). The dynamics of collective exploration and trail-formation in *Monomorium pharaonis:* Experiments and model. *Physiol Entomol* 19:291–300.

Fowler HG, Robinson SW. (1979). Foraging by *Atta sexdens* (Formicidae: Attini): Seasonal patterns, caste and efficiency. *Ecol Entomol* 4:239–47.

Franks NR, Deneubourg JL. (1997). Self-organizing nest construction in ants: Individual worker behaviour and the nest's dynamics. *Anim Behav* 54:779–96.

von Frisch K. (1967). *The Dance Language and Orientation of Bees*. Cambridge, MA: Harvard University Press.

Gotwald WH. (1995). *Army Ants—The Biology of Social Predation.* Ithaca, NY: Cornell University Press.

Grassé PP. (1986). *Termitologia, Tome 3: Comportement, Socialité, Ecologie, Evolution, Systematique.* Paris: Masson.

Grüter C, Balbuena MS, Farina WM. (2008). Informational conflicts created by the waggle dance. *Proc R Soc London B* 275:1321–27.

Hangartner W. (1969). Trail laying in the subterranean ant, *Acanthomyops interjectus. J Insect Physiol* 15:1–4.

Higashi S. (1978). Task and areal conservatism and internest drifting in a red wood ant *Formica (Formica) yessensis* Forel. *Jpn J Ecol* 28:307–17.

Hölldobler B. (1971). Recruitment behavior in *Camponotus socius* (Hym. Formicidae). *Z vergl Physiol* 75:123–42.

Hölldobler B. (1976). Recruitment behavior, home range orientation and territoriality in harvester ants, *Pogonomyrmex. Behav Ecol Sociobiol* 1:3–44.

Hölldobler B. (1981). Foraging and spatiotemporal territories in the honey ant *Myrmecocystus mimicus* Wheeler (Hymenoptera: Formicidae). *Behav Ecol Sociobiol* 9:301–14.

Hölldobler B. (1982a). Chemical communication in ants: New exocrine glands and their behavioral function. In *Proceedings of 9th International Congress of IUSSI*, Boulder, pp. 312–17.

Hölldobler B. (1982b). The cloacal gland, a new pheromone gland in ants. *Naturwissenschaften* 69:186–87.

Hölldobler B, Möglich M. (1980). The foraging system of *Pheidole militicida* (Hymenoptera: Formicidae). *Insect Soc* 27:237–64.

Hölldobler B, Wilson EO. (1978). The multiple recruitment systems of the African weaver ant *Oecophylla longinoda* (Latreille) (Hymenoptera: Formicidae). *Behav Ecol Sociobiol* 3:19–60.

Hölldobler B, Wilson EO. (1990). *The Ants.* Cambridge, MA: Harvard University Press.

Howard JJ. (2001). Costs of trail construction and maintenance in the leaf-cutting ant *Atta columbica. Behav Ecol Sociobiol* 49:348–56.

Ito F. (1993). Observation of group recruitment to prey in a primitive ponerine ant, *Amblyopone* sp. (*reclinata* group) (Hymenoptera: Formicidae). *Insect Soc* 40:163–67.

Jaffe K, Deneubourg JL. (1992). On foraging, recruitment systems and optimum number of scouts in eusocial colonies. *Insect Soc* 39:201–13.

Janson S, Middendorf M, Beekman M. (2005). Honeybee swarms: How do scouts guide a swarm of uninformed bees? *Anim Behav* 70: 349–58.

Janson S, Middendorf M, Beekman M. (2007). Searching for a new home—Scouting behavior of honeybee swarms. *Behav Ecol* 18:384–92.

Jeanson R, Ratnieks FLW, Deneubourg JL. (2003). Pheromone trail decay rates on different substrates in the Pharaoh's ant, *Monomorium pharaonis. Physiol Entom* 28:192–98.

Karsai I, Penzes Z. (1993). Comb building in social wasps: Self-organization and stigmergic script. *J Theor Biol* 161:505–25.

Lenoir A, Jaisson P. (1982). Evolution et rôle des communications antennaires chez les insectes sociaux. In Jaisson P (ed.), *Social Insects in the Tropics.* Paris : Université Paris-Nord, pp. 157–80.

Lindauer M. (1955). Schwarmbienen auf Wohnungssuche. *Z vergl Physiol* 37:263–324.

Lindauer M, Kerr WE. (1958). Die gegenseitige Verständigung bei den stachellosen Bienen. *Z vergl Physiol* 41:405–34.

López F, Acosta FJ, Serrano JM. (1994). Guerilla vs. phalanx strategies of resource capture: Growth and structural plasticity in the trunk trail system of the harvester ant *Messor barbarus. J Anim Ecol* 63:127–38.

Mallon EB, Pratt SC, Franks NR. (2001). Individual and collective decision-making during nest site selection by the ant *Leptothorax albipennis. Behav Ecol Sociobiol* 50:352–59.

Michener CD. (1974). *The Social Behavior of the Bees: A Comparative Study.* Cambridge, MA: Harvard University Press.

Michener CD. (2000). *The Bees of the World.* Baltimore: Johns Hopkins University Press.

Millor J, Pham-Delègue M, Deneubourg JL, Camazine S. (1999). Self-organized defensive behavior in honeybees. *Proc Natl Acad Sci USA* 96:12611–615.

Möglich M, Hölldobler B. (1975). Communication and orientation during foraging and emigration in the ant *Formica fusca. J Comp Physiol* 101:275–88.

Myerscough MR. (2003). Dancing for a decision: A matrix model for nest-site choice by honeybees. *Proc R Soc Lond B* 270:577–82.

Nieh JC. (2004). Recruitment communication in stingless bees (Hymenoptera, Apidae, Meliponini). *Apidologie* 35:159–82.

van Nieuwstadt MGL, Ruano ICE. (1996). Relation between size and foraging range in stingless bees (Apidae, Meliponinae). *Apidologie* 27:219–28.

Oldroyd BP, Fewell JH. (2007). Genetic diversity promotes homeostasis in insect colonies. *Trends Ecol Evol* 22:408–13.

Oldroyd BP, Rinderer TE, Buco SM. (1991). Intracolonial variance in honey bee foraging behaviour: The effects of sucrose concentration. *J Apic Res* 30:137–45.

Oldroyd BP, Wongsiri S. (2006). *Asian Honey Bees. Biology, Conservation, and Human Interactions.* Cambridge, MA: Harvard University Press.

Page RE, Erber J, Fondrk MK. (1998). The effect of genotype on response threshold to sucrose and foraging behavior of honey bees (*Apis mellifera* L.). *J Comp Physiol A* 182:489–500.

Page RE, Waddington KD, Hunt GJ, Fondrk MK. (1995). Genetic determinants of honey bee foraging behaviour. *Anim Behav* 50:1617–25.

Pasteels JM, Deneubourg JL, Verhaeghe JC, Boevé JL, Quinet Y. (1986). Orientation along terrestrial trails by ants. In Payne TL, Birch MC, Kennedy CEJ (eds.), *Mechanisms in Insect Olfaction.* Oxford: Oxford University Press, pp. 131–38.

Pratt SC, Mallon EB, Sumpter DJT, Franks NR. (2002). Quorum sensing, recruitment, and collective decision-making during colony emigration by the ant *Leptothorax albipennis. Behav Ecol Sociobiol* 52: 117–27.

Pratt SC, Sumpter DJT, Mallon EB, Franks NR. (2005). An agent-based model of collective nest choice by the ant *Temnothorax albipennis. Anim Behav* 70:1023–36.

Quinet Y, de Biseau JC, Pasteels JM. (1997). Food recruitment as a component of the trunk-trail foraging behaviour of *Lasius fuliginosus* (Hymenoptera: Formicidae). *Behav Process* 40:75–83.

Quinet Y, Pasteels JM. (1996). Spatial specialization of the foragers and foraging strategy in *Lasius fuliginosus* (Latreille) (Hymenoptera, Formicidae). *Insect Soc* 43:333–46.

Reinhard J, Srinivasan MV, Zhang S. (2004). Scent-triggered navigation in honeybees. *Nature* 427:411.

Robinson EJH, Jackson DE, Holcombe M, Ratnieks FLW. (2005). "No entry" signal in ant foraging. *Nature* 438:442.

Robinson GE, Page RE. (1989). Genetic basis for division of labor in an insect society. In Breed MD, Page RE (eds.), *The Genetics of Social Evolution.* Boulder: Westview Press, pp. 61–68.

Roces F, Núñez JA. (1993). Information about food quality influences load-size selection in recruited leaf-cutting ants. *Anim Behav* 45:135–43.

Rohrseitz K, Tautz J. (1999). Honey bee dance communication: Waggle run direction coded in antennal contacts? *J Comp Physiol A* 184:463–70.

Rosengren R, Sundström L. (1987). The foraging system of a red wood ant colony (*Formica* s. str.)—Collecting and defending food through an extended phenotype. In Pasteels JM, Deneubourg JL (eds.), *From Individual to Collective Behavior in Social Insects.* Basel: Birkhäuser, pp. 117–37.

Roubik DW. (1982). Obligate necrophagy in a social bee. *Science* 217:1059–60.

Roubik DW, Aluja M. (1983). Flight ranges of *Melipona* and *Trigona* in tropical forest. *J Kansas Entomol Soc* 56:217–22.

Schmidt VM, Schorkopf DLP, Hrncir M, Zucchi R, Barth FG. (2006). Collective foraging in a stingless bee: Dependence on food profitability and sequence of discovery. *Anim Behav* 72:1309–17.

Schneider SS. (1989). Spatial foraging patterns of the African honey bee, *Apis mellifera scutellata. J Insect Behav* 2:505–21.

Seeley TD. (1983). Division of labor between scouts and recruits in honeybee foraging. *Behav Ecol Sociobiol* 12:253–59.

Seeley TD. (1995). *The Wisdom of the Hive: The Social Physiology of Honey Bee Colonies.* Cambridge, MA: Harvard University Press.

Seeley TD. (2003). Consensus building during nest-site selection in honey bee swarms: The expiration of dissent. *Behav Ecol Sociobiol* 53:417–24.

Seeley TD, Buhrman SC. (2001). Nest-site selection in honey bees: How well do swarms implement the "best-of-N" decision rule? *Behav Ecol Sociobiol* 49:416–27.

Seeley TD, Camazine S, Sneyd J. (1991). Collective decision-making in honey bees: How colonies choose among nectar sources. *Behav Ecol Sociobiol* 28:277–90.

Seeley TD, Kühnholz S, Weidenmüller A. (1996). The honey bee's tremble dance stimulates additional bees to function as nectar receivers. *Behav Ecol Sociobiol* 39:419–27.

Seeley TD, Mikheyev AS, Pagano GJ. (2000). Dancing bees tune both duration and rate of waggle-run production in relation to nectar-source profitability. *J Comp Physiol A* 186:813–19.

Seeley TD, Tautz J. (2001). Worker piping in honey bee swarms and its role in preparing for liftoff. *J Comp Physiol A* 187:667–76.

Seeley TD, Visscher PK. (2004). Group decision making in nest-site selection by honey bees. *Apidologie* 35:101–16.

Seeley TD, Visscher PK, Passino KM. (2006). Group decision making in honey bee swarms. *Am Sci* 94:220–29.

Seeley TD, Weidenmüller A, Kühnholz S. (1998). The shaking signal of the honey bee informs workers to prepare for greater activity. *Ethology* 104:10–26.

Shepherd JD. (1982). Trunk trails and the searching strategy of a leaf-cutter ant, *Atta columbica. Behav Ecol Sociobiol* 11:77–84.

Shepherd JD. (1985). Adjusting foraging effort to resources in adjacent colonies of the leaf-cutter ant, *Atta colombica. Biotropica* 17:245–52.

Sherman G, Visscher PK. (2002). Honeybee colonies achieve fitness through dancing. *Nature* 419:920–22.

Sorensen AA, Busch TM, Vinson SB. (1985). Control of food influx by temporal subcastes in the fire ant, *Solenopsis invicta. Behav Ecol Sociobiol* 17:191–98.

Sorensen AA, Vinson SB. (1981). Quantitative food distribution studies within laboratory colonies of the imported fire ant, *Solenopsis invicta* Buren. *Insect Soc* 28:129–60.

Sumpter DJT. (2005). The principles of collective animal behaviour. *Phil Trans R Soc Lond B* 361:5–22.

Sumpter DJT, Beekman M. (2003). From non-linearity to optimality: Pheromone trail foraging by ants. *Anim Behav* 66:273–80.

Sundström L. (1993). Foraging responses of *Formica truncorum* (Hymenoptera: Formicidae); exploiting stable vs. spatially and temporally variable resources. *Insect Soc* 40:147–61.

Szlep R, Jacobi T. (1967). The mechanism of recruitment to mass foraging in colonies of *Monomorium venustum* Smith, *M. subopacum* ssp. *phoenicium* Em., *Tapinoma israelis* For. and *T. simrothi* v. *phoenicium* Em. *Insect Soc* 14:25–40.

Tautz J, Rohrseitz K. (1998). What attracts honey bees to a waggle dancer? *J Comp Physiol A* 183:661–67.

Tautz J, Rohrseitz K, Sandeman DC. (1996). One-strided waggle dance in bees. *Nature* 382:32.

Thom C, Gilley DC, Hooper J, Esch HE. (2007). The scent of the waggle dance. *PLoS Biology* 5:e228. DOI:10.1371/journal.pbio.0050228.

Topoff H, Mirenda J. (1978). Precocial behaviour of callow workers of the army ant *Neivamyrmex nigrescens*: Importance of stimulation by adults during mass recruitment. *Anim Behav* 26:698–706.

Torgerson RL, Akre RD. (1970). The persistence of army ant chemical trails and their significance in the ecitonine-ecitophile association (Formicidae: Ecitonini). *Melanderia* 5:1–28.

Traniello JFA. (1977). Recruitment behavior, orientation, and the organization of foraging in the carpenter ant *Camponotus pennsylvanicus* DeGeer (Hymenoptera: Formicidae). *Behav Ecol Sociobiol* 2: 61–79.

Traniello JFA, Fourcassié V, Graham TP. (1991). Search behavior and foraging ecology of the ant *Formica schaufussi*: Colony-level and individual patterns. *Ethol Ecol Evol* 3:35–47.

Vander Meer RK, Alvarez F, Lofgren CS. (1988). Isolation of the trail recruitment pheromone of *Solenopsis invicta. J Chem Ecol* 14:825–38.

Vander Meer RK, Lofgren CS, Alvarez F. (1990). The orientation inducer pheromone of the fire ant *Solenopsis invicta. Physiol Entomol* 15:483–88.

Verhaeghe JC. (1977). Group recruitment in *Tetramorium caespitum*. In *Proceedings of the 8th International Congress of IUSSI*, pp. 67–68.

Verhaeghe JC. (1982). Food recruitment in *Tetramorium impurum* (Hymenoptera: Formicidae). *Insect Soc* 29:67–85.

Visscher PK. (2007). Group decision making in nest-site selection among social insects. *Annu Rev Entomol* 52:255–75.

Visscher PK, Camazine S. (1999). Collective decisions and cognition in bees. *Nature* 397:400.

Visscher PK, Seeley TD. (1982). Foraging strategy of honeybee colonies in a temperate deciduous forest. *Ecology* 63:1790–801.

Waddington KD, Herbert TJ, Visscher PK, Raveret Richter M. (1994). Comparisons of forager distributions from matched honey bee colonies in suburban environments. *Behav Ecol Sociobiol* 35:423–29.

Wehner R, Harkness RD, Schmid-Hempel P. (1983). *Foraging Strategies in Individually Searching Ants, Cataglyphis bicolor (Hymenoptera: Formicidae)*. Stuttgart: Gustav Fischer Verlag.

Wilson EO. (1962a). Chemical communication among workers of the fire ant *Solenopsis saevissima* (Fr. Smith). 1. The organization of mass-foraging. *Anim Behav* 10:134–47.

Wilson EO. (1962b). Chemical communication among workers of the fire ant *Solenopsis saevissima* (Fr. Smith). 2. An information analysis of the odour trail. *Anim Behav* 10:148–58.

Wilson EO. (1962c). Chemical communication among workers of the fire ant *Solenopsis saevissima* (Fr. Smith). 3. The experimental induction of social responses. *Anim Behav* 10:159–64.

Winston ML. (1987). *The Biology of the Honey Bee*. Cambridge, MA: Harvard University Press.

Wittlinger M, Wehner R, Wolf H. (2006). The ant odometer: Stepping on stilts and stumps. *Science* 312: 1965–67.

7 Social Information Use in Foraging Insects

Ellouise Leadbeater and Lars Chittka

CONTENTS

INTRODUCTION

In 1880, the Victorian naturalist Sir John Lubbock became aware that ants can track their nestmates to rewarding food sources. Lubbock believed that ants were following scent trails, and even went as far as to suggest something approaching a form of chemical language (Lubbock 1882). While he was not so generous about bees—"[Honey] bees do not bring their friends to share any treasure they have discovered" (Lubbock 1882, p. 278)—others were less dismissive. The German pastor Ernst Spitzner wrote of honey bees: "Full of joy, they twirl in circles about those in the hive … in a few minutes, after these had made it known to the others, they came in great numbers to the place!" (Spitzner 1788; cited in Lindauer 1985, p. 192).

How surprising it is that modern science, with its scorn for anthropomorphism and overinterpretation, has borne out the enthusiastic conjecture that tiny-brained organisms should perform such extraordinary feats of communication! Several decades of rigorous research have uncovered a broad spectrum of social information use in the insects, from subtle cues provided inadvertently by other animals to astoundingly intricate, highly evolved multimodal signals (von Frisch 1967; Hölldobler and Wilson 1990; Dyer 2002; Franks et al. 2002; Nieh et al. 2004; Leadbeater and Chittka 2007a). Despite their small brains, insects show remarkably complex learning abilities (Giurfa 2003; Menzel et al. 2006), and social information often leads to the relatively long-term changes in behavior that constitute social learning. These discoveries have not only enhanced our understanding of insect behavior, but also constituted some of the most influential advances in the field of behavioral biology—a fact reflected in the award of the 1973 Nobel Prize for Physiology or Medicine to the pioneering biologist Karl von Frisch.

In this review, our major focus is the contribution that insect studies can make toward understanding the particular circumstances where social information increases biological competitiveness. Like any trait, whether social information use is favored by natural selection will depend upon its costs and benefits relative to the available alternatives, and theory predicts that the balance may often swing in favor of individual exploration (Giraldeau et al. 2002). The question of when and why social information use and social learning may be adaptive has attracted much interest over recent years (Boyd and Richerson 1988; Giraldeau and Beauchamp 1999; Giraldeau et al. 2002; Laland 2004; Kendal et al. 2005), but much of this has failed to touch upon the insect world, which in turn has not made full use of literature arising in other fields. Here, we look at old and new examples of insect social information use with the question "When is social information useful?" in mind.

USING SOCIAL INFORMATION AS A SHORTCUT TO FOOD

FINDING FORAGING BONANZAS

Using social information to locate food potentially offers an economical alternative to the arduous process of individual exploration. Even the simple presence of a foraging conspecific may provide an exploitable, up-to-date cue about a foraging bonanza. However, using social information is not an assured shortcut to success. If food is easy to find, the value of social information will be limited—particularly if gathering it incurs a time cost in itself. For example, although it might intuitively seem that a honey bee should benefit by learning from others about where they have found food, in practice the time cost of waiting for dance information might often outweigh the benefits (Dechaume-Moncharmont et al. 2005; Dornhaus et al. 2006), especially if all foragers choose to wait for social information rather than individually discovering new food patches (Giraldeau and Beauchamp 1999; Dechaume-Moncharmont et al. 2005).

Two recent studies provide an illustration of how food availability and distribution may shape social information systems. Honey bees (*Apis mellifera*), famously, communicate the location of food sources to recruits in the darkness of the hive via figure-eight dances on the vertical honeycomb. The angle between the central path of the dance and the top of the comb depicts the direction that a recruit must follow, relative to the sun's azimuth (von Frisch 1967). If a hive is placed on its side, bees are forced to dance on a horizontal surface, and thus have no directional reference point, unless an artificial light source is present—in which case, both dancers and recruits realign their reference point to the light (Figure 7.1). Comparisons of the performance of hives with oriented and disoriented dances reveal that direction communication leads to an improved foraging performance only under surprisingly limited circumstances. In one study, oriented dances led to more food being collected in the Californian winter, but not in the summer or autumn (Figure 7.2; Sherman and Visscher 2002). In another study, improved performance was found only in the tropics, but not in temperate Northern European or Mediterranean climates (Dornhaus and Chittka 2004).

Together, these findings suggest that the dance language may be a product of particular ecological circumstances. In the tropical environments where honey bees diversified, trees that produce abundant flowers over a short time window are a major source of forage. In the Californian winter, food patches are also clustered and ephemeral, and the benefits of recruitment are likely to be greater than in areas where herbs and shrubs constitute a major nectar source, as is typical of temperate summers (Heinrich 1979). The suggestion that the value of dance information reflects food patch distribution invites more direct empirical exploration, but such findings provide an enticing glimpse at how ecology may shape communication, perhaps making some headway toward the intriguing question of why no equivalent referential communication system has been found among the other social Hymenoptera. In bumble bees, for example, returning foragers make irregular runs

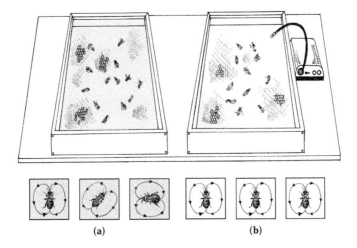

FIGURE 7.1 Examples of the orientation of dances of an individual bee (insets at bottom) within a disoriented hive (a), where the dance floor was horizontal and the hive was in complete darkness, and an oriented hive (b), where the dance floor was horizontal but a light source was available, which both dancers and recruits could use as a reference point. (Illustration by Sara Blackburn.)

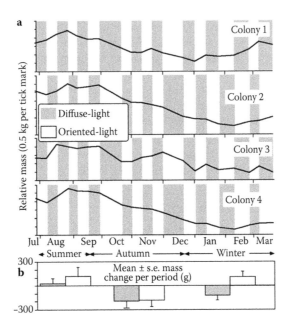

FIGURE 7.2 (a) Mass changes, as an indicator of food intake, in four honey bee colonies with successively oriented and disoriented dances. Bee colonies were aligned horizontally, and alternately allowed to perform oriented (oriented light, unshaded regions) or disoriented (diffuse light, shaded regions) dances for nineteen successive periods over the course of 9 months. (b) Mean mass changes during summer, autumn, and winter when colonies had oriented and disoriented dances. (After Sherman and Visscher 2002. With permission.)

around the nest, but this serves to distribute a pheromone that elicits foraging behavior, rather than to communicate location (Dornhaus and Chittka 1999, 2001). Stingless bees of some species also perform such irregular runs in the hive. Although the turning and spinning of the foragers in the course of their running vaguely reminds one of the honey bees' dance, it remains controversial whether this behavior, or vibrational signals produced during it, encode the location of a food source (see Chapter 11).

CHOOSING BETWEEN FLOWERS

Copying the foraging choices of others, or following their signals to find food, inevitably involves exploiting resources that have already been identified and are being harvested (Ruxton et al. 1995). If the value of social information about food location is limited in the context of searching for relatively large food patches, as is illustrated above, it follows that such information should rarely be useful at all when pollinating insects decide between individual inflorescences, which contain tiny volumes of nectar. If individual nectaries are rapidly drained by the foragers that find them, joining conspecifics will be a poor option unless the inflorescences in question contain an unusually large number of flowers. Nonetheless, an attraction to occupied inflorescences has been found in many pollinators, including bumble bees, stingless bees, honey bees, and wasps (Brian 1957; Wenner and Wells 1990; Slaa et al. 2003; Leadbeater and Chittka 2005; Kawaguchi et al. 2007). Why should an apparently maladaptive behavior be so common?

One explanation might be that individuals join each other as an aggressive response, or because they are simply attracted to conspecifics; indeed, bumble bees also tend to land beside conspecifics in a nonforaging context (Leadbeater and Chittka 2007b). Interestingly, however, a number of studies have found that joining behavior often occurs only when foragers visit flower species that they are not familiar with (Slaa et al. 2003; Leadbeater and Chittka 2005; Kawaguchi et al. 2007). Thus, perhaps joining behavior has adaptive benefits for foraging efficiency, not because it leads to individual rewarding flowers, but because it encourages sampling of rewarding flower species that might otherwise be ignored. Indeed, bumble bees that are foraging on one flower species will switch to another, more rewarding alternative more quickly if other bees are foraging there than if alone, because they are attracted to the occupied flowers (Leadbeater and Chittka 2007b). Surprisingly, there is also evidence that bumble bees can learn about rewarding flower species directly through observation of conspecific foragers. When bumble bees were permitted to watch conspecifics foraging on green flowers, and avoiding orange alternatives, observers later showed a significant preference for green when foraging alone (Worden and Papaj 2005).

TANDEM RUNNING TO LARGE FOOD SOURCES

Like honey bees in the tropics, many ants forage on foodstuffs that cannot be effectively exploited by one individual, and thus foragers use a plethora of multimodal signals to recruit their nestmates to the patches that they find (Hölldobler and Wilson 1990; Jackson and Ratnieks 2006). A behavior that features in both foraging and house hunting is tandem running, whereby a successful forager effectively leads a naïve follower to a target (Möglich et al. 1974; Hölldobler and Wilson 1990). A tandem run begins when a successful forager or scout recruits potential followers in the nest via food-offering rituals, and once antennal contact is established, the pair proceeds with the follower's antennae placed on the leader's abdomen (Figure 7.3b).

During a run, if the follower loses contact, the leader stops and adopts calling behavior until it is resumed—a feature that has led some authors to describe tandem leading as teaching (Franks and Richardson 2006). Indeed, the leader's behavior probably meets all the criteria of the commonly accepted definition of teaching provided by Caro and Hauser (1992), which states that the candidate behavior must occur only in the presence of a naïve observer, that the teacher must incur a cost (or at least no direct benefit to itself), and that the recipient must consequently acquire information more

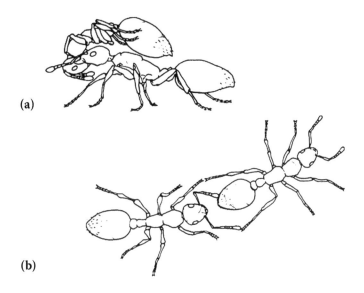

(a)

(b)

FIGURE 7.3 Ant recruitment. Transport by carrying (a) is faster than tandem running (b), but tandem running offers the advantages that recruits learn the route to the target. (Drawings by Sara Blackburn, based on photographs in Franks et al. 2002.)

quickly than it otherwise would. Tandem leaders incur a cost because the target is reached more slowly (Franks and Richardson 2006), and the follower's presence is a requirement for the pair to continue. It remains to be shown that the follower actually learns the route, rather than simply finding the food, but when tandem running occurs during house hunting, recruits repeatedly return to the target site as new tandem leaders (Möglich and Hölldobler 1974), implying that spatial learning has indeed occurred.

Under such a definition, a number of honey bee behaviors (including trophallaxis between foragers, discussed further in Chapter 10) may also qualify as teaching; however, whether it is useful to use the term *teaching* in the context of simple, declarative information is a controversial issue (Leadbeater et al. 2006; Csibra 2007; Premack 2007; Richardson et al. 2007; Thornton et al. 2007). Nonetheless, there can be little dispute that tandem running provides a fascinating example of how recruitment strategies are shaped by information requirements. Tandem pairs proceed, on average, four times more slowly than ants that return to food sources alone (Franks and Richardson 2006). Presumably, this investment is worthwhile because, unless the food source can be carried back to the nest by the two ants, the follower needs to learn the route rather than simply arrive at the food source. The trade-off between speed and the need for learning becomes clear when tandem running is considered in a house-hunting context. When scouts find potential nest sites, they lead nestmates to their chosen location via tandem runs, and the follower will then return to the nest and lead runs itself. However, once a threshold number of recruits have been led to the site, ants usually cease to act as leaders and switch to simply carrying their remaining nestmates. Carrying is much faster than tandem running, but recruits are transported upside down (Figure 7.3a), and thus (like passively displaced honey bees; Geiger et al. 1994) cannot learn the route and later become recruiters themselves (Pratt et al. 2002).

DECIDING WHEN TO FOLLOW DANCES

If animals can be choosy about when they rely on social cues and when they ignore them, then social information use may prove adaptive under a wider range of circumstances than would otherwise be the case (Laland 2004). Indeed, evidence from fish, mammals, and birds suggests that flexibility is a prominent feature of many social learning systems, and animals typically rely on social cues only when individual information is unreliable or not available (Laland 2004; Kendal et al. 2005). In insects, such hypotheses have yet to be directly tested, but a number of observations imply that they, too, do not always use social information as a blanket strategy. For example, honey bee foragers are more likely to follow dances before leaving the nest on the first few trips of their foraging career than when experienced. Experienced foragers, on the other hand, apparently rely on memorized information about food source locations rather than using spatial information provided by waggle dancers within the nest, because the simple presence of a dancing bee motivates them to fly to previously visited food patches—irrespective of vector and food odor information (Grüter et al. 2008). However, they tend to follow dances under specific circumstances where information acquired through personal experience has proved unreliable or is out of date—if their previous trip has been unsuccessful, or if they have no up-to-date personal information due to overnight inactivity (Biesmeijer and Seeley 2005).

Such flexibility can be achieved through simple associative learning. If animals modify their behavior according to the negative or positive feedback they receive, then social information use should become closely matched to environmental circumstances (Galef 1995). For example, foraging bumble bees leave chemical "footprints," derived from hydrocarbons that are passively secreted from their cuticle to prevent desiccation, on the flowers that they visit. These marks typically repel both bumble bees and other pollinators from the flowers where they are found, thus leading to avoidance of recently depleted nectar sources (Goulson et al. 1998; Stout et al. 1998; Gawleta et al. 2005; see also Chapter 13). However, if the marks are associated with high rewards, rather than with empty flowers, foragers quickly develop the opposite preference, and begin to treat them as an attractive stimulus (Saleh and Chittka 2006). The question of whether the use of scent marks as social information is entirely learned remains to be addressed, and it would be interesting to know whether naïve bees exhibit unlearned preferences or avoidance behavior toward scent-marked flowers. Nonetheless, it is clear that learning can lead to flexibility within insect social cue use, allowing for a close fit between social information use and environmental circumstances.

Insect "Traditions"

So far, we have highlighted the role of social learning as a means to rapidly track short-term changes in the environment. By nature, these behavioral changes are transient; indeed, it is precisely because environments change rapidly and unpredictably that this horizontal transmission of behavior may be adaptive (Laland et al. 1996). However, such short-term responses are notably different in kind from the long-term, cross-generational social habits that we call traditions. Traditions are the product of social learning that is not limited to individuals who utilize the same resources at the same time, but is persistent in time, maintained from generation to generation.

Many social insect colonies persist at the same site for multiple generations before relocating. Honey bee colonies, for example, survive for an average of 5.6 years at the same site under successive queens, if they make it through the first winter (Seeley 1978). Models that examine whether animals should learn behavior socially between (rather than within) generations predict that this type of occasionally changing environment should favor cross-generational social learning—in contrast to very stable environments, which should favor genetic inheritance of behavioral phenotypes, and highly unpredictable environments, where individual learning will be the most adaptive strategy (Boyd and Richerson 1985, 1988).

Traditions are rarely discussed in an invertebrate context (although see Donaldson and Grether 2007), but a series of experiments by Martin Lindauer and his student Wolfgang Kirchner provide grounds to suggest that social learning between insects of different age cohorts at least, if not different generations, may be a feature of social insect societies. In two similar studies, Lindauer (1985) and Kirchner (1987) investigated whether the temporal rhythms of a honey bee colony's foraging workforce, shaped by local environmental conditions, could be transmitted to the larval brood before emergence. They trained groups of foragers to visit highly rewarding feeders that were available for only an hour every day, either early in the morning, at midday, or in the evening. Groups of brood cells were then removed from each colony and raised in an incubator. When these bees were subsequently permitted to visit a continuously rewarding feeder, their activity patterns were similar to those of their trained predecessors, closely matching the availability of food in their mother colony. Kirchner (1987) suggests that increased levels of vibrational signaling on the dance floor above the young larvae during periods of heavy recruitment may have mediated this effect, but it seems equally possible the cue responsible may have been more frequent feeding of larvae by nurse bees during times of high food availability.

Whether this form of social learning remains stable over time remains open to question, but Lindauer's and Kirchner's studies pave the way for more detailed investigations of cross-cohort or cross-generational social learning in insects. A further interesting development in the question of invertebrate traditions in general is the suggestion that traditional behavior might arise without the need for learning at all. Donaldson and Grether (2007) argue that the cross-generational use of the same roosting sites by harvestmen, mediated by olfactory cues deposited at those locations, might also be considered traditional, especially given that the sites in question resembled a random sample of those available in terms of microclimate and ecology. The issue of whether this type of behavior, which does not require learning, constitutes a tradition remains open to debate, but the phenomenon described by Donaldson and Grether (2007) is certainly an interesting example of social information use that is unlikely to be limited to harvestmen, but may be widespread within the insects too.

SOCIAL TRANSMISSION OF FORAGING TECHNIQUES

Behavioral ecologists are interested in social information not only because it may guide decision making, but also because it can lead to the introduction of new, adaptive behaviors into an individual's repertoire. Such processes can have profound implications for a species' ecology, because socially acquired behaviors can spread through a group at an accelerating rate through positive feedback: the more "demonstrators" there are, the more naïve individuals acquire the behavior and become demonstrators themselves (Cavalli-Sforza and Feldman 1981; Pulliam 1983; Boyd and Richerson 1985). Research into how social transmission affects the dynamics of behavior has been limited to vertebrates, and well-cited examples include the social learning of pine cone stripping by black rats in Israel (Terkel 1996), and the spread of potato-washing behavior in Japanese macaques (Kawai 1965). Recently, we found that social transmission processes may be important for the dynamics of insect behavioral phenotypes, too (Leadbeater and Chittka 2008), as we discuss below.

Nectar-Robbing Bumble Bees

Nectar robbing occurs when pollinators create or reuse holes bitten through the base of a flower to extract nectar (Figure 7.4), rather than entering the corolla and hence pollinating the flower in the "legitimate" manner (Wille 1963; Inouye 1980). We noticed that broad bean plants, grown from seed in the open air, would remain intact for weeks until suddenly, every flower would be robbed over the space of only a few hours (Leadbeater and Chittka 2008). Is the spread of robbing behavior facilitated by positive feedback? The question is interesting from an evolutionary perspective, because robbers exert selection on the mutualistic relationship between flowering plants and pollinators through their varied effects on plant fitness (Maloof and Inouye 2000; Irwin 2006).

In the laboratory, we allowed naïve bees to forage either on unrobbed flowers or on flowers where robbing holes had already been cut. When bees were subsequently allowed to forage alone, on unmanipulated plants, a similar number from both groups showed a tendency to bite the flowers. However, of these, bees that had never previously used the robbing holes bit in the wrong places—around the petals and the legitimate entrance—and thus were not rewarded by finding the nectar source. In contrast, almost two-thirds of those bees that had acted as secondary robbers successfully

FIGURE 7.4 A bumble bee robbing a broad bean (*Vicia faba*) flower, mounted on a syringe needle. See color insert following page 142. (Photograph by E. Leadbeater.)

created robbing holes, becoming primary robbers. Just as many apparently complex behavior patterns can be transmitted between individuals through relatively simple inadvertent social information (Danchin et al. 2004), here, the analysis of the behavior of individual bumble bees reveals a straightforward mechanistic explanation. Bees that had experience of using the robbing holes tended to visit the base of the corolla when searching for nectar, and hence biting behavior was more likely to occur in the correct location. A common theme of insect social learning is that there is little temptation to explain social information use in terms of complex cognition; instead, behavioral change reflects simple and yet robust processes by which adaptive behavioral modifications can be achieved. We illustrate this point further in the next section.

SOCIAL INFORMATION ABOUT WHEN NOT TO FORAGE

An insect that can glean information about danger from others, and modify its foraging behavior accordingly, might bypass the risk of injury or potential death. Thus, social information use should be particularly apparent in the context of predation. In many cases, insects learn about danger through signals produced by their conspecifics; for example, aphid alarm pheromones induce nearby individuals to instantaneously drop from tree branches when their relatives are attacked (Montgomery and Nault 1977; Wohlers 1980). In other cases, however, responses to predation risk are based on information as simple as the presence or absence of conspecifics. For example, harvester ant colonies do not forage every day, but each morning a behaviorally distinct group of workers called patrollers leaves the nest and explores the colony's foraging trails (Gordon 1991). The foraging workforce does not emerge until the patrollers return safely to the nest, and the level of foraging activity that takes place depends upon the rate at which the patrollers come home (Greene and Gordon 2007).

CLASSICAL CONDITIONING IN DAMSELFLY ANTIPREDATOR RESPONSE

Some of the best evidence that insects can actually learn about their environment through the fly antipredator response behavior of others comes from the context of predation, and sometimes the learning processes involved are surprisingly complex. In areas where damselflies (*Enallagma boreale*) co-occur with predatory pike (*Esox lucius*), their larvae respond to olfactory cues from pike stimuli with antipredator behavior involving a reduction in feeding activity and movement. In contrast, larvae from allopatric populations do not respond to pike cues, but can be induced to do so if such cues are presented in combination with olfactory cues from crushed conspecifics (Wisenden et al. 1997). Wisenden and colleagues found that when the same larvae were later presented with pike cues in the absence of conspecific stimuli, they again reduced their feeding activity and movement, implying that the predation response had become conditioned to the pike stimulus.

This remarkably rapid and apparently complex learning can be explained by simple classical conditioning, although it is sometimes referred to as "observational conditioning" (Cook and Mineka 1989; Heyes 1994). Larvae from both populations showed antipredation responses to the conspecific cues alone; thus it seems likely that when the pike and conspecific cues were presented in combination, this response simply became conditioned to the new stimulus (Figure 7.5; Leadbeater and Chittka 2007a). Similar phenomena might be found in any situation where an animal shows a learned or pre-programmed response to cues from conspecifics that are the subject of predation, although little experimental investigation of this phenomenon has taken place in insects. Furthermore, there is no reason to expect that such learning should occur in response to conspecific cues alone, given that species sharing the same habitat may be at risk from the same predators. Indeed, in the damselfly study, larvae showed the same response when pike stimuli were combined with cues from injured fathead minnows—a taxonomically remote species that is hunted by the same predator (Wisenden et al. 1997).

LONG-TERM RISK ASSESSMENT IN WOOD CRICKETS

A study on wood crickets of the species *Neomobius sylvestris* (Coolen et al. 2005) provides evidence that insects can also learn about levels of predation risk, rather than predator identity, through their conspecifics' behavior. Juvenile "demonstrator" crickets placed in containers with predatory

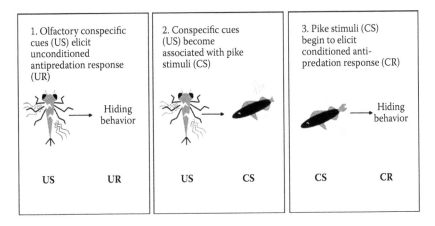

FIGURE 7.5 Classical conditioning of antipredator behavior in damselfly larvae. Larvae exhibit an initial hiding response (unconditioned response, UR) to olfactory cues from injured conspecifics (unconditioned stimulus, US). This response may become conditioned to new stimuli, such as cues produced by predators (conditioned stimulus, CS), if individuals experience both stimuli together. The hiding response thus becomes a conditioned response (CR). (From Leadbeater and Chittka 2007a. With permission.)

spiders tended to hide under the available leaf litter, even when transferred to new cages devoid of any spider cues. When naïve new companions were introduced into these new cages, they were also significantly more likely to hide under the leaves than crickets introduced into identical control cages, where demonstrators had never experienced spiders (and accordingly were not hiding). This difference in behavior persisted for 24 h after the demonstrators had been removed, suggesting that the crickets had learned indirectly about local predation risk through the behavior of their conspecifics.

CONCLUDING REMARKS

We began this chapter with the claim that social information use in the small-brained insects is surprisingly complex. Perhaps, however, one of the most consistent messages to emerge from the studies that we have discussed above is that the mechanistic basis for such behavior is surprisingly simple. When one is dealing with a relatively miniscule nervous system, there is little temptation to explain behavior in terms of advanced cognition without looking for simpler explanations first. Thus, even damselfly larvae can learn socially about predator identity through simple Pavlovian mechanisms, and ant societies can assess threat levels on the straightforward basis of whether their conspecifics make it home. Bumble bees pick up "bad habits" from others, learning to rob plants simply because they use the holes that other bees leave behind, and ants inform each other about where to find food just by walking others through the process, rather than carrying them there.

So what can these straightforward, yet highly effective social information systems offer to those interested in why social information use increases biological competitiveness? In writing this chapter, we hope to go some way toward providing a bridge between the empirical study of insect social systems and theoretical predictions about the usefulness of social information. The social insects, particularly, are unusual in that inadvertent social information has often evolved into specialized signals, and this "packaging" offers opportunities for experimental manipulation that are perhaps unparalleled in the vertebrate world. To date, few adaptive hypotheses have been tested using insect models. Yet the studies described above hint that insect social information systems have indeed been shaped by selection for reliability and cost-effectiveness, and they invite more direct development of this approach. Insects may have small brains, but they may also provide a relatively untapped system for understanding how social information use and social learning begin to evolve.

REFERENCES

Biesmeijer JC, Seeley TD. (2005). The use of waggle dance information by honey bees throughout their foraging careers. *Behav Ecol Sociobiol* 59:133–42.

Boyd R, Richerson PJ. (1985). *Culture and the Evolutionary Process*. Chicago: University of Chicago Press.

Boyd R, Richerson PJ. (1988). An evolutionary model of social learning: The effects of spatial and temporal variation. In Zentall TR, Galef BG (eds.), *Social Learning: Psychological and Biological Perspectives*. Hillsdale, NJ: Lawrence Erlbaum Associates, pp. 29–48.

Brian AD. (1957). Differences in the flowers visited by four species of bumble-bees and their causes. *J Anim Ecol* 26:71–98.

Caro TM, Hauser MD. (1992). Is there teaching in nonhuman animals? *Q Rev Biol* 67:151–74.

Cavalli-Sforza LL, Feldman MW. (1981). *Cultural Transmission and Evolution: A Quantitative Approach*. Princeton, NJ: Princeton University Press.

Cook M, Mineka S. (1989). Observational conditioning of fear to fear-relevant and fear-irrelevant stimuli in rhesus monkeys. *J Abnorm Psychol* 98:448–59.

Coolen I, Dangles O, Casas J. (2005). Social learning in noncolonial insects? *Curr Biol* 15:1931–35.

Csibra G. (2007). Teachers in the wild. *Trends Cogn Sci* 11:95–96.

Danchin E, Giraldeau LA, Valone TJ, Wagner RH. (2004). Public information: From nosy neighbours to cultural evolution. *Science* 305:487–91.

Dechaume-Moncharmont FX, Dornhaus A, Houston AI, McNamara JM, Collins EJ, Franks NR. (2005). The hidden cost of information in collective foraging. *Proc R Soc Lond B* 272:1689–95.

Donaldson ZR, Grether GF. (2007). Tradition without social learning: A scent-based communal roost formation in a Neotropical harvestman (*Prionostemma* sp.). *Behav Ecol Sociobiol* 61:801–9.

Dornhaus A, Chittka L. (1999). Evolutionary origins of bee dances. *Nature* 401:38.

Dornhaus A, Chittka L. (2001). Food alert in bumblebees (*Bombus terrestris*): Possible mechanisms and evolutionary implications. *Behav Ecol Sociobiol* 50:570–76.

Dornhaus A, Chittka L. (2004). Why do honeybees dance? *Behav Ecol Sociobiol* 55:395–401.

Dornhaus A, Klügl F, Oechslein C, Puppe F, Chittka L. (2006). Benefits of recruitment in honey bees: Effects of ecology and colony size in an individual-based model. *Behav Ecol* 17:336–44.

Dyer FC. (2002). The biology of the dance language. *Annu Rev Entomol* 47:917–49.

Franks NR, Pratt SC, Mallon EB, Britton NF, Sumpter DJT. (2002). Information flow, opinion polling and collective intelligence in house-hunting social insects. *Phil Trans R Soc Lond B* 357:1567–83.

Franks NR, Richardson T. (2006). Teaching in tandem-running ants. *Nature* 439:153.

von Frisch K. (1967). *The Dance Language and Orientation of Bees*. Cambridge, MA: Harvard University Press.

Galef BJ. (1995). Why behaviour patterns that animals learn socially are locally adaptive. *Anim Behav* 49:1325–34.

Gawleta N, Zimmermann Y, Eltz T. (2005). Repellent foraging scent recognition across bee families. *Apidologie* 36:325–30.

Geiger K, Kratzsch D, Menzel R. (1994). Bees do not use landmark cues seen during displacement for displacement compensation. *Naturwissenschaften* 81:415–17.

Giraldeau LA, Beauchamp G. (1999). Food exploitation: Searching for the optimal joining policy. *Trends Ecol Evol* 14:102–6.

Giraldeau LA, Valone TJ, Templeton JJ. (2002). Potential disadvantages of using socially acquired information. *Phil Trans R Soc Lond B* 357:1559–66.

Giurfa M. (2003). The amazing mini-brain: Lessons from a honey bee. *Bee World* 84:5–18.

Gordon DM. (1991). Behavioral flexibility and the foraging ecology of seed-eating ants. *Am Nat* 138:379–411.

Goulson D, Hawson SA, Stout JC. (1998). Foraging bumblebees avoid flowers already visited by conspecifics or by other bumblebee species. *Anim Behav* 55:199–206.

Greene MJ, Gordon DM. (2007). Interaction rate informs harvester ant task decisions. *Behav Ecol* 18:451–55.

Grüter C, Sol Balbuena M, Farina WM. (2008). Informational conflicts created by the waggle dance. *Proc R Soc B* 275:1321–27.

Heinrich B. (1979). *Bumblebee Economics*. Cambridge, MA: Harvard University Press.

Heyes CM. (1994). Social learning in animals: Categories and mechanisms. *Biol Rev* 69:207–31.

Hölldobler B, Wilson EO. (1990). *The Ants*. Cambridge, MA: Harvard University Press.

Inouye DW. (1980). The terminology of floral larceny. *Ecology* 61:1251–53.

Irwin RE. (2006). The consequences of direct versus indirect species interactions to selection on traits: Pollination and nectar robbing in *Ipomopsis aggregata*. *Am Nat* 167:315–28.

Jackson DE, Ratnieks FLW. (2006). Communication in ants. *Curr Biol* 16:R570–74.

Kawaguchi LG, Ohashi K, Toquenaga Y. (2007). Contrasting responses of bumble bees to feeding conspecifics on their familiar and unfamiliar flowers. *Proc R Soc Lond B* 274:2661–67.

Kawai M. (1965). Newly-acquired pre-cultural behavior of the natural troop of Japanese monkeys on Koshima islet. *Primates* 6:1–30.

Kendal RL, Coolen I, van Bergen Y, Laland KN. (2005). Trade-offs in the adaptive use of social and asocial learning. *Adv Stud Behav* 35:333–79.

Kirchner WH. (1987). Tradition im Bienenstaat: Kommunikation zwischen den Imagines und der Brut der Honigbiene durch Vibrationssignale. Doctoral thesis, University of Würzburg, Germany.

Laland KN. (2004). Social learning strategies. *Learn Behav* 32:4–14.

Laland KN, Richerson PJ, Boyd R. (1996). Developing a theory of animal social learning. In Heyes CM, Galef BG (eds.), *Social Learning in Animals: The Roots of Culture*. San Diego: Academic Press, pp. 129–54.

Leadbeater E, Chittka L. (2005). A new mode of information transfer in foraging bumblebees? *Curr Biol* 15:R447–48.

Leadbeater E, Chittka L. (2007a). Social learning in insects—From miniature brains to consensus building. *Curr Biol* 17:R703–13.

Leadbeater E, Chittka L. (2007b). The dynamics of social learning in an insect model, the bumblebee (*Bombus terrestris*). *Behav Ecol Sociobiol* 61:1789–96.

Leadbeater E, Chittka L. (2008). Social transmission of nectar-robbing behaviour in bumble-bees. *Proc R Soc B* 275:1669–74.

Leadbeater E, Raine NE, Chittka L. (2006). Social learning: Ants and the meaning of teaching. *Curr Biol* 16:R323–25.

Lindauer M. (1985). The dance language of honeybees: The history of a discovery. In Hölldobler B, Lindauer M (eds.), *Experimental Behavioral Ecology and Sociobiology: In Memoriam Karl von Frisch 1886–1982.* Sunderland: Sinauer, pp. 129–40.

Lubbock J. (1882). *Ants, Bees and Wasps: A Record of Observations on the Habits of the Social Hymenoptera.* New York: Appleton.

Maloof JE, Inouye DW. (2000). Are nectar robbers cheaters or mutualists? *Ecology* 81:2651–61.

Menzel R, Leboulle G, Eisenhardt D. (2006). Small brains, bright minds. *Cell* 124:237–39.

Möglich M, Hölldobler B. (1974). Social carrying behaviour and division of labor during nest moving in ants. *Psyche* 81:219–36.

Möglich M, Maschwitz U, Hölldobler B. (1974). Tandem calling: A new kind of signal in ant communication. *Science* 186:1046–47.

Montgomery ME, Nault LR. (1977). Comparative response of aphids to alarm pheromone, (E)-beta-farnesene. *Entomol Exp Appl* 22:236–42.

Nieh JC, Barreto LS, Contrera FAL, Imperatriz-Fonseca VL. (2004). Olfactory eavesdropping by a competitively foraging stingless bee, *Trigona spinipes. Proc R Soc Lond B* 271:1633–40.

Pratt SC, Mallon EB, Sumpter DJT, Franks NR. (2002). Quorum sensing, recruitment, and collective decision-making during colony emigration by the ant *Leptothorax albipennis. Behav Ecol Sociobiol* 52:117–27.

Premack D. (2007). Human and animal cognition: Continuity and discontinuity. *Proc Natl Acad Sci USA* 104:13861–67.

Pulliam HR. (1983). On the theory of gene-culture co-evolution in a variable environment. In Melgren RL (ed.), *Animal Cognition and Behavior.* Amsterdam: North-Holland Publishing Company, pp. 427–43.

Richardson TO, Sleeman PA, McNamara JM, Houston AI, Franks NR. (2007). Teaching with evaluation in ants. *Curr Biol* 17:1520–26.

Ruxton GD, Hall SJ, Gurney WSC. (1995). Attraction toward feeding conspecifics when food patches are exhaustible. *Am Nat* 145:653–60.

Saleh N, Chittka L. (2006). The importance of experience in the interpretation of conspecific chemical signals. *Behav Ecol Sociobiol* 61:215–20.

Seeley TD. (1978). Life history strategy of the honey bee, *Apis mellifera. Oecologia* 32:109–18.

Sherman G, Visscher PK. (2002). Honeybee colonies achieve fitness through dancing. *Nature* 419:920–22.

Slaa JE, Wassenberg J, Biesmeijer JC. (2003). The use of field-based social information in eusocial foragers: Local enhancement among nestmates and heterospecifics in stingless bees. *Ecol Entomol* 28:369–79.

Spitzner MJE. (1788). *Ausführliche Beschreibung der Korbbienenzucht im sächsischen Churkreise, ihrer Dauer und ihres Nutzens ohne künstliche Vermehrung nach den Gründen der Naturgeschichte und nach eigener langer Erfahrung.* Leipzig: Junius.

Stout JC, Goulson D, Allen JA. (1998). Repellent scent-marking of flowers by a guild of foraging bumblebees (*Bombus* spp.). *Behav Ecol Sociobiol* 43:317–26.

Terkel J. (1996). Cultural transmission of feeding behaviour in the black rat (*Rattus rattus*). In Heyes CM, Galef BG (eds.), *Social Learning in Animals: The Roots of Culture.* San Diego: Academic Press, pp. 17–47.

Thornton A, Raihani NJ, Radford AN. (2007). Teachers in the wild: Some clarification. *Trends Cogn Sci* 11:272–73.

Wenner AM, Wells PH. (1990). *Anatomy of a Controversy: The Question of a "Language" among Bees.* New York: Columbia University Press.

Wille A. (1963). Behavioral adaptations of bees for pollen collecting from *Cassia* flowers. *Rev Biol Trop* 11:205–10.

Wisenden BD, Chivers DP, Smith RJF. (1997). Learned recognition of predation risk by *Enallagma* damselfly larvae (Odonata, Zygoptera) on the basis of chemical cues. *J Chem Ecol* 23:137–51.

Wohlers P. (1980). Escape responses of pea aphids *Acyrthosiphon pisum* to alarm pheromones and additional stimuli. *Entomol Exp Appl* 27:156–68.

Worden BD, Papaj DR. (2005). Flower choice copying in bumblebees. *Biol Lett* 1:504–7.

8 Local Enhancement, Local Inhibition, Eavesdropping, and the Parasitism of Social Insect Communication

E. Judith Slaa and William O. H. Hughes

CONTENTS

INTRODUCTION

Food is the limiting factor for the development of social insect colonies (Hubbell and Johnson 1977; Hölldobler and Wilson 1990; Raveret Richter 2000; Eltz et al. 2002; Mattila and Seeley 2007). An efficient food exploitation system, therefore, is the key to the fitness and ecological success of a social insect colony. All social insects are central place foragers and need carbohydrates and proteins for the activity of their adults and the growth of their young. They acquire these nutritional resources from a variety of food sources. Almost all bees obtain their food from flowers (nectar and pollen) (Michener 1974), whereas most wasps obtain at least the protein part of their diet from meat, either by catching live arthropod prey or scavenging dead animals (Ross and Matthews 1991; see also Chapter 3). Most termites forage on wood or dead vegetation, with many species locating their nest within the food resource itself—a trait also true of eusocial aphids, thrips, and beetles (Costa 2006). Among all social insects, the ants utilize the broadest range of food sources, collecting, for example, meat, honeydew, fruit, nectar, or leaves, according to the respective ant species (Hölldobler and Wilson 1990).

Closely related species tend to have similar food requirements (Heithaus 1974, 1979; Roubik et al. 1986; Roubik 1989; Wilms et al. 1996) and foraging ranges of neighboring colonies within a "feeding guild" often overlap (e.g., Hubbell and Johnson 1977; Suzuki and Murai 1980; Hölldobler and Wilson 1990; Eltz et al. 2002; Slaa 2006). Consequently, and also due to the fact that many food resources are clumped in space and time (Pleasants 1983; Waser 1983; Hunter and Price 1992), both intra- and interspecific competition for food is high. This is demonstrated by the aggression and defensive behavior exhibited by social insects at and around food resources (Kerr 1959; Johnson and Hubbell 1974, 1975; Roubik 1982; Hölldobler and Wilson 1990; Nagamitsu and Inoue 1997).

Probably the most extensively studied component of the foraging biology of social insects is the way by which many species actively communicate the presence and location of food to their nest-mates. Honey bee foragers, for instance, perform "dances" inside the nest that convey this informa-tion (see Chapter 6), several bee species deposit pheromones and chemical "beacons" at the food source (see Chapters 9, 12, and 13), and many ants, termites, and some bees lay pheromone trails from the food to the nest itself (see Chapters 2, 6, and 12). These forms of active communication enable the rapid recruitment of nestmates to food sources, which plays a particularly important role in the success of most ecologically dominant species, including many invasive species of ants, wasps, and termites (Lowe et al. 2000; Holway et al. 2002). Active communication involves the use of signals, which can be defined as "stimuli that convey information and have been molded by natural selection to do so" (Seeley 1989, p. 550). Hence, active signals contrast with passive cues, which are "stimuli that contain information but have not been shaped by natural selection specifically to convey information" (Seeley 1989, p. 550). In other words, signals provide informa-tion expressly (active communication), whereas cues provide information only incidentally (passive communication).

Although most research has focused on food exploitation by social insects with group foraging strategies, many species forage solitarily. Yet, even in those species that use active recruitment com-munication during foraging, the signals are limited to aiding nestmates to locate already discov-ered food patches, which implies that at least one worker has to initially discover this food source unaided. In this case, other forms of information become paramount. One of the most reliable forms of information is the presence of collecting individuals, either from other conspecific colonies or from other species with shared food preferences. Where such individuals are present, food may be present as well. Information on the presence of other individuals can be essentially divided into two categories. The first category can be summarized as local enhancement and local inhibition. Here, information is provided by cues, such as physical presence or passively deposited odors that indicate the presence of individuals in the recent past. The second category can be denominated as eavesdropping, which means that the active signals released by a forager to recruit nestmates may potentially be detected and exploited by non-nestmates as well. In this chapter, we will focus on these little studied forms of information sources used by social insects during food exploitation. We will show that both local enhancement/inhibition and eavesdropping play an important and underap-preciated role in the food collectors' foraging decisions.

LOCAL ENHANCEMENT AND LOCAL INHIBITION

The use of passively provided cues in foraging decisions can be described as local enhancement or local inhibition. Thorpe (1956) originally defined local enhancement as "an apparent imitation resulting from directing the animal's attention to a particular object or to a particular part of the environment" (Thorpe 1956, pp. 121–22). The term is commonly used to describe the attraction of foragers to feeding individuals, or to odors left behind by them (e.g., Heyes et al. 2000). The logical converse is local inhibition, when foragers are deterred from feeding on a food patch through the presence of other individuals or through odors they had left behind (Slaa et al. 2003b). In addition, foragers may show what can be termed stimulus enhancement, when they are attracted to all objects that show the same characteristics (e.g., appearance or odor) as those on which other individuals forage (Thorpe 1956; Galef 1988; Heyes et al. 2000).

Local enhancement in social insects has sometimes been referred to as social facilitation (e.g., Parrish and Fowler 1983; Fowler 1992), or in other words, as "an increase in the frequency or inten-sity of responses or the initiation of particular responses already in an animal's repertoire, when shown in the presence of others engaged in the same behavior at the same time" (Clayton 1978, p. 374). However, *local enhancement* is a more accurate term for situations "in which an animal's attention is directed to a particular location or object by the actions or presence of conspecifics" (Raveret Richter 2000, p. 134; see also Reid et al. 1995; Raveret Richter and Tisch 1999)—which thus fits better into the context of our chapter.

The role of local enhancement in the context of social foraging may be twofold. It might (1) facilitate the finding of new food sources, and (2) facilitate the formation of feeding groups (see Pöysä 1992; Beauchamp et al. 1997). Local enhancement is best studied in vertebrates—including mammals, birds, and fish (Thorpe 1956; Heyes 1994; Galef and Giraldeau 2001; Krause and Ruxton 2002; Brown and Laland 2003)—where it seems to be the most important mechanism for social (group) foraging (Galef and Giraldeau 2001). The occurrence of local enhancement in these animals is often closely related to the benefits of social foraging, such as an enhanced feeding efficiency, an increased ability to overcome dominance by other individuals, and an enhanced predator vigilance (Pulliam and Caraco 1984; Clark and Mangel 1986; Giraldeau and Caraco 2000). Local enhancement is particularly common among group foraging birds, such as geese and gulls, in which food searching individuals are visually attracted to feeding conspecifics (Pöysä 1992). Possibly the classic example of local enhancement, though, is vultures. Most species in this group of birds are highly specialized scavengers that feed on large mammal carcasses, which on the one hand provide large food sources, but on the other hand are highly dispersed both spatially and temporally. While some individuals locate the food on their own, the majority of the individuals actually find it by following other vultures since these are far easier to detect (Jackson et al. 2008). Other examples of local enhancement in vertebrates include the attraction of agoutis to the rasping sounds made by conspecifics when feeding on nuts, or Norway rats following the foraging trails left in the undergrowth by the passage of conspecifics (Galef and Giraldeau 2001). Like local enhancement, stimulus enhancement (the attraction to objects with the same characteristics as those on which other individuals forage) is also best known from the vertebrate world. In lactating rats and rabbits, for instance, suckling pups, at weaning, show a preference for those flavors that the nursing milk contained as a consequence of the food consumed by their mothers. Similarly, the breath of rats, mice, and gerbils carries the odor of recently eaten food, which can be identified by conspecifics and causes them to prefer this kind of food (Galef and Giraldeau 2001).

Although many of the classic examples are from the vertebrate world, local enhancement is also well known from some invertebrates. The housefly *Musca domestica*, for example, commonly lands near other flies, which increases the probability of finding food (Collins and Bell 1996). Other reported examples are the males of swallow-tail butterflies, which are strongly attracted to other males in their search for salts (Otis et al. 2006), and female fruit flies (*Ceratis capitata*), which have been shown to be attracted to conspecific and heterospecific females on fruit mimics (Propoky et al. 2000).

Local enhancement and local inhibition in social insects were proposed as early as in the nineteenth century by Charles Darwin (Romanes 1884). Although it has received less attention than active recruitment, there is now good evidence that it is widespread among social insects, and that it may indeed be an important factor for the choice of food sources (Table 8.1). One of the best-studied examples is the repellent scent marks of bumble bees. When a bumble bee forager visits a flower to collect nectar or pollen, it deposits cuticular hydrocarbons from its tarsi on the petals, which then cause other bumble bees to avoid landing on this flower (Kato 1988; Goulson et al. 1998; Stout et al. 1998; Williams 1998; see also Chapter 13). These cuticular hydrocarbons have now been convincingly established to be passive cues rather than actively excreted signals (Thompson and Chittka 2001; Gawleta et al. 2005; Wilms and Eltz 2008; but see Schmidt et al. 2005). Similar footprints are probably deposited by most insects. Honey bees (Núñez 1967; Free and Williams 1983; Giurfa and Núñez 1992; Williams 1998; see also Chapter 9), some stingless bee species (Goulson et al. 2001; see also Chapter 12), and the eusocial sweat bee *Halictus aerarius* (Yokoi and Fujisaki 2007; Yokoi et al. 2007) have been demonstrated to utilize these olfactory cues as information about the nectar content of a flower. Since in most cases the nectar or pollen provided by a flower is emptied by a single bee visit, the local inhibition mediated by these footprints is advantageous to subsequent visitors because it enables them to avoid wasting time on a depleted resource (Wetherwax 1986; Kato 1988; Schmitt and Bertsch 1990; Giurfa and Núñez 1992). The repellent effect of scent marks left on flowers is not constricted to conspecifics, but works across bee families and even across insect orders (e.g., Williams 1998; Stout and Goulson 2001; Gawleta et al. 2005; Reader et al. 2005; Yokoi et al. 2007; Barth et al. 2008; see also Chapters 12 and 13).

Most records of local enhancement in social insects come from bees (Table 8.1). In honey bees and bumble bees, visual and olfactory local enhancement and local inhibition to both conspecifics and

TABLE 8.1
Compilation of Published Evidence for Local Enhancement and Local Inhibition in Social Insects

Type	Genus	Species	Local Enhancement or Inhibition	Visual or Olfactory Cues	Conspecific or Heterospecific	Reference
Bee	Halictus	aerarius	Inhibition	Olfactory	Both	[1, 2]
Bee	Apis	mellifera	Enhancement	Olfactory	Conspecific	[3–6]
Bee	Apis	mellifera	Inhibition	Olfactory	Both	[4, 7–11]
Bee	Apis	mellifera	Enhancement	Visual	Conspecific	[5, 12]
Bee	Bombus	appositus	Inhibition	Visual	Heterospecific	[13]
Bee	Bombus	diversus	Both[c]	Olfactory	Conspecific	[14, 15]
Bee	Bombus	flavifrons	Inhibition	Visual	Heterospecific	[13]
Bee	Bombus	hortorum	Inhibition	Olfactory	Both	[16]
Bee	Bombus	ignitus	Inhibition	Visual	Conspecific	[17]
Bee	Bombus	impatiens	Both	Both	Conspecific	[18, 19]
Bee	Bombus	lapidarius	Inhibition	Olfactory	Both	[8, 20, 21]
Bee	Bombus	pascuorum	Inhibition	Olfactory	Both	[16, 20–22]
Bee	Bombus	pratorum	Inhibition	Olfactory	Both	[16]
Bee	Bombus	terrestris	Inhibition	Olfactory	Both	[16, 20–25]
Bee	Bombus	terrestris	Enhancement	Both	Conspecific	[25–28]
Bee	Bombus	terrestris	Stimulus	Visual	Conspecific	[18, 29]
Bee	Bombus	vosnesenskii	Enhancement	Olfactory	Conspecific	[30]
Bee	Bombus	spp.[a]	Inhibition	Visual	Both	[31, 32]
Bee	Melipona	fasciata	Inhibition/neither[b]	Visual	Heterospecific	[33]
Bee	Melipona	seminigra	Enhancement	Olfactory	Conspecific	[34, 35]
Bee	Nannotrigona	testaceicornis	Enhancement	Olfactory	Both	[36]
Bee	Oxytrigona	mellicolor	Enhancement	Visual	Conspecific	[33, 37]
Bee	Oxytrigona	mellicolor	Inhibition/neither[b]	Visual	Heterospecific	[33]
Bee	Partamona	cupira	Enhancement	Visual	Conspecific	[33]
Bee	Partamona	cupira	Inhibition/neither[b]	Visual	Heterospecific	[33]
Bee	Plebeia	frontalis	Inhibition/neither[b]	Visual	Heterospecific	[33]
Bee	Scaptotrigona	pectoralis	Inhibition/neither[b]	Visual	Heterospecific	[33]
Bee	Trigona	amalthea	Both	Visual	Conspecific	[33]
Bee	Trigona	amalthea	Both/neither[b,d]	Visual	Heterospecific	[33]
Bee	Trigona	angustula	Enhancement	Both	Conspecific	[33, 38]
Bee	Trigona	angustula	Inhibition/neither[b]	Visual	Heterospecific	[33]
Bee	Trigona	corvina	Both/neither[b]	Visual	Heterospecific	[33]
Bee	Trigona	dorsalis	Neither	Visual	Conspecific	[33, 37]
Bee	Trigona	dorsalis	Inhibition/Neither[b]	Visual	Heterospecific	[33]
Bee	Trigona	fuscipennis	Neither	Olfactory[f]	Conspecific	[7]
Bee	Trigona	fulviventris	Inhibition	Both	Conspecific	[7, 33]

TABLE 8.1(CONTINUED)
Compilation of Published Evidence for Local Enhancement and Local Inhibition in Social Insects

Type	Genus	Species	Local Enhancement or Inhibition	Visual or Olfactory Cues	Conspecific or Heterospecific	Reference
Bee	*Trigona*	*fulviventris*	Both/neither[b]	Visual	Heterospecific	[33]
Bee	*Trigona*	*mexicana*	Enhancement	Olfactory	Conspecific	[38]
Bee	*Trigona*	*spinipes*	Enhancement	Olfactory	Heterospecific	[50]
Wasp	*Polistes*	*fuscatus*	Enhancement	Visual	Conspecific	[39]
Wasp	*Polybia*	*occidentalis*	Enhancement	Visual	Conspecific	[40, 41]
Wasp	*Polybia*	*occidentalis*	Enhancement		Heterospecific	[40]
Wasp	*Polybia*	*diguetana*	Both[e]		Heterospecific	[40]
Wasp	*Agelaia*	*pallipes*	Enhancement	Visual	Conspecific	[42]
Wasp	*Vespula*	*koreensis*	Neither	Both	Conspecific	[43]
Wasp	*Vespula*	*consobrina*	Inhibition	Visual	Heterospecific	[39]
Wasp	*Vespula*	*vidua*	Neither	Visual	Both	[39]
Wasp	*Vespula*	*flavopilosa*	Neither	Visual	Heterospecific	[39]
Wasp	*Vespula*	*fuscatus*	Enhancement	Visual	Both	
Wasp	*Vespula*	*germanica*	Enhancement	Both	Both	[39, 44–47]
Wasp	*Vespula*	*maculifrons*	Inhibition	Visual	Both	[39, 46]
Wasp	*Vespula*	*maculifrons*	Enhancement	Visual	Conspecific	[47]
Ant	*Formica*	*pratensis*	Enhancement	Visual	Heterospecific	[48]
Ant	*Formica*	*uralensis*	Enhancement		Heterospecific	[48]
Ant	*Formica*	*lugubris*	Enhancement		Heterospecific	[49]
Ant	*Formica*	*nigricans*	Enhancement		Heterospecific	[49]
Ant	*Daceton*	*armigerum*	Enhancement		Heterospecific	[49]

Note: For many of the cases suggested to involve visual cues, the concurrent use of olfactory cues cannot be ruled out, and vice versa. Empty cells indicate that the results of the respective study provide no evidence whether the local enhancement/inhibition was due to visual or olfactory cues.

[a] Various species observed, but not distinguished.

[b] Depends on species-specific interaction.

[c] *Bombus diversus*: Local enhancement if the bee is unfamiliar with the flower and local inhibition if the bee is familiar with the flower.

[d] *Trigona amalthea*: Local enhancement if naïve, inhibition if experienced.

[e] *Polybia diguetana*: Attracted to feeders with wasps but inhibited from landing.

[f] *Trigona fuscipennis*: Did not discriminate between unvisited and visited flowers with putative footprints.

[1] Yokoi et al. (2007); [2] Yokoi and Fujisaki (2007); [3] Ferguson and Free (1979); [4] Free and Williams (1983); [5] Kalmus (1954); [6] Kalmus and Ribbands (1952); [7] Goulson et al. (2001); [8] Stout and Goulson (2001); [9] Giurfa (1993); [10] Giurfa and Núñez (1992); [11] Williams (1998); [12] Wenner (1967); [13] Inouye (1978); [14] Kawaguchi et al. (2007); [15] Kato (1988); [16] Stout et al. (1998); [17] Makino and Sakai (2005); [18] Worden and Papaj (2005); [19] Saleh et al. (2006); [20] Goulson et al. (2000); [21] Stout and Goulson (2002); [22] Goulson et al. (1998); [23] Wilms and Eltz (2008); [24] Witjes and Eltz (2007); [25] Saleh and Chittka (2006); [26] Kawaguchi et al. (2006); [27] Schmitt and Bertsch (1990); [28] Schmitt et al. (1991); [29] Leadbeater and Chittka (2007); [30] Cameron (1981); [31] Thomson et al. (1987); [32] Dreisig (1995); [33] Slaa et al. (2003b); [34] Jarau et al. (2004); [35] Hrncir et al. (2004); [36] Schmidt et al. (2005); [37] Slaa et al. (2003a); [38] Villa and Weiss (1990); [39]Raveret Richter and Tisch (1999); [40] Raveret Richter (1990); [41] Hrncir et al. (2007); [42] Fowler (1992); [43] Kim et al. (2007); [44] D'Adamo et al. (2000); [45] D'Adamo et al. (2003); [46] Parrish and Fowler (1983); [47] Reid et al. (1995); [48] Reznikova (1982); [49] Hölldobler and Wilson (1990); [50] Nieh et al. (2004).

heterospecifics (or their odors) have been demonstrated (Table 8.1). Among the stingless bees, some species show local enhancement, some exhibit local inhibition, and others show neither (Slaa et al. 2003b; Table 8.1). These differences are not easily explained by differences in foraging strategy. For example, stingless bee species that show local enhancement include both group foraging (e.g., *Oxytrigona mellicolor*) and solitary foraging (e.g., *Tetragonisca angustula*) species. This parallels the fact that both bumble bees (solitary foragers) and honey bees (group foragers) show local enhancement (Kalmus 1954; Leadbeater and Chittka 2005, 2007; Kawaguchi et al. 2006; Table 8.1), which suggests that local enhancement occurs independently of the foraging strategy employed by a bee species. A number of studies have revealed the occurrence of local enhancement and local inhibition in wasps (Table 8.1), normally involving visual cues (Fowler 1992; Raveret Richter and Tisch 1999; Hrncir et al. 2007), but—at least sometimes—also olfactory cues (D'Adamo et al. 2000). Very few studies have examined the phenomenon in ants, but there are several species (see Table 8.1) in which visual enhancement does appear to occur (Reznikova 1982; Hölldobler and Wilson 1990). No studies have recorded local enhancement or local inhibition in termites, but their tunnel-based foraging systems make its occurrence seem unlikely.

Stimulus enhancement has also been described in social insects. Bumble bees, for example, copy the color choice of a demonstrator bee after a 10-min observation period (Worden and Papaj 2005). Also, floral odors perceived inside the nest can trigger a preference for this specific odor in both honey bee and bumble bee foragers when they search for food in the field (e.g., Free 1969; Jakobsen et al. 1995; Dornhaus and Chittka 1999; Arenas et al. 2007; see also Chapters 9 and 10). It has been demonstrated that insects even receive information about available food sources long before they start their foraging career. For example, adult inexperienced solitary bees of the species *Colletes fulgidus longiplumosus* prefer pollen from the plant species that they had received during their larval stages (Dobson 1987), and *Apis mellifera* preforagers have been shown to pick up and have a preference for the floral odor brought back to the hive by nestmates (Grüter et al. 2006; see also Chapter 10).

THE CONFOUNDING OF SIGHT AND SMELL

Most reported cases of local enhancement and local inhibition involve individual choices based on the visual presence of a nestmate (Table 8.1). However, in many of these experiments, odor is either present as well, or its presence cannot be ruled out. Some of the original studies on bumble bees, for example, suggested simply visual local inhibition (e.g., Inouye 1978), but more recent studies have demonstrated that scent marks also play an important role in the flower choice of these bees (Schmitt and Bertsch 1990; Schmitt et al. 1991; Goulson et al. 1998, 2000; Stout et al. 1998; Williams 1998; Stout and Goulson 2002; Wilms and Eltz 2008). Carefully controlled experiments are therefore needed, in which olfactory and visual cues are independently manipulated in order to determine to what degree either of these cues contributes to the bees' food choice in a particular system. One such study on yellow jacket wasps (D'Adamo et al. 2000) found that odor cues alone, without any visual cues of conspecifics, elicited the same response as visual and odor cues combined, whereas local enhancement to visual cues alone was significantly lower.

Local enhancement in social insects has also led to some controversies: Foragers of the European honey bee, *Apis mellifera*, are mutually attracted to each other, and Kalmus (1954) found that incoming foragers frequently settled directly on top of a bee drinking on a feeding dish. Foragers were much more hesitant to land on an empty feeder, so that the observed mutual attraction caused strong cluster formation on one of several available dishes (Kalmus 1954). This strong local enhancement among honey bee foragers caused part of the famous "dance language controversy" (see Wenner and Wells 1990) because it was disregarded in the early studies by Karl von Frisch (discussed in Johnson 1967). In these experiments (von Frisch 1948, 1954), marked foragers had been trained to feed from an experimental feeding station, and the majority of the recruited bees arrived at this experimental station rather than at any of the control stations. The results were interpreted as evidence that the recruits make use of the directional and distance information contained in the foragers' dances. However, these initial experiments lacked essential controls, one being the lack of visitation of the control feeders by foragers (while the experimental station was continuously visited). When bees, which had been trained from a different hive (which contained light-colored bees in contrast to the dark-colored bees from the experimental hive), foraged at the control stations, the results were significantly different from those of the experiments in which bees foraged only at the experimental station (Johnson 1967). In this case, the recruits were distributed more equally over the available feeding stations, with the highest number of recruits at the feeders located in the center, irrespective of whether this was the original training feeder or not. This indicates that the presence of conspecifics has a major influence on a forager's selection of food source and, consequently, on a colony's spatial foraging pattern (Johnson 1967; Wenner 1967; Wenner and Wells 1990). Similarly, unrecognized local enhancement has confounded results of recruitment experiments in wasps (Raveret Richter 2000), as well as investigations of foraging behavior in other animals (Weatherhead 1987; Mock et al. 1988). Also, a recent study that claimed that different honey bee species (*Apis cerana* and *Apis mellifera*) can understand each other's dialect (Su et al. 2008) lacked appropriate controls, and actually, further experiments are needed to test whether the foragers ended up at the "communicated" feeder because of a correct interpretation of the other species' dance language, or due to local enhancement.

THE IMPORTANCE OF CONTEXT AND EXPERIENCE

Several studies have found that the use of repellent scent marks can be facultative and context dependent. For example, bumble bee foragers are more likely to reject scent-marked flowers that are complex and require a longer handling time than scent-marked flowers that are simple (Saleh et al. 2006; see also Chapter 13). Similarly, the stingless bee *Trigona fulviventris* has been found to avoid flowers that have recently been visited by conspecifics when foraging on *Priva mexicana*, but not when robbing *Crotalaria cajanifolia* (Goulson et al. 2001; see also Chapter 12). Stout and Goulson (2002) found that in bumble bees the duration of flower avoidance after an initial visit differed among plant species, and was inversely related to nectar secretion rates (see Chapter 13). Bees are evidently able to integrate the strength of the repellent cue with their experience of the replenishment rate of the particular flower species when reaching a decision on whether to forage on it. The logical extension of this is that, under certain circumstances, scent marks may become attractive— which is indeed the case. A number of studies on honey bees (Kalmus and Ribbands 1952; Kalmus 1954), bumble bees (Cameron 1981; Schmitt and Bertsch 1990; Schmitt et al. 1991; Williams 1998; Goulson et al. 2000), and stingless bees (Villa and Weiss 1990; Aguilar and Sommeijer 2001; Nieh et al. 2003; Hrncir et al. 2004; Jarau et al. 2004; Schmidt et al. 2005; Boogert et al. 2006) have found that the scent marks deposited at food sources are attractive to foragers. Importantly, though, all of these studies involved artificial food sources with a surplus of reward, whereas all of the studies on natural flowers had found a repellent effect of scent marks. Bees are well known to be able to learn the odors associated with food. Therefore, it is not surprising that, when collecting at an unlimited resource, the bees learn to associate the scent marks deposited by previous visitors with the presence of food. This fact has indeed been confirmed experimentally by two recent studies. In one, artificial feeders were filled with small, nonreplenished amounts of reward. Under these circumstances, bumble bee foragers were found to avoid recently visited artificial food sources (Witjes and Eltz 2007). In another study, bumble bee foragers were repelled by conspecific scent marks that they encountered on nonrewarding flowers, but were attracted to those on rewarding flowers (Saleh and Chittka 2006). The latter experiments point to the fact that the same cues can both repel and attract foragers, depending on the respective context (see also Chapters 12 and 13).

The above-mentioned examples all relate to undefended food sources, at which all foragers that are attracted to then have an equal chance of feeding. In nature, however, some species may aggressively defend their food patches against both con- and heterospecifics. In these cases, local enhancement or local inhibition may occur, depending upon the relative behavioral dominance of the interacting species. Raveret Richter and Tisch (1999) found that the behaviorally dominant wasp species *Vespula germanica* and *Polistes fuscatus* tended to be visually attracted by the presence of other wasps at a food patch, whereas foragers of several other *Vespula* species, particularly those of the *V. rufa* group, tended to avoid occupied food sources. Similarly, experiments with stingless bees foraging on artificial flowers revealed that reactions to heterospecifics were often different from reactions to conspecifics, and that the exhibited behavior depended upon the respective species combination (Table 8.1; Slaa et al. 2003b): Whether foragers avoided or landed next to a feeding heterospecific depended on body size (a measure of dominance*) and aggressiveness of both the incoming and the resident bee. All nonaggressive foragers avoided aggressive foragers, and incoming bees also avoided larger resident bees. Heterospecific local enhancement was less common than heterospecific local inhibition, with the former only being exhibited by aggressive foragers (Slaa et al. 2003b). In addition, Nieh et al. (2004) reported that the stingless bee *Trigona spinipes* was attracted to scent marks left at the feeding site by *Melipona rufiventris*, and then rapidly took over the food source by driving away or killing the *M. rufiventris* foragers. Thus, heterospecific local enhancement in stingless bees appears to occur only toward a competitively submissive species. The same is true for ants—at least for the only example investigated so far: Foragers of *Formica pratensis* are attracted to the movements of *Formica cunicularia* foragers (but not to other species), over which they are behaviorally dominant, and thus are often able to usurp a food source visited by *F. cunicularia* (Reznikova 1982). Indeed, when *F. pratensis* colonies were prevented from foraging, nearby *F. cunicularia* colonies increased their food intake, and when *F. cunicularia* colonies were prevented from foraging, *F. pratensis* harvested significantly less food (Reznikova 1982). This is a clear example where heterospecific local enhancement significantly improves the foraging efficiency of a behaviorally dominant species but, on the other hand, has important negative implications for the target of the enhancement.

Finally, experience also influences local enhancement and local inhibition. In some stingless bees and bumble bees, foragers change their reactions to nestmates depending on their experience with the food source. For example, in the stingless bee *T. amalthea*, which uses pheromone trails during recruitment of nestmates, only those bees that had no experience with the food source tended to land on or close to nestmates, with the probability of doing so decreasing from 80% to 20% during the first five visits to a new food source (i.e., shifting from local enhancement to local inhibition; Slaa et al. 2003b). A similar pattern has been described for the bumble bee *Bombus diversus* (Kawaguchi et al. 2007). In this species, foragers tended to avoid occupied inflorescences when foraging on familiar flowers, whereas they showed the tendency to land on occupied inflorescences when presented with unfamiliar flowers (Kawaguchi et al. 2007). It is, in fact, important that such effects of experience are controlled for during the investigations of local enhancement/inhibition. For example, studies on wasps could suggest that local enhancement tends to occur when the food source is meat (Reid et al. 1995), whereas local inhibition occurs when it is nectar (Parrish and Fowler 1983; Raveret Richter and Tisch 1999). However, as the former study only used naïve wasps, whereas the latter studies included experienced wasps, the differences could potentially be due to experience rather than to the nature of the food resource.

THE IMPORTANCE OF LOCAL ENHANCEMENT AND LOCAL INHIBITION

As the preceding sections suggest, both local enhancement and local inhibition are evidently valuable strategies, so that multiple species of social insects have evolved these behavioral traits. Local enhancement serves to directly improve foraging efficiency by increasing a forager's probability

---

* In general, aggressive species (like *Trigona hyalinata*, *T. fuscipennis*, or *T. spinipes*) are dominant over nonaggressive species (like *Melipona* sp.). Within the aggressive and nonaggressive species, body size is a reliable measure of dominance.

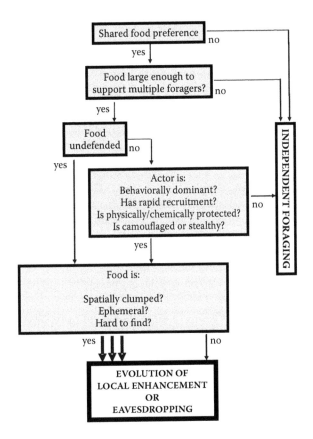

FIGURE 8.1 Flowchart of factors influencing the evolution of local enhancement or eavesdropping among non-nestmates. Note that when food sources are spatially clumped, ephemeral, or hard to find, the likelihood of local enhancement and eavesdropping evolving increases ("yes" arrows), but these factors are not essential for local enhancement and eavesdropping to evolve ("no" arrow).

of discovering food (e.g., Kawaguchi et al. 2006). On the other hand, it can also improve foraging efficiency indirectly by facilitating learning by foragers (Worden and Papaj 2005; Leadbeater and Chittka 2007). Local enhancement will be particularly important when the food is clumped or ephemeral (Figure 8.1). Yet, it is only worthwhile if the food source is large enough to support the foraging trips of multiple individuals, and if the individuals attracted by local enhancement have a reasonable chance of obtaining a share of the food (Figure 8.1). Therefore, local enhancement that involves individuals from different colonies or species tends to occur when the resident species either does not attempt to defend the food source or is unable to do so effectively against a more dominant usurper. Local inhibition, on the other hand, serves to avoid direct competition with nestmates or with foragers of other species. It also helps to avoid recently visited and thus depleted flowers, which is particularly important in cases where significant efforts are needed to access the food source, such as flowers with complex structures, or flowers a forager is unfamiliar with (Saleh et al. 2006; Kawaguchi et al. 2007).

EAVESDROPPING

Whereas local enhancement/inhibition involves passive cues, there is also the potential for foragers to detect and utilize the information conveyed by active signals that are produced by con- or heterospecifics when making their foraging decisions. Such behavior can be termed eavesdropping.

This term has been defined in different ways (e.g., Otte 1974; Wiley 1983, 1994; Bradbury and Vehrencamp 1998), but here we will follow Peake (2005), who defined eavesdropping as "the use of information in signals by individuals other than the primary target" (Peake 2005, p. 14). This definition thus excludes the use of unintentional cues such as the presence of foragers on a flower, which we categorized under local enhancement/inhibition (but see Bradbury and Vehrencamp 1998). The prerequisite for eavesdropping to be possible is that communication occurs in a network environment; i.e., signals can be received by several receivers, or receivers can receive signals, at least potentially, from several signalers at any one time (Peake 2005). Like local enhancement, eavesdropping has mostly been associated with vertebrates. Humans, for instance, can eavesdrop on the echolocation calls produced by bats, which has been utilized by biologists to study the distribution and habitat use of these animals (Fenton 1997). The Galapagos marine iguana, to give another example, has been demonstrated to eavesdrop on the alarm calls of the Galapagos mocking bird, responding to them with antipredator behavior (Vitousek et al. 2007). However, eavesdropping also occurs in invertebrates, and perhaps the best-known examples come from predator–prey relationships. For example, several predators and hymenopteran parasitoids of bark beetles are attracted to the sex or aggregation pheromones of their prey. Similarly, *Trichogramma* parasitoids use the sex pheromones of *Pieris* or *Helicoverpa* lepidopterans to locate eggs to parasitize (Stowe et al. 1995).

Peake (2005) distinguished two forms of eavesdropping: (1) eavesdropping on signals (interceptive eavesdropping) and (2) eavesdropping on signal interactions (social eavesdropping). Social eavesdropping deals with gathering information on other individuals in order to establish a relative dominance rank, for instance, which has been reported for some territorial fish and bird species (Peake 2005). Although such social eavesdropping may be important in the colonial organization of some social insects, for example, in primitive ants and wasps with behaviorally determined dominance hierarchies, it is unlikely to be relevant in social insect foraging. Interceptive eavesdropping, however, is far more likely to occur. In many social insect species, foraging areas of different colonies may overlap, creating the potential for foragers to eavesdrop on the signals left by individuals from other conspecific or heterospecific colonies, and to use the respective information when making foraging decisions. As competition for food is generally high among social insect colonies, eavesdropping may provide a beneficial strategy to find new food sources.

However, the number of conclusive records of eavesdropping in social insects is surprisingly low (Table 8.2), which is partly due to the difficulty in demonstrating it. In addition, there are several records that are to a greater or lesser extent suggestive of eavesdropping, but which could potentially be explained by co-occurring scent trails or by the coincidental independent discovery of a food resource by foragers of the different species. All of the confirmed cases involve chemical signals (pheromones), most probably because when other forms of signal are used, they generally occur only within the nest where eavesdropping will be impossible. No cases are known from termites, which may well reflect a genuine lack of trail-laying behavior in the open in this group of insects, with their mix of species living within their food and species foraging along restricted tunnels. Although termites of some species are able to detect and follow the trails of heterospecifics under laboratory conditions (Matsumura et al. 1972; Oloo and McDowell 1982; Peppuy et al. 2001), it seems very unlikely that eavesdropping occurs under natural conditions. The phenomenon of eavesdropping may be genuinely rare, because many species of social insects either do not seem to convey information about food location (e.g. wasps, bumble bees, and many solitary foraging stingless bees), or they do so within the confines of their nest, where signals will be hidden from eavesdropping (e.g., honey bees). However, many stingless bees deposit pheromonal beacons at, or in a trail toward, food resources, and eavesdropping has been recorded in two such species of *Trigona* (Table 8.2). The majority of records of eavesdropping by social insects, however, are from ants (Table 8.2). The pheromone trails laid by foragers of many ant species may overlap and constitute a communication network. Where these trails tend to lead to high-quality food sources, the value of interceptive eavesdropping will be high.

TABLE 8.2
Compilation of Published Evidence for Eavesdropping in Social Insects

	Species 1		Species 2		Eavesdropper	Reference
Ant	*Acromyrmex*	*versicolor*	*Atta*	*mexicana*	Species 1	[1]
Ant	*Atopomyrmex*	*mocquerisi*	*Camponotus*	sp. A	Coincidence	[2]
Ant	*Atopomyrmex*	*mocquerisi*	*Cataulachus*	*guineensis*	Unknown	[2]
Ant	*Atopomyrmex*	*mocquerisi*	*Crematogaster*	*castanea*	Unknown	[2]
Ant	*Atopomyrmex*	*mocquerisi*	*Polyrhachis*	*weissi*	Unknown	[2]
Ant	*Camponotus*	*beebei*	*Azteca*	*chartifex*	Species 1	[3]
Ant	*Camponotus*	*femoratus*	*Crematogaster*	*levior*	Species 1	[4, 5]
Ant	*Camponotus*	*lateralis*	*Crematogaster*	*scutellaris*	Species 1	[3]
Ant	*Camponotus*	*rufifemur*	*Crematogaster*	*modiglianii*	*parabiosis*	[6]
Ant	*Camponotus*	sp. 1	*Crematogaster*	*coriaria*	Mutual	[7]
Ant	*Dolichoderus*	*bispinosus*	*Crematogaster*	*carinata*	Parabiosis	In [5]
Ant	*Dolichoderus*	*debilis*	*Crematogaster*	*carinata*	Species 1	[5]
Ant	*Dolichoderus*	*inermis*	*Crematogaster*	*carinata*	Parabiosis	In [4]
Ant	*Odontomachus*	*panamensis*	*Crematogaster*	*carinata*	Parabiosis	In [4]
Ant	*Odontomachus*	*mayi*	*Crematogaster*	*carinata*	Parabiosis	[8]
Ant	*Odontomachus*	*mayi*	*Dolichoderus*	*debilis*	Parabiosis	In [5]
Ant	*Oecophylla*	*longinoda*	*Cataulachus*	*guineensis*	Coincidence	[2]
Ant	*Oecophylla*	*longinoda*	*Crematogaster*	*castanea*	Unknown	[2]
Ant	*Polyrhachis*	*armata*	*Polyrhachis*	*bihamata*	Species 1	[9]
Ant	*Polyrhachis*	*rufipes*	*Gnamptogenys*	*menadensis*	Species 1	[10]
Ant	*Polyrhachis*	sp.	*Campontotus*	spp.	Species 1	[11]
Ant	*Cephalotes*	*maculatus*	*Azteca*	*trigona*	Species 1	[12]
Bee	*Trigona*	*amalthea*	*Trigona*	*fulviventris*	Species 1	[13, 14]
Bee	*Trigona*	*spinipes*[a]	*Melipona*	*rufiventris*	Species 1	[15]

Note: Several additional cases of ant species that share trails are more probably due to co-occurring trails rather than eavesdropping, and have therefore been excluded.

[a] *T. spinipes* was attracted to scent marks left by *M. rufiventris*, which was described by the authors as eavesdropping, but might rather be local enhancement (see text).

Cephalotes maculates: Previously known as *Zacryptocercus maculates*.
Crematogaster carinata: Previously known as *Crematogaster limata parabiotica* (Longino 2003, Zootaxa 151:1–150).
Crematogaster levior: Previously known as *Crematogaster limata parabiotica* (Longino 2003, Zootaxa 151:1–150).
Dolichoderus debilis: Previously known as *Monacis debilis* and *Monacis rufescens*.
Dolichoderus bispinosus: Previously known as *Monacis bispinosa*.

[1] Mintzer (1980); [2] Dejean (1996); [3] Wilson (1965); [4] Vantaux et al. (2007); [5] Swain (1980); [6] Menzel et al. (2008); [7] Menzel and Blüthgen (2006); [8] Orivel et al. (1997); [9] Starr (1981); [10] Gobin et al. (1998); [11] Davidson et al. (2007); [12] Adams (1990); [13] Johnson (1983); [14] Johnson (1987); [15] Nieh et al. (2004).

EAVESDROPPING IN GARDEN ANTS AND THE TRADE-OFF BETWEEN DISCOVERY RATE AND BEHAVIORAL DOMINANCE

Many of the ant species that have been observed to be involved in eavesdropping are "garden ants." These are arboreal ants that are mutualists with epiphytes. Frequently, two ant species share the same nest, a phenomenon known as parabiosis. Where colonies of two species are so closely integrated, encounters with the trails of the other parabiotic partner will be extremely common and eavesdropping particularly likely. In some of these parabiotic associations, eavesdropping may be bidirectional and amicable (Table 8.2). However, many other relationships are not so benign. In the association of *Crematogaster levior* and *Dolichoderus debilis* (previously known as *C. limata para-biotica* and *Monacis debilis*, respectively), it was first reported that the two species use common odor trails and foraged peacefully together on the same plants (Forel 1898). This was also observed for *Crematogaster levior* and *Camponotus femoratus* (Wheeler 1921). However, later studies showed that both *D. debilis* and *C. femoratus* follow the trails laid by *Cr. levior*, and displace them from food sources (Swain 1980; Vantaux et al. 2007). It appears that the smaller, more numerous *Cr. levior* is generally quicker to discover prey, and that the other species eavesdrop on their trails to improve their own discovery rate of food. Although *Cr. levior* is also able to follow *C. femoratus* trails, this occurs less frequently, and *Cr. levior* are only able to obtain a small proportion of the food sources discovered by *C. femoratus* (Vantaux et al. 2007). This trade-off between discovery rate and behavioral dominance has also been found in a guild of woodland ants (Fellers 1987; but see also Davidson 1998), between *Camponotus cylindricus* and *Polyrhachis* (Davidson et al. 2007), *Polyrhachis rufipes* and *Gnamptogenys menadensis* (Gobin et al. 1998), and in guilds of stingless bees (Hubbell and Johnson 1978; Nagamitsu and Inoue 1997; Slaa 2003). If this is a common trait in social insects, we would expect that heterospecific trail following is biased toward dominant species following trails of subordinate species. Although only few studies report on heterospecific trail following, those on stingless bees are in accordance to the above expectation. Observations by Johnson (1983, 1987) suggest that the stingless bee *Trigona amalthea* may exploit the ability of *Trigona ful-viventris* to quickly find new resources by attraction to its scent marks, and Slaa (2003) found that *T. amalthea* competitively displaces *T. fulviventris* from high-quality nectar sources. Similar olfactory eavesdropping and displacement have also been reported for the stingless bee *T. spinipes*, which is behaviorally dominant over *M. rufiventris* (Nieh et al. 2004). However, the origin of the scent marks left at the food source by both latter species had not been investigated. Yet, since it is very likely that these scent marks were unintentional footprints, they consequently were cues rather than signals. Under our definition (see above), the behavior described by Nieh et al. (2004) is therefore classified as local enhancement.

Eavesdropping appears to play a particularly important role in the foraging biology of the arboreal ant *Cephalotes maculatus* (previously referred to as *Zacryptocerus maculates*). Nests of *C. maculatus* are small, and in Panama often occur in territories of *Azteca trigona*, which is also arboreal with large nests and territories, which it aggressively defends against most other ant species. Foragers of *C. mac-ulatus* are apparently able to follow the pheromone trails of *A. trigona*, which substantially increases their probability of discovering food (Adams 1990). Unlike the previous examples, *C. maculatus* is not behaviorally dominant. It is generally ignored by *A. trigona* and, if attacked, utilizes its unusual, dorsoventrally flattened morphology to protect itself by flattening itself against the substrate (Adams 1990). In fact, the use of alien scent signals may be widespread among species of these two genera, with *C. foliaceus* and *C. umbraculatus* having also been observed at baits discovered by *A. trigona*, and *A. velox* was observed to share baits with *C. umbraculatus* (Adams 1990).

THE IMPORTANCE OF EAVESDROPPING

The value to the eavesdropper is obvious. Much of the food fed on by social insects is clumped, with detection requiring a considerable search effort. A recruitment signal left by another individual acts as a reliable sign as to the location of food, so following it to the food—and provided that this food can then be utilized—will reduce search effort and increase foraging efficiency. It is also important to consider the effect on the individual being eavesdropped. The latter individual has produced a signal to help its nestmates locate the food it has discovered, thereby increasing the efficiency with which its colony collects food, and thus ultimately increasing its own inclusive fitness. If an individ-ual from another colony eavesdrops on a forager's signal and utilizes this to gather some or all of the food resource (thereby inevitably leaving less for individuals from the forager's own colony), then the effect on the signaling forager can only be negative. Eavesdropping during foraging is therefore in a sense parasitizing the communication system of other individuals (Nuechterlein 1981) and has been described as "chemical espionage" (Vinson 1984).

The impact of eavesdropping on either party depends upon their foraging strategies and behavioral dominance. In many of the interactions recorded, particularly those associated with a parabiotic relationship, it appears that the two species share the food resource amicably. Where the parabiosis is mutualistic (i.e., with both species benefiting from nesting together), then the negative effect of being eavesdropped may be offset because the "stolen" food does at least benefit a mutualist partner, and therefore to some extent the eavesdropped individuals themselves. Unless the eavesdropped colony has a surplus of food, this benefit will not be enough to make being eavesdropped upon neutral or beneficial, but it may well reduce the cost to a level at which attempting to avoid eavesdropping (e.g., by aggressive dominance of the food) is unprofitable. Eavesdropping will also cause very limited costs when the species doing the eavesdropping has few foragers or a solitary foraging strategy, or in cases where the food resource is very large. Under these circumstances the loss of food stolen may be insignificant. An example for this is the interaction between *Cephalotes maculatus* and *Azteca trigona*. The former species eavesdrops on the trails of the latter species, but has far fewer foragers, and so probably has relatively little impact on the fitness of the latter species (Adams 1990). However, in many other interactions where the eavesdropper is behaviorally dominant over the eavesdropped species, the negative impact may be significant. Such dominant eavesdroppers may aggressively prevent individuals of the eavesdropped species from utilizing the food they originally discovered, and so cause significant loss of resources to the latter species, particularly when the search effort involved in locating the food is considered.

The question of how widespread eavesdropping is, is harder to answer. Although there are a number of records of eavesdropping from ants and bees (Table 8.2), they are relatively few in number. To some extent this could reflect a general rarity of eavesdropping. Many ants are well known for their strongly territorial behavior throughout their foraging area, which will greatly limit the potential for eavesdropping to occur. Within the stingless bees, while some species will feed peacefully next to noncolony members, others are very aggressive toward con- and heterospecifics (Kerr 1959; Johnson and Hubbell 1974, 1975; Roubik 1982; Nagamitsu and Inoue 1997; Slaa 2003). It seems likely, however, that eavesdropping is far more widespread than current data suggest. All of the current records relate to heterospecific eavesdropping, which is relatively straightforward for the human observer to detect. However, it is conspecific eavesdropping that is, in fact, most likely to occur, both because individuals will share food requirements and because they will share the ability to detect and interpret the signals. Identifying conspecific eavesdropping, though, is problematic, since it requires detailed and simultaneous observation of multiple individuals. Such laborious investigations will probably not be conducted unless detecting eavesdropping is the specific aim. It seems likely that such studies may prove fruitful.

CONCLUDING REMARKS

This review shows that social insect foragers use the little-studied information systems of *local enhancement*, *local inhibition*, and *eavesdropping* to locate food sources, in addition to active recruitment systems, which are much better studied. As the best-studied cases of local enhancement/ inhibition show, the respective behaviors are potentially complex, being dependent on context and experience of the involved individuals. The number of reported cases is still low, which might reflect the fact that these information systems are easily overlooked, but they might be more important than we have thought. At the very least, it is critical to consider them as possibilities when studying other aspects of social insect foraging and recruitment; otherwise, they may confound the aspect being investigated. Clearly, more studies are needed to confirm their importance and frequency, but it is already obvious that local enhancement/inhibition and eavesdropping can play important roles in guiding the foraging choices of social insects.

REFERENCES

Adams ES. (1990). Interaction between the ants *Zacryptocerus maculatus* and *Azteca trigona*: Interspecific parasitization of information. *Biotropica* 22:200–6.

Aguilar I, Sommeijer M. (2001). The deposition of anal excretions by *Melipona favosa* foragers (Apidae: Meliponinae): Behavioural observations concerning the location of food sources. *Apidologie* 32:37–48.

Arenas A, Fernández VM, Farina WM. (2007). Floral odor learning within the hive affects honeybees' foraging decisions. *Naturwissenschaften* 94:218–22.

Barth FG, Hrncir M, Jarau S. (2008). Signals and cues in the recruitment behavior of stingless bees (Meliponini). *J Comp Physiol A* 194:313–27.

Beauchamp G, Bélisle M, Giraldeau LA. (1997). Influence of conspecific attraction on the spatial distribution of learning foragers in a patchy habitat. *J Anim Ecol* 66:671–82.

Boogert NJ, Hofstede FE, Monge IA. (2006). The use of food source scent marks by the stingless bee *Trigona corvina* (Hymenoptera : Apidae): The importance of the depositor's identity. *Apidologie* 37:366–75.

Bradbury JW, Vehrencamp SL. (1998). *Principles of Animal Communication.* Sunderland: Sinauer Associates.

Brown C, Laland KN. (2003). Social learning in fishes: A review. *Fish Fish* 4:280–88.

Cameron SA. (1981). Chemical signals in bumble bee foraging. *Behav Ecol Sociobiol* 9:257–60.

Clark CW, Mangel M. (1986). The evolutionary advantages of group foraging. *Theor Popul Biol* 3:45–75.

Clayton DA. (1978). Socially facilitated behavior. *Quart Rev Biol* 53:373–92.

Collins RD, Bell WJ. (1996). Enhancement of resource finding efficiency by visual stimuli in *Musca domestica* (Diptera: Muscidae). *J Kans Entomol Soc* 69:204–7.

Costa JT. (2006). *The Other Insect Societies.* Cambridge, MA: Belknap Press of Harvard University Press.

D'Adamo P, Corley J, Sackmann P, Lozada M. (2000). Local enhancement in the wasp *Vespula germanica*. Are visual cues all that matter? *Insect Soc* 47:289–91.

D'Adamo P, Lozada M, Corley J. (2003). Conspecifics enhance attraction of *Vespula germanica* (Hymenoptera: Vespidae) foragers to food baits. *Ann Entomol Soc Am* 96:685–88.

Davidson DW. (1998). Resource discovery versus resource domination in ants. *Ecol Entomol* 23:484–90.

Davidson DW, Lessard JP, Bernau CR, Cook SC. (2007). The tropical ant mosaic in a primary Bornean rain forest. *Biotropica* 39:468–75.

Dejean A. (1996). Trail sharing in African arboreal ants (Hymenoptera: Formicidae). *Sociobiology* 27:1–10.

Dobson HEM. (1987). Role of flower and pollen aromas in host-plant recognition by solitary bees. *Oecologia* 72:618–23.

Dornhaus A, Chittka L. (1999). Evolutionary origins of bee dances. *Nature* 401:38

Dreisig H. (1995). Ideal free distributions of nectar foraging bumblebees. *Oikos* 72:161–72.

Eltz T, Brühl CA, van der Kaars S, Linsenmair KE. (2002). Determinants of stingless bee nest density in low-land dipterocarp forests of Sabah, Malaysia. *Oecologia* 131:27–34.

Fellers JH. (1987). Interference and exploitation in a guild of woodland ants. *Ecology* 68:1466–78.

Fenton MB. (1997). Science and the conservation of bats. *J Mammal* 78:1–14.

Ferguson AW, Free JB. (1979). Production of forage-marking pheromone by the honeybee. *J Apic Res* 18:128–35.

Forel A. (1898). La parabiose chez les fourmis. *Bull Soc Vaud Sci Nat* 34:380–84.

Fowler HG. (1992). Social facilitation during foraging in *Agelaia* (Hymenoptera: Vespidae). *Naturwissenschaften* 79:424.

Free JB. (1969). Influence of the odour of a honeybee colony's food stores on the behaviour of its foragers. *Nature* 222:778

Free JB, Williams IH. (1983). Scent-marking of flowers by honeybees. *J Apic Res* 22:86–90.

von Frisch K. (1948). Solved and unsolved problems of bee language. *Bull Anim Behav* 9:2–25.

von Frisch K. (1954). *The Dancing Bees: An Account of the Life and Senses of the Honey Bee.* London: Methuen.

Galef BG. (1988). Imitation in animals: History, definition, and interpretation of data from the psychological laboratory. In Zentall TR, Galef BG (eds.), *Social Learning: Psychological and Biological Perspectives.* Hillsdale, NJ: Lawrence Erlbaum Associates, pp. 3–28.

Galef BG, Giraldeau LA. (2001). Social influences on foraging in vertebrates: Causal mechanisms and adaptive functions. *Anim Behav* 61:3–15.

Gawleta N, Zimmerman Y, Eltz T. (2005). Repellent foraging scent recognition across bee families. *Apidologie* 36:325–30.

Giraldeau LA, Caraco T. (2000). *Social Foraging Theory*. Princeton, NJ: Princeton University Press.

Giurfa M. (1993). The repellent scent-mark of the honeybee *Apis mellifera ligustica* and its role as communication cue during foraging. *Insect Soc* 40:59–67.

Giurfa M, Núñez JA. (1992). Honeybees mark with scent and reject recently visited flowers. *Oecologia* 89:113–17.

Gobin B, Peeters C, Billen J, Morgan ED. (1998). Interspecific trail following and commensalism between the ponerine ant *Gnamptogenys menadensis* and the formicine ant *Polyrhachis rufipes*. *J Insect Behavior* 11:361–69.

Goulson D, Chapman JW, Hughes WOH. (2001). Discrimination of unrewarding flowers by bees: Direct detection of rewards and use of repellent scent marks. *J Insect Behav* 14:669–78.

Goulson D, Hawson SA, Stout JC. (1998). Foraging bumblebees avoid flowers already visited by conspecifics or by other bumblebee species. *Anim Behav* 55:199–206.

Goulson D, Stout JC, Langley J, Hughes WOH. (2000). The identity and function of scent marks deposited by foraging bumblebees. *J Chem Ecol* 26:2897–911.

Grüter C, Acosta LE, Farina WM. (2006). Propagation of olfactory information within the honeybee hive. *Behav Ecol Sociobiol* 60:707–15.

Heithaus ER. (1974). The role of plant-pollinator interactions in determining community structure. *Ann Mo Bot Gard* 61:675–91.

Heithaus ER. (1979). Flower visitation records and resource overlap of bees and wasps in northwest Costa Rica. *Brenesia* 16:9–52.

Heyes CM. (1994). Social learning in animals—Categories and mechanisms. *Biol Rev* 69:207–31.

Heyes CM, Ray ED, Mitchell CJ, Nokes T. (2000). Stimulus enhancement: Controls for social facilitation and local enhancement. *Learn Motiv* 31:83–98.

Holway DA, Lach L, Suarez AV, Tsutsui ND, Case TJ. (2002). The causes and consequences of ant invasions. *Ann Rev Ecol Syst* 33:181–233.

Hölldobler B, Wilson EO. (1990). *The Ants*. Cambridge, MA: Belknap Press of Harvard University Press.

Hrncir M, Jarau S, Zucchi R, Barth FG. (2004). On the origin and properties of scent marks deposited at the food source by a stingless bee, *Melipona seminigra* Friese 1903. *Apidologie* 35:3–13.

Hrncir M, Mateus S, Nascimento FS. (2007). Exploitation of carbohydrate food sources in *Polybia occidentalis*: Social cues influence foraging decisions in swarm-founding wasps. *Behav Ecol Sociobiol* 61:975–83.

Hubbel SP, Johnson LK. (1977). Competition and nest spacing in a tropical stingless bee community. *Ecology* 58:949–63.

Hubbel SP, Johnson LK. (1978). Comparative foraging behavior of six stingless bee species exploiting a standardized resource. *Ecology* 59:1123–36.

Hunter MD, Price PW. (1992). Playing chutes and ladders: Heterogeneity and the relative roles of bottom-up and top-down forces in natural communities. *Ecology* 73:724–32.

Inouye DW. (1978). Resource partitioning in bumble bees: Experimental studies of foraging behaviour. *Ecology* 59:672–78.

Jackson AL, Ruxton GD, Houston DC. (2008). The effect of social facilitation on foraging success in vultures: A modelling study. *Biol Lett* 4:311–13.

Jakobsen HB, Kristjánsson K, Rohde B, Terkildsen M, Olsen CE. (1995). Can social bees be influenced to choose a specific feeding station by adding the scent of the station to the hive air? *J Chem Ecol* 21:1635–48.

Jarau S, Hrncir M, Ayasse M, Schulz C, Francke W, Zucchi R, Barth FG. (2004). A stingless bee (*Melipona seminigra*) marks food sources with a pheromone from its claw retractor tendons. *J Chem Ecol* 30:793–804.

Johnson DL. (1967). Honey bees: Do they use the direction information contained in their dance maneuver? *Science* 155:844–47.

Johnson LK. (1983). Foraging strategies and the structure of stingless bee communities in Costa Rica. *Soc Insects Trop* 2:31–58.

Johnson LK. (1987). Communication of food source location by the stingless bee *Trigona fulviventris*. In Eder J, Rembold H (eds.), *Chemistry and Biology of Social Insects*. München: Peperny, pp. 689–99.

Johnson LK, Hubbell SP. (1974). Aggression and competition among stingless bees: Field studies. *Ecology* 55:120–27.

Johnson LK, Hubbell SP. (1975). Contrasting foraging strategies and coexistence of two bee species on a single resource. *Ecology* 56:1398–406.

Kalmus H. (1954). The clustering of honeybees at a food source. *Br J Anim Behav* 2:63–71.

Kalmus H, Ribbands CR. (1952). The origin of the odours by which honeybees distinguish their companions. *Proc R Soc Lond B* 140:50–59.

Kato M. (1988). Bumblebee visits to *Impatiens* spp.: Pattern and efficiency. *Oecologia* 76:364–70.

Kawaguchi LG, Ohashi K, Toquenaga Y. (2006). Do bumble bees save time when choosing novel flowers by following conspecifics? *Funct Ecol* 20:239–44.

Kawaguchi LG, Ohashi K, Toquenaga Y. (2007). Contrasting responses of bumble bees to feeding conspecifics on their familiar and unfamiliar flowers. *Proc R Soc Lond B* 274:2661–67.

Kerr WE. (1959). Bionomy of Meliponids. VI. Aspects of food gathering and processing in some stingless bees. In *Symposium on Food Gathering Behavior of Hymenoptera*, Cornell University, Ithaca, NY, pp. 24–31.

Kim KW, Noh S, Choe JC. (2007). Lack of field-based recruitment to carbohydrate food in the Korean yellowjacket, *Vespula koreensis*. *Ecol Res* 22:825–30.

Krause J, Ruxton GD. (2002). *Living in Groups*. Oxford: Oxford University Press.

Leadbeater E, Chittka L. (2005). A new mode of information transfer in foraging bumblebees? *Curr Biol* 15:R447–48.

Leadbeater E, Chittka L. (2007). The dynamics of social learning in an insect model, the bumblebee (*Bombus terrestris*). *Behav Ecol Sociobiol* 61:1789–96.

Lowe SJ, Browne M, Boudjelas S, De Poorter M. (2000). *100 of the World's Worst Invasive Alien Species: A Selection from the Global Invasive Species Database*. Auckland: The Invasive Species Specialist Group (ISSG).

Makino TT, Sakai S. (2005). Does interaction between bumblebees (*Bombus ignitus*) reduce their foraging area? Bee-removal experiments in a net cage. *Behav Ecol Sociobiol* 57:617–22.

Matsumura F, Jewett DM, Coppel HC. (1972). Interspecific response of termite to synthetic trail-following substances. *J Econ Entomol* 65:600–2.

Mattila HR, Seeley TD. (2007). Genetic diversity in honey bee colonies enhances productivity and fitness. *Science* 317:362–64.

Menzel F, Blüthgen N. (2006). *Crematogaster-Camponotus* associations in a tropical rainforest: Mechanisms and specificity of interspecific recognition. Paper presented at Proc GTO Conference, Kaiserlautern, Germany.

Menzel F, Linsenmair KE, Blüthgen N. (2008). Selective interspecific tolerance in tropical *Crematogaster-Camponotus* associations. *Anim Behav* 75:837–46.

Michener CD. (1974). *The Social Behavior of Bees: A Comparative Study*. Cambridge, MA: Harvard University Press.

Mintzer A. (1980). Simultaneous use of foraging trails by two leafcutter ant species in the Sonoran desert. *J NY Entomol Soc* 88:102–5.

Mock DW, Lamey TC, Thompson DBA. (1988). Falsifiability and the information centre hypothesis. *Ornis Scand* 19:231–48.

Nagamitsu T, Inoue T. (1997). Aggressive foraging of social bees as a mechanism of floral resource partitioning in an Asian tropical rainforest. *Oecologia* 110:432–39.

Nieh JC, Barreto LS, Contrera FAL, Imperatriz-Fonseca VL. (2004). Olfactory eavesdropping by a competitively foraging stingless bee, *Trigona spinipes*. *Proc R Soc Lond B* 271:1633–40.

Nieh JC, Ramírez S, Nogueira-Neto P. (2003). Multi-source odormarking of food by a stingless bee, *Melipona mandacaia*. *Behav Ecol Sociobiol* 54:578–86.

Nuechterlein GL. (1981). Information parasitism in mixed colonies of western grebes and Forster terns. *Anim Behav* 29:985–89.

Núñez JA. (1967). Sammelbienen markieren versiegte Futterquellen durch Duft. *Naturwissenschaften* 54: 322–23.

Oloo GW, McDowell PG. (1982). Interspecific trail-following and evidence of similarity of trails of trinervitermes species from different habitats. *Insect Sci Appl* 3:157–61.

Orivel J, Errard C, Dejean A. (1997). Ant gardens: Interspecific recognition in parabiotic ant species. *Behav Ecol Sociobiol* 40:87–93.

Otis GW, Locke B, McKenzie NG, Cheung D, MacLeod E, Careless P, Kwoon A. (2006). Local enhancement in mud-puddling swallowtail butterflies (*Battus philenor* and *Papilio glaucus*). *J Insect Behav* 19: 685–98.

Otte D. (1974). Effects and functions in the evolution of signalling systems. *Annu Rev Ecol Syst* 5:385–417.

Parrish MD, Fowler HG. (1983). Contrasting foraging related behaviours in two sympatric wasps (*Vespula maculifrons* and *V. germanica*). *Ecol Entomol* 8:185–90.

Peake TM. (2005). Eavesdropping in communication networks. In McGregor PK (ed.), *Animal Communication Networks.* Cambridge, MA: Cambridge University Press, pp. 13–37.

Peppuy A, Robert A, Sémon E, Bonnard O, Son NT, Bordereau C. (2001). Species specificity of trail pheromones of fungus-growing termites from northern Vietnam. *Insect Soc* 48:245–50.

Pleasants JM. (1983). Structure of plant and pollinator communities. In Jones CE, Little RJ (eds.), *Handbook of Experimental Pollination Biology.* New York: Van Nostrand Reinhold, pp. 375–93.

Pöysä H. (1992). Group foraging in patchy environments: The importance of coarse-level local enhancement. *Ornis Scand* 23:159–66.

Propoky RJ, Miller NW, Duan JJ, Vargas RI. (2000). Local enhancement of arrivals of *Ceratitis capitata* females on fruit mimics. *Entomol Exp Appl* 97:211–17.

Pulliam HR, Caraco T. (1984). Living in groups: Is there an optimal group size? In Krebs JR, Davies NB (eds.), *Behavioural Ecology: An Evolutionary Approach.* 2nd ed. Oxford: Blackwell Scientific Publications, pp. 122–47.

Raveret Richter M. (1990). Hunting social wasp interactions: Influence of prey size, arrival order, and wasp species. *Ecology* 71:1018–30.

Raveret Richter M. (2000). Social wasp (Hymenoptera: Vespidae) foraging behavior. *Annu Rev Entomol* 45:121–50.

Raveret Richter M, Tisch VL. (1999). Resource choice of social wasps: Influence of presence, size and species of resident wasps. *Insect Soc* 46:131–36.

Reader T, MacLeod I, Elliott PT, Robinson OJ, Manica A. (2005). Inter-order interactions between flower-visiting insects: Foraging bees avoid flowers previously visited by hoverflies. *J Insect Behav* 18:51–57.

Reid BL, MacDonald JF, Ross DR. (1995). Foraging and spatial dispersion in protein-scavenging workers of *Vespula germanica* and *V. maculifrons* (Hymenoptera: Vespidae). *J Insect Behav* 8:315–30.

Reznikova JI. (1982). Interspecific communication between ants. *Behaviour* 80:84–95.

Romanes GJ. (1884). The Darwinian theory of instinct. *The Nineteenth Century* 91:434–50.

Ross KG, Matthews RW (eds.). (1991). *The Social Biology of Wasps.* Ithaca, NY: Comstock Publishing Associates of Cornell University Press.

Roubik DW. (1982). Ecological impact of Africanized honeybees on native neotropical pollinators. In Jaisson P (ed.), *Social Insects in the Tropics 2.* Paris: Université Paris-Nord, pp. 233–47.

Roubik DW. (1989). *Ecology and Natural History of Tropical Bees.* Cambridge, MA: Cambridge University Press.

Roubik DW, Moreno JE, Vergara C, Wittmann D. (1986). Sporadic food competition with the African honey bee: Projected impact on neotropical social bees. *J Trop Ecol* 2:97–111.

Saleh N, Chittka L. (2006). The importance of experience in the interpretation of conspecific chemical signals. *Behav Ecol Sociobiol* 61:215–20.

Saleh N, Ohashi K, Thomson JD, Chittka L. (2006). Facultative use of the repellent scent mark in foraging bumblebees: Complex versus simple flowers. *Anim Behav* 71:847–54.

Schmidt VM, Zucchi R, Barth FG. (2005). Scent marks left by *Nannotrigona testaceicornis* at the feeding site: Cues rather than signals. *Apidologie* 36:285–91.

Schmitt U, Bertsch A. (1990). Do foraging bumblebees scent-mark food sources and does it matter? *Oecologia* 82:137–44.

Schmitt U, Lübke G, Francke W. (1991). Tarsal secretion marks food sources in bumblebees (Hymenoptera: Apidae). *Chemoecology* 2:35–40.

Seeley TD. (1989). The honey bee colony as a superorganism. *Am Sci* 77:546–53.

Slaa EJ. (2003). Foraging ecology of stingless bees: From individual behaviour to community ecology. Doctoral thesis, University of Utrecht, The Netherlands.

Slaa EJ. (2006). Population dynamics of a stingless bee community in the seasonal dry lowlands of Costa Rica. *Insect Soc* 53:70–79.

Slaa EJ, Tack AJM, Sommeijer MJ. (2003a). The effect of intrinsic and extrinsic factors on flower constancy in stingless bees. *Apidologie* 34:457–68.

Slaa EJ, Wassenberg J, Biesmeijer JC. (2003b). The use of field-based social information in eusocial foragers: Local enhancement among nestmates and heterospecifics in stingless bees. *Ecol Entomol* 28:369–79.

Starr C. (1981). Trail-sharing by two species of *Polyrhachis* (Hymenoptera: Formicidae). *Philipp Entomol* 5:5–8.

Stout JC, Goulson D. (2001). The use of conspecific and interspecific scent marks by foraging bumblebees and honeybees. *Anim Behav* 62:183–89.

Stout JC, Goulson D. (2002). The influence of nectar secretion rates on the responses of bumblebees (*Bombus* spp.) to previously visited flowers. *Behav Ecol Sociobiol* 52:239–46.

Stout JC, Goulson D, Allen JA. (1998). Repellent scent-marking of flowers by a guild of foraging bumblebees (*Bombus* spp.). *Behav Ecol Sociobiol* 43:317–26.

Stowe MK, Turlings TCJ, Loughrin JH, Lewis WJ, Tumlinson JH. (1995). The chemistry of eavesdropping, alarm, and deceit. *Proc Natl Acad Sci USA* 92:23–28.

Su S, Cai F, Si A, Zhang S, Tautz J, Chen S. (2008). East learns from West: Asiatic honeybees can understand dance language of European honeybees. *PLoS ONE* 3:e2365.

Suzuki H, Murai M. (1980). Ecological studies of *Roparidia fasciata* in Okinawa Island I. Distribution of single- and multiple-foundress colonies. *Res Popul Ecol* 22:184–95.

Swain RB. (1980). Trophic competition among parabiotic ants. *Insect Soc* 27:377–90.

Thomson JD, Chittka L. (2001). Pollinator individuality: When does it matter? In Chittka L, Thomson JD (eds.), *Cognitive Ecology of Pollination.* Cambridge, MA: Cambridge University Press, pp. 191–213.

Thomson JD, Peterson SC, Harder LD. (1987). Response of traplining bumble bees to competition experiments: Shifts in feeding location and efficiency. *Oecologia* 71:295–300.

Thorpe WH. (1956). *Learning and Instinct in Animals.* 1st ed. London: Methuen.

Vantaux A, Dejean A, Dor A, Orivel J. (2007). Parasitism versus mutualism in the ant-garden parabiosis between *Camponotus femoratus* and *Crematogaster levior. Insect Soc* 54:95–99.

Villa JD, Weiss MR. (1990). Observations of the use of visual and olfactory cues by *Trigona* spp. foragers. *Apidologie* 21:541–45.

Vinson, SB. (1984). Parasite-host relationships. In Bell WJ, Cardé RT (eds.), *Chemical Ecology of Insects.* London: Chapman & Hall, pp. 205–33.

Vitousek MN, Adelman JS, Gregory NC, St. Clair JJH. (2007). Heterospecific alarm call recognition in a non-vocal reptile. *Biol Lett* 3:632–34.

Waser NM. (1983). Competition for pollination and floral character differences among sympatric plant species: A review of the evidence. In Jones CE, Little RJ (eds.), *Handbook of Experimental Pollination Ecology.* New York: Van Nostrand Reinhold, pp. 277–93.

Weatherhead PJ. (1987). Field tests of information transfer in communally roosting birds. *Anim Behav* 35:614–15.

Wenner AM. (1967). Honey bees: Do they use distance information contained in their dance maneuver? *Science* 155:847–49.

Wenner AM, Wells PH. (1990). *Anatomy of a Controversy: The Question of a "Language" among Bees.* New York: Columbia University Press.

Wetherwax PB. (1986). Why do honeybees reject certain flowers? *Oecologia* 69:567–70.

Wheeler WM. (1921). A new case of parabiosis and the "ant gardens" of British Guiana. *Ecology* 2:89–103.

Wiley RH. (1983). The evolution of communication: Information and manipulation. In Halliday TR, Slater PJB (eds.), *Animal Behavior.* Vol. 2. *Communication.* New York: WH Freeman, pp. 156–89.

Wiley RH. (1994). Errors, exaggeration, and deception in animal communication. In Real LA (ed.), *Behavioral Mechanisms in Evolutionary Ecology.* Chicago: University of Chicago Press, pp. 157–89.

Williams CS. (1998). The identity of the previous visitor influences flower rejection by nectar-collecting bees. *Anim Behav* 56:673–81.

Wilms J, Eltz T. (2008). Foraging scent marks of bumblebees: Footprint cues rather than pheromone signals. *Naturwissenschaften* 95:149–53.

Wilms W, Imperatriz-Fonseca VL, Engels W. (1996). Resource partitioning between highly-social bees and possible impact of the introduced Africanized honey bee on native stingless bees in the Brazilian Atlantic rain forest. *Stud Neotrop Fauna Environ* 31:137–51.

Wilson EO. (1965). Trail sharing in ants. *Psyche* 72:2–7.

Witjes S, Eltz T. (2007). Influence of scent deposits on flower choice: Experiments in an artificial flower array with bumblebees. *Apidologie* 38:12–18

Worden BD, Papaj DR. (2005). Flower choice copying in bumblebees. *Biol Lett* 1:504–7.

Yokoi T, Fujisaki K. (2007). Repellent scent-marking behaviour of the sweat bee *Halictus (Seladonia) aerarius* during flower foraging. *Apidologie* 38:474–81.

Yokoi T, Goulson D, Fujisaki K. (2007). The use of heterospecific scent marks by the sweat bee *Halictus aerarius. Naturwissenschaften* 94:1021–24.

9 The Role of Scents in Honey Bee Foraging and Recruitment

Judith Reinhard and Mandyam V. Srinivasan

CONTENTS

For all insects as a class, have, thanks to the species of odor correlated with nutrition, a keen olfactory sense of their proper food from a distance, even when they are very far away from it; such is the case with the bees.

Aristotle, 330 BC

INTRODUCTION

Scents play a crucial role in a honey bee's life. Both in the darkness of the hive and in the outside environment bees encounter an overwhelming array of different scents, from which they retrieve information. The scents they come across are either produced by other bees (pheromones) or originate from the bees' environment, such as the food processed inside the hive and the flowering plants that bees use as food sources. The role of honey bee pheromones during social activities such as mating, reproduction, brood care, kin recognition, swarming, alarm, and defense has been researched and reviewed extensively (Free 1987). Here, we will focus on the role of pheromones as well as floral scents in honey bee foraging and recruitment. The first section of this chapter introduces the honey bee's sense of smell: it explains how odors are detected, goes on to describe the bee's olfactory anatomy, how olfactory information is processed, and how bees learn and discriminate scents. The next section introduces scents that typically occur in the honey bee foraging environment and gives some background with respect to the scents' origin and chemistry. In this section, we also briefly discuss how honey bees might perceive and interpret the complex scents they encounter. The main section of this book chapter describes the ways that honey bees make use of scents during foraging and recruitment, and discusses the kind of information bees can obtain and transfer via scents.

HONEY BEE OLFACTION

MECHANISM OF ODOR DETECTION

The sense of smell is the most ancient, most basic, universal sense. It plays a vital role in all animal species in detecting and processing information regarding virtually every aspect of life. Across the animal kingdom, chemosensory systems, including the honey bee's, are remarkably similar, sharing a number of fundamental mechanisms (Hildebrand and Shepherd 1997; Strausfeld and Hildebrand 1999). Odorants are detected by olfactory sensory neurons. Odorant molecules reach the dendrites of these olfactory neurons through an aqueous medium via odorant binding proteins (Figure 9.1) (Pelosi 2001; Nagnan-Le Meillour and Jacquin-Joly 2003). Once they have reached the dendrite membrane, the odorant molecules bind to olfactory receptor proteins. This initiates the odorant signal transduction cascade, resulting in opening of ion channels, and culminating in the generation of electric activity that is transmitted down the neuron axon to the olfactory neuropil and higher brain centers for processing (Stengl et al. 1999; Breer 2003; Zwiebel 2003; Ruetzler and Zwiebel 2005).

THE HONEY BEE'S OLFACTORY SYSTEM

The main olfactory organ of insects is their antennae, which are covered with a variety of sensory structures (sensilla), that mostly come in the shape of hairs or pegs (Schneider 1964; Keil 1999). In honey bees, an unusual sensillum type, the placoid sensilla (pore plates), is predominantly responsible for scent detection (Figure 9.2) (Goodman 2003). Thousands of such pore plates are found on each honey bee antenna, and five to thirty-five sensory neurons innervate each pore plate (Schneider and Steinbrecht 1968). In the worker honey bee, some forty-eight thousand olfactory neurons run

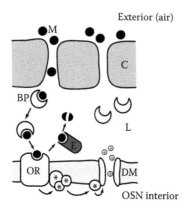

FIGURE 9.1 Concept of perireceptor events, odor detection, and transduction in an insect. An airborne odor molecule (M) passes through pores in the sensillum cuticle (C) into the sensillum lymph (L), where it is captured by a binding protein (BP), which transports it to an odorant receptor protein (OR), a transmembrane protein embedded on the dendrite membrane (DM) of an olfactory sensory neuron (OSN). This initiates a signal transduction cascade (*) within the OSN, resulting in opening of ion channels and generation of electric activity. The odorant is then degraded by a specific enzyme (E), freeing the OR for the next odorant.

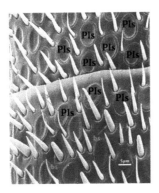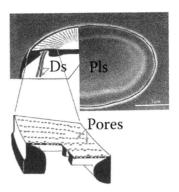

FIGURE 9.2 Left: Magnification of two flagellar segments of the honey bee worker antenna showing placoid sensilla (pore plates) (Pls). Right: Pore plate (Pls) showing rim of pores and dendrites (Ds) of olfactory sensory neurons underneath. (Modified from Goodman 2003. With permission.)

through the antenna and extend their axons along the antennal nerve to the first center of olfactory processing, the antennal lobes (Figure 9.3). In the honey bee worker each antennal lobe possesses about 160 glomeruli (Arnold et al. 1985; Flanagan and Mercer 1989; Galizia et al. 1999a). From the antennal lobes, the olfactory information is passed on to higher brain centers, such as the mushroom bodies, which are involved in multimodal sensory processing, learning, and memory (Figure 9.3) (Mobbs 1982; Strausfeld 2002).

How is olfactory information encoded in this system? Each olfactory neuron carries one type of olfactory receptor protein responding to a particular kind of odor. Some olfactory receptors are highly specific, responding to only one particular odorant, as is the case for pheromone receptors. Other olfactory receptors, like the ones detecting floral scents, are more broadly tuned, and respond to a range of chemically similar odorants (Goodman 2003). Each olfactory neuron projects onto one glomerulus within the antennal lobe (Brockmann and Brückner 1995; Kelber et al. 2006). Neurons that carry the same olfactory receptor protein project onto the same glomerulus. Neurons that carry different receptors project onto different glomeruli, even if the neurons come from the same sensillum (Kelber et al. 2006). Thus, the approximately 160 olfactory receptor proteins found in the

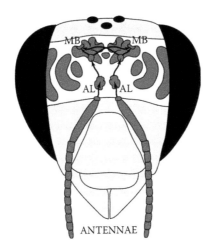

FIGURE 9.3 Schematic drawing of a honey bee head showing the exposed parts of the bee brain (MB, mushroom bodies; AL, antennal lobes) and the olfactory processing pathway (arrows).

worker honey bee (Robertson and Wanner 2006) correspond to the 160 glomeruli found in each antennal lobe. When a honey bee detects a scent, which is usually a mixture of different odorants (see further below), several different olfactory receptors respond to the different components. This leads to a spatiotemporal pattern of activity in the antennal lobe that is specific for this particular scent: it encodes information on odor composition, concentration, and ratio of components, and is also affected by prior olfactory experience, that is, the bee's scent memories (Joerges et al. 1997; Faber et al. 1999; Galizia et al. 1999b; Sachse et al. 1999; Galizia and Menzel 2000; Faber and Menzel 2001). The plasticity of the olfactory system is made possible by a dense network of excitatory and inhibitory neural connections within the antennal lobes and their projections to higher brain centers, which continues to be the subject of many investigations (Hansson and Anton 2000; Laurent 2002; Müller et al. 2002; Sachse and Galizia 2002; Deisig et al. 2006). This plasticity ensures that olfactory information is detected and processed in context of the specific situation in which the bee finds itself.

ODOR DISCRIMINATION AND LEARNING IN HONEY BEES

The worker honey bee's sophisticated sense of smell and her capacity to discriminate between different fragrances have been known for a long time (von Frisch 1919). Numerous behavioral studies have explored odor discrimination in honey bees, demonstrating that honey bees are able to recognize and discriminate between hundreds of components with exquisite sensitivity and specificity (e.g., Laska et al. 1999; Laloi et al. 2000; Laska and Galizia 2001). Bees distinguish odorants based on their carbon chain length, and the type, position, and number of functional groups; e.g., they can easily tell an alcohol from an aldehyde or an ester.

The brain of a honey bee contains only about 960,000 neurons and is 1 mm^3 in size (Menzel and Giurfa 2001). However, it supports well-developed learning and memory capacities (Menzel 1990; Hammer and Menzel 1995; Giurfa 2007). Learning of odors in honey bees has been investigated for many decades, using behavioral assays with both restrained and freely flying bees (Kuwabara 1957; von Frisch 1993). Olfactory learning is a fast and robust process; in most cases a single learning trial is sufficient for bees to learn to associate an odor with a food reward (Koltermann 1969; Menzel 1993). Within 24 h this associative experience is consolidated to form a robust, long-term memory. The bee is able to recognize a learned scent and recall the association for days even without reinforcement of the experience (Beekman 2005).

THE HONEY BEE SCENT ENVIRONMENT

SCENT ORIGIN AND CHEMISTRY

The honey bee world is a world of scents. Every time a bee enters the hive she is confronted with a rich and informative array of odors, many of which are pheromones, i.e., produced by other bees. These can range from mixtures of just a handful of terpenoid components to complex blends of dozens of esters, fatty acids, alcohols, and aliphatics. Pheromones inside the hive have a variety of functions: they contain information about the status and needs of the colony regarding relatedness, brood care, disease, nourishment, and reproductive matters (for comprehensive reviews see Winston 1987 and Free 1987). While all these pheromones are omnipresent in the hive, they are not directly relevant to the tasks involved in foraging.

Scents inside the hive that are of importance to honey bee foraging and recruitment emanate from the colony's food stores, as well as from the food brought in by returning foragers. Nectar, pollen, and honey all release distinctive bouquets based on their floral origin. The original sources of floral scents are the flowers in the outside environment that a honey bee encounters every time she leaves the hive in search for food. Flowers produce scents in either floral scent glands, also called osmophores, or other epidermal tissues of the inflorescences (Effmert et al. 2006). The volatile scent compounds are emitted through the epidermal membranes in order to attract pollinators. Floral odors are highly complex—they can be composed of hundreds of different chemical components (Knudsen et al. 2006). The majority of floral odorants are terpenoids, but we also find large numbers of alcohols, aldehydes, ketones, and esters. Among them, there are twelve odorants that occur in over 50% of all floral bouquets analyzed and are hence regarded as typical floral odorants: limonene, (E)-β-ocimene, myrcene, linalool, α-pinene, β-pinene, benzaldehyde, methyl salicylate, benzyl alcohol, 2-phenyl ethanol, caryophyllene, and 6-methyl-5-hepten-2-one. Of course, different flower species have different scents. The difference in fragrance can be due to a difference in chemical composition, but also in intensity, i.e., concentration and ratio of the components. Furthermore, a floral bouquet can also vary within a species, depending on the environmental conditions, such as the location of an individual flower, time of day, pollination status, nectar content, and age of the flower.

While floral scents are major players in the olfactory world outside the hive, honey bees also encounter pheromones in their exterior environment. One of them is the Nasonov pheromone, which is released into the air and serves as orientation and attraction pheromone for nestmates (Free 1987; Winston 1987). It is composed of seven terpenoids, some of which are also found in scents of many flowers: geraniol, citral, and nerol. There are also a number of other social scents and pheromones that honey bees encounter during foraging and recruitment, about which only little is known, such as potential trail pheromones and marking pheromones (see below).

PERCEPTION OF COMPLEX SCENTS

As outlined above, the scents honey bees encounter during foraging are highly complex, often composed of many different odorants. Furthermore, quantitative and qualitative changes, however subtle they might be, occur frequently among natural scents such as floral bouquets. From numerous behavioral studies we know that honey bees have amazing abilities to learn, discriminate, and recognize such scents in spite of their complexity and variability. However, it is still unclear exactly how bees perceive and interpret information contained in complex scents. Some studies have shown that bees that were trained to a mixture of floral odorants responded to some of the odorants better than to others when the odorants were presented alone (Pham-Delègue et al. 1993; Wadhams et al. 1994; Le Métayer et al. 1997; Laloi et al. 2000). The authors concluded that bees learned floral scents by learning individual key odorants as representative for the entire mixture. This sparse-code strategy would enable honey bees to recognize a floral scent based on just a few of its components. Alternatively, it has been suggested that bees might interpret a complex scent as a Gestalt, that is, as a unit. In this case, a scent would only be recognized if it completely matches the learned template in both composition and concentration. Some studies suggest that bees might employ this alternative strategy when interpreting cuticular hydrocarbon patterns, which serve as nestmate recognition cues (Page et al. 1991; Breed et al. 2004). While there have been a number of behavioral as well as physiological studies investigating the question of complex scent perception (e.g., Joerges et al. 1997; Galizia and Menzel 2000; Deisig et al. 2002, 2006), it is still not clear how bees truly interpret natural scents. Perception and processing of complex scents is a topic with many unknowns, the discussion of which exceeds the focus of this chapter. For the interested reader, more detailed information on honey bee learning, recognition, and discrimination of floral odors can be found in an excellent review by Smith et al. (2006).

SCENT INFORMATION USED DURING HONEY BEE FORAGING AND RECRUITMENT

When a foraging bee searches for a potential food source, for example, a flower field or a blossoming tree, she is attracted by visual signals such as colors, patterns, and shapes (Gould 1993). Successful foragers recruit nestmates by performing a "dance" on return to the hive, which provides information about the distance and direction of a food source (von Frisch 1993). The recruited bees use visual information such as the position of the sun, the polarized light pattern in the sky, and the amount of optic flow experienced en route to locate the advertised food source (Lindauer 1963; Wehner 1982; Wehner and Rossel 1985; von Frisch 1993; Esch and Burns 1995, 1996; Srinivasan et al. 1996, 1997, 2000; Esch et al. 2001). While visual information undoubtedly plays an important role in honey bee foraging and recruitment, honey bees also rely heavily on olfactory signals and cues to detect and recognize their food sources (Kriston 1973; Menzel and Erber 1978). Here, we will review which scents honey bees use as information source during foraging and recruitment, and what kind of information they can obtain from these scents.

FLORAL SCENTS AS ATTRACTANTS

Every honey bee researcher who has trained bees to an artificial feeder with sugar water will know that application of some drops of a flowery scent on the feeder increases the feeder's attractive-ness, luring foraging bees toward it. The attractive effect of floral scents on pollinators such as the honey bee has been known for a long time (von Frisch 1919). When there is no wind, and the air is perfectly still, the floral scent that is released by a flower will, in theory, diffuse uniformly in all directions away from a flower, resulting in a scent concentration that is highest at the flower and which decreases smoothly with increasing distance from the flower. Under such conditions, a bee can, in principle, home in on the flower by flying "uphill" along the resulting concentration gradient. However, such conditions rarely prevail. More commonly, the floral scents that are released by the flower are taken up by air currents and transported downwind. The resulting odor plume is usually turbulent, and its structure changes with the speed and direction of the wind (Greenfield 2002). Under such conditions, the plumes can look like wavy ribbons or filaments, mirroring the way in which the wind has changed direction in the recent past (Figure 9.4). How do honey bees make use of such floral scent plumes to find the target? Wenner and co-workers (1991) suggested that after detecting odor molecules in the air, honey bees locate the source by flying upwind when in contact with odor and by casting transversely in a zigzag motion upon losing contact, until they find the source (Figure 9.4)—similar to the pheromone-searching flights of moths (Baker 1986). This kind of chemically induced, upwind orientation (anemotaxis) does not rely on the detection and process-ing of potential concentration gradients within an odor plume in order to find its source. It is believed that by using this tactic, insects can home in on an odor source from over 100 m away (Greenfield 2002). Considering the honey bee's olfactory capacity, we can assume that bees are, in theory, able to use floral scent plumes as guides for long-distance foraging flights. However, given the constantly changing nature of a scent plume in moving air, this kind of olfactory-guided foraging seems rather time- and energy-consuming—both of which are crucial factors to take into account for a foraging bee—if it is the only means of navigating to the target. Furthermore, odor-based navigation will not work when the bee is upstream of the food source: no scent plumes will reach the bee. Therefore, it is more likely that for the long-distance part of a foraging flight, honey bees rely more on visual

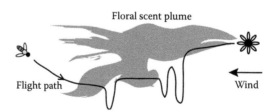

FIGURE 9.4 Theoretical flight path of a foraging honey bee within a floral scent plume (overhead view). After detection of scent molecules the bee orients upwind. If contact with the odor is lost, the bee casts in a zigzag motion until she is once again within the odor plume.

information to orient toward potential food sources, such as familiar landmarks, a prominent tree, or a colorful flower field.

Floral scents become increasingly important once a foraging bee approaches a potential food site. Presented with an array of blossoms or flowers, which might all look equally attractive, the bee has to find the ones that actually produce nectar or pollen. The floral scent emitted by the productive flowers helps a foraging bee to locate them quickly and accurately, and saves her from a time-consuming trial-and-error search. Floral odor even serves as orientation cue at very close range once the bee has landed on the plant. The fragrance-emitting areas of a blossom or flower are located in close proximity to the actual nectaries and often congruent with the so-called nectar guides at the base of the flower (Effmert et al. 2006). These are characteristic markings, visible in the UV, which guide the bees to the nectaries. The odor-emitting areas of a blossom serve as an additional olfactory nectar guide, directing the forager to the precise location of the food.

Floral scent helps not only the foragers but also the recruits to find food. In a number of field studies, Wenner and co-workers observed that honey bee recruits were not able to find a food source unless it carried scent (Wenner et al. 1969; Friesen 1973). They reported that recruited bees did not necessarily show up at the food sites indicated by the dances of returning foragers, when the feeders on offer were unscented. Rather, the recruits came to a third, scented feeder that was placed in relative proximity to the unscented ones. Wenner concluded that honey bees are recruited exclusively through olfaction; i.e., recruits locate a food source merely by following floral scent plumes, and not based on information encoded in the dance. However, several later studies (e.g., Esch et al. 2001) have shown that recruits can be directed to locations that are signaled solely by a forager's dance—even when the target offers no scent cues. By now, the controversy between Wenner's olfaction hypothesis and the dance hypothesis of honey bee recruitment has largely been resolved, with the research community widely agreeing that recruitment through dance communication and recruitment through olfactory search behavior are two strategies that complement rather than exclude each other (Kirchner and Grasser 1998). How so? When following the dance of a returning forager in the hive, a honey bee recruit obtains information about the distance and direction of the advertised food source. However, this positional information is not 100% precise. It does not direct the bee to a specific flower, but rather to an area within which the relevant flowers, bushes, or trees are located. After arriving in the advertised area, a new recruit needs to locate the actual food source. If the food source is large and diffuse, like a field of simultaneously productive flowers, as in an agricultural planting, finding a nectar- or pollen-carrying flower is not a problem. However, if the food source is small and highly localized, or not all of the flowers in the area are productive, the situation would present a challenge to the new recruit. It is then that the floral scent emitted by the food source serves as crucial orientation guide, helping the arriving recruits home in on the food source.

In addition to this olfactory information, the recruits are also offered some visual help by the foragers. Foragers are known to synchronize their return flights to the food source they are advertising, and to form a small swarm circling briefly over the site before landing (Pflumm 1969). The sight of these circling foragers seems to additionally assist recruits in locating the food source by what is known as local enhancement (reviewed by Slaa and Hughes in Chapter 8). However, if the food is scented, the visual aid provided by foragers is of much less importance to recruits than the directional information provided by floral scent (Tautz and Sandeman 2003).

Clearly, floral scents contribute significantly to successful honey bee foraging and recruitment, acting as attractants and orientation guides. But they can also have an indirect effect on recruitment. In a simple field experiment, Lindauer (1948) used two identical sugar water feeders, one scented with an extract of flowers that bees typically visit, such as roses, and the other scented with an extract of flowers that honey bees do not use as food source, such as holly. He observed that bees that had foraged on the feeder scented with rose extract danced with higher intensity, leading to increased recruitment, than bees that had foraged on the holly-scented feeder—the latter danced with a lower intensity. Bees obviously used the chemical information contained in the respective scents to regulate the intensity of their dances, thus directing a greater number of recruits to the more highly valued feeder.

Lindauer's observation indicates that floral scents not only act as simple attractants for foraging bees, but also are a rich source of information. What kind of information can bees extract from a floral scent? To start with, it is a signal that the flower or blossom emitting the scent is currently producing nectar or pollen, or both. The intensity, i.e., quantity of scent emitted, is in direct correlation to the amount of nectar or pollen produced, and thus a decisive cue for a foraging bee (Dudareva and Piechersky 2006). Flowers that have been depleted of their nectar stores by a pollinator have a markedly reduced scent emission, with emission rates declining by almost 90% (Theis and Raguso 2005; Dudareva and Piechersky 2006). Hence, low scent emission informs an arriving forager that this flower has already been visited. It drastically lowers the attractiveness of these flowers to foraging bees, thus directing them to the unpollinated flowers of the plant or species.

But it is not only the intensity of the scent that is important to bees. The quality of the scent, i.e., its chemical composition, provides crucial information on the flower's identity, and with that, information on the quality of the nectar and pollen that is produced by this particular species. An experienced forager uses this qualitative information to decide whether the flower species emitting the scent is worth exploiting. This has been elegantly demonstrated in Lindauer's experiments described above (Lindauer 1948). Importantly, a flower's scent composition can change over time: senescent flowers smell different than young, more productive flowers of the same species (Schade et al. 2001), informing foraging bees which flower or blossom is worth visiting.

It is clear that honey bees are attracted by floral scents and use both quantitative and qualitative information contained in a floral bouquet to make foraging decisions. However, one aspect has not yet been taken into account—that floral bouquets do not provide stable information. The scent emitted by an individual flower can vary depending on the time of day, temperature, sun exposure, and nectar content (see above), to mention just a few factors. How do honey bees deal with this variability? As described above (in "Perception of Complex Scents" section), honey bees seem to learn complex floral scents by learning individual key compounds as representative for the entire scent (Pham-Delègue et al. 1993; Wadhams et al. 1994; Le Métayer et al. 1997; Laloi et al. 2000). In combination with their well-known capacity to generalize between similar olfactory stimuli (Smith 1991; Pham-Delègue et al. 1993), key compound learning enables bees to also respond to floral scents that are not identical to the original scent, but only share some components. When given a choice, bees will prefer the scent that matches the originally learned one most closely, but if need be, they can even respond to floral scents they have never encountered before, as long as the novel scents share a certain extent of similarity with the original scent. Thus, the combination of key compound learning and scent generalization allows bees to find profitable food sources in spite of fluctuations in their volatile emissions. Put simply, a foraging bee will recognize a flower or flower species, even if it does not smell exactly the same as during the bee's previous foraging trip.

SCENT-ASSOCIATED MEMORIES OF FOOD SOURCES

As outlined above, honey bees have an impressive capacity to learn floral scents. They form associative memories of scents and food sources, when smelling an odor shortly before or concurrently with receiving a food reward (Menzel et al. 1993). These scent-food associations can be formed in the field when forager bees visit flowers, but also inside the hive when bees experience the scent and taste of incoming food from another bee, during trophallactic interactions. Trophallaxis occurs under a variety of circumstances in a bee colony, for example, when successful foragers pass the collected food on to food processor bees, or when returning foragers distribute samples of the collected food to potential recruits following the dance. During such contacts a receiver bee learns the scent of the food via the food samples she receives, and via the floral odor clinging to the returning forager's body (von Frisch 1993; Gil and De Marco 2005; Farina et al. 2007). The significance of associative learning of food odors by honey bees inside the nest during trophallactic interactions with foragers is reviewed in detail by Walter Farina and Christoph Grüter in Chapter 10.

There are a number of ways in which scent-associated food memories aid honey bee foraging and recruitment. In the previous section we have discussed how floral scent generally helps a recruit pinpoint the actual food source after arriving in the area advertised by the dance. But what if the area is a flower field with numerous different flower species all emitting attractive scents? How can a recruit locate the specific flower advertised by the forager without making time-consuming mistakes? When following a dancing forager in the hive, a recruit bee obtains olfactory information on the flower the forager is advertising—be it through the nectar samples passed on by the dancer or via the floral scent clinging to the dancer's body. The recruit bee stores this olfactory information in her memory. On arrival in the area advertised by the dance, she is able to recognize the correct scent among the array of floral scents she might encounter, by matching it with the scent in her memory (Koltermann 1969). Thus, short-term memories of food-related scents can help honey bee recruits locate an advertised food source quickly and accurately, which greatly facilitates food exploitation.

Scent-associated memories of food sources are not only short-term, but can become stable long-term memories when honey bees are repeatedly exposed to the same food scent (Menzel and Erber 1978; Menzel 1993). For example, scents experienced by bees inside the hive via incoming food or via the food stores can condition them on a long-term basis: foragers often show preferences for scents they have experienced within the hive, be it as preforaging individuals or later in life (Pham-Delègue et al. 1990; Arenas et al. 2007). This fact is frequently exploited by beekeepers: they condition honey bee colonies by feeding them scented sugar solution, thus inducing the colony to preferentially collect from specific floral sources, which in turn produce highly valued monofloral honeys (von Frisch 1943, 1993).

Olfactory memories can be simple, as is the case for a recruit that forms a scent–reward association between the floral scent that she smells when following a dancer and the nectar sample that she receives at the same time. In this case, all the recruit bee does when flying out is to recall the scent in order to locate the correct flower and receive the expected food reward. However, bees can also form more complex scent-associated memories of food sources. When arriving at a nectar source, a foraging bee will smell the flower's scent shortly before landing and while feeding from it. The forager will not only associate the scent with the nectar reward, but also learn the flower's color and shape (Menzel 1993). Furthermore, she will form memories of the flower's location and the navigational path (Menzel 1993). Why would a bee invest in storing all this information in her memory? We ran a series of experiments to investigate this question (Reinhard et al. 2004a, 2004b). We trained individually marked bees to forage at two sugar feeders placed at two different locations some 100 m from the hive. One of the feeders was scented with rose scent, the other with lemon scent. After letting the bees forage at both feeder locations for several days, we replaced the scented training feeders with unscented, empty ones. We then blew the training scents into the hive one after the other. When rose scent was blown into the hive, the vast majority of trained bees immediately flew out to the location where the rose-scented feeder used to be. When lemon scent was blown into the hive, the same bees all flew to the lemon feeder location (Figure 9.5). As the feeders were unscented during this test and cleaned of any potential scent residues, the bees could not have simply homed in on the scent. Instead, the familiar scent the bees perceived in the hive triggered recall of the specific visual and navigational memories associated with this scent, such as the distance and direction of the food source, and possibly landmarks encountered en route. These memories guided the bees straight back to the food source without the need for distance and direction information as usually provided by the dance. We showed that this cross-modal associative recall also works to link scents with colors rather than locations. Thus, a familiar scent can cause bees to recall not only the location of the associated feeder, but also its visual appearance (Reinhard et al. 2004b). Furthermore, we showed that bees can form and recall complex foraging memories of up to three different locations, colors, and scents (Reinhard et al. 2006). Thus, the odor and taste of nectar samples distributed by returning foragers in the hive, and the odor clinging to the forager's body can trigger recall of navigational memories associated with the food site in experienced recruits, and facilitate their navigation back to the site. For example, if a previously exploited food source suddenly reappears, the familiar smell brought back by

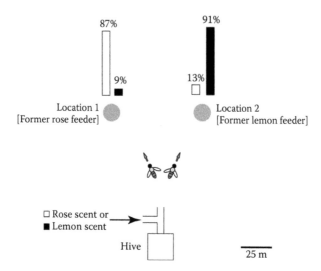

FIGURE 9.5 Experiment investigating scent-triggered recall of navigational memories in foraging honey bees (Reinhard et al. 2004a, 2004b). Individually marked bees were trained to visit a rose-scented feeder at location 1 and a lemon-scented feeder at location 2. During the test the feeders were removed and the bees were tested by blowing rose or lemon scent into the hive in turn. The bars show the percentage of trained bees visiting each location after injecting rose scent (white) or lemon scent (black) into the hive. Injection of rose scent causes the majority of marked bees to visit location 1; injection of lemon scent has the opposite effect. (Figure modified from Srinivasan et al. 2006, data from Reinhard et al. 2004b. With permission.)

the first foragers will reactivate other experienced foragers without the need for dance information. Interestingly, a recent study showed that in cases where the dance creates an informational conflict (for example, when dancing foragers carry the scent of a familiar flower species but advertise a new and unfamiliar location), experienced recruits actually prefer to use their own scent-associated navigational memories over the new navigational information provided by the dancing foragers (Grüter et al. 2008). Scent–food associative memories and their recall is therefore a recruitment mechanism used by honey bees in addition to the dance language to navigate successfully and repeatedly to attractive food sources, enhancing the foraging efficiency of the colony.

THE NASONOV PHEROMONE

Floral scents are not the only odors that play a role in honey bee foraging and recruitment; there are also a number of pheromones that provide information. The most well studied is the Nasonov pheromone, a mixture of biosynthetically related terpenoid compounds (geraniol, (E)-citral, (Z)-citral, geranic acid, nerolic acid, (E,E)-farnesol, nerol) produced in the Nasonov gland, which is found on the dorsal surface of the seventh abdominal tergum (Figure 9.6) (Free 1987; Winston 1987). Interestingly, the Nasonov compounds are also commonly found in floral scents (Knudsen et al. 2006). Bees release the pheromone by lifting their abdomen and exposing the Nasonov gland to let the pheromone evaporate into the air, often assisted by fanning of the wings (Figure 9.6). The highly volatile pheromone attracts bees to the point of release and induces them to land. It acts as an orientation cue in a variety of situations, including finding the nest entrance, forage marking, and swarming (Free 1987). Because of the differences in the volatility of the components, their relative proportions will change with distance from the gland (Pickett et al. 1981). It is conceivable that a bee that is orienting to Nasonov pheromone follows a gradient of component concentrations. Unfortunately, there is no information on the distance from which a bee can perceive the pheromone and direct its flight appropriately.

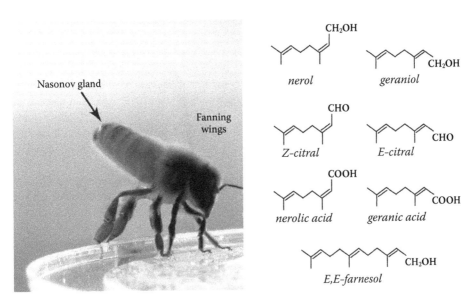

FIGURE 9.6 Left: Honey bee forager at an artificial food source releasing Nasonov pheromone as orientation cue for other foragers. Arrow indicates exposed Nasonov gland; note the fanning wings used to assist evaporation and distribution of the pheromone. (Photo courtesy P. Kraft.) Right: Nasonov pheromone compounds.

Newly emerged workers produce very little Nasonov pheromone, but the amounts increase rapidly with age, peaking when the bee is about 28 days old, roughly coinciding with the onset of foraging activity (Winston 1987). In a foraging context, bees release Nasonov pheromone after landing on a profitable food or water source in order to aid arriving recruits in pinpointing the location. The Nasonov pheromone thus has the same function as a floral scent in providing directional information to searching recruits. Interestingly, bees rarely release the pheromone when landing on natural flowers, but do so when landing on artificial feeders and to advertise water sources (Free 1987). Presumably, the floral scent emanating from natural flowers suffices to guide searching recruits to them, while unscented food sources like sugar solution feeders or water require pheromone release, as recruits are otherwise unlikely to find these sites. However, honey bees do not always release Nasonov pheromone at unscented sugar feeders: it depends on the concentration of the sugar water offered, the forager's previous experience with the food source and its expected reward, and the amount of food the colony has in store (Winston 1987; Fernández and Farina 2001). Evidently, honey bee workers adjust their pheromone marking behavior according to the quality of the food imbibed in combination with the condition of their colony's food stores, thus controlling the number of foragers directed to a specific food source.

Nasonov pheromone is not only released after landing at a food source. As mentioned above, experienced foragers that arrive at a profitable food source undertake circular flights around the target for several seconds before landing (Pflumm 1969). These flights are thought to provide visual orientation cues for other foragers following behind. Depending on the profitability of the food source, the forager's previous experience with it, and the condition of the colony's food stores, bees are known to also release Nasonov pheromone during these circular flights (Pflumm 1969), which presumably provides an additional orientation cue for newly arriving bees. Indeed, Tautz and Sandeman (2003) showed that recruit bees approaching an unscented food source fly after a moving lure if the lure is accompanied by the release of geraniol, the main component of the Nasonov pheromone. Thus, these circular flights might provide visual as well as olfactory orientation cues for arriving recruits.

SCENT MARKS ON FOOD SOURCES

In addition to the Nasonov pheromone, which is released into the air, honey bees also deposit scent marks directly onto the food source. Scent marks can be attractive, directing recruits to a profitable food source, or repellent, when left on flowers whose nectar sources have been depleted—helping to avoid unproductive visits by other foragers. These scent marks have also been named footprint pheromone or forage-marking pheromone (Free 1987). However, due to our lack of knowledge about their origin and chemistry, as well as our limited understanding of their function and interplay with other chemical signals, we suggest that, at least for the present, these scents be included under the more general term *scent mark*.

It is well known that artificial feeders at which bees have been foraging are more attractive than new artificial feeders that have never been visited (Free 1987). It is only necessary for a bee to land briefly on a feeder, and not even obtain food from it, for the feeder to become attractive to subsequent foragers. It has therefore been suggested that bees leave some sort of scent mark behind when landing and walking on a food source. This scent mark was reported to be long-lasting, retaining its attractiveness for many hours (Ferguson and Free 1979). Interestingly, the same attractive scent mark seems to be deposited near the nest entrance, and is thus not an exclusive foraging pheromone, but a more general orientation cue for honey bees. In a foraging context, it acts synergistically with the Nasonov pheromone, attracting and directing foragers to a profitable food source. From where in the body do honey bees release these scent marks? Unfortunately, neither the source nor the chemistry of attractive scent marks has been elucidated in honey bees. The bee's tarsal Arnhart glands have been suggested as a source, with the bee leaving scent marks behind while walking on a flower, thus the term *footprint pheromone* (Free 1987; Winston 1987). But as extracts of other body parts proved to be more attractive than extracts of the legs, it is more likely that the attractive substance is produced elsewhere in or on the bee's body. Considering the long-lasting effect of attractive scent marks it is possible that they are composed of cuticular hydrocarbons, which are not actively deposited, but simply rub off from a bee's feet and her body onto the flower's petals while she collects nectar and pollen. Indeed, it has been shown that many chemical compounds found on the cuticle of honey bee workers are volatile enough to be detected over distance, and that "nestmate odor" triggers orientation flights in other bees when offered in a wind tunnel (Chaffiol et al. 2005; Schmitt et al. 2007).

There is also evidence that honey bees avoid flowers that have been recently visited by themselves or conspecifics, if these flowers have been exhausted and are not yet replenished with nectar (Free 1987; Giurfa 1993). It was concluded that foraging bees deposit repellent scent marks on food sources after feeding on them, not only to ensure that they themselves do not revisit it, but also to inform subsequent foragers that this food source's nectar stores are currently depleted. Like the attractive scent marks, the glandular origin and the chemistry of the repellent scent marks in honey bees are not well studied. It is likely that the repellent scent mark is a rather short-lived, volatile substance. 2-heptanone, which is secreted by the mandibular glands and known to repel foragers from food sources, has been suggested as a constituent of the repellent signal, but this has never been confirmed (Butler 1966; Ferguson and Free 1979).

One question comes to mind: If a honey bee leaves behind an attractive scent mark on each flower she visits by inadvertently rubbing against the flower petals, and then marks the same flower as repellent after feeding from it, how can the flower become attractive again to new foragers? It is believed that the balance between the long-lasting attractive scent mark and the short-lived repellent scent mark gives the ultimate message. For a certain time after a flower has been visited and depleted of nectar the deterrent scent with which it has been marked will override any attractive scent left on it. As the deterrent chemicals evaporate, the influence of the long-lasting attractive scent mark becomes dominant, and the flower will again become attractive to new foragers, by which time the supply of nectar is likely to have been replenished. This change from repellent to attractive happens within 20–60 min, during which time nectar replenishes in flowers (Stout and Goulson 2001). One would expect that the longevity of a repellent scent mark is tuned to match the rate at which the flower replenishes its nectar store. It would be interesting to investigate how good this match is, and to explore whether bees vary the amount or concentration of repellent marks depending upon their perception of the level of the flower's nectar reserves, and upon the species of the flower being visited, if different species are known to replenish their nectar at different rates.

Given the honey bee's capacity for associative learning of odors (see "Scent-Associated Memories of Food Sources" section), it is conceivable that honey bee foragers learn to associate the scent marks left on food sources by previous visitors with the reward level they encounter, i.e., rich vs. depleted food source, and accordingly adjust their acceptance or rejection behavior during subsequent encounters with the same scent. This was recently shown for bumble bees (see Chapter 13) and stingless bees (see Chapter 12), but data for honey bees are not available yet.

Scent marks are indeed a common means of chemical communication among stingless bees (reviewed in Chapter 12) and bumble bees (reviewed in Chapter 13). Given that many of these pollinators share the same food sources, it would be interesting to know whether honey bees use the chemical information left behind by other species and vice versa. Stout and Goulson (2001) reported that honey bees indeed make use not only of their own scent marks, but also of repellent marks left behind by heterospecifics such as bumble bees, avoiding all previously visited and emptied flowers. This capacity to use heterospecific information is intriguing, considering that bumble bee and stingless bee repellent scents have a different chemical composition than honey bee repellent scents: the former are composed of cuticular hydrocarbons (see Chapters 12 and 13), while the latter are thought to contain a short-lived ketone. Unfortunately, too little is known about these scent marks to fully understand how honey bees use heterospecific cues. Evidently, the biological activity, chemical nature, and origin of scent marking in honey bees need considerable clarification, but it is at least clear that honey bee workers produce and use a number of food marking signals as additional information sources to facilitate foraging and recruitment.

TRAIL PHEROMONES AND SCENT TRAILS

Various stingless bees do not use a dance language to communicate the location of profitable food sources, but recruit nestmates by laying odor trails between the nest and the foraging site (reviewed in Chapter 12). Do honey bees also use trail pheromone communication as an additional recruitment mechanism? Butler (1967) found that bees he had trained to forage in a darkened arena produced an odor trail between their hive and the dish of sugar water. However, in this experimental setup it is more likely that the bees walked instead of flying, leaving behind on the floor a trail of attractive scent marks from their body, as described above.

Trail pheromones of flying insects can either be deposited at regular intervals on plant substrates (Nieh 2004; see also Chapter 12) or be released into the air while flying. The resulting pheromone plume is followed using the same mechanism described above for following floral scent plumes to home in on the source. There are no reports of foraging honey bees landing at regular intervals on plant surfaces during their homeward flight, and it is therefore unlikely that they deposit pheromone trails similar to those known from stingless bees (Nieh 2004). However, it is theoretically possible that bees release Nasonov pheromone during their foraging flights, similar to the pheromone release during the circular "buzzing" flights mentioned above. Also, Wenner and co-workers suggested that foragers returning from a food source leave a trail of floral scent behind from the floral odor clinging to their bodies (Wenner et al. 1991). They proposed that, over time, with numerous bees using the same flight path and repeated foraging trips to the same food source, this floral scent trail would become an "odor tunnel," serving as a directional guide for recruits. While there have been no further investigations on potential honey bee trail pheromones, it has often been observed that foragers and recruits follow one another when flying to food sources (Tautz and Sandeman 2003). It was assumed that they follow each other using exclusively visual cues. However, considering how sparse our knowledge is on this topic, we should not exclude the possibility that these bees might also use chemical cues to follow one another—be they pheromones released into the air or floral odors evaporating from the foragers' bodies.

THE DANCE PHEROMONE

The honey bee dance has been studied for centuries, with von Frisch demonstrating that it conveys information to nestmates about distance and direction of a profitable food source (von Frisch 1993). The floral scent on the dancer's body and the food samples passed on to bees following the dance contain further information for potential recruits, as discussed above. However, it has recently been shown that there is yet another scent connected to the honey bee dance, a pheromone (Thom et al. 2007). Successful foragers performing a waggle dance in the hive release a mixture of two alkanes (tricosane, pentacosane) and two alkenes (Z-(9)-tricosene, Z-(9)-pentacosene) onto their abdomens, from where the compounds evaporate into the airspace above, possibly assisted by the relatively high body temperature of dancers. In contrast, nondancing foragers produce these substances in only minute quantities, significantly less than dancers. The four compounds are cuticular hydrocarbons. As such, they are produced subcutaneously and are not stored within a gland. Thom and colleagues (2007) therefore suggested that the waggle dance scent is best classified as a semiochemical or nonglandular pheromone.

When an artificially prepared mixture of the pheromone was blown into the hive, it increased foraging activity, with significantly more bees exiting the hive (Thom et al. 2007). The exact mechanism underlying this observed effect is still unclear, but the authors suggest that the pheromone acts as an attractant, luring inactive foragers toward the dance floor and assisting them in locating the dancers themselves, which could enhance recruitment. As it is not specific for particular food sources, it could also generally alert inactive but experienced foragers, informing them of good foraging conditions, for instance, at the beginning of a day or after a spell of bad weather. In addition to alerting foragers, the dance scent might also have an effect on food–processing bees. A high concentration of the pheromone could alert them to the fact that food is coming in and attract them in large numbers to the dance floor to unload the returning foragers.

Interestingly, some recent work has shown that, together with numerous other scents, the four compounds from the dance pheromone are also present in the air surrounding foragers at a feeder (Schmitt et al. 2007). Possibly foraging honey bees do not release this pheromone only inside the hive while dancing, but also outdoors to alert and attract foraging nestmates to a profitable food source. Thus, this newly described honey bee dance pheromone might have a complementary or possibly even synergistic function to the Nasonov pheromone, and the attractive scent marks described above.

CONCLUDING REMARKS

Scents play a major part in a honey bee's life, and this chapter hopefully demonstrates how important both floral scents and social odors are to all aspects of honey bee foraging and recruitment. However, and in spite of this chapter's emphasis on the role of scents as information source, it is crucial to keep the following in mind: Scents are not the only information channel for a foraging bee. She also receives visual, auditory, mechanosensory, gravitational, and even magnetic input. While scents are sometimes crucial for foraging and recruitment success, at other times depending on availability and reliability of additional information, scents might only play a fractional role. It is the interplay between all the information channels that makes honey bee foraging and recruitment a robust and evolutionary successful biological phenomenon that has fascinated scientists since the time of Aristotle.

ACKNOWLEDGMENTS

The work on honey bee associative learning and scent-triggered navigation by the authors was supported by the Alexander von Humboldt Foundation (FLF-1093207) and the Australian Research Council (DP0450535, DP0208683).

REFERENCES

Arenas A, Fernández VM, Farina WM. (2007). Floral odor learning within the hive affects honeybees' foraging decisions. *Naturwissenschaften* 94:218–22.

Arnold G, Masson C, Budharugsa S. (1985). Comparative study of the antennal lobes and their afferent pathway in the worker bee and the drone (*Apis mellifera*). *Cell Tissue Res* 242:593–605.

Baker TC. (1986). Pheromone-modulated movements of flying moths. In Payne TL, Birch MC, Kennedy CEJ (eds.), *Mechanisms in Insect Olfaction*. Oxford: Claredon Press, pp. 39–48.

Beekman M. (2005). How long will honey bees (*Apis mellifera* L.) be stimulated by scent to revisit past-profitable forage sites? *J Comp Physiol A* 191:1115–20.

Breed MD, Diaz PH, Lucero KD. (2004). Olfactory information processing in honeybee, *Apis mellifera*, nestmate recognition. *Anim Behav* 68:921–28.

Breer H. (2003). Olfactory receptors: Molecular basis for recognition and discrimination of odors. *Anal Bioanal Chem* 377:427–33.

Brockmann A, Brückner D. (1995). Projection pattern of poreplate sensory neurones in honey bee worker, *Apis mellifera* L. (Hymenoptera: Apidae). *Int J Ins Morph Embryol* 24:405–11.

Butler CG. (1966). Mandibular gland pheromone of worker honey bee. *Nature* 212:530.

Butler CG. (1967). Insect pheromones. *Biol Res* 42:42–87.

Chaffiol A, Laloi D, Pham-Delègue MH. (2005). Prior classical olfactory conditioning improves odour-cued orientation of honey bees in a wind tunnel. *J Exp Biol* 208:3731–37.

Deisig N, Giurfa M, Lachnit H, Sandoz JC. (2006). Neural representation of olfactory mixtures in the honeybee antennal lobe. *Eur J Neurosci* 24:1161–74.

Deisig N, Lachnit H, Giurfa M. (2002). The effect of similarity between elemental stimuli and compounds in olfactory patterning discriminations. *Learn Mem* 9:112–21.

Dudareva N, Piechersky E. (2006). Floral scent metabolic pathways: Their regulation and evolution. In Dudareva N, Pichersky E (eds.), *Biology of Floral Scent*. Boca Raton, FL: CRC Press, pp. 55–78.

Effmert U, Buss D, Rohrbeck D, Piechulla B. (2006). Localization of the synthesis and emission of scent compounds within the flower. In Dudareva N, Pichersky E (eds.), *Biology of Floral Scent*. Boca Raton, FL: CRC Press, pp. 105–124, pp. 105–24.

Esch HE, Burns JE. (1995). Honeybees use optic flow to measure the distance of a food source. *Naturwissenschaften* 82:38–40.

Esch HE, Burns JE. (1996). Distance estimation by foraging honeybees. *J Exp Biol* 199:155–62.

Esch HE, Zhang SW, Srinivasan MV, Tautz J. (2001). Honeybee dances communicate distances measured by optic flow. *Nature* 411:581–83.

Faber T, Joerges J, Menzel R. (1999). Associative learning modifies neural representation of odors in the insect brain. *Nature Neurosci* 2:74–78.

Faber T, Menzel R. (2001). Visualizing mushroom body response to a conditioned odor in honeybees. *Naturwissenschaften* 88:472–76.

Farina WM, Grüter C, Acosta L, McCabe S. (2007). Honeybees learn floral odors while receiving nectar from foragers within the hive. *Naturwissenschaften* 94:55–60.

Ferguson AW, Free JB. (1979). Production of a forage-marking pheromone by the honeybee. *J Apic Res* 18:128–35.

Fernández PC, Farina WM. (2001). Changes in food source profitability affect Nasonov gland exposure in honeybee foragers *Apis mellifera* L. *Insect Soc* 48:366–71.

Flanagan D, Mercer AR. (1989). An atlas and 3-D reconstruction of the antennal lobes in the worker honey bee, *Apis mellifera* L. (Hymenoptera: Apidae). *Int J Insect Morphol Embryol* 18:145–59.

Free JB. (1987). *Pheromones of Social Bees*. Ithaca, NY: Cornell University Press.

Friesen LJ. (1973). The search dynamics of recruited honeybees, *Apis mellifera ligustica* Spinola. *Biol Bull* 144:107–31.

von Frisch K. (1919). Über den Geruchssinn der Biene und seine blütenbiologische Bedeutung. *Zool Jb Physiol* 37:1–238.

von Frisch K. (1943). Versuche über die Lenkung des Bienenfluges durch Duftstoffe. *Naturwissenschaften* 31:445–60.

von Frisch K. (1993). *The Dance Language and Orientation of Bees*. Cambridge, MA: Harvard University Press.

Galizia CG, McIlwrath SL, Menzel R. (1999a). A digital three-dimensional atlas of the honeybee antennal lobe based on optical sections acquired by confocal microscopy. *Cell Tissue Res* 295:383–94.

Galizia CG, Menzel R. (2000). Odour perception in honeybees: Coding information in glomerular patterns. *Curr Op Neurobiol* 10:504–510.

Galizia CG, Sachse S, Rappert A, Menzel R. (1999b). The glomerular code for odor representation is species specific in the honeybee *Apis mellifera*. *Nat Neurosci* 2:473–78.

Gil M, De Marco RJ. (2005). Olfactory learning by means of trophallaxis in *Apis mellifera*. *J Exp Biol* 208:671–80.

Giurfa M. (1993). The repellent scent-mark of the honeybee *Apis mellifera ligustica* and its role as communication cue during foraging. *Insect Soc* 40:59–67.

Giurfa M. (2007). Behavioral and neural analysis of associative learning in the honeybee: A taste from the magic well. *J Comp Physiol A* 193:801–24.

Goodman L. (2003). *Form and Function in the Honey Bee*. Cardiff: IBRA.

Gould JL. (1993). Ethological and comparative perspectives on honey bee learning. In Papaj DR, Lewis AC (eds.), *Insect Learning*. New York: Chapman & Hall, pp. 18–50.

Greenfield MD. (2002). *Signalers and Receivers: Mechanisms and Evolution of Arthropod Communication*. Oxford: Oxford University Press.

Grüter C, Balbuena MS, Farina WM. (2008). Informational conflicts created by the waggle dance. *Proc R Soc B* 275:1321–27.

Hammer M, Menzel R. (1995). Learning and memory in the honeybee. *J Neurosci* 15:1617–30.

Hansson BS, Anton S. (2000). Function and morphology of the antennal lobe: New developments. *Ann Rev Entomol* 45:203–31.

Hildebrand JG, Shepherd GM. (1997). Mechanisms of olfactory discrimination: Converging evidence for common principles across phyla. *Ann Rev Neurosci* 20:595–631.

Joerges J, Küttner A, Galizia CG, Menzel R. (1997). Representation of odours and odour mixtures visualized in the honeybee brain. *Nature* 387:285–88.

Keil TA. (1999). Morphology and development of the peripheral olfactory organs. In Hansson BS (ed.), *Insect Olfaction*. Berlin: Springer Verlag, pp. 5–47.

Kelber C, Rössler W, Kleineidam CJ. (2006). Multiple olfactory receptor neurons and their axonal projections in the antennal lobe of the honeybee *Apis mellifera*. *J Comp Neurol* 496:395–405.

Kirchner WH, Grasser A. (1998). The significance of odor cues and dance language information for the food search behavior of honeybees (Hymenoptera: Apidae). *J Insect Behav* 11:169–78.

Knudsen JT, Eriksson R, Gershenzon J, Ståhl B. (2006). Diversity and distribution of floral scent. *Bot Rev* 72:1–120.

Koltermann R. (1969). Lern- und Vergessensprozesse bei der Honigbiene - aufgezeigt anhand von Duftdressuren. *Z vergl Phsyiol* 63:310–34.

Kriston I. (1973). Die Bewertung von Duft- und Farbsignalen als Orientierungshilfen an der Futterquelle durch *Apis mellifera* L. *J Comp Physiol* 84:77–94.

Kuwabara M. (1957). Bildung des bedingten Reflexes von Pavlovs Typus bei der Honigbiene, *Apis mellifica*. *J Fac Sci Hokkaido Univ Ser VI Zool* 13:458–64.

Laloi D, Bailez O, Blight MM, Roger B, Pham-Delegue MH, Wadhams LJ. (2000). Recognition of complex odors by restrained and free-flying honeybees, *Apis mellifera*. *J Chem Ecol* 26:2307–19.

Laska M, Galizia CG. (2001). Enantioselectivity of odor perception in honeybees (*Apis mellifera carnica*). *Behav Neurosci* 115:632–39.

Laska M, Galizia CG, Giurfa M, Menzel R. (1999). Olfactory discrimination ability and odor structure—Activity relationships in honeybees. *Chem Senses* 24:429–38.

Laurent G. (2002). Olfactory network dynamics and the coding of multidimensional signals. *Nature Rev Neurosci* 3:884–95.

Le Métayer M, Marion-Poll F, Sandoz JC, Pham-Delègue MH, Blight MM, Wadhams LJ, Masson C, Woodcock CM. (1997). Effect of conditioning on discrimination of oilseed rape volatiles by the honeybee: Use of a combined gas chromatography-proboscis extension behavioural assay. *Chem Senses* 22:391–98.

Lindauer M. (1948). Über die Einwirkung von Duft- und Geschmacksstoffen sowie anderer Faktoren auf die Tänze der Bienen. *Z vergl Physiol* 31:348–412.

Lindauer M. (1963). Kompassorientierung. *Ergebn Biol* 26:158–81.

Menzel R. (1990). Learning, memory, and cognition in honeybees. In Kesner RP, Olton DS (eds.), *Neurobiology of Comparative Cognition*. Hillsdale, NJ: Lawrence Erlbaum, pp. 237–92.

Menzel R. (1993). Associative learning in honey bees. *Apidologie* 24:157–68.

Menzel R, Erber J. (1978). Learning and memory in bees. *Sci Am* 239:102–9.

Menzel R, Giurfa M. (2001). Cognitive architecture of a mini-brain: The honeybee. *Trends Cogn Sci* 5: 62–71.

Menzel R, Greggers U, Hammer M. (1993). Functional organization of appetitive learning and memory in a generalist pollinator, the honey bee. In Papaj DR, Lewis AC (eds.), *Insect Learning*. New York: Chapman & Hall, pp. 79–125.

Mobbs PG. (1982). The brain of the honeybee *Apis mellifera*. I. The connections and spatial organization of the mushroom bodies. *Phil Trans R Soc Lond B* 298:309–54.

Müller D, Abel R, Brandt R, Zöckler M, Menzel R. (2002). Differential parallel processing of olfactory information in the honeybee, *Apis mellifera* L. *J Comp Physiol A* 188:359–70.

Nagnan-Le Meillour P, Jacquin-Joly E. (2003). Biochemistry and diversity of insect odorant-binding proteins. In Blomquist GJ, Vogt RG (eds.), *Insect Pheromone Biochemistry and Molecular Biology*. London: Elsevier, pp. 509–37.

Nieh JC. (2004). Recruitment communication in stingless bees (Hymenoptera, Apidae, Meliponini). *Apidologie* 35:159–82.

Page RE Jr, Metcalf RA, Metcalf RL, Erickson JEH, Lampman RL. (1991). Extractable hydrocarbons and kin recognition in honeybees (*Apis mellifera* L.). *J Chem Ecol* 17: 745–56.

Pelosi P. (2001). The role of perireceptor events in vertebrate olfaction. *CMLS* 58:503–9.

Pflumm W. (1969). Beziehungen zwischen Putzverhalten und Sammelbereitschaft bei der Honigbiene. *Z vergl Physiol* 64:1–36.

Pham-Delègue MH, Bailez O, Blight MM, Picard-Nizou AL, Wadhams LJ. (1993). Behavioural discrimination of oilseed rape volatiles by the honeybee *Apis mellifera* L. *Chem Senses* 18:483–94.

Pham-Delègue MH, Roger B, Charles R, Masson C. (1990). Effet d'une pré-exposition olfactive sur un comportement d'orientation en olfactomètre dynamique à quatre voies chez l'abeille (*Apis mellifera* L.). *Insect Soc* 37:181–87.

Pickett JA, Williams IH, Smith MC, Martin AP. (1981). Nasonov pheromone of the honey bee, *Apis mellifera* L. (Hymenoptera, Apidae). Part III. Regulation of pheromone composition and production. *J Chem Ecol* 7:543–54.

Reinhard J, Srinivasan MV, Guez D, Zhang SW. (2004b). Floral scents induce recall of navigational and visual memories in honeybees. *J Exp Biol* 207:4371–81.

Reinhard J, Srinivasan MV, Zhang SW. (2004a). Scent-triggered navigation in honeybees. *Nature* 427:411.

Reinhard J, Srinivasan MV, Zhang SW. (2006). Complex memories in honeybees: Can there be more than two? *J Comp Physiol A* 192:409–16.

Robertson HM, Wanner KW. (2006). The chemoreceptor superfamily in the honey bee, *Apis mellifera*: Expansion of the odorant, but not gustatory, receptor family. *Genome Res* 16:1395–403.

Ruetzler M, Zwiebel LJ. (2005). Molecular biology of insect olfaction: Recent progress and conceptual models. *J Comp Physiol A* 191:777–90.

Sachse S, Galizia CG. (2002). Role of inhibition for temporal and spatial odor representation in olfactory output neurons: A calcium imaging study. *J Neurophysiol* 87:1106–17.

Sachse S, Rappert A, Galizia CG. (1999). The spatial representation of chemical structures in the antennal lobe of honeybees: Steps towards the olfactory code. *Eur J Neurosci* 11:3970–82.

Schade F, Legge RL, Thompson JE. (2001). Fragrance volatiles of developing and senescing carnation flowers. *Phytochemistry* 56:703–10.

Schmitt T, Herzner G, Weckerle B, Schreier P, Strohm E. (2007). Volatiles of foraging honeybees *Apis mellifera* L. (Hymenoptera: Apidae) and their potential role as semiochemicals. *Apidologie* 38:164–70.

Schneider D. (1964). Insect antennae. *Annu Rev Entomol* 9:103–22.

Schneider D, Steinbrecht RA. (1968). Checklist of insect olfactory sensilla. *Symp Zool Soc Lond* 23: 279–97.

Smith BH. (1991). The olfactory memory of the honeybee *Apis mellifera*: I. Odorant modulation of short- and intermediate-term memory after single-trial conditioning. *J Exp Biol* 161:367–82.

Smith BH, Wright GA, Daly KC. (2006). Learning-based recognition and discrimination of floral odors. In Dudareva N, Pichersky E (eds.), *Biology of Floral Scent*. Boca Raton, FL: CRC Press, pp. 263–95.

Srinivasan MV, Zhang SW, Altwein M, Tautz J. (2000). Honeybee navigation: Nature and calibration of the 'odometer.' *Science* 287:851–53.

Srinivasan MV, Zhang SW, Bidwell N. (1997). Visually mediated odometry in honeybees. *J Exp Biol* 200:2513–22.

Srinivasan MV, Zhang SW, Lehrer M, Collett TS. (1996). Honeybee navigation en route to the goal: Visual flight control and odometry. *J Exp Biol* 199:237–44.

Srinivasan MV, Zhang SW, Reinhard J. (2006). Small brains, smart minds: Vision, perception, navigation, and 'cognition' in insects. In Warrant E, Nilsson DA (eds.), *Invertebrate Vision*. Cambridge, MA: Cambridge University Press, pp. 462–93.

Stengl M, Ziegelberger G, Boekhoff I, Krieger J. (1999). Perireceptor events and transduction mechanisms in insect olfaction. In Hansson BS (ed.), *Insect Olfaction*. Berlin: Springer Verlag, pp. 49–66.

Stout JC, Goulson D. (2001). The use of conspecific and interspecific scent marks by foraging bumblebees and honeybees. *Anim Behav* 62:183–89.

Strausfeld NJ. (2002). Organization of the honey bee mushroom body: Representation of the calyx within the vertical and gamma lobes. *J Comp Neurol* 450:4–33.

Strausfeld NJ, Hildebrand JG. (1999). Olfactory systems: Common design, uncommon origins. *Curr Op Neurobiol* 9:634–39.

Tautz J, Sandeman DC. (2003). Recruitment of honeybees to non-scented food sources. *J Comp Physiol A* 189:293–300.

Theis N, Raguso RA. (2005). The effect of pollination on floral fragrance in thistles. *J Chem Ecol* 31:2581–600.

Thom C, Gilley DC, Hooper J, Esch HE. (2007). The scent of the waggle dance. *PloS Biology* 5:e228.

Wadhams LH, Blight MM, Kerguelen V, Le Métayer M, Marion-Poll F, Masson C, Pham-Delègue MH, Woodcock CM. (1994). Discrimination of oilseed rape volatiles by honey bee: Novel combined gas chromatographic-electrophysiological behavioral assay. *J Chem Ecol* 20:3221–31.

Wehner R. (1982). Himmelsnavigation bei Insekten. Neurophysiologie und Verhalten. *Neujahrsbl Naturforsch Ges Zürich* 184:1–132.

Wehner R, Rossel S. (1985). The bee's celestial compass—A case study in behavioral neurobiology. *Fortschr Zool* 31:11–53.

Wenner AM, Meade DE, Friesen LJ. (1991). Recruitment, search behavior, and flight ranges of honey bees. *Am Zool* 31:768–82.

Wenner AM, Wells PH, Johnson DJ. (1969). Honey bee recruitment to food sources: Olfaction or language? *Science* 164:84–86.

Winston ML. (1987). *The Biology of the Honey Bee*. Cambridge, MA: Harvard University Press.

Zwiebel LJ. (2003). The biochemistry of odor detection and its future prospects. In Blomquist GJ, Vogt RG (eds.), *Insect Pheromone Biochemistry and Molecular Biology*. London: Elsevier, pp. 371–90.

10 Trophallaxis
A Mechanism of Information Transfer

Walter M. Farina and Christoph Grüter

CONTENTS

INTRODUCTION

Trophallaxis is the exchange of liquid material between individuals, mostly members of the same colony. Wheeler (1918) was the first to propose the term *trophallaxis* for describing these interactions between nestmates in ant colonies. He interpreted them as being "clearly cooperative and mutualistic relationships," and thereby separated his term from the term *trophobiosis*, which had been suggested earlier by Roubaud (1916) to indicate oral food transfers based on a "trophic exploitation" in social wasps (for review and historical background, see Sleigh 2002).

There are two main kinds of intraspecific liquid food transfer in social insect nests: In the first one, adults exchange liquids with larvae (they imbibe larval saliva from the brood and transfer glandular secretions, honey, and pollen to the larvae). In the second one, the liquid is transferred between two adults. The stomodeal (or oral) trophallaxes are the most common ones. Here, donors regurgitate a drop of food from their crops while one or more receivers drink the liquid (Wilson 1971; Michener 1974). During these mouth-to-mouth contacts, intensive antennation (Montagner and Galliot 1982; Hölldobler 1985; Goyret and Farina 2003; Hrncir et al. 2006; McCabe et al. 2006), occasional foreleg movements of both partners (Hölldobler et al. 1974; McCabe et al. 2006), or as in stingless bees, the transmission of pulsed vibrations (Hrncir et al. 2006) accompany the oral contacts. Abdominal or anal trophallaxes are also used to distribute material such as intestinal symbionts for wood digestion in termite colonies (Grassé and Noirot 1945).

The evolutionary origin of trophallaxis might have been related to the regulation of aggression in group-living insects (Roubaud 1916). Indeed, aggressive behaviors often stimulate trophallaxis, and the aggression ceases after a food offering (Wilson 1971; Hölldobler 1977; Wcislo and

González 2006). However, adult–adult food sharing may also enhance the chance of survival of a colony where unfavorable weather conditions prevent foraging over longer periods of time (Wcislo and González 2006).

The occurrence and high frequency of trophallaxes among adult individuals are a common characteristic of highly social insects (Michener 1969; Wilson 1971). While eusocial insects, such as honey bees and many other bee, ant, termite, and wasp species, engage in frequent oral interactions inside their nests, this form of contact is rare in communal insects (e.g., halictine bees, Kukuk and Crozier 1990).

During the course of evolution, trophallaxis probably became more important in species for which it considerably improved the efficiency of the performance of vital tasks like food collection or nest construction. The partitioning of tasks is assumed to increase overall colony task performance (Ratnieks and Anderson 1999). Once the material transfer became an important aspect of work organization, it offered an opportunity for both food donors and receivers to acquire information about internal and external environmental parameters via incidental cues, such as searching delays (Seeley 1995) and numbers of receivers (Farina 2000), and about olfactory (Gil and De Marco 2005; Farina et al. 2007) as well as gustatory (Pankiw et al. 2004; Martínez and Farina 2008) cues contained in the transferred material.

The frequency of trophallaxis varies greatly between the different social insect species. It is particularly high in honey bees (Michener 1974; Hölldobler 1977), where trophallaxes occur between bees of all castes and ages (Free 1957; Moritz and Hallmen 1986; Crailsheim 1998; Grüter and Farina 2007). In experimental hives, bees performing the trophallaxes are easily identified: the food donor opens her mandibles and regurgitates the food, whereas the recipients protrude the proboscis to contact the donor's prementum. Since the mechanism of food transfer in other social insects, such as ants and wasps, differs from that of the honey bees, the distinction between donors and receivers has to be made by differences in either the head position or the antennation between the trophallactic partners (Figure 10.1).

As central place foragers, honey bees perform successive feeding trips to a profitable food source, interrupted by hive stays. This allows the researcher to analyze their trophallactic behavior in observation hives in the context of foraging (von Frisch 1967; Núñez 1970; Seeley 1986, 1989; Farina 1996; De Marco and Farina 2001; Farina and Wainselboim 2001a). In this chapter we will discuss the role of trophallaxis for the transfer of information in the context of nectar foraging in honey bees. We will denote all mouth-to-mouth contacts as trophallaxes, because even short interactions with a low probability of an effective food transfer potentially provide important chemosensory information, such as the odor and taste of the exploited food source.

a

b

FIGURE 10.1 Trophallaxis in the honey bee *Apis mellifera* (a) and the carpenter ant *Camponotus mus* (b). In honey bees, the roles of the donor (D) and the receiver (R) of food are easily distinguishable. The receiver bee protrudes her proboscis and contacts the donor's prementum, which causes different head position of the trophallactic partners. In the carpenter ant *C. mus,* the position of the heads and differences in the antennation intensity provide information about the trophallactic roles. (Photographs by Christoph Grüter (a) and Sofia McCabe (b).)

RETURNING TO THE NEST AFTER FORAGING

Division of labor, i.e., the formation of groups specialized in different tasks, is supposed to improve the efficiency of collective activities in a constantly changing environment (Wilson 1971; Michener 1974). In honey bees, young and middle-aged workers perform in-hive duties (e.g., cleaning cells, caring for brood, grooming, receiving and processing of nectar), while older workers forage outside (Rösch 1925; Lindauer 1952; Seeley 1982). Nectar foraging in honey bees is a partitioned task: foragers collect nectar in the field and, inside the nest, transfer the food to bees of middle age (often called food processors, receivers, or food storers) in the delivery area close to the hive entrance (Park 1925; Lindauer 1954; von Frisch 1967; Seeley 1995). These bees of middle age are then mainly responsible for processing the nectar into honey and storing it in cells (Park 1925).

Foragers returning from a profitable food source sometimes display dance maneuvers (von Frisch 1967; Seeley 1995), which encode the location of a food source (von Frisch 1967; Riley et al. 2005). The dance also communicates the existence of an attractive food source and increases the attention and activity of bees in the dancer's vicinity and, subsequently, the disposition of inactive foragers to leave the hive and search for the food source (von Frisch 1923, 1967; Božič and Valentinčič 1991; Thom et al. 2007). However, the dance display attracts not only potential foragers but also food processor bees, which unload the forager (Farina 2000). Foragers not only provide information for other individuals, but also receive information during their interactions with hive bees. They seem to use both the time to find a bee for unloading and the number of unloading bees as cues to adjust their dance behavior after unloading (Lindauer 1954; Seeley 1992; Kirchner and Lindauer 1994; Seeley and Tovey 1994; Farina 2000). Thus, social interactions on the delivery area provide the foragers with information that helps them to adjust the dance levels according to the general interest of hive bees in a particular food source and the availability of food processor bees (Seeley 1995).

Waggle dances are often performed immediately after arrival, and prior to food unloading (Thom 2003). This shows that unloading delays depend not only on the availability of food processor bees but also on the motivation of the foragers themselves, which in turn depends, for example, on the recently experienced profitability of the food source (von Frisch 1967; Seeley 1986; Seeley et al. 2000; De Marco and Farina 2001). Laboratory studies similarly suggested an important role of the donor: the delays to initiate trophallaxis between worker pairs in experimental arenas increased when the crop load of donors was reduced (Farina and Núñez 1993) or when the sugar concentration decreased (Tezze and Farina 1999). Thus, the interplay between hive-internal and hive-external information available to the forager determines the amount of dancing (see also Seeley 1995).

Previous experiences of donors and receivers might be another factor affecting queuing delays and the social feedback inside the hive. Experiments with arenas demonstrated that if bees had prior odor information, new scents present in the crop of the donors negatively affected the occurrence of trophallaxis (Gil and Farina 2003). Similarly, food processor bees show a preference to unload foragers that had collected sugar solution with a previously experienced food scent (Goyret and Farina 2005a).

THE OFFERING BEHAVIOR OF ACTIVE FORAGERS

Foragers often perform several food-offering contacts with a wide range of durations after returning from a nectar source (from milliseconds up to more than 60 s in some cases; De Marco and Farina 2001; Farina and Wainselboim 2001a, 2005) (Figure 10.2a). In general, nectar foragers perform one or two trophallaxes that last for more than 2–3 s per hive stay, and a much more variable number of shorter offering contacts (De Marco and Farina 2001). During the interactions with a duration of at least 2 s, food is effectively transferred between foragers and receivers (Farina and Wainselboim 2001a). Under constant reward conditions the frequency of food-offering contacts (short and long trophallaxes) is fairly constant between different foraging trips (De Marco and Farina 2001), and it is similar for different reward rates offered at the feeder (Fernández et al. 2003).

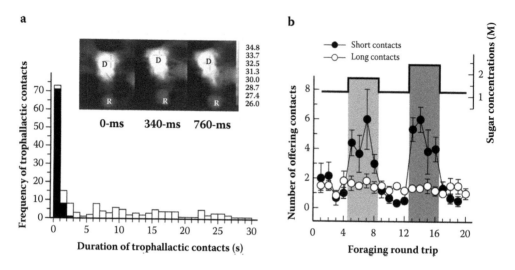

FIGURE 10.2 Trophallactic offering behavior of active foragers. (a) Frequency of offering contacts performed by foragers in relation to the contact duration. The bees collected 50% w/w sucrose solution from a feeder, which provided a constant flow rate of 8.2 µl/min. Given are the contacts without increase in the receiver's proboscis temperature (black bars) and contacts with an increase in proboscis temperature (white bars). A temperature increase represents an effective transfer of the liquid food among partners. Inset: Thermograms showing the surface temperatures of a donor forager (D) and a receiver (R) at three different times during one trophallaxis. The proboscis temperature of the receiver changed rapidly during the trophallactic interaction, whereas the donor's head and thorax temperature remained at 39.4°C. Grey-value scale indicates the measured temperature ranges. (After Farina and Wainselboim 2001b. With permission.) (b) The offering contacts by incoming foragers during their hive stays were categorized according to their length: contacts shorter than 2 s (filled circles) or longer than 2 s (empty circles), respectively. Given is the number of short and long offering contacts when foragers collected at a feeder with constant flow rate (5 µl of sugar solution per minute) during twenty consecutive round-trips. The sucrose concentration (line above plot) changed after every fourth foraging bout. (After De Marco and Farina 2001. With permission.)

Under such conditions, most of the short offering contacts occur at the beginning of a hive stay (De Marco and Farina 2001). However, when reward conditions fluctuate, an adjustment of the number of short trophallactic contacts occurs (Farina 1996; De Marco and Farina 2001). An increase in profitability causes a rapid increase in the number of the short interactions (Figure 10.2b). In a changing environment, therefore, there is a clear and positive relation between the profitability of a food source (i.e., sugar concentration) and the number of short contacts as well as the dance duration (De Marco and Farina 2001).*

In honey bees, short offering contacts can take place before, during, and after dancing (Park 1925; von Frisch 1967; De Marco and Farina 2001; Farina and Wainselboim 2005; Díaz et al. 2007), which leads to a rather equal distributing of contacts during hive stays, at least after an increase in food profitability (De Marco and Farina 2001). Hence, short offering contacts are not simply a failure to unload all the food at once due to the forager's high motivation for dancing. But why do foragers perform several short offering contacts if the chance to transfer food during these interactions is very low (Farina and Wainselboim 2001a)? Actually, the context in which they occur suggests that they might play a role in providing information about fluctuating resources, such as the taste and scent of the exploited food source (for more details, see the "Odor Learning Through Trophallaxis" section below).

---

* A correlation between offering contacts and spinning behaviors was also observed in the stingless bee *Melipona beecheii* (Hart and Ratnieks 2002). This study, however, did not investigate the effect of food quality (sugar concentration) on these behaviors.

While the number of short-offering contacts is highly variable, the number of long trophallaxes (longer than 2–3 s) is quite constant (Figure 10.2b) and does not seem to depend on the amount of food collected by the forager (Fernández et al. 2003) but rather on the colony's nectar influx (Huang and Seeley 2003; Gregson et al. 2003). In addition, a mismatch between the crop loads of foragers and the crop capacities of hive bees receiving the nectar seems to explain why foragers perform more than one long unloading trophallaxis (Gregson et al. 2003; Huang and Seeley 2003).

DYNAMICS OF FOOD TRANSFER

Bees modulate their crop-loading behavior at the feeding place according to the food source profitability (Núñez 1966, 1970, 1982). Similarly, they adjust their crop-unloading behavior during the long trophallaxes according to a food source's profitability (Farina and Núñez 1991). In observation hives, the estimated transfer rate increased with higher crop loads, which in turn depended on the reward rates of the food source (Farina 1996; Farina and Wainselboim 2001a). In experiments using small interaction arenas, the transfer rate was further affected by the sugar concentration of the transferred food (Farina and Núñez 1991; Tezze and Farina 1999) and the reward rate experienced at a food source by a food donor (Wainselboim and Farina 2000a, 2000b). In addition, disturbances of foragers during food collection also affected the transfer rate. In summary, bees seem to evaluate the profitability of the food source by integrating an overall flow rate throughout the entire visit, instead of measuring only the current flow rate delivered at the feeder (Wainselboim et al. 2003; Wainselboim and Farina 2003). Furthermore, foragers also seem to be able to detect sudden changes in the delivered flow of solution within a single foraging bout, and subsequently adjust the transfer rate within the hive in relation to these changes (Wainselboim et al. 2002; Wainselboim and Farina 2003).

But does the modulation of trophallactic behavior by donors actually modify the behavior of food-receiver bees? Infrared thermal analysis of foraging bees showed that if a feeder offered food with a higher reward rate, foragers initiated unloading inside the hive at higher thoracic temperatures, compared to low reward rates. During the food transfer, the receivers actively heat up their thoraxes. Their heating rate positively correlates with the foragers' thoracic temperatures and with the reward rate experienced by the donor at the feeder (Farina and Wainselboim 2001a; Figure 10.3a). These heating rates also depend on the orientation of the receiver toward the donor. Receivers positioned frontally to the donor forager warm up more quickly and attain higher proboscis temperatures than those positioned laterally (Farina and Wainselboim 2001a; Figure 10.4). These differences in proboscis temperature indicate that the unloading bees received different portions of sugar solution.

Receiving bees (mostly food processors) also adjust their nectar processing behavior in accordance to the profitability of the nectar source. After receiving nectar, food processors perform offering contacts or cell inspections, and often even both behaviors, prior to returning to the delivery area (Pírez and Farina 2004). When performing only one of these tasks, the occurrence of cell inspections increases or, alternatively, the amount of offering contacts decreases when the highest reward rate is offered to the donor forager (Pírez and Farina 2004). These results strongly suggest that first-order receivers acquire quantitative information about the nectar source exploited by foragers.

Another factor, which correlates with aspects of trophallaxis, is the intensive antennal contacts performed by trophallactic partners during food transfer (Montagner and Galliot 1982). In honey bees, the antennal movements of both donor and receiver, which are rapid during the food transfer (mean frequency of 13 Hz), vary according to the reward rate experienced by the food donor and show a positive correlation between both trophallactic partners (Goyret and Farina 2003; Figure 10.3b). In addition to information about the reward rate of the food source, information about the food supply of a colony might also be encoded in the tactile stimulation during trophallaxis, as has been suggested for the carpenter ant *Camponotus mus* (McCabe et al. 2006).

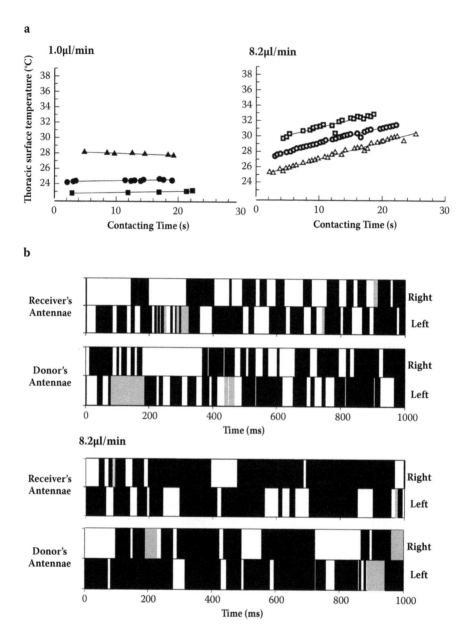

FIGURE 10.3 Thermal behavior and antennation during trophallaxis in honey bees. (a) Thoracic temperature of food receivers in relation to the duration of the trophallactic contact with a forager that collected either 1.0 or 8.2 µl of a 50% sucrose solution per minute at a feeding station. Different symbols represent different food-receiving bees. (After Farina and Wainselboim 2001a. With permission.) (b) Example of the antennal contacts (strokes, black bars) of a donor and a single receiver during a long trophallaxis inside the nest. The donor forager returned from a rate feeder offering either 1.0 or 8.2 µl of a 50% sucrose solution per minute. Temporal resolution: 5 ms. Grey bars indicate moments where the position of the antennae could not be precisely determined. Under these experimental conditions, the mean values of the thoracic temperature of the donor foragers at the beginning of the trophallaxis were 31.8°C for a reward rate of 1 µl/min and 37.5°C for 8.2 µl/min, respectively, and the estimated transfer rates of the donors for these reward rates were 1.1 and 2.1 µl/s. (After Goyret and Farina 2003. With permission.)

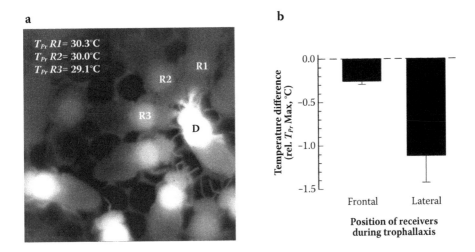

FIGURE 10.4 Trophallaxis with multiple food receivers. (a) Thermogram showing the food donor (D), which was an active forager trained to collect at a rate feeder that offered 8.2 µl of a 50% sucrose solution per minute. Receivers in front of the donor (R1 and R2) attained higher proboscis temperatures (T_{Pr}) than receiver R3, which contacted the donor from its side. The heating of the frontal receivers was independent of the number of simultaneous receivers. (b) Mean differences between the receivers' proboscis temperatures and the maximum proboscis temperature (T_{Pr} Max) observed during trophallaxis with an incoming forager. Receivers in front of the donor had smaller temperature differences than receivers positioned laterally to the forager. (Farina, unpublished data.)

CHANGING THE TROPHALLACTIC ROLE

After unloading their crop, foragers walk across the delivery area to the hive entrance, thereby often protruding their proboscis and touching the mouthparts of their nestmates. It has been suggested that these begging contacts are refueling events for the forthcoming foraging trip (Beutler 1950; von Frisch 1967). After leaving the hive, foragers carry more food if they do not know the feeding site well (Brandstetter et al. 1988) or if they collect far from the nest (Istomina-Tsvetkova 1960). This could be explained either by refueling inside the hive or by unloading only a part of the collected crop. However, begging behavior can also be observed in foragers that constantly collect food over a longer period of time at a feeder located close to the hive (Núñez 1970; Farina 1996; De Marco and Farina 2001). So, why do these bees beg for food? It would seem much more efficient if experienced bees simply retained the amount necessary for the flight to the food source during the unloading event.

Food source profitability affects the forager's begging behavior, just as it affects the short offering contacts, yet in the opposite direction: nectar foragers increase the frequency of begging contacts after food unloading when they return from a low-profit source (Farina 1996; De Marco and Farina 2001) or if the diversity of odor cues and food qualities encountered in the exploited food patch is high (De Marco and Farina 2003). These begging contacts often last less than 1 s (De Marco and Farina 2003), which indicates that the probability of an actual food transfer is very low (Farina and Wainselboim 2001b). Hence, it is unlikely that foragers are refueled during these begging interactions.

An alternative hypothesis proposes that begging might be a means of information acquisition (Núñez 1970; Farina 1996; De Marco and Farina 2001, 2003). Active foragers may direct their begging behavior toward other employed nectar foragers in order to obtain chemosensory information about other resources, which facilitates the reevaluation of their own food source. Consequently, employed foragers could decide whether to continue exploiting their food source, whether to switch to a previously exploited one that reappeared again (indicated by the presence of its scent in the hive), or whether to stop foraging and stay in the hive. However, to date there is no evidence for these hypotheses, and begging contacts remain a puzzling phenomenon.

Experiments conducted under low-reward-rate conditions can help us to understand communication strategies in a more natural context. Natural flowers normally offer minute amounts of nectar with variable flow rates (Núñez 1977; Vogel 1983), and bees often visit several hundred flowers per foraging trip (Ribbands 1949). Many of the modulatory effects described above become apparent only when bees collect food at a low reward rate.

THE DISTRIBUTION OF THE NECTAR INSIDE THE HIVE

After receiving the nectar from foragers, a majority of processor bees offer it to other bees, sometimes large parts of their load, on their way to the honey cells (von Frisch 1923; Park 1925; Seeley 1989; Pírez and Farina 2004; Grüter and Farina 2007). Whereas workers that are not involved in food processing have only 0.25–0.75 trophallactic contacts per 10 min (Istomina-Tsvetkova 1953a, 1953b, cited in Free 1959; Grüter and Farina 2007), food processor bees perform between 4.3 and 10.5 contacts during the same period of time (Seeley 1989), which highlights their role in the rapid distribution of the collected food among the hive bees. The workers that receive the food from processors (second-order receivers) are mainly nurse bees, but they can also be foragers and other food processors (Grüter and Farina 2007). These second-order receivers perform about four contacts per 10 min, most of them being offering contacts. As a consequence, the incoming nectar is rapidly distributed among bees of all ages. Nixon and Ribbands (1952) measured this food distribution within honey bee nests using a radioactive tracer in the sugar solution (^{32}P). They fed between five and nine foragers belonging to colonies of different population sizes 10–20 ml of radioactive solution and found that within 4 h a majority of the hive bees had received samples of this food. Similarly, DeGrandi-Hoffman and Hagler (2000) found a rapid distribution of food among young hive bees by using a protein marker.

Several characteristics of food processing, like the rate of trophallactic events or the food storing behavior of the hive bees, depend on a variety of factors, such as the nutritional state of the colony, the amount of brood, the workers' genotype, the amount of incoming nectar, or the season (Free 1959; Kloft et al. 1976; Hillesheim 1986; Seeley 1989; Crailsheim 1998). Whereas experiments with caged bees also showed an effect of age on the trophallactic activity (Moritz and Hallmen 1986), the trophallactic activity of workers within beehives seems to depend more on the performed task than on the bees' age (Free 1957).

Another interesting aspect of trophallaxis in honey bees is the fact that the transfer rates of subsequent trophallaxes are positively correlated (Goyret and Farina 2005b). In other words, trophallactic experiences of bees affect their trophallactic behavior in the immediate future in similar ways as nectar flow rates affect the unloading rate of foragers. Consequently, food receiver bees that are not directly unloading foragers might still be able to acquire information about the colony's foraging situation. Given the extensive sharing of food among bees of all ages, cues present in the collected food that convey information about food source characteristics can reach most hive members in a relatively short time. Information available to most or all individuals of a colony, or "global" information (Mitchell 2006), potentially affects the behavior of the majority of the nestmates, thereby causing a global response (Moritz and Southwick 1992; Pankiw et al. 2004). For example, the sugar concentration of incoming nectar affects the sugar response thresholds (SRTs) of nectar receivers (Martínez and Farina 2008) and even of young hive bees, which are not involved in foraging (Pankiw et al. 2004) and have little direct contact with engaged foragers (Seeley 1995). These results not only indicate a fine-tuning of sensory thresholds in hive bees, but also highlight the role of trophallaxis as a mechanism to transfer gustatory information in honey bees (Martínez and Farina 2008).

ODOR LEARNING THROUGH TROPHALLAXIS

Bees are excellent learners and readily establish associations between odors (or other cues) and a reward, such as a sugar solution (e.g., von Frisch 1967; Koltermann 1969; Menzel 1999). During olfactory conditioning the sugar solution functions as an unconditioned stimulus (US), while the odor becomes the conditioned stimulus (CS).

Olfactory learning has a strong effect on foraging decisions (see Chapter 9). In a series of simple and elegant experiments, von Frisch (1923) showed that bees recruited by a forager showed a strong preference for food with the odor brought back by the recruiting bee. In the meantime, these findings have been confirmed in other social insects like bumble bees (Dornhaus and Chittka 1999), stingless bees (Lindauer and Kerr 1960; see also Chapter 12), wasps (Maschwitz et al. 1974; Jandt and Jeanne 2005), and ants (Roces 1990). It has been suggested that recruits learn food odors while receiving food samples from foragers, i.e., during trophallaxis (von Frisch 1967). Here, the transferred food samples could function as a reward for learning. Dirschedl (1960) found that more than 95% of all recruits arriving at a feeder that offered stained sugar solution had received small samples of this food inside the hive prior to leaving the colony.

The proboscis extension response (PER) test has been used with great success to study associative learning under controlled laboratory conditions, and to analyze the physiology and memory processes underlying learning in honey bees (Kuwabara 1957; Bitterman et al. 1983; reviewed in Menzel et al. 1993; Menzel 1999). Bees extend their proboscis when chemoreceptors on their antennae, tarsi, or proboscis are stimulated with sucrose solution (US). If an odor (CS) is presented simultaneously, it will subsequently elicit the proboscis extension (conditioned response, CR) if presented alone, often even after a single learning trial.

More recently, the PER assay was also used to investigate learning processes during trophallaxis. Honey bees associatively learn food odors while receiving food from other bees in a variety of behavioral contexts (e.g., within an experimental arena, Gil and De Marco 2005; within an observation hive, Farina et al. 2007). Several experiments have demonstrated that increasing the concentration of either the CS or the US results in better learning during trophallaxis (Gil and De Marco 2005). Interestingly, successful learning does not seem to depend much on the duration of the oral contact (e.g., trophallaxes as short as 1.2 s led to learning; Gil and De Marco 2005). For this kind of learning it should be of little importance whether the receivers perceive the odor in the solution or on the foragers' bodies. Most important, however, is the contiguity between CS and US (Rescorla 1988; Menzel et al. 1993).

Furthermore, PER assays revealed that bees that are recruited to a specific food source associatively learn the food's odors inside the colony (Farina et al. 2005; Grüter et al. 2006; Figure 10.5). Similarly, food processor bees show the PER for a particular scent with an elevated probability after having unloaded food containing this scent from foragers (Farina et al. 2007; Figure 10.5a). Grüter et al. (2006) fed scented sucrose solution to marked foragers for about a week and at the same time measured the PER of bees at the age of 4–9 days and 12–16 days, and of a sample of foragers (Figure 10.5b). During the feeding period, the proportion of bees extending their proboscises upon presentation of the solution's scent increased in all age groups, which indicates that olfactory information about the flower species exploited by foragers propagates within the entire colony as a consequence of food sharing. Potentially, this has long-term consequences, since olfactory information acquired inside the hive can be stored in the bees' long-term memory (LTM) (Farina et al. 2005; Gil and De Marco 2006; Arenas et al. 2007), a form of memory that affects the behavior of bees for several days (e.g., Menzel 1999).

In the experiments with hive bees, the PER frequencies were lower than expected when comparing them to learning performances under controlled laboratory conditions, where a single learning trial often already leads to about 50% of bees responding to the odor (Menzel et al. 1993). However, there are several problems with such comparisons. First, in the experiments with hive bees, the individuals were tested in a context that differed from the context they had experienced when learning

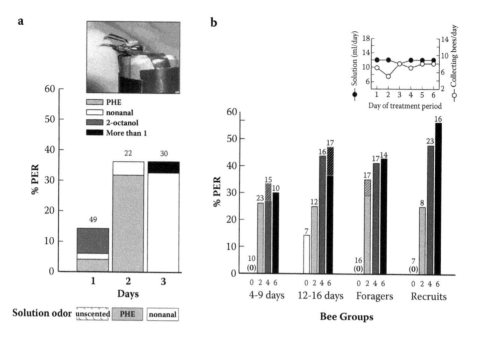

FIGURE 10.5 Olfactory memories formed within the hive during the sharing of scented food. (a) Percentage of food-receiver bees that extended the proboscis on the first presentation of an odor in a proboscis extension response (PER) assay. In the course of the experiment, foragers collected a 2.0-M sucrose solution that was either unscented (day 1), scented with phenylacetaldehyde (PHE) (day 2), or scented with nonanal (day 9). Responses for PHE (grey), nonanal (white), 2-octanol (dark grey), and for more than one test odor (black) are shown. Numbers of tested bees are given above the bars. (After Farina et al. 2007. With permission.) (b) PER on the first presentation of a treatment odor (linalool) and a novel odor (2-octanol) in 4- to 9-day-old bees, 12- to 16-day-old bees, foragers, and recruits. PER frequencies were measured on days 0, 2, 4, and 6 after starting to feed the colony with a linalool-scented sucrose solution. Bars indicate the percentage of bees showing a spontaneous PER to linalool (filled bars), or to both linalool and 2-octanol (hatched bars). The inset shows the daily quantity of the collected sucrose solution in milliliters (filled circles) and the number of trained foragers (empty circles). (After Grüter et al. 2006. With permission.)

the odor. This can cause a considerable reduction in performance (Bouton and Moody 2004), which, as a consequence, might lead to an underestimation of information acquisition. Second, within the hive, bees probably not always perceive an odor associated with a reward, but often without it, e.g., when attending or following a dancing bee without receiving a food sample. These CS-only experiences (retrieval trials after acquisition) often result in a reduction of the conditioned response (CR) (Stollhoff et al. 2005).

Given the rapid sharing of food among all bees of a colony (Farina et al. 2005, Grüter et al. 2006), the propagation of the information about the odor of a food source would be much more extensive if the scent were present in the nectar itself rather than just clinging to the forager's body. But are nectars scented at all? Unfortunately, very little is known about the presence of scents in nectars of most plant species visited by bees. Apparently, many types of nectars contain scents, but more systematic studies on the distribution of scented floral nectars (reviewed in Raguso 2004) are needed to evaluate their potential role as an information source in communication and learning processes in bees.

The propagation of olfactory information among hive bees has important behavioral and ecological implications (see also Chapter 9). Novice foragers without foraging experience leave the hive with socially acquired olfactory information about the food plants, which can help them to discover new food patches of the same plant species (i.e., other patches than those exploited by the recruiting foragers). The bees' preference for odors they had experienced in the hive while searching for food in the field is also used for crop production: feeding honey bee colonies with sugar solution scented with crop plant odors is the basic principle of the attempts to guide bees to certain plant species in order to increase the visitation rates and seed production (von Frisch 1943, 1967).

As discussed above, olfactory memory also affects unloading decisions of food processor bees (Goyret and Farina 2005a). Consequently, the flower constancy observed in honey bee foragers (von Frisch 1967) can already be found in nectar processors during unloading, yet to a lesser degree. Thus, a forager returning with a new odor might be received with less interest by nectar processors, which, in turn, affects the forager's motivation to perform dances (Goyret and Farina 2005a). However, more studies are needed to investigate the consequences of olfactory experiences for these in-hive interactions. In addition, research should be extended to other social insect species in order to better understand the general importance of trophallaxis for learning in these organisms.

In the context of olfactory learning, the bee dance might have an amplifying effect. As mentioned earlier, dancers attract both foragers and food processors, and thereby increase the number of trophallactic interactions (Farina 2000). In this way, the dance creates an environment for the acquisition of olfactory information (von Frisch 1923, 1967; Díaz et al. 2007).

CONCLUDING REMARKS

The attempt to understand how a honey bee colony coordinates the single individuals in order to efficiently obtain and process food in a system lacking central control is a fascinating challenge. Trophallaxis seems to be one of the means by which individuals that belong to worker groups carrying out different tasks are rapidly informed about characteristics of the collected resources. Hence, trophallaxis is an important mechanism not only to transfer food, but also to inform individuals about fluctuating foraging opportunities, to adjust in-hive tasks related to the foraging process, and to create information networks that connect different groups of workers.

ACKNOWLEDGMENTS

We are indebted to R. Josens and F. Segers for valuable comments and suggestions on an early version of this manuscript. This study was supported by funds from ANPCYT (01-12319), CONICET, and the University of Buenos Aires (X 036) to WMF. CG is supported by the Dr. De Giacomi Stiftung, Basler Stiftung für biologische Forschung, and the Janggen-Pöhn Stiftung.

REFERENCES

Arenas A, Fernández VM, Farina WM. (2007). Floral odor learning within the hive affects honeybees' foraging decisions. *Naturwissenschaften* 94:218–22.
Beutler R. (1950). Zeit und Raum im Leben der Sammelbiene. *Naturwissenschaften* 37:102–5.
Bitterman ME, Menzel R, Fietz A, Schäfer S. (1983). Classical-conditioning of proboscis extension in honeybees (*Apis mellifera*). *J Comp Psychol* 97:107–19.
Bouton ME, Moody EW. (2004). Memory processes in classical conditioning. *Neurosci Biobehav Rev* 28:663–74.
Božič J, Valentinčič T. (1991). Attendants and followers of honey bee waggle dances. *J Apicult Res* 30:125–31.
Brandstetter M, Crailsheim K, Heran H. (1988). Provisioning of food in the honeybee before foraging. In Nachtigall W (ed.), *Biona Report 6: The Flying Honeybee*. Stuttgart: Gustav Fischer Verlag, pp. 129–48.
Crailsheim K. (1998). Trophallactic interactions in the adult honeybee (*Apis mellifera* L.). *Apidologie* 29:97–112.
DeGrandi-Hoffman G, Hagler J. (2000). The flow of incoming nectar through a honey bee (*Apis mellifera* L.) colony as revealed by a protein marker. *Insect Soc* 47:302–6.

De Marco RJ, Farina WM. (2001). Changes in food source profitability affect the trophallactic behavior of forager honeybees (*Apis mellifera*). *Behav Ecol Sociobiol* 50:441–49.

De Marco RJ, Farina WM. (2003). Trophallaxis in forager honeybees (*Apis mellifera*): Resource uncertainty enhances begging contacts? *J Comp Physiol A* 189:125–34.

Díaz PC, Grüter C, Farina WM. (2007). Floral scents affect the distribution of hive bees around dancers. *Behav Ecol Sociobiol* 61:1589–97.

Dirschedl H. (1960). Die Vermittlung des Blütenduftes bei der Verständigung im Bienenstock. Doctoral thesis, University of Munich, Germany.

Dornhaus A, Chittka L. (1999). Evolutionary origins of bee dances. *Nature* 401:38.

Farina WM. (1996). Food-exchange by foragers in the hive—A means of communication among honey bees? *Behav Ecol Sociobiol* 38:59–64.

Farina WM. (2000). The interplay between dancing and trophallactic behavior in the honey bees *Apis mellifera*. *J Comp Physiol A* 186:239–45.

Farina WM, Grüter C, Diaz PC. (2005). Social learning of floral odours inside the honeybee hive. *Proc R Soc London B* 272:1923–28.

Farina WM, Grüter C, Acosta L, McCabe S. (2007). Honeybees learn floral odors while receiving nectar from foragers within the hive. *Naturwissenschaften* 94:55–60.

Farina WM, Núñez JA. (1991). Trophallaxis in the honeybee, *Apis mellifera* (L.) as related to the profitability of food sources. *Anim Behav* 42:389–94.

Farina WM, Núñez JA. (1993). Trophallaxis in honey bees: Transfer delay and daily modulation. *Anim Behav* 45:1227–31.

Farina WM, Wainselboim AJ. (2001a). Changes in the thoracic temperature of honey bees while receiving nectar from foragers collecting at different reward rates. *J Exp Biol* 204:1653–58.

Farina WM, Wainselboim AJ. (2001b). Thermographic recordings show that honeybees may receive nectar from foragers even during short trophallactic contacts. *Insect Soc* 48:360–62.

Farina WM, Wainselboim AJ. (2005). Trophallaxis within the dancing context: A behavioral and thermographic analysis in honeybees. *Apidologie* 36:43–47.

Fernández PC, Gil M, Farina WM. (2003). Reward rate and forager activation in honeybees: Recruiting mechanisms and temporal distribution of arrivals. *Behav Ecol Sociobiol* 54:80–87.

Free JB. (1957). The transmission of food between worker honeybees. *Br J Anim Behav* 5:41–47.

Free JB. (1959). The transfer of food between the adult members of a honeybee community. *Bee World* 40:193–201.

von Frisch K. (1923). Über die "Sprache" der Bienen. Eine tierpsychologische Untersuchung. *Zool Jahrb, Abt Physiol* 40:1–186.

von Frisch K. (1943). Versuche über die Lenkung des Bienenfluges durch Duftstoffe. *Naturwissenschaften* 31:445–60.

von Frisch K. (1967). *The Dance Language and Orientation of Bees*. Cambridge, MA: Harvard University Press.

Gil M, De Marco RJ. (2005). Olfactory learning by means of trophallaxis in *Apis mellifera*. *J Exp Biol* 208:671–80.

Gil M, De Marco RJ. (2006). *Apis mellifera* bees acquire long-term olfactory memories within the colony. *Biol Lett* 2:98–100.

Gil M, Farina WM. (2003). Crop scents affect the occurrence of trophallaxis among forager honeybees. *J Comp Physiol A* 189:379–82.

Goyret J, Farina WM. (2003). Descriptive study of antennation during trophallactic unloading contacts in honeybees *Apis mellifera carnica*. *Insect Soc* 50:274–76.

Goyret J, Farina WM. (2005a). Non-random nectar unloading interactions between foragers and their receivers in the honeybee hive. *Naturwissenschaften* 92:440–43.

Goyret J, Farina WM. (2005b). Trophallactic chains in honeybees: A quantitative approach of the nectar circulation amongst workers. *Apidologie* 36:595–600.

Grassé PP, Noirot C. (1945). La transmission des flagelles symbiotiques et les aliments des termites. *Bull Biol Fr Belg* 79:273–92.

Gregson AM, Hart AG, Holcombe M, Ratnieks FLW. (2003). Partial nectar loads as a cause of multiple nectar transfer in the honey bee (*Apis mellifera*): A simulation model. *J Theor Biol* 222:1–8.

Grüter C, Acosta LE, Farina WM. (2006). Propagation of olfactory information within the honeybee hive. *Behav Ecol Sociobiol* 60:707–15.

Grüter C, Farina WM. (2007). Nectar distribution and its relation to food quality in honeybee (*Apis mellifera*) colonies. *Insect Soc* 54:87–94.

Hart AG, Ratnieks FLW. (2002). Task-partitioned nectar transfer in stingless bees: Work organisation in a phylogenetic context. *Ecol Entomol* 27:163–68.

Hillesheim E. (1986). Trophallaxis von Arbeiterinnen in Kleingruppen und im Volk (*Apis mellifera* L.). *Apidologie* 17:343–46.

Hölldobler B. (1977). Communication in social Hymenoptera. In Sebeok TA (ed.), *How Animals Communicate*. Bloomington: Indiana University Press, pp. 418–71.

Hölldobler B. (1985). Liquid food transmission and antennation signals in ponerine ants. *Israel J Entomol* 19:89–99.

Hölldobler B, Möglich M, Maschwitz U. (1974). Communication by tandem running in the ant *Camponotus sericeus*. *J Comp Physiol A* 90:105–27.

Hrncir M, Schmidt VM, Schorkopf DLP, Jarau S, Zucchi R, Barth FG. (2006). Vibrating the food receivers: A direct way of signal transmission in stingless bees (*Melipona seminigra*). *J Comp Physiol A* 192:879–87.

Huang MH, Seeley TD. (2003). Multiple unloadings by nectar foragers in honey bees: A matter of information improvement or crop fullness? *Insect Soc* 50:330–39.

Istomina-Tsvetkova KP. (1953a). New data on the behavior of bees. *Pchelovodstvo* 30:15–23.

Istomina-Tsvetkova KP. (1953b). Reciprocal feeding between bees. *Pchelovodstvo* 30:25–29.

Istomina-Tsvetkova KP. (1960). Contribution to the study of trophic relations in adult worker bees. In *Proceedings of the XVII International Beekeeping Congress*, Bologna-Roma, Italy, Vol. 2, pp. 361–68.

Jandt JM, Jeanne RL. (2005). German yellowjacket (*Vespula germanica*) foragers use odors inside the nest to find carbohydrate food sources. *Ethology* 111:641–51.

Kirchner WH, Lindauer M. (1994). The causes of the tremble dance of the honeybee *Apis mellifera*. *Behav Ecol Sociobiol* 35:303–8.

Kloft WJ, Djalal AS, Drescher W. (1976). Untersuchung der unterschiedlichen Futterverteilung in Arbeiterinnengruppen verschiedener Rassen von *Apis mellifica* L. mit Hilfe von ^{32}P als Tracer. *Apidologie* 7:49–60.

Koltermann R. (1969). Lern- und Vergessensprozesse bei der Honigbiene—aufgezeigt anhand von Duftdressuren. *Z vergl Physiol* 63:310–34.

Kukuk PF, Crozier RH. (1990). Trophallaxis in a communal halictine bee *Lasioglossum (Chilalictus) erythrurum*. *Proc Natl Acad Sci USA* 87:5402–4.

Kuwabara M. (1957). Bildung des bedingten Reflexes von Pavlovs Typus bei der Honigbiene, *Apis mellifica*. *J Fac Sci Hokkaido Univ Ser VI Zool* 13:458–64.

Lindauer M. (1952). Ein Beitrag zur Frage der Arbeitsteilung im Bienenstaat. *Z vergl Physiol* 34:299–345.

Lindauer M. (1954). Temperaturregulierung und Wasserhaushalt im Bienenstaat. *Z vergl Physiol* 36:391–432.

Lindauer M, Kerr WE. (1960). Communication between the workers of stingless bees. *Bee World* 41:29–41, 65–71.

Martínez A, Farina WM. (2008). Honeybees modify gustatory responsiveness after receiving nectar from foragers within the hive. *Behav Ecol Sociobiol* 62:529–35.

Maschwitz U, Beier W, Dietrich I, Keidel W. (1974). Futterverständigung bei Wespen der Gattung *Paravespula*. *Naturwissenschaften* 61:506.

McCabe S, Farina WM, Josens RB. (2006). Antennation of nectar-receivers encodes colony needs and food-source profitability in the ant *Camponotus mus*. *Insect Soc* 53:356–61.

Menzel R. (1999). Memory dynamics in the honeybee. *J Comp Physiol A* 185:323–40.

Menzel R, Greggers U, Hammer M. (1993). Functional organization of appetitive learning and memory in a generalist pollinator, the honey bee. In Papaj D, Lewis AC (eds.), *Insect Learning: Ecological and Evolutionary Perspectives*. New York: Chapman & Hall, pp. 79–125.

Michener CD. (1969). Comparative social behavior of bees. *Annu Rev Entomol* 14:299–342.

Michener CD. (1974). *The Social Behavior of Bees: A Comparative Study*. Cambridge, MA: Harvard University Press.

Mitchell M. (2006). Complex systems: Network thinking. *Artif Intell* 170:1194–212.

Montagner H, Galliot G. (1982). Antennal communication and food exchange in the domestic bee *Apis mellifera* L. In Breed MD, Michener CD, Evans ME (eds.), *The Biology of Social Insects*. Boulder: Westview Press, pp. 302–6.

Moritz RFA, Hallmen M. (1986). Trophallaxis of worker honeybees (*Apis mellifera* L.) of different ages. *Insect Soc* 33:26–31.

Moritz RFA, Southwick EE. (1992). *Bees as Superorganisms: An Evolutionary Reality*. Berlin: Springer-Verlag.

Nixon HL, Ribbands CR. (1952). Food transmission within the honeybee community. *Proc R Soc Lond B* 140:43–50.

Núñez JA. (1966). Quantitative Beziehungen zwischen den Eigenschaften von Futterquellen und dem Verhalten von Sammelbienen. *Z vergl Physiol* 53:142–64.

Núñez JA. (1970). The relationship between sugar flow and foraging and recruiting behavior of honeybees (*Apis mellifera* L.). *Anim Behav* 18:527–38.

Núñez JA. (1977). Nectar flow by melliferous flora and gathering flow by *Apis mellifera ligustica*. *J Insect Physiol* 23:265–75.

Núñez JA. (1982). Honeybee foraging strategies at a food source in relation to its distance from the hive and the rate of sugar flow. *J Apicult Res* 21:139–50.

Pankiw T, Nelson M, Page RE, Fondrk MK. (2004). The communal crop: Modulation of sucrose response thresholds of pre-foraging honey bees with incoming nectar quality. *Behav Ecol Sociobiol* 55:286–92.

Park W. (1925). The storing and ripening of honey by honeybees. *J Econ Entomol* 18:405–10.

Pírez N, Farina WM. (2004). Nectar-receiver behavior in relation to the reward rate experienced by foraging honeybees. *Behav Ecol Sociobiol* 55:574–82.

Raguso RA. (2004). Why are some floral nectars scented? *Ecology* 85:1486–94.

Ratnieks FLW, Anderson C. (1999). Task partitioning in insect societies. *Insect Soc* 46:95–108.

Rescorla RA. (1988). Behavioral studies of pavlovian conditioning. *Annu Rev Neurosci* 11:329–52.

Ribbands CR. (1949). The foraging method of individual honey-bees. *J Anim Ecol* 18:47–66.

Riley JR, Greggers U, Smith AD, Reynolds DR, Menzel R. (2005). The flight paths of honeybees recruited by the waggle dance. *Nature* 435:205–7.

Roces F. (1990). Olfactory conditioning during the recruitment process in a leaf-cutting ant. *Oecologia* 83:261–62.

Rösch GA. (1925). Untersuchungen über die Arbeitsteilung im Bienenstaat. 1. Teil: Die Tätigkeiten im normalen Bienenstaate und ihre Beziehungen zum Alter der Arbeitsbienen. *Z vergl Physiol* 2: 571–631.

Roubaud E. (1916). Recherches biologiques sur les guêpes solitaires et social d'Afrique: la genèse de l'instinct maternel chez les vespides. *Annal Sci Nat* 10e(1):1–160.

Seeley TD. (1982). Adaptive significance of the age polyethism schedule in honeybee colonies. *Behav Ecol Sociobiol* 11:287–93.

Seeley TD. (1986). Social foraging by honeybees: How colonies allocate foragers among patches of flowers. *Behav Ecol Sociobiol* 19:343–54.

Seeley TD. (1989). Social foraging in honey bees: How nectar foragers assess their colony's nutritional status. *Behav Ecol Sociobiol* 24:181–99.

Seeley TD. (1992). The tremble dance of the honey bee: Message and meanings. *Behav Ecol Sociobiol* 31:375–83.

Seeley TD. (1995). *The Wisdom of The Hive: The Social Physiology of Honey Bee Colonies*. Cambridge, MA: Harvard University Press.

Seeley TD, Mikheyev AS, Pagano GJ. (2000). Dancing bees tune both duration and rate of waggle run production in relation to nectar-source profitability. *J Comp Physiol A* 186:813–19.

Seeley TD, Tovey CA. (1994). Why search time to find a food-storer bee accurately indicates the relative rates of nectar collecting and nectar processing in honey bee colonies. *Anim Behav* 47:311–16.

Sleigh C. (2002). Brave new worlds: Trophallaxis and the origin of society in the early twentieth century. *J Hist Behav Sci* 38:133–56.

Stollhoff N, Menzel R, Eisenhardt D. (2005). Spontaneous recovery from extinction depends on the reconsolidation of the acquisition memory in an appetitive learning paradigm in the honeybee (*Apis mellifera*). *J Neurosci* 25:4485–92.

Tezze A, Farina WM. (1999). Trophallaxis in the honeybee, *Apis mellifera* (L.): The interaction between viscosity and sucrose concentration of the transferred solution. *Anim Behav* 57:1319–26.

Thom C. (2003). The tremble dance of honey bees can be caused by hive-external foraging experience. *J Exp Biol* 206:2111–16.

Thom C, Gilley DC, Hooper J, Esch HE. (2007). The scent of the waggle dance. *PLoS Biol* 5:e228.

Vogel S. (1983). Ecophysiology of zoophilic pollination. In Lange OL, Nobel PS, Osmond CB, Ziegler H (eds.), *Physiological Plant Ecology III (Encyclopedia of Plant Physiology)*. Berlin: Springer-Verlag, pp. 559–624.

Wainselboim AJ, Farina WM. (2000a). Trophallaxis in filled-crop honeybees (*Apis mellifera* L.): Food-loading time affects unloading behavior. *Naturwissenschaften* 87:280–82.

Wainselboim AJ, Farina WM. (2000b). Trophallaxis in the honeybee, *Apis mellifera* (L.): The interaction between flow of solution and sucrose concentration of the exploited food sources. *Anim Behav* 59:1177–85.

Wainselboim AJ, Farina WM. (2003). Trophallaxis in the honeybees, *Apis mellifera* (L.) as related to their past experience at the food source. *Anim Behav* 66:791–95.

Wainselboim AJ, Roces F, Farina WM. (2002). Honey bees assess changes in nectar flow within a single foraging bout. *Anim Behav* 63:1–6.

Wainselboim AJ, Roces F, Farina WM. (2003). Assessment of food source profitability in honeybees (*Apis mellifera*): How does disturbance of foraging activity affect trophallactic behaviour? *J Comp Physiol A* 189:39–45.

Wcislo WT, González VH. (2006). Social and ecological contexts of trophallaxis in facultatively social sweat bees, *Megalopta genalis* and *M. ecuadoria* (Hymenoptera, Halictidae). *Insect Soc* 53:220–225.

Wheeler WM. (1918). A study of some young ant larvae with a consideration of the origin and meaning of social habits among insects. *Proc Am Phil Soc* 57:293–343.

Wilson EO. (1971). *The Insect Societies*. Cambridge, MA: Belknap Press of Harvard University Press.

11 Mobilizing the Foraging Force
Mechanical Signals in Stingless Bee Recruitment

Michael Hrncir

CONTENTS

INTRODUCTION

Already at school, probably all of us have learned that bees perform dances that map the position of a food source. This is certainly true—but only for the honey bees (Apini; about ten species), which merely constitute a meager 1% of all eusocial bee species within the Apidae family. So far, we do not have the least evidence for dances encoding either the direction or the distance of a food source among the remaining 99%, comprising the stingless bees (Meliponini; more than 400 species) and the bumble bees (Bombini; about 250 species)—not to mention the primitively eusocial colony stages found in several species within the tribes Allodapini and Ceratini, or within the family Halictidae (Michener 2000).

Among the stingless bees, we encounter a broad spectrum of signals involved in foraging and recruitment processes (Lindauer and Kerr 1958, 1960; Nieh 2004; Barth et al. 2008). In this bee group, both the search for and the choice of a specific food patch are strongly biased by signals and cues in the field (field-based information). In some Meliponini, for instance, collecting foragers deposit pheromone marks at and near the food source to guide newcomers to the goal. Consequently, the foraging processes in colonies of these species are extremely directed toward one particular target area (see Chapter 12). Apart from such genuine signals, field-based social cues have been shown to affect the foraging decisions of searching individuals. Within a food patch, a bee's choice of where to collect is influenced by scent marks left on flowers (see Chapter 12), and by the physical presence of con- or heterospecific foragers (Slaa et al. 2003a, 2003b; Biesmeijer and Slaa 2004; see also Chapter 8).

Yet, the use of field-based information is not the entire story of success in the Meliponini. Foragers that return from an extremely profitable food source run agitatedly through the nest, thereby jostling their nestmates (Lindauer and Kerr, 1958, 1960; Nieh 1998; Hrncir et al. 2000) (Figure 11.1). In addition to these excited movements, the foragers generate thoracic vibrations ("sounds") in the course of their stay in the hive, predominantly during trophallactic interactions with hive bees (Lindauer and Kerr 1958, 1960; Hrncir et al. 2006a, 2006b) (Figure 11.1). The agitated jostling run

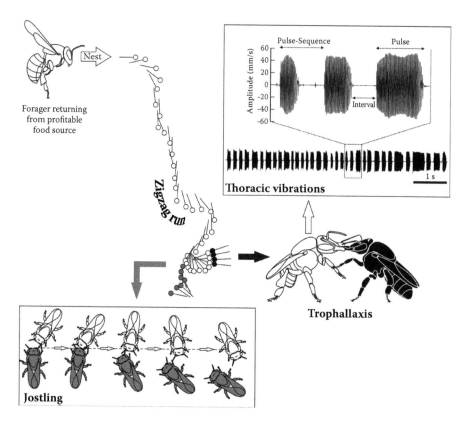

FIGURE 11.1 Nest-internal behaviors displayed by stingless bee foragers. When a forager (white bees and symbols) returns from a high-profit food source, she excitedly runs through the colony (zigzag run), thereby jostling her nestmates (grey bees and symbols). While running, but predominantly during trophallaxis with hive bees (black bee and symbols), the forager generates pulsed thoracic vibrations.

as well as the sound production attracted the attention of scientists because both behaviors were considered potential meliponine equivalents to the dance "language" of the honey bees (Esch et al. 1965; Nieh 1998; Dyer 2002). The present chapter gives an overview of what is currently known about the potential message(s) and meaning(s) of these putative signals.

ECOLOGICAL CONSIDERATIONS

Similar to honey bees, stingless bees live in perennial colonies, which typically comprise a few dozen to several thousand individuals (Schwarz 1948; Michener 1974, 2000; Roubik 1989). As in other social insects, the survival of a colony greatly depends on the success of foraging individuals in finding and collecting carbohydrates (usually nectar) and proteins (usually pollen). Both are stored within the nest to guarantee a permanent food supply and, consequently, to ensure the survival of the colony during possible events of resource scarcity. Since stingless bees are mainly found in the tropical regions around the globe, where periods without a considerable number of plant species in bloom are insignificant (Bawa 1983), the acquisition of food occurs all year round. Nevertheless, meliponine colonies adjust their collecting activity to the availability of resources in the environment by increasing both the individual foraging activity and the foraging force as soon as a surplus of food is available (Roubik et al. 1986; Eltz et al. 2001; Nagamitsu and Inoue 2002; Hofstede 2006; Hofstede and Sommeijer 2006).

RESOURCE AVAILABILITY

In the tropics, we encounter two distinct patterns of flowering: one in which the individual plant produces small numbers of flowers over an extended period of time ("steady state"), and the second in which an individual produces a large amount of new flowers each day over a short period of time, often less than a week ("big bang" or "mass flowering") (Gentry 1974; Augspurger 1980, 1983). Within a plant population, mass-flowering individuals of one species bloom synchronously in most cases (Figure 11.2a). Asynchronous blooming, which also occurs, results in an extended blooming period on the population level (Bawa 1983).

These two modes of flowering constitute two distinct resource types for tropical bees:

1. *Low profitability—long-lived resources.* Since steady-state plants offer only a small number of flowers, bees occasionally have to move long distances between conspecific specimens to encounter an open flower (increased search costs and risk of predation). However, as soon as a bee knows all (or many) individuals of a steady-state species within her flight range, these plants represent a reliable food source due to a flowering period that often exceeds her foraging lifetime.

2. *High profitability—short-lived resources.* Since mass-flowering plants produce an excess of flowers,[*] bees foraging on this kind of resource have low search costs and can rapidly fill their crop, or load pollen, during a single foraging trip.[†]

The two flowering patterns, steady state and mass flowering, should result in different foraging strategies of eusocial bee colonies exploiting these food sources. Steady-state resources provide a continuous but low food influx into the colony. Due to the small number of available flowers within the foraging range of a colony, such resources can be exploited by solitary foraging. And since the value (profitability) of the resource is low, recruitment to these plants should be low or inexistent. By contrast, in order to efficiently exploit mass-flowering plants that offer a great opportunity to hoard

---

[*] Mori and Piploy (1984) estimated up to five hundred thousand open flowers per day and individual for the tropical tree *Miconia minutiflora*, Melastomaceae.

[†] An example for three *Melipona* species collecting nectar on the mass-flowering shrub *Hybanthus prunifolius*, Violaceae, is given by Roubik and Buchmann (1984).

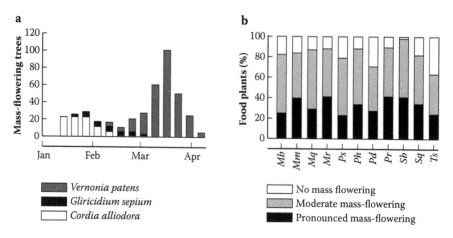

FIGURE 11.2 Availability of mass-flowering plants and their significance for stingless bees. (a) Synchronous blooming of mass-flowering trees. Given is the number of individuals of three tree species in bloom between January and April 2004, within a circumference of 500 m around stingless bee hives (*Melipona beecheii*) in a Neotropical forest in El Salvador (data from Hofstede 2006). (b) Food plants used by stingless bees in the Brazilian Atlantic rainforest. Given is the proportion of plants with pronounced (black bars) or moderate (grey bars) mass flowering, or without mass flowering (white bars), on which bees of the respective species foraged. *Mb, Melipona bicolor; Mm, M. marginata; Mq, M. quadrifasciata; Mr, M. rufiventris; Ps, Paratrigona subnuda; Ph, Partamona helleri; Pd, Plebeia droryana; Pr, Plebeia remota; Sb, Scaptotrigona bipunctata; Sq, Schwarziana quadripunctata; Ts, Trigona spinipes.* (Data from Wilms et al. 1996.)

large amounts of nectar or pollen for bee colonies, the entire potentially available foraging force should be quickly activated to visit the respective food source.

In fact, mass-flowering plants are the predominant source of nectar and pollen for stingless bees (Absy et al. 1980, 1984; Wilms et al. 1996; Wilms and Wiechers 1997; Ramalho 2004) (Figure 11.2b). In a detailed survey of the floral resources used by native stingless bees in the Brazilian Atlantic rainforest, Wolfgang Wilms and co-workers (Wilms et al. 1996; Wilms and Wiechers 1997) and Mauro Ramalho (2004) found that mass-flowering plants contributed by up to 90% to the annual nutritional input into the colonies. The peaks in collecting activity during mass-flowering events found in meliponine colonies (Roubik et al. 1986; Hofstede 2006) are a result of both an increased activity of the individuals and a boost in the total foraging force (Hofstede 2006).

COMPETITION OVER FOOD SOURCES

Stingless bees are opportunistic generalists (Biesmeijer and Slaa 2004; Nagamitsu and Inoue 2005). Therefore, since the foraging ranges of different meliponine colonies in a community highly overlap (Hubbell and Johnson 1977, 1978; Ramalho 2004), a particular resource will not be in private hands of a single colony. Instead, it will be shared by both conspecific non-nestmates and heterospecifics (Biesmeijer and Slaa 2004). This generalized utilization of a common resource will inevitably result in exploitive competition over valuable food sources, such as mass-flowering plants (Nagamitsu and Innoue 2005).

Among the stingless bees we encounter different strategies to deal with the competition over food sources. *Competition winners* are those species that are able to quickly guide large groups of individuals to a specific location and, in addition, aggressively chase away conspecific non-nestmates or heterospecifics from the food sources. This kind of foraging strategy, referred to as "extirpation" by Leslie K. Johnson (1983), has been described for some species of the genera *Trigona* and *Oxytrigona* (Kerr and Cruz 1961; Kerr 1969; Johnson and Hubbell 1974, 1975; Hubbell and Johnson

1978; Roubik 1980; Johnson 1983; Nagamitsu and Inoue 1997; Slaa 2003; Biesmeijer and Slaa 2004; Nieh et al. 2004). Through aggressive group foraging, foragers from a single colony manage to monopolize clumped resources. The trade-off for this elevated competitive ability, however, is a reduced capacity to independently discover new food sources or even neighboring food patches (Hubbell and Johnson 1978; Biesmeijer and Slaa 2004).

Competition avoiders, by contrast, should be able to quickly switch to a new food patch as soon as they are dislodged by the aggressive species. Here, recruitment mechanisms with little or no location specificity, relying, however, on the simultaneous discovery of various patches of the same food source, would be an advantage. A colony whose foraging force leaves the nest only with information on the odor of a food source is capable of discovering most or all of the existent food patches within its foraging area (Hubbell and Johnson 1978). Certainly, field-based information, such as olfactory or visual cues, biases the distribution of the foragers toward those patches that have not been discovered—or are not yet dominated—by aggressive species (Slaa et al. 2003b; Nieh et al. 2004). In cases where recruitment occurs, the activation of new foragers is sustained as long as the food source remains profitable, or as long as potential recruits are available within the nest. The important advantages of competition avoiders over competition winners are their ability to discover new food patches independently and their capacity to switch the colony's foraging focus to another food patch. Due to this increased agility in their foraging, in contrast to the aggressive group foraging species, the competition avoiders are able to capitalize on highly valuable food sources, but only as long as they keep one step ahead of the extirpators.

A third strategy to benefit from rich food sources is *competition indifference*. Here, solitary foragers, or foraging groups, collect food at the same patches as other species, yet without physical interference with other bees (Johnson 1983; Biesmeijer and Slaa 2004).

These three strategies of how to deal with resource competition are not species specific (Biesmeijer and Slaa 2004), but rather depend on the composition of the foraging community present at the food patch. Some species, for instance, which act in a competition-indifferent manner when other not-aggressive species are present, can switch to competition avoidance and abandon the patch as soon as an aggressive species shows up.

INFERENCES ON RECRUITMENT MECHANISMS

Which conclusions can now be drawn regarding the nest-internal recruitment mechanisms employed by stingless bees? In nature, recruitment of foragers occurs predominantly during "big bang" events (Hofstede 2006). Since all individuals of a mass-flowering species bloom in synchrony, we often encounter a high spatial density and abundance of a single type of food source (Augspurger 1980; Bawa 1983; Mori and Piploy 1984; Wilms and Wiechers 1997; Hofstede 2006) (Figure 11.2a). Hence, a rapid activation of foragers, informed solely about the scent of the resource prior to leaving the nest, would be adequate for a colony to quickly discover many or all patches within its foraging area. The conspicuous odor emitted by the multitude of flowers, which can be detected even by human noses over long distances (Mori and Piploy 1984; da Silva and Pinheiro 2007), is certainly sufficient to guide the foragers toward a patch. For colonies of not-aggressive Meliponini, chiefly for competition avoiders, a quick detection of many food patches and a quick activation of all available foragers are necessary to get a head start before being dislodged from some of the patches. Aggressive species, on the other hand, need to rapidly mobilize huge numbers of foragers in order to chase other species away from a food patch, and to subsequently defend this patch against other aggressive colonies. In these species, as far as is known, the guidance of the group toward a specific location is accomplished by pheromone marks at and near the food patch itself (see Chapter 12). Although competition winners and competition avoiders employ fundamentally different foraging strategies, a quick activation of unemployed foragers is required in both cases. Consequently, nest-internal recruitment mechanisms in stingless bees should predominantly aim at mobilizing the foraging force.

OF JOSTLING AND BUZZING: THE EARLY FINDINGS BY LINDAUER AND KERR

In the first experimental analyses of the recruitment abilities of stingless bees, Martin Lindauer and Warwick Estevam Kerr investigated the nest-internal behaviors of foragers in eleven species of Meliponini that showed evident differences in their success to activate nestmates to food sources (Lindauer 1956; Lindauer and Kerr 1958, 1960). In some species, for instance, five collecting bees recruited only one or a few additional nestmates to a feeding dish during 1 h (e.g., *Frieseomelitta silvestrii*, one recruit, food source 25 m from the nest). Other species, by contrast, were even more successful than honey bees (e.g., *Scaptotrigona postica*, 107 recruits; *Apis mellifera*, 77 recruits; both food sources 150 m from the nests) (Lindauer and Kerr 1958, 1960). Inside the nest, the authors observed two conspicuous behaviors performed by the foragers upon their return from "a good source of food" (von Frisch 1993, p. 307): *jostling in zigzag runs* and *buzzing sounds*. Both were postulated to occur in all stingless bee species, in some more accentuated than in others (Lindauer and Kerr 1958, 1960; Kerr 1969): "To Lindauer's surprise, the behavior of the foragers on the combs did not reveal any significant differences, whether he was dealing with species that had little success in arousing their comrades or with others that had much" (von Frisch 1993, p. 309).

WHY HAD FORAGER SOUNDS NOT BEEN NOTICED BEFORE 1953?

That stingless bee foragers emit sounds on their return from a profitable food source was first noticed by Warwick E. Kerr: "Some idea of the intensity of this buzzing is given by the fact that it was heard for the first time by Kerr in 1953, produced in a colony of *M. quadrifasciata* in an observation hive $1^{1}/_{2}$ m away; even at this distance the sound was strong enough to attract attention" (Lindauer and Kerr 1960, p. 34).

By the time Lindauer and Kerr published their findings on the recruitment abilities of stingless bees, and on the communication mechanisms that are potentially involved in that process (Lindauer and Kerr 1958, 1960), several detailed accounts on the biology of the Meliponini existed (e.g., von Buttel-Reepen 1903; von Ihering 1904; Schwarz 1948). Interestingly, none of these studies ever mentioned sounds or buzzing within the nests during food gathering processes, although nest-internal behaviors as well as foraging had been surveyed in some detail. How could something as conspicuous as sounds, which are audible from a distance of more than a meter away, have escaped the attention of the biologists before 1953?

A possible answer to this puzzle is the use of artificial food sources (feeding dishes) for the systematic study of recruitment behavior in bees, which was a novelty in stingless bee research in the mid-twentieth century. The method was introduced in 1919 by the Austrian Karl von Frisch for the study of honey bees (see von Frisch 1993, p. 18), and finally adopted with success* in the late 1940s by the Brazilians Warwick E. Kerr and Paulo Nogueira-Neto for their research on stingless bees. In fact, when the sounds of *Melipona quadrifasciata* were registered for the first time, the foragers were indeed collecting sugar solution at a feeding dish (Kerr and Esch 1965). Compared to natural food sources, the time and energy spent by the foragers to find the food and fill their crop is extremely reduced when collecting at artificial food sources that offer sugar or honey solutions *ad libitum*. Consequently, feeding dishes constitute "nectar sources" of unnaturally high profitability. Since in stingless bees the temporal pattern of the pulsed forager sounds as well as the foragers' disposition to produce them depend to a high degree on the profitability of a food source (see below), it stands to reason that both the sounds and their occurrence during foraging processes differ between natural food sources (observations before 1950) and artificial food sources (Lindauer and Kerr 1958, 1960, and subsequent studies until today). Yet, it would be wrong to assume that the generation of sounds by stingless bee foragers is a behavioral artifact induced by the use of feeding dishes. Sounds are, *de facto*, also produced by foragers that collect at natural nectar and pollen sources. In these

---

* Lutz (1933) reported an unsuccessful attempt to train *Trigona cressoni parastigma* (=Frieseomelitta paupera) to several kinds of syrup during a stay in Panama.

cases, however, if sounds occur at all, these are generated at a lower rate (Kerr et al. 1963) and with lower amplitude (Hrncir, unpublished data) than sounds produced by foragers collecting at artificial food sources.

THE ALERTING HYPOTHESIS

Lindauer and Kerr (1958, 1960) considered both the excited zigzag runs and the buzzing of stingless bee foragers "methods of alerting" nestmates about the existence of a valuable food source—comparable, actually, to the round dance in *Apis mellifera* (Lindauer and Kerr 1960, p. 68). The authors observed that the behavior of the returning foragers caused a general agitation within the colony, which was followed by an increasing number of newcomers arriving at the feeding dishes. In those species, however, that showed a less vigorous nest-internal forager behavior (*Frieseomelitta silvestrii, Plebeia droryana*) the number of collecting bees hardly increased at all (Lindauer and Kerr 1958, 1960).

The publications by Lindauer and Kerr (1958, 1960), unfortunately, do not provide any details to what extent a correlation between the nest-internal behavior of foragers and spatial parameters of the food source was systematically analyzed. But given the extraordinary observation experience of these two scientists, it was certainly not without cause when they declared: "Neither the 'buzzes' nor the zigzag runs gave any support of the idea that the alerting system included an indication of distance or direction" (Lindauer and Kerr 1960, p. 36; see also Lindauer 1956).

THE RECRUITER'S EXCITEMENT

From the pioneering studies on the recruitment behavior in stingless bees by Lindauer and Kerr (Lindauer 1956; Lindauer and Kerr 1958, 1960), we learn the following about the nest-internal recruitment signals in Meliponini: (1) Both their occurrence and their vigor depend on the profitability of a food source. (2) The signaling behavior of the foragers goes along with both an elevated excitement within the colony and an increased number of nestmates arriving at the food source. (3) Neither zigzag runs nor buzzing seem to depend on spatial parameters (distance, direction) of the food source.

Techniques and tools to observe, record and analyze the behavior of bees have considerably advanced during the last 50 years. How much progress, however, has our knowledge about the message(s) and the meaning(s) of the nest-internal recruitment signals in stingless bees really made since these early findings by Lindauer and Kerr?

THE FOOD SOURCE'S VALUE AND THE RECRUITER'S EXCITEMENT

Foraging eusocial bees evaluate the current value of a food source for their decision whether to continue or stop collecting (Biesmeijer et al. 1998). In the case of nectar foraging, important parameters that indicate the profitability of a food patch are the sugar concentration of the collected nectar and the availability of exploitable flowers. Both are subject to sometimes striking variations in the course of a collecting period. The availability of exploitable flowers within a specific food patch, in addition to the plant-specific alterations in the number of open flowers throughout the day, is strongly affected by the density of collecting con- and heterospecific foragers. Here, the quick usurpation of a patch by aggressive competition winners can result in a rapid and unpredicted decline of available flowers for competition avoiders.

The food sources' profitability not only affects a bee's foraging decision, but also influences both the thoracic temperature of foragers (honey bees, Stabentheiner and Hagmüller 1991; Stabentheiner et al. 1995; Stabentheiner 2001; Farina and Wainselboim 2001; stingless bees, Nieh and Sánchez 2005) and their "excitement" when returning to the colony (honey bees, Waddington 1982, 2001; Waddington and Kirchner 1992; Seeley et al. 2000; stingless bees, Hrncir et al. 2004a, 2004b).

These measurable forager qualities can be scientifically exploited to investigate which parameters of a collecting trip are used by the bees to evaluate the profitability of a food source (Chittka 2004). In stingless bees, so far, the following parameters have experimentally been determined: the sugar concentration of the collected food (Aguilar and Briceño 2002, Hrncir et al. 2002, 2004b; Nieh et al. 2003; Nieh and Sánchez 2005; Schmidt et al. 2008), the energetic costs during foraging (Hrncir et al. 2002, 2004a), the distance of a food patch from the nest (Nieh and Sánchez 2005), and the past foraging experience of the foragers (Schmidt et al. 2006b).

THE AGITATION THRESHOLD

Agitated movements and thoracic vibrations of foragers can only be observed when the profitability of a food source is above a certain threshold. The profitability value necessary to elicit the bees' excitement is certainly not identical for all foragers, and it also changes—at the colony level—in the course of a year. Thus, researchers are sometimes confronted with the fact that meliponine foragers do not show any form of nest-internal agitation, even if collecting highly concentrated sugar solutions. In an analogous manner, Lindauer (1948) described for the dances of the honey bee: "Es gab Tage, an denen ich mit einer $^1/_8$ mol Zuckerlösung Tänze auslösen konnte, während zu anderer Zeit, wo eine 2 mol Lösung am Futterplatz aufgestellt war, kaum noch Tänze zu verzeichnen waren" [On some days, I could trigger dances with a $^1/_8$ mol sugar solution, whereas on others, even when a 2 mol solution was offered at the feeding place, almost no dances could be registered] (Lindauer 1948, p. 352). Lindauer (1948) found that the dance threshold in honey bees greatly depends on the season (high threshold in spring, low threshold in late summer and autumn), but also, to a minor extent, on the weather situation and the colony's food reserves (see also Michener 1974). In stingless bees, the season might play a similarly important role for the foragers' excitement, since on several occasions—and in different species—no obvious nest-internal recruitment activity by bees that collected at highly rewarding artificial food sources occurred during the spring and early summer months in Brazil (September–November/December) (for *Melipona quadrifasciata, M. scutellaris, M. seminigra*, MH, unpublished observations; for *Nannotrigona testaceicornis*, Allerstorfer 2004; for *Scaptotrigona* aff. *depilis*, Veronika M. Schmidt, personal communication). So far, however, potential factors that influence and determine these periods of stagnation in the foragers' excitement have not been systematically studied in the Meliponini.

EXCITEMENT AND MOTION

Probably the most eye-catching behavior by returning foragers is their excited running through the nest, or at least through parts of it, which can be observed in bees collecting sugar solution from feeding dishes as well as in bees gathering nectar or pollen at natural sources (Sommeijer et al. 1983). So far, however, all detailed investigations of these excited movements have been performed with foragers trained to artificial food sources, and their movement patterns have been studied on "unloading platforms" (Nieh 1998; Nieh et al. 2003) or in "observation boxes" (Hrncir et al. 2000; Samwald 2000; Schmidt et al. 2006b, 2008) placed between the entrance/exit and the actual hive (Figure 11.3). Although such platforms and boxes are no ideal copy of the natural nest entrance, which is usually a batumen-lined tube, funnel, or similarly spatially restricted structure (Wille and Michener 1973; Nieh 1998; Morawetz 2007), they prove to be essential for the observation of the nest-internal forager behavior. For the interpretation of the results it should be kept in mind, however, that the foragers' movements in natural nests are strongly influenced and determined by the complex, narrow, and curved environment, and that they certainly differ from the motion patterns exhibited by the bees on the flat and spacious entrance substitutes.

In recent years, the "zigzag runs," as the foragers' excited movements had been denominated by Lindauer and Kerr (1960), were meticulously dissected into the following subunits: running path, turning behavior, and spinning behavior. Typically, a bee that returns from a collecting trip enters

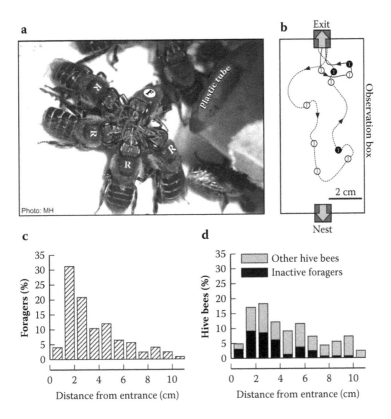

FIGURE 11.3 Foragers entering the hive after a foraging trip. (a) Foragers (F) distribute the collected food to receivers (R) predominantly close to the nest entrance (shown is the entrance tube of an observation box). (b) Running path of a forager in an observation box (positioned between the entrance/exit and the nest) on two subsequent returns to the colony (*Melipona seminigra*, recorded in March 2005). Arrowheads indicate the running direction. Open circles, J, locations of jostling contacts; filled circles, T, locations of trophallaxis. (c) Unloading position (distance from the nest entrance) of ten active *M. panamica* foragers during multiple visits back to the nest. (Data from Nieh 1998.) (d) Position of inactive *M. panamica* foragers (black bars) and other hive bees (grey bars) in relation to the nest entrance. (Data from Nieh 1998.)

the nest and runs around until she finds a hive bee to unload the food (Figure 11.3). After unloading her crop, the forager performs excited spinning movements on the spot (observed in *M. favosa*, Pereboom and Sommeijer 1993; *M. panamica*, Nieh 1998; *M. quadrifasciata, M. scutellaris*, Hrncir et al. 2000; *M. beecheii*, Hart and Ratnieks 2002). While running to the nest exit to leave for the next foraging bout, "the zigzag run was interrupted by a sharp turn in a semicircle; then the bee gave up another portion of syrup and immediately continued the zigzag movement" (Lindauer and Kerr 1960, p. 33).

Running Path

The movement pattern of foragers varies considerably. A bee collecting at a feeder might unload her food directly at the entrance without even properly entering the nest, but run excitedly through the colony the next time she returns from the feeding site (Hrncir et al. 2000; Samwald 2000) (Figure 11.3). Already from this simple observation, it can be concluded that the actual length of a forager's running path is not influenced by features of the food source, like its position or its quality. Most probably it is the presence and availability of food-unloading bees around the entrance that determines how far a forager enters into the nest (Nieh 1998) (Figure 11.3).

Turning and Spinning

Although the sudden turnings during the zigzag runs of stingless bee foragers vaguely remind one of the honey bee's waggle movements, neither the number nor the angles of the turnings correlate with either the distance or the direction of the food source (*M. quadrifasciata*, Hrncir et al. 2000). Similarly, the forager's spinning movements after food unloading are independent of the position of the feeding site (*M. panamica*, Nieh 1998; *M. quadrifasciata*, *M. scutellaris*, Hrncir et al. 2000). In *Melipona panamica*, James C. Nieh (1998) observed an interesting feature of the spinning: although the foragers started their spinning at random orientation with respect to the nest exit, they stopped their movement, on average, facing it (Nieh 1998). A similar, yet more narrative description had already been given by Martin Lindauer in 1956 for *T. iridipennis*:

> Gleichzeitig macht die Trachtbiene deutliche Schüttelbewegungen mit dem ganzen Körper, was die umstehenden Bienen veranlaßt, sich ihr zuzuwenden, sie mit den Antennen zu betasten und ihr ein Stück zu folgen. Besonders auffallend ist, daß die Sammlerin, wenn sie auf solche Weise ein Gefolge von 3-4 Bienen hinter sich hat, sich plötzlich umdreht und mit betonten Schüttelbewegungen zum Flugloch weist. [At the same time, the forager performs conspicuous shaking movements with her entire body, which causes the surrounding bees to turn towards her, to touch her with the antennae, and to follow her part of the way. It is particularly eye-catching that, as soon as the collecting bee has 3–4 followers behind her, she suddenly turns around and points towards the exit with pronounced shaking movements]. (Lindauer 1956, pp. 553–54)

Although it is tempting to interpret this feature of the foragers' spinning as a signal for the nest-mates to leave the nest, it remains to be investigated whether and how the hive bees perceive that the forager is facing toward the exit.

EXCITEMENT AND JOSTLING RUNS

An important parameter, which often appears to be ignored when nest-internal behaviors of bees are analyzed and interpreted, is the complete darkness inside a bee hive (with exception of the open-nesting *Apis* species), which impedes the use of visual information. Therefore, all intranidal recruitment signals by foragers—like the waggle movement of the honey bees or the agitated running of the stingless bees—require unequivocal mechanical or chemical components to be perceptible for hive bees.

In stingless bees, one possible way for prospective recruits to perceive the presence of excited foragers within the nest is the amount of contacts with this forager during a certain time interval. As had already been pointed out by Lindauer and Kerr (1958, 1960), agitated foragers jostle their nestmates while running through the colony. Recent studies indicate that this jostling activity of the foragers depends on the value of the visited food source. In *Nannotrigona testaceicornis*, for instance, the number of jostling contacts by the foragers correlated directly with the sugar concentration of the collected food (Schmidt et al. 2008) (Figure 11.4a). Interestingly, in *M. quadrifasciata* (Hrncir 1998; Hrncir et al. 2000), the jostling activity of foragers was found to be highest immediately after they started collecting at a food source (Figure 11.4b). In light of the fact that *Melipona* bees are competition avoiders (Biesmeijer and Slaa 2004), and that the risk of a sudden detection and usurpation of the food patch by competition winners increases with time, the value of a food patch for these bees is certainly highest right after its discovery. Therefore, it may be assumed that in *Melipona*, too, the individually experienced value of a food source influences the jostling activity of the foragers. In line with the findings for *Nannotrigona* and *Melipona*—although appearing contradictory at first sight—are the findings for *Scaptotrigona* aff. *depilis* (Schmidt et al. 2006b). In this species, the foragers' jostling activity remained constant when they collected at a food source of steadily increasing quality. It significantly decreased, however, when the bees experienced a successive decline in food profitability in the course of an experiment (Schmidt et al. 2006b). Apparently, the foragers of *Scaptotrigona* aff. *depilis* include their past foraging experience when evaluating a food source. As long as the exploitation of the food source seems

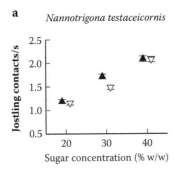
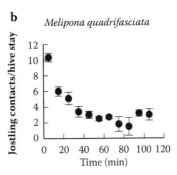

FIGURE 11.4 Jostling contacts of returning foragers. (a) In *Nannotrigona testaceicornis*, the number of jostling contacts directly correlated with the sugar concentration of the collected food. Given are mean values ±1 SE of the foragers' jostling contacts when the sugar concentration at the food source steadily increased (filled triangles; N = 8 bees) or decreased (open triangles; N = 8 bees). (Data from Schmidt et al. 2008; raw data kindly provided by Veronika M. Schmidt.) (b) In *Melipona quadrifasciata*, the amount of jostling contacts by the foragers decreased with time (mean values ±1 SE; N = 10 bees). (Taken from Hrncir 1998.)

"worthwhile" (constant or increasing sugar concentration at the food patch), the foragers' nest-internal recruitment behavior remains more or less the same. The foragers "lose interest", however, when experiencing a gradually decreasing sugar concentration at a food patch. In consequence of such a successive depletion, or a decline in profitability, the foragers show a reduced excitement inside the nest.

For an unemployed hive bee, a high amount of contacts with foragers potentially indicates the existence and availability of a valuable food patch. In addition, the amount of jostling contacts could as well provide important information to active foragers. One prerequisite for an efficient exploitation of nectar sources in eusocial bees is a balance between food gatherers and food processors (Seeley 1992; Anderson and Ratnieks 1999). Therefore, if during periods of high resource abundance the nectar influx exceeds the food processing capacity of the colony, nectar receiving should be reinforced or, as an alternative, the colony's foraging activity should be reduced. Yet, the premise for any form of control mechanism to be actuated is the detection of the imbalance between the food intake rate and the nectar processing capacity. In honey bees, nectar foragers perceive a shortage of food receivers through long search times inside the nest when trying to unload their crop (Seeley 1992; Seeley and Tovey 1994; Kirchner and Lindauer 1994). In stingless bees, additionally, nectar foragers could perceive the presence and abundance of hive bees close to the nest entrance by the amount of contacts experienced in the course of their jostling runs. Thus, food collectors could assess both the colony's current nectar processing capacity and the availability of inactive recruitable foragers—which predominantly cluster around the nest entrance (Nieh 1998) (Figure 11.3)—during their stay in the nest.

EXCITEMENT AND THORACIC VIBRATIONS

Without doubt, the best-studied nest-internal recruitment behavior in stingless bees is the generation of thoracic vibrations ("sounds") by foragers that collect at high-profit nectar sources—or rather at nectar source substitutes. "Immediately after feeding had started, and the returning foragers began to enter the hive, a persistent high-pitched buzz could be heard, consisting of long and short components, for instance: ·· — — · — · · · — · — — —" (Lindauer and Kerr 1960, p. 34). The pulsed structure of these vibrations, which reminds one of a Morse code, gave birth to the idea that information about the food source could be encoded within the temporal pattern of the sounds. From what we know today, however, the assumption that stingless bees use a symbolic "language" similar to that of the honey bee (Esch et al. 1965; Esch 1967), or even a more sophisticated one (Nieh and Roubik 1998), has to be reconsidered.

The key to decoding the message of a putative signal is the unequivocal identification of the factors that influence and shape this communication behavior (Seeley 1992). From this point of view, the first attempts to decipher the message of the thoracic vibrations in stingless bees (Esch et al. 1965; Esch 1967; Nieh and Roubik 1998) had drawn overhasty conclusions about the existence of a referential location communication. These studies solely focused on a correlation analysis between the sounds and the spatial parameters of a food source relative to the nest, but did not consider at all numerous additional criteria that could potentially influence the temporal pattern of the sounds. In fact, the foragers' thoracic vibrations have now been shown to correlate with a variety of factors. Thus, their actual message is not easily detected. Yet, with every new finding it becomes more and more evident that the thoracic vibrations reflect the excitement of a collecting bee rather than actually encode spatial information about the visited food patch.

Threshold

In accordance with their agitated movements, foragers do not generate thoracic vibrations as long as the value of a food source is below a certain threshold. Since in all available studies the bees had been trained to feeding dishes that offered sugar solution *ad libitum*, the threshold was governed by the energetic gains in terms of the sugar concentration of the collected solution (Esch 1967; Hrncir et al. 2000; Schmidt et al. 2006b, 2008). Yet, another important determinant for the excitement of a forager is the individually experienced profitability of the visited food patch. Consequently, alterations in sugar concentration–independent parameters of profitability, like search and collecting times, lead to changes in the bee's disposition to generate thoracic vibrations inside the nest.*

Energetic Gains

The energetic gains during a foraging trip, which, in bees, are strongly influenced by the sugar concentration of the collected food (Roubik and Buchmann 1984; Roubik et al. 1995; Hrncir et al. 2004b; Schmidt et al. 2006b, 2008), are probably one of the principal factors used by foragers to determine the profitability of a food patch. In honey bees, the liveliness of the waggle and round dances, as well as the temporal pattern of the concomitant sounds generated by the foragers, strongly depends on the imbibed sugar concentration (Esch 1963; Waddington 1982, 2001; Waddington and Kirchner 1992; Seeley et al. 2000; De Marco and Farina 2001). Although the importance of food quality for the honey bee's communication has been known for more than 80 years (von Frisch 1923), the possible relation between nest-internal recruitment signals and the food profitability in stingless bees has only recently been investigated (*Melipona costaricensis*, Aguilar and Briceño 2002; *M. bicolor, M. mandacaia*, Nieh et al. 2003; *M. seminigra*, Hrncir et al. 2002, 2004a, 2004b; *M. rufiventris*, Hrncir et al. 2006a; *Nannotrigona testaceicornis*, Allerstorfer 2004; Schmidt et al. 2008; *Scaptotrigona* aff. *depilis*, Schmidt et al. 2006b).

In all studied *Melipona* species, as well as in *N. testaceicornis*, the temporal pattern of the foragers' pulsed thoracic vibrations is significantly related to the energetic gains at the food patch. As soon as the value of the food exceeds the agitation threshold, the duration of the pulses increases and the interval between the pulses decreases with increasing sugar concentration (*M. costaricensis*, Aguilar and Briceño 2002; *M. bicolor, M. mandacaia*, Nieh et al. 2003; *M. rufiventris*, Hrncir et al. 2006a; *M. seminigra*, Hrncir et al. 2002, 2004a, 2004b; *N. testaceicornis*, Allerstorfer 2004; Schmidt et

---

* During an experiment in which three *M. seminigra* foragers collected a 40% w/w (≈ 1.4 mol/l) sugar solution at a feeding dish, the bees suddenly stopped to produce sounds when returning to the colony. The fact that the foragers continued to deliver food to their nestmates indicated that the artificial food patch still provided food. Yet, several dozen *Trigona recursa* foragers had usurped the feeder. Due to this mass occupation by another species, the foragers of *M. seminigra* only occasionally managed to land on the feeding dish and collect sugar solution (MH, unpublished observation). In this case, apparently, the cessation of the foragers' recruitment activity was not related to a decrease in sugar concentration below the agitation threshold, but to a sudden decay in the overall profitability of the food patch due to an increase in search and collecting times.

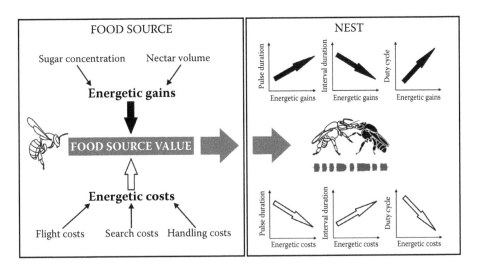

FIGURE 11.5 The value of a food source influences the thoracic vibrations of stingless bee foragers. Several energetic parameters determine the value of a food source for collecting bees and, consequently, influence the temporal pattern of the foragers' thoracic vibrations: Increasing energetic gains at the food patch result in longer pulses, shorter intervals, and consequently, an increasing duty cycle (duty cycle = pulse duration/[pulse duration + interval duration]). Increasing energetic costs, by contrast, result in shorter pulses, longer intervals, and a decreasing duty cycle.

2008) (Figure 11.5). In *Scaptotrigona* aff. *depilis*, by contrast, the collected sugar concentration had only a small effect on the bees' thoracic vibrations (Schmidt et al. 2006b). Similarly to what has been found for the jostling activity in this bee species (see above), the past foraging experience strongly influenced the temporal pattern of the foragers' thoracic vibrations. The sound pattern remained more or less the same when the foragers collected a constant or an increasing sugar concentration at the food patch. Yet, their excitement decreased (decreasing pulse duration, increasing interval duration) when experiencing a gradual reduction of the food's profitability (Schmidt et al. 2006b).

Flight Costs

Energetic costs during foraging might be another important factor for the foragers' evaluation of the food source's profitability. Consequently, and in accordance with the findings for energetic gains, increased energetic expenses during a collecting trip should reduce the agitation of a forager. And indeed, in *Melipona seminigra* the effect that elevated flight costs had on the temporal pattern of the foragers' thoracic vibrations was exactly the opposite of that of increased energetic gains* (Hrncir et al. 2004a, 2004b) (Figure 11.5).

---

* Foragers had been trained to collect sugar solution first at the entrance, and then at the far end of flight tunnels that varied in their width and in the visual pattern of their lateral walls (black-and-white stripes either perpendicularly or parallel to the flight direction of the bees). In fact, these experiments had been designed to study the possible influence of the visual flow (lateral image motion)—which foragers of *M. seminigra* use to estimate their flight distances (Hrncir et al. 2003)—on potential recruitment signals of these bees. It turned out that the thoracic vibrations were not at all affected by differences in the visual flow (induced through different tunnel widths) experienced by the bees. Even so, the temporal pattern of the foragers' vibrations differed between the feeding situations at the tunnel entrance and at the far end of the tunnel, respectively. As soon as the bees had to pass through a tunnel to get to the food source, the duration of the pulses decreased significantly, and the interval between the pulses increased (Hrncir et al. 2004a). A possible explanation for the observed reduction in the foragers' excitement is an increase in energetic costs as soon as the bees collected at the far end of the tunnels. The foragers' flight speed inside all tunnels was about 0.3 m/s, and therefore considerably reduced compared to their flight velocity in a natural environment (ca. 4.25 m/s) (Hrncir et al. 2004a). Since the energetic expenditures of flying bees are substantially higher at low flight velocities (<1 m/s) than at velocities of about 4 m/s (measured in honey bees, Nachtigall et al. 1995), the foragers of *M. seminigra* quite possibly spent more energy when collecting at the far end of the tunnel than when colleting at the tunnel entrance.

Natural Nectar Sources

When stingless bee foragers collect at natural food sources, their foraging trips are considerably longer than when they collect at feeding dishes (Allerstorfer 2004). Since sugar solution foragers do not have to search for a rewarding flower, the round-trip times mainly depend on the distance of the feeding dish from the nest. Foraging times of bees that collect at natural nectar sources, by contrast, are composed of the time to reach a food patch plus the time to find and exploit rewarding flowers.* In addition to increased collecting times, bees that collect at natural resources often return with only a partly filled crop (Roubik and Buchmann 1984). Therefore, due to elevated energetic expenses and reduced energetic gains, the profitability of natural nectar sources is certainly lower than that of sugar solution feeders—even if both sources provided food of the same sugar concentration. Consequently, we should expect less excitement when foragers collect at natural nectar sources than when they collect at feeding dishes.

To this day, the thoracic vibrations of stingless bee foragers that collect at natural food sources have not been studied (or at least results have not been published). In *M. seminigra*, however, two studies tentatively simulated a natural feeding situation in order to investigate the possible influence of a reduced nectar intake together with increased search times on the foragers' vibrations (Hrncir et al. 2002; Hrncir, unpublished data). The bees collected sugar solution at feeder arrays consisting of either six (Hrncir et al. 2002) or nine (MH, unpublished data; Figure 11.6) small cups, of which only three contained a small reward. This collecting situation, which caused elevated search times and reduced energetic gains for the bees, indeed resulted, as predicted, in a reduced excitement of the foragers inside the nest (Figure 11.6). Although the foragers' collecting times were almost two times longer at the cup feeder arrays than at common feeding dishes (collecting time: 62 s at cup feeder, 30 s at feeding dish; MH unpublished data), they still were considerably shorter than foraging times at natural food sources.* We can, consequently, venture the prognosis that bees foraging at natural nectar sources show even less nest-internal excitement than those collecting at the artificial "mass flowers."

MOBILIZING THE FORAGING FORCE

Due to the high competition over food sources, a quick activation of the colonies' foraging force to a profitable resource would be advantageous for several meliponine species. For unemployed but food source–experienced foragers, scent/taste information provided by returning foragers that matches their own experience is sufficient to make them return to the food patch they previously visited (Biesmeijer et al. 1998; see also Chapters 9 and 10). On the other hand, those bees that have no previous experience with the resource in question start their collecting activity only when their foraging motivation is above a certain threshold (Biesmeijer et al. 1998). Therefore, any recruitment signal that aims at activating new foragers should primarily modulate the foraging readiness of inexperienced nestmates by increasing their foraging motivation.

The Excitement of Potential Newcomers

Research on the nest-internal recruitment communication in stingless bees, so far, has focused on the behavior of the recruiters. In order to understand a communication system in-depth, however, both the sender and the receiver of a signal have to be considered. In most cases, the sender—in the present case the employed forager—is easily identified. By contrast, it often proves difficult to pinpoint the receivers, due to sometimes very subtle changes of their behavior in response to the signal.

---

* Roubik and Buchmann (1984) calculated foraging times between 7 and 12 min for three *Melipona* species collecting nectar at a mass-flowering shrub. In compliance with these calculations, de Bruijn and Sommeijer (1997) observed that nectar foragers of *M. costaricensis* spent up to 10 min on a foraging trip, and Allerstorfer (2004) described natural foraging times of even up to 50 min in *N. testaceicornis*.

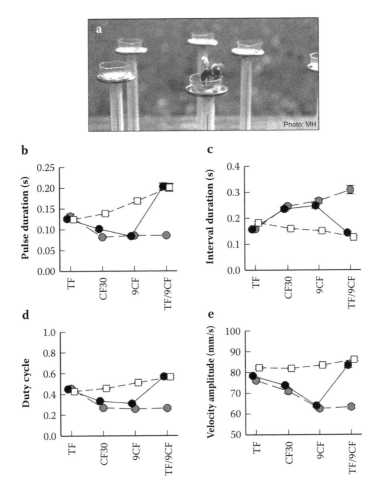

FIGURE 11.6 Thoracic vibrations of foragers collecting at an artificial "mass flower". Foragers of *Melipona seminigra* were trained to collect sugar solution (50% w/w) from a feeding dish with unlimited solution flow (TF). In the subsequent experimental step, the feeding dish was replaced by a small plastic cup containing 30 μl of sugar solution (CF30), which is less than the average full crop load of the foragers (about 40 μl; Hrncir et al. 2004b). Thereafter, 30 μl of sugar solution was distributed among three randomly chosen cups in a nine-cup feeder array (9CF), which resulted in elongated search times for the foragers. In the last experimental step, half of the foragers (grey-filled circles; N = 6 bees) continued foraging at the cup feeder array, whereas the other half (black-filled circles; N = 6 bees) returned to collecting at the training feeder. Each experimental step lasted 30 min. A third group (open squares; N = 6 bees) collected at TF for 120 min to control for eventual changes in the thoracic vibrations along with the duration of an experiment. The thoracic vibrations of all foragers were recorded during trophallaxis inside an observation box placed between the entrance/exit tube and the nest. (a) *M. seminigra* forager collecting at a cup feeder array. (b–e) There were significant changes in the temporal pattern of the thoracic vibrations in the course of an experiment (given are mean values ± 1 SE): the duration of the pulses (b), the duty cycle (d), and the velocity amplitude (e) decreased, and the duration of the intervals (c) increased when the collected reward decreased below the full crop load of the foragers (TF → CF30). The thoracic vibrations changed even more as soon as the foragers' search times increased (CF30 → 9CF). The comparison with the control group demonstrates that the observed decrease in the foragers' excitement was a consequence of reduced energetic gains and increased search costs at the food patch (Hrncir, unpublished data).

In the case of recruitment communication, the only receivers that can unequivocally be determined are those recruits that reach an advertised food patch during an ongoing foraging process.

Studies on the nest-internal case history of individually marked recruits demonstrated that, before hive bees initiated their foraging activity, their interactions with active foragers steadily increased (*Melipona quadrifasciata*, Hrncir 1998; Hrncir et al. 2000; *M. seminigra*, Kronberger 2000). Prospective newcomers to a food source had trophallactic interactions with the foragers, but also attended trophallactic food transfers without receiving food samples, staying close to the forager's thorax-abdominal region (Hrncir 1998; Hrncir et al. 2000; Samwald 2000). In both these situations the forager's thoracic vibrations are putatively transmitted to the recruits (Hrncir et al. 2006b, 2008). During such interactions, therefore, hive bees are potentially informed about the agitation level of a forager—expressed in the temporal pattern of its thoracic vibrations—and consequently, about the profitability of the food source. This information is important for the hive bees' decision whether or not to forage (Biesmeijer et al. 1998). Some bees, apparently, require only little input (interactions with active foragers) to initiate their collecting activity (low foraging threshold), whereas others need repeated information from the active foragers (high foraging threshold). In *M. quadrifasciata* and *M. seminigra*, for example, the studied prospective newcomers received between one and fifteen food samples, and attended between two and thirteen trophallaxes, respectively, before leaving the nest for the food source (Figure 11.7).

Considering that communication is characterized as "the process in which actors use specially designed signals or displays to modify the behaviour of reactors" (Krebs and Davies 1993, p. 349), we have to ask whether and to which degree prospective newcomers change their behavior after their interactions with active foragers. One behavioral parameter that can be used for these investigations is the agitation level of inactive foragers. Lindauer and Kerr (1960) observed: "Until the first bees reached the feeding table, the bees in the hive were quiet, inactive and scattered about. But they started running about excitedly when the first bees returned from the feeder" (Lindauer and Kerr 1960, p. 33). This agitation inside the hive immediately after the onset of a collecting process is certainly, in parts, due to the raised activity of nectar unloaders and food processors. Yet, an elevated agitation can also be observed in inactive foragers. In *Melipona seminigra*, for instance, prospective newcomers significantly increased their jostling activity after their first contacts with active foragers (Kronberger 2000; Figure 11.7c). This sudden excitement of inactive foragers can be taken as an indicator for their elevated foraging motivation in response to the interactions with the food collectors, and it supports the idea that the nest-internal behaviors of active foragers are indeed communication signals.

Recruiter's Excitement and Recruitment Success

A fact that corroborates the alerting hypothesis of Lindauer and Kerr (1958, 1960) is the correlation between the profitability of a food source and the recruitment success (amount of newcomers per unit time) of the bees collecting there.

As long as the value of the food is below the agitation threshold of a forager (neither excited movements nor sounds inside the nest), no newcomers arrive at the food source (Jarau et al. 2000).* If, however, the food source's profitability exceeds the agitation threshold, the foragers' recruitment success correlates with the sugar concentration of the collected food. This fact has been demonstrated for several *Melipona* species (*M. bicolor*, *M. mandacaia*, Nieh et al. 2003; *M. panamica*, Nieh and Sánchez 2005) and for *N. testaceicornis* (Schmidt et al. 2008) (Figure 11.8). Since in these species, however, the recruiters' excitement directly correlates with the energetic gains at the food source (see above), it is difficult to ascertain whether the recruitment success depends on the sugar

---

* Researchers have made use of this circumstance when trying to avoid the initiation of recruitment processes while training foragers to a specific feeding site (e.g., Jarau et al. 2000, 2003; Schmidt et al. 2006a, 2008).

Interactions with foragers

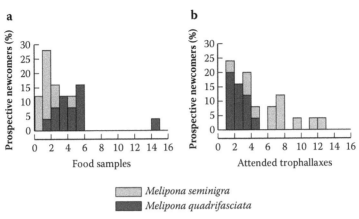

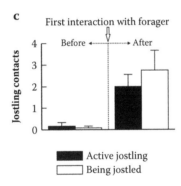

FIGURE 11.7 Behavior of prospective newcomers inside the nest prior to reaching a food source. Prospective newcomers of *Melipona quadrifasciata* (dark grey; N = 13 bees) and *M. seminigra* (light grey; N = 12 bees) received between one and fifteen food samples (a) or attended between two and thirteen trophallaxes without receiving food (b) before initiating their collecting activity. (c) The jostling activity of prospective newcomers (means + 1 SE for twelve individuals of *M. seminigra*) increased significantly after their first interactions with active foragers. (Data for *M. quadrifasciata* from Hrncir 1998; data for *M. seminigra* from Kronberger 2000.)

concentration of the collected and distributed food per se, or whether it is related to the foragers' nest-internal agitation. Here, the possible solution is provided by experiments with *Scaptotrigona* aff. *depilis*, in which the recruiter's excitement (jostling activity as well as thoracic vibrations) was shown to depend on the past foraging experience rather than on the current food profitability (see above). When offering a steadily increasing sugar concentration at the food source, the foragers' nest-internal excitement did not change, and neither did the recruitment success (Schmidt et al. 2006b) (Figure 11.8). Hence, the quality of the received food samples did not influence the foraging motivation of the hive bees. Yet, when the profitability of the food source continuously decreased, both the recruiters' excitement and their recruitment success diminished (Schmidt et al. 2006b) (Figure 11.8). From these findings, it can be concluded that—at least in this bee species—the foraging motivation of inexperienced bees does not depend on the quality of the food brought in by the foragers but, indeed, on the excitement of the recruiters.

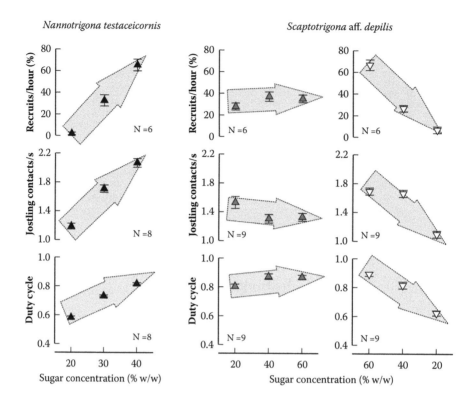

FIGURE 11.8 Recruitment success compared to the excitement of the recruiters. In *Nannotrigona testacei-cornis*, both the amount of recruited bees (100% = total number of recruits per experiment) and the excitement of the foragers—expressed in their jostling activity and the temporal pattern of their thoracic vibrations (duty cycle given as example; duty cycle = pulse duration/[pulse duration + interval duration])—correlated with the sugar concentration of the collected food. In *Scaptotrigona* aff. *depilis*, by contrast, the amount of recruited bees and the recruiters' excitement did not change much when the foragers experienced a steadily increasing food quality (dark-grey-filled triangles). Only when the food quality decreased in the course of time (open triangles) did the recruitment success and the bees' nest-internal excitement diminish. These findings suggest that the recruitment success depends on the foragers' excitement rather than on the actual sugar concentration of the food brought into the colony. Graphs represent mean values ± 1 SE; numbers of experiments (recruitment success) and foragers (jostling contacts and duty cycle) are indicated in the respective plots. (Data from Schmidt et al. 2006b, 2008; raw data kindly provided by Veronika M. Schmidt.)

CONCLUDING REMARKS

The scientific interest in stingless bees has steadily increased during the past decades, since from this ecologically and biologically highly diverse group of eusocial bees much can be learned about the organization of insect societies. Nevertheless, we are still a far way behind what is known about honey bees, which is probably due to the fact that in the Meliponini we have to cope with more than four hundred species. Yet, it becomes more and more evident that stingless bees are far from simply being a group of "more primitively organized" honey bees (von Frisch 1993, p. 306).

Regarding the foraging and recruitment strategies of stingless bees, we encounter a fascinating variety and diversity among the different species (Lindauer and Kerr 1958, 1960; Nieh 2004; Barth et al. 2008). So far, however, we have only managed to grasp a small fragment of this meliponine world of scents and sounds.

To this day, the significance of the behaviors displayed by employed foragers inside the nest is not entirely unscrambled. In addition to the alerting hypothesis introduced by Lindauer and Kerr

(1958, 1960; see also Hrncir et al. 2000, 2004a, 2004b; Jarau et al. 2000), several studies claimed a meliponine equivalent to the honey bee's dance language, postulating that *Melipona* foragers encode the distance of a food source from the nest—and even its height above ground level—within the temporal pattern of their thoracic vibrations. Consequently, the recruits were thought to leave the nest with detailed information about the position of the visited food patch (Esch et al. 1965; Esch 1967; Nieh and Roubik 1998; Aguilar and Briceño 2002; Nieh et al. 2003). However, in spite of putative correlations between the duration of the vibratory pulses and the position of a food source (distance or height), convincing evidence is lacking that the foragers' sounds really constitute a form of referential location communication comparable to the honey bees' dance. Nieh and Roubik (1998) performed experiments in which they removed the recruiters—trained foragers (either ten or twenty individuals) collecting at an experimental feeder—during an ongoing recruitment process. From the observation that, after capturing the foragers, more newcomers arrived at the experimental feeder than at a control feeder (either at a different distance from the nest or a different height above the ground), the authors concluded that these bees had found the food source based on information provided by foragers inside the nest (Nieh and Roubik 1998). It remained unclear, however, for how long the newcomers had already been searching for the food source prior to alighting at the experimental feeder. Thus, it cannot be excluded that the bees had actually left the nest long before the experienced foragers were removed, and that their search had been biased by the presence of up to twenty collecting nestmates—as has been observed in honey bees (Friesen 1973).

So far, in fact, experiments have only undoubtedly demonstrated that the foragers' signals are followed by an increasing number of bees leaving the nest (Lindauer and Kerr 1958; Esch 1967; Pereboom and Sommeijer 1993) or by an increasing number of bees arriving at a food source (Lindauer and Kerr 1958, 1960; Nieh and Roubik 1995; Jarau et al. 2000, 2003; Nieh et al. 2003; Schmidt et al. 2006b, 2008). Furthermore, the recent findings that foragers of *M. seminigra* do not encode the visually measured distance of a food source in the temporal pattern of their thoracic vibrations (Hrncir et al. 2004a) actually provide strong evidence *against* a referential location communication. In fact, all available experimental data corroborate the alerting hypothesis by demonstrating that (1) the recruiters' excitement strongly depends on the individually perceived value of a food source, and (2) the foraging motivation of bees depends on the nest-internal behavior of the recruiters. In addition, ecological considerations indicate that a quick activation of foragers is of considerable advantage for various species of stingless bees in order to exploit valuable food sources efficiently. It has to be kept in mind, however, that so far only a small percentage of the Meliponini has been investigated with regard to their recruitment communication, and that many surprises still might await discovery.

ACKNOWLEDGMENTS

I thank Friedrich G. Barth and Ronaldo Zucchi, who paved my way to stingless bee research, as well as Stefan Jarau, Dirk L. P. Schorkopf, Veronika Schmidt and Sidnei Mateus for continuous help, discussions, and a lot of fun. This work was supported by FAPESP-grants P06/50809-7 and P06/53839-4.

REFERENCES

Absy ML, Bezerra EB, Kerr WE. (1980). Plantas nectaríferas utilizadas por duas espécies de *Melipona* da Amazônica. *Acta Amazônica* 10:271–81.
Absy ML, Camargo JMF, Kerr WE, Miranda IP de A. (1984). Espécies de plantas visitadas por Meliponinae (Hymenoptera; Apoidea), para coleta de pólen na região do médio Amazonas. *Rev Bras Biol* 44: 227–37.

Aguilar I, Briceño D. (2002). Sounds in *Melipona costaricensis* (Apidae: Meliponini): Effect of sugar concentration and nectar source distance. *Apidologie* 33:375–88.

Allerstorfer S. (2004). Rekrutierung und Kommunikation bei *Nannotrigona testaceicornis* Lep. (1836) (Hymenoptera; Apidae; Meliponini). Diploma thesis, University of Vienna, Austria

Anderson C, Ratnieks FLW. (1999). Worker allocation in social insect societies: Coordination of nectar foragers and nectar receivers in honey bee (*Apis mellifera*) colonies. *Behav Ecol Sociobiol* 46:73–81.

Augspurger CK. (1980). Mass-flowering in a tropical shrub (*Hybanthus prunifolius*): Influence on pollinator attraction and movement. Evolution 34:475–88.

Augspurger CK. (1983). Phenology, flowering synchrony, and fruit set of six neotropical shrubs. *Biotropica* 15:257–67.

Barth FG, Hrncir M, Jarau S. (2008). Signals and cues in the recruitment behavior of stingless bees (Meliponini). *J Comp Physiol A* 194:313–27.

Bawa KS. (1983). Patterns of flowering in tropical plants. In Jones CE, Little RJ (eds.), *Handbook of Experimental Pollination Biology*. New York: Van Nostrand Reinhold, pp. 394–410.

Biesmeijer JC, Slaa EJ. (2004). Information flow and organization of stingless bee foraging. *Apidologie* 35:143–57.

Biesmeijer JC, van Nieuwstadt MGL, Lukács S, Sommeijer MJ. (1998). The role of internal and external information in foraging decisions of *Melipona* workers (Hymenoptera: Meliponinae). *Behav Ecol Sociobiol* 42:107–16.

de Bruijn LLM, Sommeijer MJ. (1997). Colony foraging in different species of stingless bees (Apidae, Meliponinae) and the regulation of individual nectar foraging. *Insect Soc* 44:35–47.

von Buttel-Reepen HB. (1903). Die phylogenetische Entstehung des Bienenstaates, sowie Mitteilungen zur Biologie der solitären und sozialen Apiden - Fortsetzung. *Biologisches Centralblatt* 23:129–54.

Chittka L. (2004). Dances as windows into insect perception. *PLoS Biol* 2:898–900.

De Marco RJ, Farina WM. (2001). Changes in food source profitability affect the trophallactic and dance behavior of forager honeybees (*Apis mellifera* L.). *Behav Ecol Sociobiol* 50:441–49.

Dyer FC. (2002). The biology of the dance language. *Annu Rev Entomol* 47:917–49.

Eltz T, Brühl CA, van der Kaars S, Chey VK, Linsenmair KE. (2001). Pollen foraging and resource partitioning of stingless bees in relation to flowering dynamics in a Southeast Asian tropical rainforest. *Insect Soc* 48:273–79.

Esch H. (1963). Über die Auswirkung der Futterplatzqualität auf die Schallerzeugung im Werbetanz der Honigbiene (*Apis mellifica*). *Zool Anz* (Suppl 26):302–9.

Esch H. (1967). Die Bedeutung der Lauterzeugung für die Verständigung der stachellosen Bienen. *Z vergl Physiol* 56:199–220.

Esch H, Esch I, Kerr WE. (1965). Sound: An element common to communication of stingless bees and to dances of the honey bee. *Science* 149:320–21.

Farina WM, Wainselboim AJ. (2001). Changes in the thoracic temperature of honeybees while receiving nectar from foragers collecting at different reward rates. *J Exp Biol* 204:1653–58.

Friesen LJ. (1973). The search dynamics of recruited honeybees, *Apis mellifera ligustica* Spinola. *Biol Bull* 144:107–31.

von Frisch K. (1923). Über die "Sprache" der Bienen, eine tierphysiologische Untersuchung. *Zool Jahrb Abt Physiol* 40:1–186.

von Frisch K. (1993). *The Dance Language and Orientation of Bees*. Cambridge, MA: Harvard University Press.

Gentry AH. (1974). Flowering phenology and diversity in tropical Bigoniaceae. *Biotropica* 6:54–68.

Hart AG, Ratnieks FLW. (2002). Task-partitioned nectar transfer in stingless bees: Work organisation in a phylogenetic context. *Ecol Entomol* 27:163–68.

Hofstede FE. (2006). Foraging task specialisation and foraging labour allocation in stingless bee colonies. Doctoral thesis, Utrecht University, The Netherlands.

Hofstede FE, Sommeijer MJ. (2006). Effect of food availability on individual foraging specialisation in the stingless bee *Plebeia tobagoensis* (Hymenoptera, Meliponini). *Apidologie* 37:387–97.

Hrncir M. (1998). Futterplatzrekrutierung und Kommunikation bei *Melipona quadrifasciata* Lep. (1836) (Hymenoptera; Apidae; Meliponinae). Diploma thesis, University of Vienna, Austria.

Hrncir M, Barth FG, Tautz J. (2006a). Vibratory and airborne-sound signals in bee communication (Hymenoptera). In Drosopoulos S, Claridge MF (eds.), *Insect Sounds and Communication: Physiology, Behaviour, and Evolution*. Boca Raton, FL: CRC Press, Taylor & Francis Group, pp. 421–36.

Hrncir M, Jarau S, Zucchi R, Barth FG. (2000). Recruitment behavior in stingless bees, *Melipona scutellaris* and *M. quadrifasciata*. II. Possible mechanisms of communication. *Apidologie* 31:93–113.

Hrncir M, Jarau S, Zucchi R, Barth FG. (2003). A stingless bee (*Melipona seminigra*) uses optic flow to estimate flight distances. *J Comp Physiol A* 189:761–68.

Hrncir M, Jarau S, Zucchi R, Barth FG. (2004a). Thorax vibrations of a stingless bee (*Melipona seminigra*). I. No influence of visual flow. *J Comp Physiol A* 190:539–48.

Hrncir M, Jarau S, Zucchi R, Barth FG. (2004b). Thorax vibrations of a stingless bee (*Melipona seminigra*). II. Dependence on sugar concentration. *J Comp Physiol A* 190:549–60.

Hrncir M, Schmidt VM, Schorkopf DLP, Jarau S, Zucchi R, Barth FG. (2006b). Vibrating the food receivers: A direct way of signal transmission in bees (*Melipona seminigra*). *J Comp Physiol A* 192: 879–87.

Hrncir M, Schorkopf DLP, Schmidt VM, Zucchi R, Barth FG. (2008). The sound field generated by tethered stingless bees (*Melipona scutellaris*): Inferences on its potential as a recruitment mechanism inside the hive. *J Exp Biol* 211:686–98.

Hrncir M, Zucchi R, Barth FG. (2002). Mechanical recruitment signals in *Melipona* vary with gains and costs at the food source. Anais do V Encontro sobre Abelhas, Ribeirão Preto, SP, Brazil, pp. 172–75.

Hubbell SP, Johnson LK. (1977). Competition and nest spacing in a tropical stingless bee community. *Ecology* 58:949–63.

Hubbell SP, Johnson LK. (1978). Comparative foraging behavior of six stingless bee species exploiting a standardized resource. *Ecology* 59:1123–36.

von Ihering H. (1904). Biologie der stachellosen Honigbienen Brasiliens. *Zool Jahrb Abt Syst Geogr Biol Tiere* 19:179–287, tables 10–22.

Jarau S, Hrncir M, Schmidt VM, Zucchi R, Barth FG. (2003). Effectiveness of recruitment behavior in stingless bees (Apidae, Meliponini). *Insect Soc* 50:365–74.

Jarau S, Hrncir M, Zucchi R, Barth FG. (2000). Recruitment behavior in stingless bees, *Melipona scutellaris* and *M. quadrifasciata*. I. Foraging at food sources differing in direction and distance. *Apidologie* 31:81–91.

Johnson LK. (1983). Foraging strategies and the structure of stingless bee communities in Costa Rica. In Jaisson P (ed.), *Social Insects in the Tropics*. Paris: Université Paris-Nord, pp. 31–58.

Johnson LK, Hubbell SP. (1974). Aggression and competition among stingless bees: Field studies. *Ecology* 55:120–27.

Johnson LK, Hubbell SP. (1975). Contrasting foraging strategies and coexistence of two bee species on a single resource. *Ecology* 56:1398–406.

Kerr WE. (1969). Some aspects of the evolution of social bees (Apidae). *Evol Biol* 3:119–75.

Kerr WE, Cruz C da C. (1961). Funções diferentes tomadas pela glândula mandibular na evolução das abelhas em geral e em *Trigona* (*Oxytrigona*) *tataira* em especial. *Rev Bras Biol* 21:1–16.

Kerr WE, Esch H. (1965). Comunicação entre as abelhas sociais brasileiras e sua contribuição para o entendimento da sua evolução. *Ciên Cult* 17:529–38.

Kerr WE, Ferreira A, Mattos NS de. (1963). Communication among stingless bees—Additional data (Hymenoptera: Apidae). *J NY Entomol Soc* 71:80–90.

Kirchner W, Lindauer M. (1994). The causes of the tremble dance of the honeybee, *Apis mellifera*. *Behav Ecol Sociobiol* 35:303–8.

Krebs JR, Davies NB. (1993). *An Introduction to Behavioural Ecology*. 3rd ed. Oxford: Blackwell Science.

Kronberger E. (2000). Futterplatzrekrutierung bei *Melipona seminigra merillae* CKLL. (1919) (Hymenoptera; Apidae; Meliponinae). Diploma thesis, University of Vienna, Austria.

Lindauer M. (1948). Über die Einwirkung von Duft- und Geschmackstoffen sowie anderer Faktoren auf die Tänze der Bienen. *Z vergl Physiol* 31:348–412.

Lindauer M. (1956). Über die Verständigung bei indischen Bienen. *Z vergl Physiol* 38:521–57.

Lindauer M, Kerr WE. (1958). Die gegenseitige Verständigung bei den stachellosen Bienen. *Z vergl Physiol* 41:405–34.

Lindauer M, Kerr WE. (1960). Communication between workers of stingless bees. *Bee World* 41:29–41, 65–71.

Lutz FE. (1933). Experiments with "stingless bees" (*Trigona cressoni parastigma*) concerning their ability to distinguish ultraviolet patterns. *Am Mus Novit* 641:1–26.

Michener CD. (1974). *The Social Behavior of the Bees: A Comparative Study*. Cambridge, MA: Harvard University Press.

Michener CD. (2000). *The Bees of the World*. Baltimore: Johns Hopkins University Press.

Morawetz L. (2007). Reichweite und Übertragung vibratorischer Signale bei der Kommunikation stachelloser Bienen. Diploma thesis, University of Vienna, Austria.

Mori SA, Piploy JJ. (1984). Observations on the big bang flowering of *Miconia minutiflora* (Melastomaceae). *Brittonia* 36:337–41.

Nachtigall W, Hanauer-Thieser U, Mörz M. (1995). Flight of the honey bee. VII. Metabolic power versus flight speed relation. *J Comp Physiol B* 165:484–89.

Nagamitsu T, Inoue T. (1997). Aggressive foraging of social bees as a mechanism of floral resource partitioning in an Asian tropical rainforest. *Oecologia* 110:432–39.

Nagamitsu T, Inoue T. (2002). Foraging activity and pollen diets of subterranean stingless bee colonies in response to general flowering in Sarawak, Malaysia. *Apidologie* 33:303–14.

Nagamitsu T, Inoue T. (2005). Floral resource utilization by stingless bees (Apidae, Meliponini). In Roubik DW, Sakai S, Karim AAH (eds.), *Pollination Ecology and the Rain Forest—Sarawak Studies*. Berlin: Springer, pp. 73–88.

Nieh JC. (1998). The food recruitment dance of the stingless bee, *Melipona panamica*. *Behav Ecol Sociobiol* 43:133–45.

Nieh JC. (2004). Recruitment communication in stingless bees (Hymenoptera, Apidae, Meliponini). *Apidologie* 35:159–82.

Nieh JC, Barreto LS, Contrera FAL, Imperatriz-Fonseca VL. (2004). Olfactory eavesdropping by a competitively foraging stingless bee, *Trigona spinipes*. *Proc R Soc Lond B* 271:1633–40.

Nieh JC, Contrera FAL, Rangel J, Imperatriz-Fonseca VL. (2003). Effect of food location and quality on recruitment sounds and success in two stingless bees, *Melipona mandacaia* and *Melipona bicolor*. *Behav Ecol Sociobiol* 55:87–94.

Nieh JC, Roubik DW. (1995). A stingless bee (*Melipona panamica*) indicates food location without using a scent trail. *Behav Ecol Sociobiol* 37:63–70.

Nieh JC, Roubik DW. (1998). Potential mechanisms for the communication of height and distance by a stingless bee, *Melipona panamica*. *Behav Ecol Sociobiol* 43:387–99.

Nieh JC, Sánchez D. (2005). Effect of food quality, distance and height on thoracic temperature in the stingless bee *Melipona panamica*. *J Exp Biol* 208:3933–43.

Pereboom JJM, Sommeijer MJ. (1993). Recruitment and flight activity of *Melipona favosa*, foraging on an artificial food source. *Proc Exp Appl Entomol* 4:73–78.

Ramalho M. (2004). Stingless bees and mass flowering trees in the canopy of Atlantic forest: A tight relationship. *Acta Bot Bras* 18:37–47.

Roubik DW. (1980). Foraging behavior of competing Africanized honeybees and stingless bees. *Ecology* 61:836–45.

Roubik DW. (1989). *Ecology and Natural History of Tropical Bees*. New York: Cambridge University Press.

Roubik DW, Buchmann SL. (1984). Nectar selection by *Melipona* and *Apis mellifera* (Hymenoptera: Apidae) and the ecology of nectar intake by bee colonies in a tropical forest. *Oecologia* 61:1–10.

Roubik DW, Moreno JE, Vergara C, Wittmann D. (1986). Sporadic food competition with the African honey bee: Projected impact on neotropical social bees. *J Trop Ecol* 2:97–111.

Roubik DW, Yanega D, Aluja SM, Buchmann SL, Inouye DW. (1995). On optimal nectar foraging by some tropical bees (Hymenoptera: Apidae). *Apidologie* 26:197–211.

Samwald U. (2000). Mechanismen der Futterplatzrekrutierung bei *Melipona seminigra merillae* CCKL. (1919) (Hymenoptera; Apidae; Meliponinae). Diploma thesis, University of Vienna, Austria.

Schmidt VM, Hrncir M, Schorkopf DLP, Mateus S, Zucchi R, Barth FG. (2008). Food profitability affects intranidal recruitment behaviour in the stingless bee *Nannotrigona testaceicornis*. *Apidologie* 39:260–72.

Schmidt VM, Schorkopf DLP, Hrncir M, Zucchi R, Barth FG. (2006a). Collective foraging in a stingless bee: Dependence on food profitability and sequence of discovery. *Anim Behav* 72:1300–17.

Schmidt VM, Zucchi R, Barth FG. (2006b). Recruitment in a scent trail laying stingless bee (*Scaptotrigona* aff. *depilis*): Changes with reduction but not with increase of the energy gain. *Apidologie* 37:487–500.

Schwarz HF. (1948). Stingless bees (Meliponidae) of the western hemisphere. *Bull Am Mus Nat Hist* 90:1–546.

Seeley TD. (1992). The tremble dance of the honey bee: Message and meanings. *Behav Ecol Sociobiol* 31:375–83.

Seeley TD, Mikheyev AS, Pagano GP. (2000). Dancing bee tunes both duration and rate of waggle-run production in relation to nectar-source profitability. *J Comp Physiol A* 186:813–19.

Seeley TD, Tovey CA. (1994). Why search time to find a food-storer bee accurately indicates the relative rates of nectar collecting and nectar processing in honey bee colonies. *Anim Behav* 47:311–16.

da Silva ALG, Pinheiro MCB. (2007). Biologia floral e da polinização de quatro espécies de *Eugenia* L. (Myrtaceae). *Acta Bot Bras* 21:235–47.

Slaa EJ. (2003). Foraging ecology of stingless bees: From individual behaviour to community ecology. Doctoral thesis, Utrecht University, The Netherlands.

Slaa JE, Tack AJM, Sommeijer MJ. (2003a). The effect of intrinsic and extrinsic factors on flower constancy in stingless bees. *Apidologie* 34:457–68.

Slaa EJ, Wassenberg J, Biesmeijer JC. (2003b). The use of field-based social information in eusocial foragers: Local enhancement among nestmates and heterospecifics in stingless bees. *Ecol Entomol* 28:369–79.

Sommeijer MJ, De Rooy GA, Punt W, De Bruijn LLM. (1983). A comparative study of foraging behavior and pollen resources of various stingless bees (Hym., Apidae) in Trinidad, West-Indies. *Apidologie* 14: 205–24.

Stabentheiner A. (2001). Thermoregulation of dancing bees: Thoracic temperature of pollen and nectar foragers in relation to profitability of foraging and colony need. *J Insect Physiol* 47:385–92.

Stabentheiner A, Hagmüller K. (1991). Sweet food means hot dancing in honeybees. *Naturwissenschaften* 78: 471–73.

Stabentheiner A, Kovac H, Hagmüller K. (1995). Thermal behavior of round and wagtail dancing honeybees. *J Comp Physiol B* 165:433–44.

Waddington KD. (1982). Honey bee foraging profitability and round dance correlates. *J Comp Physiol A* 148: 297–301.

Waddington KD. (2001). Subjective evaluation and choice behavior by nectar- and pollen-collecting bees. In Chittka L, Thomson JD (eds.), *Cognitive Ecology of Pollination*. Cambridge, MA: Cambridge University Press, pp. 41–60

Waddington KD, Kirchner WH. (1992). Acoustical and behavioral correlates of profitability of food sources in honey bee round dances. *Ethology* 92:1–6.

Wille A, Michener CD. (1973). The nest architecture of stingless bees with a special reference to those of Costa Rica (Hymenoptera, Apidae). *Rev Biol Trop* 21(Suppl 1):1–278.

Wilms W, Imperatriz-Fonseca VL, Engels W. (1996). Resource partitioning between highly eusocial bees and possible impact of the introduced Africanized honey bee on native stingless bees in the Brazilian Atlantic rainforest. *Stud Neotrop Fauna Environm* 31:137–51.

Wilms W, Wiechers B. (1997). Floral resource partitioning between native *Melipona* bees and the introduced Africanized honey bee in the Brazilian Atlantic rain forest. *Apidologie* 28:339–55.

12 Chemical Communication during Food Exploitation in Stingless Bees

Stefan Jarau

CONTENTS

INTRODUCTION

Chemical compounds play a key role in the communication systems of many living organisms. Information-carrying molecules, known as *semiochemicals*, have been classified into several groups, depending on the identities of the sender and recipient organisms and on who benefits from the transmitted information (reviewed, e.g., in Howse et al. 1998; Wyatt 2003). In short, chemical signals that act within species, to induce either a specific behavior (*releaser pheromones*) or a developmental process (*primer pheromones*), are known collectively as *pheromones*, whereas semiochemicals that act between species are known as *allelochemicals*. There are three main types of allelochemicals, which can either (1) benefit the receiver at the cost of the sender (*kairomones*), (2) benefit the sender at the cost of the receiver (*allomones*), or (3) benefit both sender and receiver alike (*synomones*). In addition to pheromones and allelochemicals, other *chemical cues,* which are not primarily emitted for the purpose of communication, can also act as a source of information for many animals, that way influencing their behavior. Social insects are one of the best examples of a group of organisms whose daily lives depend heavily on chemical communication. Wilson (1990) estimated that for the vast majority, or if not all species of social insects, at least 90% of communication is mediated principally by chemicals. Or, to express it in Murray S. Blum's more poetic words, "the road to insect sociality was paved with pheromones" (Blum 1974a, p. 197).

Apart from honey bees (Hymenoptera, Apidae, Apini), stingless bees (Hymenoptera, Apidae, Meliponini) are the only other group of bees that have developed an advanced eusocial colonial organization (Michener 1974). Depending on the species, the population of a stingless bee colony can range from a few dozen to many tens of thousands of adult workers (Michener 2000). As with all eusocial insects, not all individuals within the colony leave the nest at the same time to forage for food. Therefore, for a colony to survive, it is critical that a limited number of foraging workers can collect enough food, at any one time, to nourish the entire population of the nest. Important dietary components that help to sustain a colony include pollen, or animal proteins (in a few necrophageous species) to feed the developing larvae, as well as nectar or other carbohydrate sources to provide energy for adult bees. Stingless bees have evolved a variety of communication mechanisms for the transfer of information among workers, on both the nature and locality of a food source, to increase the overall foraging efficiency of the colony (reviewed in Nieh 2004; Barth et al. 2008). Chemical signals, including pheromones and allelochemicals, as well as indirect chemical cues are of great importance in this context to stingless bees.

In this chapter I review the existing literature on the role of semiochemicals in the foraging ecology of stingless bees and divide the main volatile compounds into the following four categories: (1) food odors, (2) food source marking volatiles, (3) trail pheromones, and (4) the chemicals used by robber bees and casual thieves during nest plundering.

FOOD ODORS

Food odors, such as flower volatiles or compounds emanating from carcasses, are used by many animals to detect and orient toward food sources. In social insects, foraging information can also be transferred to the colony by food scent trapped on the body of a returning worker. The importance of food odors for the recruitment of workers remains largely unstudied in stingless bees, but it has been investigated in great detail in honey bees (see Chapters 9 and 10). However, the existing data collected for some species of stingless bees indicate that food odors play an important role both for the flower constancy of individual bees and for the recruitment of fellow workers within the nest.

Slaa et al. (1998) tested the flower constancy of three species of stingless bees (*Trigona fuscipennis*, *T. fulviventris*, and *Frieseomelitta nigra*) in relation to floral scent using an experimental arrangement in which the bees were exposed to equal numbers of artificial flowers of the same color and shape, but scented with two distinct fragrances (eighteen rosewood scented and eighteen peppermint scented). In all three species, individual foragers displayed a distinct preference for a single floral scent, with 78–87% of individuals returning to flowers with the same floral scent during consecutive foraging bouts. Furthermore, most recruited individuals preferred the same floral scent as the first foraging bee to return to the nest during each experiment (Slaa et al. 1998). A similar constancy was also displayed by all three species of bees in relation to flower color. Thus, both visual and chemical cues appear to be equally important for the flower constancy of *Trigona fuscipennis*, *T. fulviventris*, and *Frieseomelitta nigra* (Slaa et al. 1998). This was also reported for *Oxytrigona mellicolor* and *Tetragona dorsalis* (Slaa et al. 2003). However, flower constancy in the latter two species significantly increased when the tested flowers differed in both color and odor (Slaa et al. 2003).

The first evidence that stingless bee foragers transmit food odor to fellow workers inside the nest, and that they use this information to locate food in the field came from an experiment conducted by Lindauer (1956) on the island of Sri Lanka. He trained foraging bees of *Trigona iridipennis* to an artificial food source scented with peppermint oil, and found that all bees that were subsequently recruited by the foragers were exclusively attracted to the same scent, and ignored another feeder scented with lavender oil placed next to it. However, when both feeders were scented with peppermint oil, the newly recruited bees no longer discriminated between the two food sources

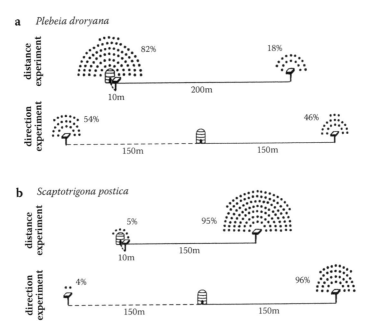

a *Plebeia droryana*

distance experiment

82%   18%

200m

10m

direction experiment

54%   46%

150m   150m

b *Scaptotrigona postica*

distance experiment

5%   95%

150m

10m

direction experiment

4%   96%

150m   150m

FIGURE 12.1 Recruitment to artificial feeders containing peppermint-scented sugar solution. The recruiting foragers were allowed to regularly collect food at the feeder at the end of the solid line, whereas any bee arriving at the control feeder at the end of the broken line was immediately captured. (a) In *Plebeia droryana*, a species that does not communicate the location of a particular food source, the majority of the recruited bees landed on the feeder closest to the nest in the distance experiment, and their distribution was almost equal in the direction experiment. (b) In *Scaptotrigona postica*, a species that precisely guides recruits to the feeder by means of pheromone spots deposited along the route between the food source and the nest, the great majority of recruits always arrived at the feeder visited by the foragers. The black dots denote individual recruits, and the percentages give their distribution. (After Lindauer and Kerr 1958. With permission.)

(Lindauer 1956). Similar results were also recorded for a second species, *Plebeia tica*, which, like *Trigona iridipennis*, lacks a precise mechanism for the communication of food localities (Aguilar 2004; Aguilar et al. 2005). In addition, 80% of the *P. tica* recruits landed on a control feeder close to the nest (5 m) when it had the same peppermint odor as the experimental feeder, which was visited by the recruiting foragers, at a distance of 50 m (Aguilar et al. 2005). This number is very similar to that reported by Lindauer and Kerr (1958, 1960) for a related species, *P. droryana*, in which 82% of the recruits landed on the scented feeder positioned closest to the nest (Figure 12.1a). However, when both feeders were positioned at equal distances from the nest (150 m), but in opposite directions, almost equal numbers of *P. droryana* recruits landed on them both (Figure 12.1a). Apparently, the recruits left the nest with a knowledge of the food scent, but searched for the source of the scent at random (Lindauer and Kerr 1958, 1960). Food odor is also likely to influence the localization of sugar water feeders by recruited bees of the species *Trigona carbonaria*, as demonstrated in a study by Nieh et al. (1999/2000). In this study more recruits landed on the scented experimental feeder, which was situated upwind from the nest and visited by the initial foragers, than on a control feeder treated with the same scent, which was placed at an equal distance, but in the opposite direction. When the experimental feeder was placed downwind, however, only a slight majority of new bees came to it. In experiments with the control feeder placed at 50 m and in the same direction as the experimental feeder (150 m), the distribution of newcomers on the two food sources was almost equal when they were both treated with the same scent. Interestingly, when unscented feeders were used, significantly more newcomers alighted at the control feeder

(Nieh et al. 1999/2000). It is not clear, in this case, whether the bees were attracted by visual stimuli associated with the feeding device (which they had never seen before), or whether the scent of the sugar water alone (Adrian Wenner, personal communication) was sufficient to attract them. However, it is clear that *T. carbonaria* recruits can use food odors brought back to the hive by their foraging nestmates to locate the same food resource in the field—although not necessarily the same individual plants. Kerr (1969) mentioned that the orientation of recruited bees of two species (*Nannotrigona testaceicornis* and *Axestotrigona ferruginea tescorum*) improved considerably when odor-impregnated filter paper was added to the feeding device visited by the recruiting foragers. However, no experimental data were presented to support this observation. Esch et al. (1965) reported that foragers of *Melipona quadrifasciata* revisited a former feeding site, 300 m from the nest, when exposed to playback recordings of sounds produced by bees returning to the hive from the same food source together with the associated food scent. However, in contrast to the food odor, the auditory cues alone were not sufficient to induce experienced bees to leave the nest (Esch et al. 1965). Although these observations are only anecdotal, without giving underlying experimental data, they suggest that the mechanism involved in quickly reactivating stingless bee foragers to visit a known food source by providing the food's odor within the nest may be similar to what is known from honey bees (see Chapter 9).

The studies reviewed above clearly demonstrate that food odor can help experienced bees to relocate a specific food source, thereby promoting constancy toward particular flowers or artificial feeders. Furthermore, recruits apparently learn the odor inside the nest and use the information when searching for food in the field. The observations that prospective recruits of *M. quadrifasciata* and *M. scutellaris*, within the nest, preferentially touch the thoraxes and abdomens of returning foragers before they take up food samples (Hrncir et al. 2000), along with the recent demonstration that workers of *M. quadrifasciata* can be conditioned to floral odors through associative learning (McCabe et al. 2007), suggest that the food odor clinging to a forager's body in connection with a delivered nectar sample (reward) is an important element for food odor learning during the recruitment process. Again, these potential communication mechanisms have been studied in much greater detail in honey bees (see Chapter 10).

FOOD SOURCE MARKING VOLATILES

ATTRACTIVE SCENT MARKS

The deposition of volatile organic compounds at food resources and the function of these scent marks in the foraging behavior of stingless bees have mainly been investigated through the use of artificial feeders that dispense a sugar solution. The compounds that accumulate on the feeders with each visit by a foraging bee attract both the depositor herself and other incoming bees that arrive at the food source.

To test for the possible effects of scent marking at food sources, Villa and Weiss (1990) trained foragers of *Tetragonisca angustula* to sugar water feeders of a particular color (yellow or blue). The trained bees were then exposed to two identically colored feeders, only one of which had been visited by foragers before. These experiments demonstrated that feeders, which were previously visited by foraging bees, were significantly more attractive—presumably due to chemical marks left on them. Indeed, even when the color of the visited feeder was changed, the bees still continued to prefer it to an identical, but unmarked feeder of the original color, indicating that the scent marks are more important than color in feeder constancy of *T. angustula* (Villa and Weiss 1990). The authors also found that the volatiles deposited by foraging bees were attractive to conspecifics from a different colony. Likewise, *Scaptotrigona mexicana* workers landed equally on two clean feeders, but exclusively chose scent-marked feeders, previously visited by conspecifics, when given the choice (Villa and Weiss 1990). A similar result has also been recorded for *Plebeia tica* (Aguilar 2004). *Melipona panamica* newcomers (feeder inexperienced bees) exclusively chose a marked feeder over a clean one when they were placed 1 m apart, regardless of whether the food itself was scented, when they arrived while experienced foragers were present on the feeder (Nieh 1998). The newcomers' choice in these experiments likely was influenced by visual local enhancement (see Chapter 8), but it only decreased to ca. 91% of landings on the used feeder when the experienced foragers were caught prior to testing the newcomers' choice (Nieh 1998). Likewise, a significant majority of *M. mandacaia* (Nieh et al. 2003b) and *M. rufiventris* (Nieh et al. 2004a) foragers had chosen a putatively odor marked paper disc from a previously visited feeder over a clean control paper in two-feeder-choice experiments that excluded visual local enhancement. In *Trigona corvina* a significant majority of the foragers chose a previously used (scent-marked) feeder over a clean feeder, too, when the odors were deposited by the same bee that was tested, by its nestmates, or by foragers of a different conspecific colony (Boogert et al. 2006). However, feeders previously visited by foragers of another species, *Melipona beecheii*, did not attract *T. corvina* workers (Boogert et al. 2006). By contrast, *T. spinipes* foragers, which were significantly more attracted to a feeder bearing their nestmates' scent marks over a feeder scent-marked by *Melipona rufiventris* workers when both were presented at the *T. spinipes* training site, preferred the heterospecifics' odor over their own scent marks at the *Melipona* feeding site only 2 m away (Nieh et al. 2004a). Furthermore, *T. spinipes* aggressively drove away *M. rufiventris* and took over the food sources. Nieh et al. (2004a) interpreted these findings as olfactory eavesdropping used by *T. spinipes* to locate resources that had been encountered by competitors (see Chapter 8). They further speculated that olfactory eavesdropping could have driven the evolution of a concealed symbolic communication of food location inside the nest in some stingless bee species (i.e., *Melipona*), and that such referential location information replaced odor trail information. This, however, would imply that pheromone trails are the ancestral mode of location communication in stingless bees, which is unlikely. For example, many species within the genera *Scaptotrigona* and *Trigona*, which are derived rather than basal taxa (Michener 2000; Costa et al. 2003), use trail pheromones to communicate the precise location of a food source to their nestmates. Furthermore, a conclusive demonstration that any stingless bee species indeed uses a referential location information communicated inside the nest to locate a food source is still lacking to this day (Hrncir et al. 2006; Barth et al. 2008; see also Chapter 11).

More detailed studies investigating the influence of the number of forager visits at a food source on the subsequent choice behavior of the bees (experienced foragers and newly arriving individuals) have been carried out with *Melipona seminigra* (Hrncir et al. 2004) and *Nannotrigona testaceicornis* (Schmidt et al. 2005). Artificial feeders only became attractive to *M. seminigra* foragers after bees had landed and fed on them on forty occasions, after which up to 72% of the bees subsequently tested chose the used feeder over a clean control feeder. Under similar experimental conditions, the number of visits necessary to attractively mark a feeder for *N. testaceicornis* foragers was only twenty. In this species approximately 86% of the bees chose the marked feeder when it was presented alongside a clean control feeder (Schmidt et al. 2005). In both species, the feeders previously visited by bees attracted more newcomers than clean feeders, even when the food used was itself scented. However, discrimination between the two food sources was significantly higher when both feeders contained unscented sugar solution. Apparently the bees were attracted by the food odor itself, but the information from the deposited scent marks was not lost. The attractiveness of feeders that had been visited by foragers on twenty occasions was also demonstrated for *Scaptotrigona* aff. *depilis* (Schmidt et al. 2003). In this species the majority of bees always landed on the putatively marked feeders rather than on clean control feeders in two-feeder-choice experiments, regardless of whether the sugar solution was 0.75 M, 1.5 M, or 3 M, and whether the feeders were separated by a distance of 20 or 170 cm. *Scaptotrigona* bees recruit nestmates by means of scent trails, which lead them toward the food (Lindauer and Kerr 1958, 1960; Kerr et al. 1963; Schmidt et al. 2003; see below). The attractive scent marks at the feeders, which apparently differ from the actual trail pheromone markings, can help the bees to locate the endpoint of the odor trail. This conjecture is based on the observation that scent-marked feeder plates displaced from their original food site—and set up 20 m beyond it or half the way toward the nest—still attracted the bees, whereas clean feeders did not (Schmidt et al. 2003). The high precision of recruitment to a particular food source found in *Scaptotrigona* aff. *depilis*, where 97.5–100% of the recruited bees arrived at the feeder visited by the recruiting foragers, likely can be explained by the combination of a scent trail and the attractive food-marking volatiles (Schmidt et al. 2003). A similarly high precision in the arrival of bees at used and potentially marked artificial food sources was found in *S. mexicana*, both at a larger scale with control feeders placed 5 and 10 m away from the experimental feeder and in a within-patch situation with the control feeders arranged 0.5 and 1 m around it (Sánchez et al. 2004, 2007). It seems likely that scent marks left on the experimental feeder by the recruiting foragers, as well as visual cues caused by the foragers' presence while taking up sugar solution (local enhancement; see Chapter 8), accounted for the attraction of the newly arriving bees to the particular experimental feeder in *S. mexicana*. Scent marks deposited by bees on a sugar water feeder also accounted for the precise recruitment of experienced foragers to this food source when it was presented together with four unvisited, clean feeders in *Melipona panamica* and *Partamona peckolti* (Contrera and Nieh 2007).

The attractiveness of the scent marks deposited by foragers at feeders persisted for 70 min in *Melipona mandacaia* (Nieh et al. 2003b). Feeders marked by *M. seminigra* attracted bees for up to 2 h, and their active range was restricted to approximately 1 m (Hrncir et al. 2004). When the distance between the scent-marked feeder and a control feeder was 1.5 m or greater the bees' choice behavior did not significantly differ from their choice between two clean feeders. Experiments with *M. panamica*, using similar sugar solution feeders, showed that the effective radius of forager-deposited scent marks lay somewhere between 6 and 12 m (Nieh 1998). Probably the larger radius of attraction in *M. panamica* was due to a much larger quantity of accumulated scent on the feeders. Nieh (1998) allowed twenty foragers to freely visit the feeders for at least 30 min prior to the experiments (which probably resulted in a minimum of 250 visits), whereas a fixed number of forty forager visits was allowed in the case of *M. seminigra* (Hrncir et al. 2004).

Repellent Scent Marks

Studies carried out with bumble bees (Chapter 13), honey bees (Chapter 9), and the social sweat bee *Halictus aerarius* (Yokoi and Fujisaki 2007; Yokoi et al. 2007), as well as with a few species of solitary bees (e.g., *Anthophora plumipes*, Gilbert et al. 2001; *Anthidium manicatum*, Gawleta et al. 2005), have demonstrated that scent marks left on natural flowers (as opposed to artificial feeders) by foraging bees can be used by subsequent visitors to quickly reject a depleted food source, which is likely to significantly increase foraging efficiency. In stingless bees the only observations of workers foraging on natural flowers, within the context of scent marking, were reported for *Trigona fulviventris* (Goulson et al. 2001). Individuals that foraged on *Priva mexicana* (Verbenaceae), which has small tubular flowers, accepted ca. 76% of the flowers that had not been visited by other bees for at least 45 min. By contrast, only 40% of flowers were landed on and probed if they had been visited by another bee within the previous 2 min. Recently visited flowers were still rejected even when the depleted nectar was artificially replenished, using nectar from other flowers. This, therefore, indicates that the bees' decisions were based on the presence or absence of scent marks rather than direct assessment of the reward (Goulson et al. 2001). However, in the case of the plant species *Byrsonima crassifolia* (Malpighiaceae), foragers of *T. fulviventris* equally accepted both recently visited and unvisited flowers, which are simple in structure and have easily accessible nectaries (Goulson et al. 2001). Therefore, although *T. fulviventris* workers apparently can detect and respond to scent marks when foraging, they do not necessarily use such information when flower handling time is short and visiting depleted flowers not costly, as might be the case with *B. crassifolia*. This conjecture is supported by recent experiments with bumble bees foraging on artificial flowers carried out by Saleh et al. (2006). The authors found that the bees responded to scent marks more strongly (higher rejection rate) when they were encountered on long flowers—which require a longer handling time—than on short flowers.

Another interesting observation of a repellent effect of volatiles deposited by bees was reported by Nieh et al. (2004a). This study demonstrated that workers of *Melipona fulviventris* show a strong aversion to feeders that had previously been visited by foragers of *Trigona spinipes*, which is known to aggressively defend food sources. The avoidance behavior suggests that *M. fulviventris* can detect interspecific scent marks at food sources, and probably retreats from them to avoid encounters with aggressive competitors. Whether this avoidance behavior is a general response to foreign odors or the result of prior experience with *T. spinipes* remains to be determined (Nieh et al. 2004a).

ORIGIN AND CHEMICAL COMPOSITION OF SCENT MARKS

Kerr and Rocha (1988) hypothesized that the volatiles used to mark food sources by foragers of *M. rufiventris* and *M. compressipes* come from the anal liquids that the bees sometimes excrete at or near a sugar water feeder after food uptake. However, this conjecture was only based on the observation of defecation behavior, without demonstrating that bees are actually attracted by the anal droplets. Aguilar and Sommeijer (1996, 2001) subsequently observed that in *M. favosa* the percentage of foragers that deposited anal droplets on and around sugar solution feeders, as well as the total number of droplets, increased in relation to the distance between the food source and the nest. The total numbers of anal droplets deposited by *M. favosa* foragers that fed on sugar solutions of both low and high concentrations were similar (Aguilar and Sommeijer 2001). In *M. panamica*, however, the deposition rate increased at feeders containing less concentrated sugar solution, i.e., at poor food sources (Nieh 1998). In both species, the volume of the droplets decreased as the sugar concentration increased. In *M. mandacaia*, the foragers deposited twice as many droplets when feeding at 1.25-M sugar solution feeders than at 2.5-M solution feeders (Nieh et al. 2003b). As argued by Nieh et al. (2003b), the putative scent from the anal excretions would therefore be stronger for low-quality food sources than for high-quality food resources. This seriously undermines the hypothesis that anal droplets function as attractive food-marking substances. Nevertheless, Aguilar and Sommeijer (2001) found that the visitation rate of *M. favosa* foragers at feeding tables with anal droplets was higher than that at clean feeders, and that a larger proportion of bees (ca. 63%) landed on a feeder with anal droplets than on a simultaneously offered clean control feeder. However, the authors did not exclude the possibility that other volatiles, in addition to anal excretions, were also deposited on the feeders when the bees were foraging. Therefore, the results of this study have to be interpreted with caution. In two other studies that controlled for other potential sources of attractant, workers showed no preference for feeders baited with anal deposits over clean feeders (*M. panamica*, Nieh 1998; *M. seminigra*, Hrncir et al. 2004). Compounds excreted from a bee's anus thus can be excluded as the source for attractive food-marking odors. In the same way, the sugar water upon which the bees have fed, as well as their mandibular gland secretions, can also be excluded (Nieh 1998; Hrncir et al. 2004). Nieh et al. (2003b) reported that *M. mandacaia* foragers were attracted to feeders baited with anal droplets collected on filter papers and to feeders baited with filter papers simply held below the abdomen of foragers—without even touching them. They assumed that these papers were impregnated with a "ventro-abdominal odor" of a rather mystic origin. However, the preference for the marked papers is only significant if the data for all experiments are pooled, as within the original study (Nieh et al. 2003b). Separate comparisons, on the data recorded from each of the individual experiments (e.g., with chi-square tests; data given in Table 1 in Nieh et al. 2003b), reveal a nonrandom forager distribution for only one of the five experiments that were conducted using both types of scent marks. Therefore, the conclusion drawn by Nieh et al. (2003b) that *M. mandacaia* foragers are attracted to abdominal droplets and ventro-abdominal odors should be taken with caution.

Compelling evidence for the importance of footprint secretions in the food-marking context—and against any other potential origin of the volatiles—comes from experiments carried out with *M. seminigra* by Hrncir et al. (2004). These authors covered the tips of the foragers' legs with nail polish, thus sealing any potential gland opening on the legs' distal segments. As a consequence of

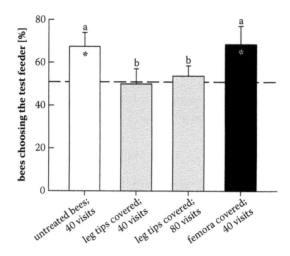

FIGURE 12.2 Choice behavior of *Melipona seminigra* workers between a clean (unvisited) sugar solution feeder and an identical feeder, which was previously visited by nestmates that were either untreated (open bar; N = 12 experiments) or had their leg tips (gray bars; N = 4 each) or femora (black bar; N = 4) covered with nail polish. The bars and whiskers give means + 1 SD. Asterisks indicate statistically significant differences in the proportion of bees choosing the previously visited feeder compared to the proportion of bees choosing one of two identical and clean feeders during a series of control experiments (50.8 ± 1.2%; N = 6; dashed line = mean); different letters denote significant differences between bars (one-way ANOVA and Tukey pair-wise comparisons; minimum significance level $P < 0.05$). Obviously, the foragers were unable to leave attractive volatiles at the feeder when their leg tips were covered, which likely is due to the sealing of glands that open there and account for the deposition of footprint substances. (After Hrncir et al. 2004. With permission.)

this treatment, the feeders the foragers had visited did not become attractive to other bees even after forty or eighty visits (Figure 12.2). By contrast, feeders visited forty times by untreated foragers, or by bees with their femora covered with nail polish, were significantly preferred over clean feeders by newly arriving bees (Figure 12.2; Hrncir et al. 2004). These experiments clearly demonstrated that the volatiles deposited while a bee is sitting or running on a feeding plate are secreted from glands within their legs that open at the legs' tips, rather than from any other part of the bee, including the general cuticle surface.

The most obvious glands that could account for the secretion of footprint substances are the tarsal glands, which are situated in the fifth tarsomeres of a hymenopteran's legs (Dahl 1885; Arnhart 1923). However, a histological study on the species *M. seminigra* has shown that these glands lack any openings to the outside (Jarau et al. 2005), as is the case in other stingless bees (Cruz-Landim et al. 1998), honey bees (Lensky et al. 1985; Federle et al. 2001), and bumble bees (Pouvreau 1991). Therefore, they are unlikely to be involved in the production of chemical footprints (Jarau et al. 2005). The contradiction between the apparent use of attractive footprint secretions by *M. semini-gra* foragers at food sources on the one hand and the lack of openings of the tarsal glands on the other was resolved by the discovery of a different system of glands within the bees' legs (Jarau et al. 2004b). Each claw retractor tendon, which runs from a leg's femur through its tibia and tarsus and connects to the base of the pretarsus, has specialized glandular epithelia within the femur and tibia (Figure 12.3). The glands' products are secreted into the tendons, which form hollow tubes and serve as the excretory canals leading to the legs' tips, where they are secreted as footprints. Feeding tables baited with extracts of the tendon glands, dissected from *M. seminigra* foragers, attracted the bees in the same way as feeders naturally marked by foragers (Jarau et al. 2004b), thus providing strong evidence that the secretions of these glands account for the attraction of bees to footprints.

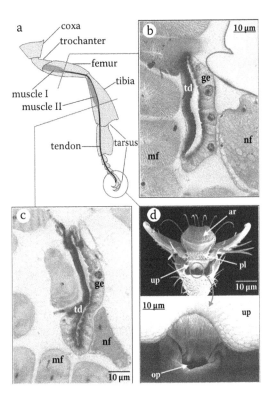

FIGURE 12.3 The tendon gland of *Melipona seminigra*, which produces the footprint secretions that are released at a leg's tip. (a) Diagram of a metathoracic leg illustrating the location of the tendon and its associated muscles. (b, c) Cross sections through the tendon gland in the leg's femur and tibia showing the cuticular part of the hollow tendon (*td*) and the large cells of the glandular epithelium (*ge*). (d) Ventral view of the fifth tarsomere and the pretarsus (top) and of an enlargement (bottom) of the opening (*op*) of the tendon at the base of the unguitractor plate (*up*). Further abbreviations: *ar*, arolium; *mf*, muscle fibers; *nf*, nerve fibers; *pl*, planta. See color insert following page 142. (After Jarau et al. 2004b. With permission.)

The chemical structures of the compounds that are deposited by stingless bees at food sources have so far been elucidated for only one species (*Melipona seminigra*). The scent marks collected from artificial feeders that were previously visited by foragers consisted of 12 alkanes, 8 alkenes, 1 methyl alkane, and 1 aldehyde (Jarau et al. 2004b). The main compounds, each constituting ≥10% of the total amount of the identified volatiles, were pentacosane, heptacosane, and the corresponding alkenes 7-(Z)-pentacosene and 7-(Z)-heptacosene. The same compounds were also detected in extracts collected from the tendon glands of *Melipona seminigra* (see above) as well as from its last tarsomeres. These extracts also contained a further forty-one compounds, mainly esters, acids, and methyl alkanes (Jarau et al. 2004b). Interestingly, the hydrocarbons identified from *M. seminigra* scent marks are similar to the compounds reported from bumble bee scent marks (e.g., Schmitt et al. 1991; Eltz 2006; Saleh et al. 2007), which have a similar effect on the behavior of foraging bees (see Chapter 13).

FOOD-MARKING CUES OR TRUE SIGNALS?

One important question that arises from the discovery that bees are attracted to food sources that have been previously visited by other foragers is whether the deposited substances are true signals or merely cues. Signals have been specifically shaped by natural selection to broadcast information and are released on purpose, whereas cues can contain useful information, but have not especially

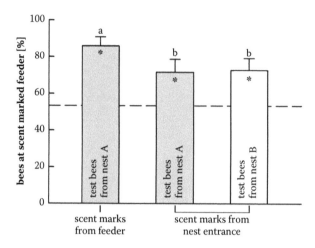

FIGURE 12.4 Choices of *Nannotrigona testaceicornis* workers between a feeder bearing scent marks and a clean (unmarked) but otherwise identical feeder. The scent marks originated from bees of nest A that either landed and sat on the feeder 25 m away from the colony in order to collect sugar solution, or simply ran over it directly in front of the nest's flight hole. The bars and whiskers give means + 1 SD (N = 6 experiments each). Asterisks indicate statistically significant differences in the proportion of bees choosing the scent-marked feeder compared to the proportion of bees choosing one of two identical and clean feeders during a series of control experiments (53.4 ± 3.6%; N = 6; dashed line = mean); different letters denote significant differences between bars (one-way ANOVA and Tukey pair-wise comparisons; minimum significance level $P < 0.05$). Apparently, the attractive volatiles are neither nest specific nor specifically deposited while foraging at a food source. (After Schmidt et al. 2005. With permission.)

been molded by natural selection to convey them (Seeley 1989). Such cues are incidentally left on the substrate by foragers, and the accumulating odor could then be interpreted by other bees as, for example, an indicator for the presence of food.

Schmidt et al. (2005) recently investigated this question in *Nannotrigona testaceicornis*. In experiments with artificial food sources the bees had to choose between two feeder plates: one that was clean and unvisited by bees and one (which was optically identical and contained the same sugar solution) that was scent-marked by other bees. Regardless of whether the scent marks were deposited by twenty foragers that collected food from the test feeder or twenty bees that simply ran over its surface to reach the entrance to their nest, the bees significantly preferred the marked feeder over the control (Figure 12.4). A significantly larger proportion of bees chose the feeders marked at feeding sites than feeders marked by workers returning to their nest, but this difference can probably be explained by variation in the total amounts of deposited volatiles between the two locations. Whereas each forager spent ca. 20 s on the feeder, the time required for a bee to cross it at the entrance to the nest was less than 4 s. There is good reason to assume that the approximately fivefold time the bees spent on the feeders compared to the nest entrance plates has resulted in the deposition of more scent. The compounds left on the substrate at the nest entrance, presumably as footprints, were not nest specific, but also attracted conspecific bees of another colony (Figure 12.4; Schmidt et al. 2005). The substances that bees apparently leave on any substrate simply by walking on it, regardless of the behavioral context, which act as an attractant for other bees, thus are cues that carry the information only incidentally. This conclusion can also explain the observation, reported by Nieh et al. (2003b), that *M. mandacaia* foragers deposited attractive volatiles on rich food sources (2.5-M sugar solution), to which they recruited nestmates, as well as on poor food sources (1.25 M), to which recruitment did not occur. Recent experiments have demonstrated that in bumble bees, too, the chemical compounds left on substrates are passive cues rather than deliberately deposited signals (Saleh et al. 2007; Wilms and Eltz 2008; see also Chapter 13).

Additional evidence that food-marking volatiles are context-dependent cues rather than fixed signals containing specific information on the nature or quality of a food resource comes from recent experiments conducted with *Scaptotrigona mexicana*. Foragers of this species learned to positively or negatively associate scent marks with the quality of the food in choice experiments using artificial feeders (Sánchez et al. 2008). When the bees had to choose between two feeders, one of which was scent-marked and the other a clean control, they landed on both at the same frequency, when each feeder contained the same rewarding sugar solution. However, the bees showed a preference for the clean feeder when the contents of the scent-marked feeder were replaced with pure water (no sugar reward). The opposite occurred when the rewards were switched, so that only the scent-marked feeder contained sugar (Sánchez et al. 2008). Likewise, *Melipona scutellaris* foragers also use volatiles to avoid revisiting feeders that are depleted after a single visit (Roselino et al. 2007), but they are attracted to scent-marked feeders if they contain an even higher concentrated sugar solution during a second visit (Rodrigues et al. 2008). Apparently, stingless bees can learn to associate forager-deposited volatiles with the quality of a food resource and adapt their feeding behavior during subsequent visits. Volatiles deposited by bees can, therefore, provide information on a food source that can be quickly and effectively learned in the same way as other floral cues, such as flower odor, color, or shape (Menzel 1999). Recently, the ability to learn and predict food reward levels from volatile food markings has also been demonstrated in bumble bees (Saleh and Chittka 2006; Witjes and Eltz 2007; see also Chapter 13), and may represent a general feature of the foraging behavior of bees.

TRAIL PHEROMONES

The use of trail pheromones, which guide recruited foragers from the nest to a food resource, has been observed in several species of stingless bees within the genera *Trigona*, *Scaptotrigona*, *Geotrigona*, *Cephalotrigona*, and *Oxytrigona* (Lindauer and Kerr 1958, 1960; Kerr 1960, 1969, 1973; Kerr and Cruz 1961; Kerr et al. 1963, 1981; Hebling et al. 1964; Cruz-Landim and Ferreira 1968; Blum et al. 1970; Johnson 1987; Noll 1997; Schmidt et al. 2003; Nieh et al. 2003a, 2004b; Jarau et al. 2004a, 2006; Sánchez et al. 2004; Aguilar et al. 2005; Schorkopf et al. 2007). This remarkable behavior belongs to the most precise and effective communication mechanisms for the exploitation of food found within the social bees (Lindauer and Kerr 1958, 1960; Jarau et al. 2003). As a result of the high numbers of workers that quickly and precisely gather at a particular food source, following its discovery by scouts, some species can easily compete with the recruitment efficiency of honey bees (Lindauer and Kerr 1958, 1960). Furthermore, trail pheromone communication allows the precise recruitment of bees not only at a particular direction and distance from the nest (Figure 12.1b), but also to a particular height above ground level (Lindauer and Kerr 1958, 1960; Nieh et al. 2004b). Height, which is not communicated in the honey bees' dance language (von Frisch 1965), is an important dimension for stingless bees that live in tropical forests.

Kerr and Rocha (1988) and Kerr (1994) also proposed the use of short odor trails by *Melipona* bees, based on observations that foragers of several species (*M. rufiventris*, *M. compressipes*, *M. scutellaris*, *M. bicolor*, and *M. quadrifasciata*) sometimes landed and walked on leaves (for several centimeters) on their return flights from a feeding station. However, so far there is no proof that the foragers deposit scent and that recruited bees search for scent marks when orienting toward a food source. Furthermore, the infrequent occurrence of interrupted return flights from a food source in *M. scutellaris* and *M. quadrifasciata* (0.25% and 0.62%, respectively, of ca. 6,000 observations per species) found by Hrncir et al. (2000) points to the insignificance of this behavior for the guidance of recruits to a food source in *Melipona* bees. Likewise, the actual existence and use of an "aerial odor tunnel," which is assumed to be created by *Melipona* (Kerr 1994) and *Partamona* (Kerr 1969) foragers through the release of a pheromone during their flight toward a food source and followed by recruited nestmates, has yet to be scientifically demonstrated.

SCENT MARKING BEHAVIOR AND THE SPATIAL DISTRIBUTION OF PHEROMONE MARKS

The deposition of trail pheromone marks, even when deposited at the food source itself, markedly differs from the passive transfer of odor cues to food sources described above. It is always connected with a very conspicuous behavior exhibited by the foragers that deposit the scent marks. This behavior was first described in detail by Lindauer and Kerr (1958, 1960), who trained foragers of *Scaptotrigona postica* to feed at sugar solution feeders at a distance of 50 m from the nest. The trained bees returned several times to the feeders to collect food and flew directly back to the nest without showing any special behavior. After a couple of visits, however, the bees appeared excited, briefly left the feeding table in a hectic flight, alighted on it again, flew up and landed on nearby blades of grass or sticks, and finally left in the direction of the nest. On their return flight they again landed on leaves or sticks at intervals of 1–2 m. About 8 m before reaching the nest they usually turned around and flew back to the food (Figure 12.5a). Occasionally, a forager also entered the nest, presumably to alert and recruit nestmates. The hectic behavior displayed by the foragers is clearly connected with the deposition of scent marks that are subsequently used by the recruits to find the food: When the nest and feeder were placed on the opposite sides of a small lake, denying the bees any solid substrate for the deposition of pheromones, no newly recruited bees found the food resource (Lindauer and Kerr 1958, 1960). Recruitment took place, however, when the foragers

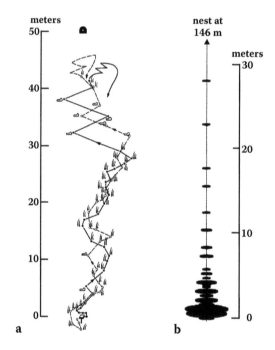

FIGURE 12.5 Spatial distribution of pheromone marks deposited (a) by *Scaptotrigona postica* foragers on stones and ground-proximate vegetation and (b) by *Trigona hyalinata* workers along a rope with leaves fastened to it at 1 m intervals, which extended for 100 m from the food source toward the nest (the leave-bedecked rope was the only possibility for bees to deposit scent marks during the experiment). Whereas recruiting foragers of *S. postica* deposit pheromone marks along most of the distance covered by their return flight to the nest (two forager flights shown; dots denote a scent mark, arrows indicate the flight direction), the scent marks are most concentrated at the food source and along the first few meters toward the nest in *T. hyalinata* (area of ellipses correspond to the number of scent marks deposited, the smallest ones being equal to 1 mark; in sum, 103 marks are shown). The food sources (sugar solution feeding tables) were located at 0 m; note the different scales and distances to the nest in (a) and (b). The black hive symbol in (a) denotes the location of the bees' nest. (a, after Lindauer and Kerr 1958; b, after Nieh et al. 2003a. With permission.)

were provided with a substrate in the form of a rope decorated with twigs and leaves tightened over the water surface, where they could land and deposit pheromone marks on their way back to the nest (Lindauer and Kerr 1958, 1960).* A rapid change from normal feeding to the hectic scent marking behavior of recruiting foragers has also been observed in *Scaptotrigona* aff. *depilis* (Schmidt et al. 2003), *S. mexicana* (Sánchez et al. 2004), *Trigona fulviventris* (Johnson 1987), *T. hyalinata* (Nieh et al. 2003a), *T. recursa* (Jarau et al. 2004a), and *T. spinipes* (Nieh et al. 2004b). To deposit scent marks, foragers of *T. recursa* run for a short duration of about 0.6 s across the substrate, thereby covering a distance of approximately 1 cm (Jarau et al. 2004a). An equally short duration (ca. 0.7 s) of scent mark deposition was also reported for *T. hyalinata* and *T. spinipes* (Nieh et al. 2003a, 2004b).

The trail of scent observed in the species *S. postica*, which extended for almost the entire distance from a food source toward the nest (Lindauer and Kerr 1958, 1960), appears not to be the general rule in stingless bees.* In the species *T. amalthea*, distribution of scent marks was non-uniform, with the highest concentration of scent deposited on the food source and on the first 5 m of the trail toward the nest, and with larger intervals between the marks thereafter (Kerr et al. 1963). A similar distribution of scent marks could have accounted for the correct approach of *S. postica* recruits at a feeder located at the most distant end of the trail made by their nestmates (Kerr et al. 1963). Likewise, Johnson (1987) reported that foragers of *T. fulviventris* concentrate the scent marks they deposit at the food source or within 1 m around it (51% and 43% of 277 observed scent marking events, respectively). Only 6% of the scent marks were deposited on leaves 1–2 m away from the feeder in the direction of the nest. Thus, *T. fulviventris* does not lay an entire trail of pheromone spots from a food source to the nest—at least when the food resource is within 20 m of the nest (Johnson 1987). Instead, the bees concentrate the odor marks around the food source itself. More recently, Nieh et al. (2003a, 2004b) studied the distribution of scent marks in two further *Trigona* species. During their experiments the trained foragers could only deposit their pheromone on leaves, which were attached, at 1–m intervals, along ropes that extended from the feeders to the nest in an otherwise unnatural environment (a terrace paved with tiles). The authors found that both species laid only short trails and deposited the majority of their scent marks on the feeding table itself or only a few meters away from it. In *T. hyalinata*, the scent marks were mainly deposited within 5 m of the feeder and formed a gradient of decreasing scent concentration that extended for a maximum of 27 m from the food source to the nest, which was situated 146 m away (Figure 12.5b; Nieh et al. 2003a). Likewise, *T. spinipes* foragers deposited trail scent marks at a maximum distance of 29 m from the food source in the direction of the nest, which was up to 225 m away, with 95% of the marks placed within 3 m of the feeder (Nieh et al. 2004b). The much larger concentration of scent marks close to the feeders resulted in a polarization of the trails. Foragers of both species used the odor gradient to orient toward the end of the trail and preferred a feeder at this position compared to a feeder within the trail or at the end of the trail that was closest to the nest, even if the trails (on the ropes) were displaced from their original positions (Nieh et al. 2003a, 2004b). In *T. spinipes* the concentration of the scent marks was also highest on and around the feeder when it was placed on top of a 12-m-high water tower, with a decreasing number of scent marks laid toward the tower's base (Nieh et al. 2004b). Recruited *T. spinipes* workers were attracted to the scent trail for 8 min (Kerr et al. 1963) to 25 min (Nieh et al. 2004b). A similar time span (ca. 15 min) during which recruits followed scent trails was found in *Scaptotrigona postica* and *S. bipunctata* when the deposition of further scent spots by foraging bees was experimentally hindered (Lindauer and Kerr 1960; Kerr et al. 1963). However, under normal conditions, trails are rapidly reinforced by the scent deposits of recruiting bees (Schmidt et al. 2006), and are likely to remain effective over much longer periods of time. In *S. postica* and *T. corvina* the orientation of newly recruited bees along a scent

---

* Dirk Louis P. Schorkopf and colleagues recently repeated the lake experiments with *S. postica*, and gathered strong evidence that an entire scent trail is not imperative for the recruitment success of this species (Schorkopf, personal communication). The publication of their results, not yet available at the preparation of this chapter, likely will change our view of stingless bee trail pheromones.

trail probably is aided by the recruiting foragers, which were observed to guide small groups of bees on their first flight to the food source (Lindauer and Kerr 1958, 1960; Aguilar et al. 2005).

GLANDULAR ORIGIN OF TRAIL PHEROMONES

The origin of the pheromones deposited by stingless bees when marking scent trails was long considered to be their mandibular glands. This idea has been perpetuated by virtually every textbook or review article dealing in some way with stingless bees, trail pheromones, or animal behavior (e.g., Blum 1970, 1974b; Wilson 1971; Blum and Brand 1972; Shorey 1973; Michener 1974; Wille 1983; Haynes and Birch 1985; Free 1987; Roubik 1989; Morgan 1990; Traniello and Robson 1995; Alcock 2001; Wyatt 2003; Nieh 2004). This is astounding since no experiment, within the primary literature, has ever demonstrated that the trail pheromone of any stingless bee species is secreted from its mandibular glands (see below). In fact, in their seminal study on stingless bee recruitment, Lindauer and Kerr (1958, 1960) had already reported to the contrary, that *Scaptotrigona postica* recruits *never* followed experimental trails, made from mandibular gland secretions, when these trails deviated away from the bees' natural scent trail! However, because the bees somehow "examined" mandibular gland marks when the authors placed them between the natural scent marks deposited by foragers, and because a scent marking bee alights on the substrate and runs a short stretch on it with her mandibles opened, Lindauer and Kerr (1958, 1960) concluded that the trail pheromone is produced in the mandibular glands. In several later studies, researchers merely assumed that trail pheromones in stingless bees are produced in their mandibular glands without conducting specific experiments, such as bioassays that employed artificial scent trails, to test this assumption (Kerr and Cruz 1961; Kerr et al. 1963; Cruz-Landim and Ferreira 1968; Blum et al. 1970). The most recent study heavily influenced by the "mandibular gland trail pheromone paradigm" was published by Nieh et al. (2003a). These authors presented extracts at feeding tables, prepared from mandibles with their appendicular glands plus additional tissues pulled out of the heads of *T. hyalinata* foragers, and observed the behavior of newcomers approaching them. For about 8 min the arriving bees showed aggressive behavior, even attacking the vials containing the extracts, and did not land (Nieh et al. 2003a). Only after this deterrent effect had ceased did the bees show a preference for the baited feeders when compared to unbaited control feeders. Although these findings stand in clear contrast to the observed synchronization of the foragers' scent marking behavior and the newcomers' landing at the feeders—and despite the fact that bees never attack food sources naturally scent-marked by foragers prior to landing—Nieh et al. (2003a, p. 2194) concluded that recruiting *T. hyalinata* foragers "evidently use mandibular gland secretions to create the short odour trail to guide nestmates." As argued by Jarau et al. (2004a), however, it is likely that the repellent effect of the extracts prepared by Nieh et al. (2003a) was caused by mandibular gland compounds, whereas the attraction of bees after 8 min can probably be attributed to compounds from the bees' labial glands that had been unknowingly extracted by the authors due to the crude method of mandibular gland dissection. In fact, neat preparation and extraction methods are imperative for any study on the role of gland secretions. Indeed, when mandibular and labial glands were carefully separated, Jarau et al. (2004a) found in experiments with *T. recursa* that extracts from the former gland were purely repellent, whereas extracts from the latter were clearly attractive to bees arriving at artificial feeders. In choice experiments with two feeders, one baited with gland extract and the other with pure solvent (control), ca. 84% of the bees landed on the feeder bearing labial gland components, whereas in the case of the mandibular gland treatment, 73% of the bees chose the control (Jarau et al. 2004a). Importantly, the attraction of recruited bees to labial gland secretions was not limited to the food source alone. When artificial scent trails were created using labial gland secretions, large numbers of newly recruited bees deviated from their natural scent trails in order to follow them, whereas mandibular gland extracts and pure solvent pentane (control experiments) had no effect (Jarau et al. 2006). Therefore, the trail pheromone of *T. recursa* clearly is produced in its labial glands. Recently, it was unequivocally shown, by similar artificial scent trail bioassays, that the trail pheromones

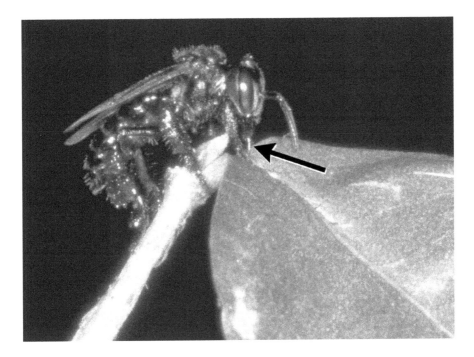

FIGURE 12.6 A forager of *Trigona recursa* lands on a leaf in order to deposit a pheromone mark from her labial glands on her way back from a food source to the nest. The arrow points to the extended glossa, at the base of which the glands' excretory duct opens to release their secretions. (Photograph by Stefan Jarau.)

of *T. corvina* (Dambacher 2006; Dambacher et al. 2007), *T. spinipes* (Schorkopf et al. 2007), and *Scaptotrigona pectoralis* (Hemmeter 2008) are also produced in the foragers' labial glands. In all three species none of the recruited bees followed artificial scent trails made of mandibular gland extracts, whereas labial gland extracts induced trail-following behavior in a significantly larger proportion of bees than the solvent control (*T. corvina*, 51% vs. 0.5% [labial vs. solvent]; *T. spinipes*, 90% vs. 10% [different experimental setup]; *S. pectoralis*, 35% vs. 0.5%). Indirect evidence for the importance of labial gland secretions for foraging workers also comes from a histological study published by Cruz-Landim and Puga (1967), who found that the labial glands of *S. postica* develop with age and are most developed and productive within the oldest bees (i.e., the foragers).

The labial gland secretions are released at the base of a bee's glossa (Snodgrass 1956; Cruz-Landim 1967). To deposit a pheromone mark from these glands, foragers of *T. recursa* extend their mouthparts before landing on the substrate and then rub their glossa against the surface as they run across it (Figure 12.6; Jarau et al. 2004a). Nieh et al. (2003a, 2004b) reported that foragers of *T. hyalinata* and *T. spinipes* also frequently extended their proboscises when depositing a scent mark. Interestingly, Simpson and Riedel (1964) found that honey bee workers are able to discharge their labial gland secretions even when their tongues are almost completely retracted. Assuming that stingless bees probably have a similar method of discharge, foragers may be able to apply their labial gland secretions to the substrate by rubbing the dorsal surface of the almost completely retracted proboscis, which is held between the mandibles. As already suggested by Jarau et al. (2004a), this behavior could have misleadingly been interpreted as "rubbing the mandibles on the substrate" in earlier studies.

CHEMICAL STRUCTURES OF TRAIL PHEROMONE COMPOUNDS

The first bioassays with a synthetic compound, which aimed to verify its function as a trail pheromone in a stingless bee, were reported by Blum et al. (1970). These authors tested the reaction of *Geotrigona subterranea* workers toward small wooden blocks treated with citral, which was identified as the main compound of their mandibular glands (constituting ca. 95% of the detectable volatiles). When the citral source was placed on a feeding table, all the arriving bees landed on the wooden block, grasped it lightly, and only after releasing it crawled to the syrup and fed. When citral-impregnated blocks were placed on the ground halfway between the nest and a feeder placed only 80 cm away from it, the bees interrupted their flight in order to alight on them (Blum et al. 1970). The authors' conclusion that citral must be regarded as the primary trail pheromone employed by *G. subterranea* workers because it acts as a powerful attractant is, however, questionable. Real trail-following behavior was not demonstrated by the experiments, and grasping the food source by the arriving bees prior to feeding does not occur at resources naturally scent-marked by foragers. In addition, Blum et al. (1970) also found a clear alarm reaction of the workers when a high concentration of citral was placed at the entrance to a nest. It seems likely that the observed behavior of the bees toward the less concentrated citral sources placed 40 cm away from the nest or at the feeding table was a somehow weaker alarm reaction. Blum (1970) and Blum and Brand (1972) cite unpublished experiments in which workers of *Scaptotrigona tubiba* and *S. postica* were strongly attracted to benzaldehyde alone or in combination with 2-tridecanone and 2-pentadecanone, respectively, and conclude that these chemicals function as critical volatile elements of the trail pheromones. However, they do not give details on how and where the tests were conducted, and do not describe whether benzaldehyde and the ketones induced trail-following behavior rather than mere attraction. Later, Kerr et al. (1981) trained approximately forty workers of *T. spinipes* to a sugar solution feeder at a distance of 36 m from their nest. When the authors reinstalled the feeder on the following day and simultaneously provided an artificial scent trail made of drops of synthetic 2-heptanol, the main compound identified from head extracts of *T. spinipes*, six bees arrived and landed on the feeder within 20 min. Kerr et al. (1981) deduced from this result that 2-heptanol guided these bees to the food source, and thus may be the major constituent of the trail pheromone of *T. spinipes*. However, as already argued by Jarau et al. (2004a), it is much more likely that the six workers visiting the feeder on the second day were not newly recruited bees that were attracted by the 2-heptanol trail, but inspectors (Biesmeijer and de Vries 2001) that spontaneously checked the feeding site they had visited the day before. In fact, the ability of foragers to remember the location of a food source and the time of day when it is available, along with the demonstration that the bees make use of their memory to revisit known feeding sites, was reported for honey bees 80 years ago (Beling 1929) and, more recently, has also been shown for several species of stingless bees (*Melipona quadrifasciata* and *Partamona cupira*, Freitas and Raw 1996; *Trigona amalthea*, Breed et al. 2002; *M. scutellaris, M. seminigra, M. compressipes*, Schorkopf et al. 2004).

Recently, compounds from the trail pheromones of three *Trigona* species were identified by means of both (1) chemical analyses of extracts prepared from foragers' labial glands, the source of trail pheromones in stingless bees (see above), and (2) bioassays verifying the function of single synthetic compounds or compound mixtures as releasers of actual trail-following behavior in newly recruited bees. The main compound of labial gland extracts prepared from *T. recursa* foragers is hexyl decanoate, which makes up about 72% of its constituents (Jarau et al. 2006). Artificial scent trails that branched away from *T. recursa*'s natural scent trails and which were baited with this ester diverted a significant number of recruits from their original flight direction, whereas control trails made of pure solvent did not. However, hexyl decanoate alone was less effective at releasing trail-following behavior than the natural labial gland extract, indicating that the entire trail pheromone of *T. recursa* is a blend containing additional constituents that enhance the activity of the signal that is dominated by this specific ester (Jarau et al. 2006). Octyl decanoate, hexyl decanoate, isopropyl octadecenoate (double bond position unknown), and two unidentified compounds were the next most abundant volatiles in the labial glands of *T. recursa*, followed by a variety of minor constituents (Jarau et al. 2006). It is not yet known which of these compounds act synergistically with hexyl decanoate to release full trail-following behavior of this species. Dambacher (2006) and Dambacher et al. (2007) analyzed labial gland extracts prepared from *T. corvina* and also carried out gaschromatographic analyses coupled to electroantennographic detection (GC-EAD) with these extracts and worker antennae. Of the twelve compounds that released a detectable response in the chemoreceptors on the bees' antennae (indicating that the bees can perceive them), seven could be identified (one terpene ester and six carboxylic acid alkyl esters). The main component in the labial gland extracts was octyl octanoate, which did not release trail-following behavior in recruited *T. corvina* workers along artificial scent trails when it was applied alone (Dambacher 2006). A synthetic blend of five of the identified compounds (octyl hexanoate, octyl octanoate, geranyl octanoate, decyl octanoate, and octyl decanoate; making together ca. 83% of the detected volatiles), which were mixed in approximately natural relative proportions, did, however, successfully lead the bees along experimental scent trails (Dambacher 2006; Dambacher et al. 2007). Thus, the trail pheromone of *T. corvina* is a blend of esters, with its main component alone being inactive. Interestingly, *T. spinipes* differs from *T. recursa* and *T. corvina* in that a single compound, octyl octanoate, was just as effective at inducing trail-following behavior in food-searching bees as the natural labial gland extract (Schorkopf et al. 2007). In experiments that used these odors to create artificial trails branching off the bees' natural path, about 90% of the recruits followed them rather than a trail that was simultaneously provided and marked with the pure solvent pentane. Not surprisingly, the main compound of labial gland extracts prepared from *T. spinipes* foragers was identified as octyl octanoate (making ca. 74% of the volatile constituents). This compound was also detected on substrates previously scent-marked by bees (Figure 12.7). Furthermore, mandibular gland extracts, which completely failed to induce trail-following behavior in recruited bees, contained no trace of octyl octanoate (Schorkopf et al. 2007).

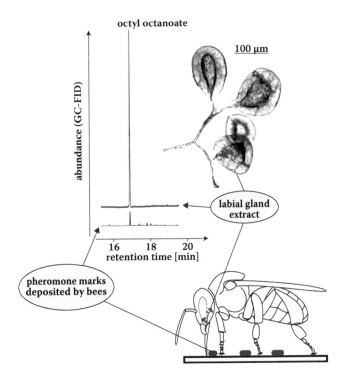

FIGURE 12.7 The trail pheromone compound of *Trigona spinipes*, octyl octanoate, is deposited on the substrate by recruiting foragers. Chemical analyses by means of gas chromatography (GC-FID, signals shown) and gas chromatography coupled to mass spectrometry (GC-MS, not shown) revealed that octyl octanoate is produced in the bees' labial glands and constitutes about 74% of the total amount of their secretion's volatile components. (After Schorkopf et al. 2007. With permission.)

TABLE 12.1
Trail Pheromone Compounds Identified from Labial Gland Secretions of Stingless Bees[a]

Species	Compound	Molecular Formula	Molecular Weight	Estimated Boiling Point[b]	Reference for Trail Bioassays
Trigona corvina	Octyl hexanoate	$C_{14}H_{28}O_2$	228	280°C	Dambacher (2006)
	Octyl octanoate	$C_{16}H_{32}O_2$	256	320°C	
	Geranyl octanoate	$C_{18}H_{32}O_2$	280	360°C	
	Decyl octanoate	$C_{18}H_{36}O_2$	284	360°C	
	Octyl decanoate	$C_{18}H_{36}O_2$	284	360°C	
Trigona recursa	Hexyl decanoate	$C_{16}H_{32}O_2$	256	320°C	Jarau et al. (2006)
Trigona spinipes	Octyl octanoate	$C_{16}H_{32}O_2$	256	320°C	Schorkopf et al. (2007)

[a] Only components are considered that have been identified by means of both chemical analyses and behavioral bioassays with synthetic compounds. For a critique on the remaining literature reports of proposed trail pheromone compounds, see the text.

[b] Boiling points are roughly estimated by the rule of thumb "number of C atoms × 20°C."

To summarize, all trail pheromone compounds so far identified from gland extracts of stingless bee foragers that have also proved to be effective in triggering trail-following behavior belong to the chemical class of carboxylic acid alkyl esters, with the exception of one terpene ester (Table 12.1). The compounds have similarly low molecular weights (MW 228 to 284), estimated boiling points of between 280°C and 360°C, and show medium to low polarity. These properties are likely to make the volatility of these esters well suited for attracting bees over long distances. For example, one of the most frequently occurring compounds in moth sex pheromones is (Z)-9-tetradecenyl acetate (Schulz 2001), which belongs to the same chemical class of compounds and has a similar molecular weight (MW 254). Its ideal physicochemical behavior as a volatile signal is well known.

The similarity of the properties of the single compounds of trail pheromone blends may be important to maintain species or nest specific compositions (including relative proportions) following their application to the substrate by foragers. The same compounds, as well as other esters, were also found in labial gland extracts of *Scaptotrigona pectoralis* (Hemmeter 2008) and in head extracts of several other species of stingless bees (Kerr et al. 1981; Francke et al. 1983, 2000; Johnson et al. 1985; Engels et al. 1987), some of which communicate the location of food sources to their nestmates by means of scent trails. The function of single compounds or mixtures of the detected esters has not yet been investigated by means of bioassays in these species, but from the results obtained from *Trigona* species it may be predicted that at least some of them may be used for scent trail communication too. This assumption, however, awaits examination in future studies.

NEST SPECIFICITY OF TRAIL PHEROMONES

Despite the potential importance of nest specific trail pheromones for competitor avoidance and resource partitioning, only a few studies have addressed the question of whether trail pheromones in stingless bees are primarily, or even exclusively, followed by the nestmates of a trail-laying forager. Lindauer and Kerr (1958, 1960) described an experiment in which foragers from two different nests of *Scaptotrigona postica* were placed beside each other and trained to feed at separate sugar solution feeders in such a way that the paths of their respective scent trails crossed. After recruitment took place for 1 h they recorded the colony identity of the bees at the feeders and found that workers from both nests collected sugar solution at the two food sources. However, because all bees were allowed to return to their nest after food uptake throughout the entire experiment, it cannot be

TABLE 12.2
Nest Specific Effectiveness of Trail Pheromones in Two Species of Stingless Bees

	Percentage of Recruits That Followed Artificial Trails Made of:		
	Own Nest's Labial Gland Extract	Foreign Nest's Labial Gland Extract	Solvent
Scaptotrigona pectoralis	34.7 ± 21.5^a (472)	12.4 ± 5.9^b (766)	0.6 ± 1.4^c (238)
Trigona corvina	51.1 ± 24.5^a (1,338)	6.8 ± 7.3^b (625)	0.4 ± 1.1^b (265)

Note: The percentage of the recruited bees that followed artificially baited scent trails that branched off from the bees' natural pheromone trails is given as mean $\pm$ 1 SD (100% = total number of recruits that followed the natural plus the artificial trails). Different letters denote statistically significant differences in the proportion of bees of a given species misled by the artificial trails (U-tests and Bonferroni corrections; overall $P < 0.05$). Values in parentheses give the numbers of the tested bee individuals. (Data for *S. pectoralis* are taken from Hemmeter 2008; data for *T. corvina* from Dambacher 2006.)

absolutely excluded that foragers from both nests recruited to either feeder. Recently, the nest specific effectiveness of labial gland extracts was clearly demonstrated in *Trigona corvina* (Dambacher 2006; Dambacher et al. 2007) and *S. pectoralis* (Hemmeter 2008). In both species, labial gland extracts from nestmates were more attractive to recruits than those made from foragers from a foreign nest (Table 12.2). This effect was more evident in *T. corvina*, where the proportion of bees diverted by the foreign labial gland extracts did not differ statistically from experiments using pure solvent. In both species foragers taken from different nests could clearly be separated by means of principal component- and discriminant function analyses based on the composition of the compounds contained in their labial glands (Dambacher 2006; Dambacher et al. 2007, Hemmeter 2008). This demonstrates the nest specific relative compositions of the glands' secretions and explains their nest specific effectiveness.

The observation that *S. pectoralis* recruits are more likely to follow conspecific scent trails made by foragers of a foreign nest than is the case in *T. corvina* likely can be explained by differences in the foraging strategies and aggressiveness of these two species. *T. corvina* is an aggressive group foraging species, and encounters with foragers from other nests always lead to costly combats between the foragers of the involved colonies (Johnson 1983; Slaa 2003). Thus, the avoidance of scent trails laid by *T. corvina* foragers of foreign nests, instead of following them in order to reach the food source already discovered by them, is advantageous, because it diminishes intraspecific competition leading to the death of many of a colony's foragers. As a consequence, natural selection likely has favored *T. corvina* colonies whose foragers did not follow foreign trails. *S. pectoralis* workers, on the other hand, forage peacefully at resources alongside other bees without displaying any aggression (Johnson 1983; Slaa 2003). Therefore, following the trail of a foreign nest is not as disastrous in this species as it would be in *T. corvina* (for similar findings in ants, see Chapter 8).

CHEMICALS INVOLVED IN NEST PLUNDERING

The species belonging to the genera *Lestrimelitta* in South and Central America and *Cleptotrigona* in Africa have evolved a highly specialized mode of food acquisition. They are exclusively cleptobiotic. Instead of visiting flowers they raid the nests of other stingless bee species, or even honey bee colonies, in order to steal food along with other resources, such as building materials and stored resins, which are then transported back to their own nests (Müller 1874; Portugal-Araújo 1958; Sakagami and Laroca 1963; Bego et al. 1991; Sakagami et al. 1993; Nogueira-Neto 1997; Quezada-Euán and González-Acereto 2002). The workers of these robber bees have shiny and sparsely haired bodies, and the structures for pollen collection (corbicula, penicillum, rastellum) are reduced, which presumably is related to the robbing behavior and the lack of flower-visiting habits (Michener 2000). Apparently, robber bees prefer to raid the nests of certain host species. Quezada-Euán and González-Acereto (2002) observed that *Lestrimelitta niitkib* on the Yucatan Peninsula in Mexico mainly raided the nests of *Frieseomelitta nigra* and *Nannotrigona perilampoides*. In Brazil and Panama *L. limao* preferentially attacked nests of several species of *Scaptotrigona*, *Nannotrigona*, and *Plebeia* (Sakagami and Laroca 1963; Roubik 1989; Bego et al. 1991; Sakagami et al. 1993). Less frequently, *Lestrimelitta* also invades conspecific nests, colonies of *Melipona*, *Paratrigona*, and *Tetragonisca*, or hives of *Apis mellifera* (Sakagami and Laroca 1963; Sakagami et al. 1993; Quezada-Euán and González-Acereto 2002). In Angola, Portugal-Araújo (1958) observed raids of *Cleptotrigona cubiceps* mainly on *Hypotrigona braunsi* colonies.

Typical raids, which usually do not entirely destroy the victim's nests, involve several dozen or hundreds of robber bees that invade a host colony within minutes, then take over and guard the nest entrance while simultaneously removing building materials, larval provisions from the younger brood combs, and food from the storage pots. Within the nest, most host workers and the queen retreat to the older brood combs or to the nest's periphery where they remain immobile, while host foragers that return from the field do not enter their nest. Researchers who have witnessed the raids of *Lestrimelitta* and *Cleptotrigona* have all noticed a strong, lemon-like odor emanating from the entrances of attacked nests, which are occupied by the robber bees—especially during the early phase of a raid (Portugal-Araújo 1958; Sakagami and Laroca 1963; Wille 1983; Bego et al. 1991; Sakagami et al. 1993; Nogueira-Neto 1997; Quezada-Euán and González-Acereto 2002). This strong odor emanates from the bees' mandibular glands, which are filled with secretions, particularly in workers more than 15 days old (Cruz-Landim and Camargo 1970). The two main components of *Lestrimelitta limao*'s mandibular gland secretion were identified as the two stereoisomers of citral, i.e., geranial and neral, in a ratio of ca. 2:1 (Blum 1966). In head extracts of the same species, several additional components belonging to the chemical classes of alcohols, acetates, esters, carboxylic acids, ketones, and hydrocarbons (probably extracted from the cuticle) were also identified (Wittmann et al. 1990; Francke et al. 2000). However, due to the method of extracting whole worker heads, the exact glandular origin of these compounds remains unknown. Citral may be used by the first robber bee foragers arriving at the entrance of a host nest both as a pheromone to attract further nestmates and, in high concentrations, as an allomone that irritates the host workers and disrupts their defensive organization (Blum et al. 1970; Michener 1974; Roubik 1989).

In bioassays with citral-impregnated wooden blocks that were introduced into the nest entrances of different bee colonies, Blum et al. (1970) found that workers of *Paratrigona subnuda*, *Plebeia droryana*, and *Nannotrigona testaceicornis* became highly agitated. Furthermore, the organized colonial activities of the nest ceased, and many bees left the hive and did not return for about 20 min. Likewise, in *Scaptotrigona pectoralis*, *S. mexicana*, and *Trigona fulviventris*, living or freshly killed *L. limao* workers, extracts prepared from their heads, or pure citral presented in front of the nest mainly released strong excitement in the bees at the entrance and hindered returning foragers to enter the colony (Weaver et al. 1975). The reaction of the test colonies to citral in these experiments indeed resembles the observed behavior of host workers during natural raids by *Lestrimelitta* (Sakagami and Laroca 1963; Sakagami et al. 1993). By contrast, workers of *Frieseomelitta varia* started to cover a citral-impregnated wooden block with resin 3 min after its introduction to the nest entrance, and *S. postica* guards removed the citral source from their nest (Blum et al. 1970). Wittmann (1985) found that the guards of *Tetragonisca angustula*, which constantly hover in front of the entrance of their nest, immediately attack live *L. limao* workers brought into their frontal visual field, and hypothesized that the unique defense behavior of this species evolved under the pressure of *Lestrimelitta* attacks. The robber bees are apparently recognized and distinguished from returning *T. angustula* foragers by their specific odor bouquet, with citral and 6-methyl-5-hepten-2-one acting as key volatiles that trigger the defense reaction (Wittmann et al. 1990). Additional minor compounds identified from head extracts of *L. limao* had a synergistic effect when added to the citral and 6-methyl-5-hepten-2-one mixture, but did not release attacks in *Tetragonisca* guards when presented alone (Wittmann et al. 1990). Markedly different reactions to citral were also found in two *Melipona* species. In *M. quadrifasciata*, colonial activities totally broke down, and many workers left the nest and did not return as long as citral was present near or in the nest entrance, whereas *M. rufiventris* workers were apparently indifferent to citral (Blum et al. 1970; Pompeu and Silveira 2005). Interestingly, a colony of *M. rufiventris* was also observed to successfully repel a real attack by *L. limao* (Pompeu and Silveira 2005). The species-specific differences in the reactions to citral may contribute to the known host preferences of the cleptobiotic bees (Blum et al. 1970, Michener 1974), but raids against host nests may also depend on their conditions, with weak colonies being more susceptible to attacks (Pompeu and Silveira 2005). In relation to both of these ideas, an interesting but as yet unanswered question is how robber bee foragers select the colonies they are about to attack. It could be that *Lestrimelitta* and *Cleptotrigona* workers have an innate preference for specific host species, but they may also be attracted to host odors they already know from previous raids, or even learn to avoid odors from species where they had experienced severe resistance by guards. There is good reason to believe that the volatiles emanating from the host nests are important for their localization by robber bees and probably also for their selection. Future research, monitoring which species are raided by single *Lestrimelitta* or *Cleptotrigona* colonies over longer periods of time, could help to answer this question.

The chemical compounds that account for the disruption of the social organization within raided nests remain to be identified. Citral seems to be a good candidate (Blum et al. 1970; Michener 1974), but its typical lemon-like odor was absent in most host colonies housed in observation boxes that were opened during robber bee attacks by Portugal-Araújo (1958) and Sakagami et al. (1993). Therefore, other volatiles that are released during raids should also be considered instead of citral. Furthermore, Sakagami et al. (1993) found indications that citral, which they released at the entrance of *Scaptotrigona* nests that were raided by *L. limao* workers, triggered the retreat of the robber bees and the mass return of host workers. Thus, citral probably signifies a "robber retreat message" to end a raid, rather than serving host disruption.

Aside from the obligate robber bees, several species belonging to the genera *Melipona*, *Scaptotrigona*, *Trigona*, *Tetragona*, *Tetragonisca*, and particularly *Oxytrigona* occasionally steal food from nests of other stingless bees and from colonies of *Apis mellifera* (von Ihering 1904; Schwarz 1948; Roubik 1983, 1989; Bian et al. 1984; Roubik et al. 1987; Rinderer et al. 1988; Nogueira-Neto 1997; Souza et al. 2007). *Oxytrigona tataira* is even regarded as a pest species by beekeepers in certain areas in northeastern Brazil (Souza et al. 2007). The mandibular glands of *Oxytrigona* workers are extensively developed (Kerr and Cruz 1961), and their caustic secretions are primarily used as a powerful defense against vertebrate predators due to the skin burns they produce when chewed into the cutis (Kerr and Cruz 1961; Wille 1961, 1983; Michener 1974). Head extracts of *O. mellicolor* and *O. daemonica* (Bian et al. 1984; Roubik et al. 1987) as well as mandibular gland extracts of *O. mediorufa* (Cruz-López et al. 2007) are unique in relation to other stingless bees due to the large numbers and high quantities of saturated and unsaturated mono- and diketones (2-heptanone, heptan-2,5-dione, 3-hepten-2-one, 5-hepten-2-one, *E*- and *Z*-3-hepten-2,5-dione, nonan-2,5-dione, 5-nonen-2-one, *E*- and *Z*-3-nonen-2,5-dione) they contain, in addition to several other compounds (hydrocarbons, carboxylic acids, carboxylic acid alkyl esters). Rinderer et al. (1988) conducted bioassays with several synthetic compounds, which were applied to rubber septa and introduced into the observation hives of honey bees, and found that *E*-3-hepten-2,5-dione and *E*-3-nonen-2,5-dione had a clear, but localized repellent effect on the bees on the comb. Furthermore, the honey bees' defensive behavior was significantly reduced when the diketones were presented in the nest compared to the presentation of clean rubber septa or septa treated with 2-heptanone, 2-decanone, 3-hepten-2-one, pentadecane, or with various acetate esters (all compounds previously identified from *Oxytrigona* head extracts; Bian et al. 1984). The effect of *E*-3-hepten-2,5-dione and *E*-3-nonen-2,5-dione on honey bee workers thus is similar to the observed absence of organized resistance against *Oxytrigona* foragers that invade *Apis* colonies in order to rob their resources (Bian et al. 1984). The latter observation is interesting, because a ketone, 6-methyl-5-hepten-2-one, was also identified from *Lestrimelitta limao* head extracts (Wittmann et al. 1990; Francke et al. 2000), where it may serve the same function, i.e., as a repellent or appeasement allomone released inside host nests during raids (see above).

CONCLUDING REMARKS

The studies reviewed on the preceding pages clearly show that chemicals derived from food sources or released by foragers are of the utmost importance for the localization and exploitation of food, as well as for the recruitment of stingless bees. Orientation toward food odors by foraging animals is surely an ancestral characteristic that can be found throughout the entire animal kingdom. Beyond this, in eusocial insects the odor of a food source, which is passed on from one worker of a colony to others, and which then biases the food-searching behavior of these individuals in the field, also functions as social information. However, we are only just beginning to understand the importance of food odors in this context, which is also true of odor cues left by visitors at food sources that influence a forager's decision whether to visit them. Exciting new studies with stingless bees and bumble bees have shown that the message conveyed by these volatiles is not fixed but depends on an individual bee's past experience. Since foraging efficiency is increased by the use of these chemical cues, similar mechanisms are also likely to be at work in other insects—social and solitary species alike. However, this remains to be studied.

Among all social insects whose workers forage on the wing, some species of stingless bees are unique by depositing pheromone spots on solid substrates at and around food sources, as well as along the route back to their nest. These pheromone marks are true signals involved in recruitment communication and strongly enhance the attraction of the recruits to a particular food source, increasing the efficiency of its exploitation. However, the use of pheromones to create continuous scent trails has to be questioned, as recent studies provide accumulating evidence for short odor trails that are basically deposited in the immediate vicinity of the food. Instead, the pheromone marks may function as a "strong and extended food odor," which is easily encountered by recruits that search for it after leaving the nest. It can also contain information about the identity (at both the species and colony level) of other foragers that exploit a resource, which in turn could influence a forager's decision whether to land and feed on it. Only recently we have identified the actual glands that produce these pheromones and studied the chemical structure of some of the compounds. This, finally, allows us to address more specific questions. For example: How do recruits orient toward the deposited pheromones? Or how does the specific pheromone composition of foragers from different nests contribute to competitor avoidance? We already have a clearer idea of the roles played by volatile chemicals in many aspects of the foraging ecology of stingless bees, but there can be no doubt that many intriguing details await discovery in years to come.

ACKNOWLEDGMENTS

I thank Michael Hrncir, Robert Hodgkison, and Wittko Francke for their valuable comments that helped to improve this chapter. I am grateful to Dirk Louis P. Schorkopf for letting me know the details of his recent, not yet published, investigations on the importance of scent trails in *Scaptotrigona postica*.

REFERENCES

Aguilar I. (2004). Communication and recruitment for the collection of food in stingless bees: A behavioral approach. Ph.D. thesis, Utrecht University, The Netherlands.
Aguilar I, Fonseca A, Biesmeijer JC. (2005). Recruitment and communication of food source location in three species of stingless bees (Hymenoptera, Apidae, Meliponini). *Apidologie* 36:313–24.

Aguilar I, Sommeijer MJ. (1996). Communication in stingless bees: Are the anal substances deposited by *Melipona favosa* scent marks? *Proc Exper Appl Entomol* 7:57–63.

Aguilar I, Sommeijer M. (2001). The deposition of anal excretions by *Melipona favosa* foragers (Apidae: Meliponinae): Behavioural observations concerning the location of food sources. *Apidologie* 32:37–48.

Alcock J. (2001). *Animal Behavior. An Evolutionary Approach.* 7th ed. Sunderland: Sinauer Associates.

Arnhart L. (1923). Das Krallenglied der Honigbiene. *Arch Bienenk* 5:37–86.

Barth FG, Hrncir M, Jarau S. (2008). Signals and cues in the recruitment behavior of stingless bees (Meliponini). *J Comp Physiol A* 194:313–27.

Bego LR, Zucchi R, Mateus S. (1991). Notas sobre a estratégia alimentar (cleptobiose) de *Lestrimelitta limao* Smith (Hymenoptera, Apidae, Meliponinae). *Naturalia* 16:119–27.

Beling I. (1929). Über das Zeitgedächtnis der Bienen. *Z vergl Physiol* 9:295–338.

Bian Z, Fales HM, Blum MS, Jones TH, Rinderer TE, Howard DF. (1984). Chemistry of cephalic secretion of fire bee *Trigona (Oxytrigona) tataira*. *J Chem Ecol* 10:451–61.

Biesmeijer JC, de Vries H. (2001). Exploration and exploitation of food sources by social insect colonies: A revision of the scout-recruit concept. *Behav Ecol Sociobiol* 49:89–99.

Blum MS. (1966). Chemical releasers of social behavior. VIII. Citral in the mandibular gland secretion of *Lestrimelitta limao* (Hymenoptera: Apoidea: Melittidae [sic]). *Ann Ent Soc Am* 59:962–64.

Blum MS. (1970). The chemical basis of insect sociality. In Beroza M (ed.), *Chemicals Controlling Insect Behavior*. New York: Academic Press, pp. 61–94.

Blum MS. (1974a). Pheromonal bases of social manifestations in insects. In Birch MC (ed.), *Pheromones*. Amsterdam: North-Holland Publishing Company, pp. 190–99.

Blum MS. (1974b). Pheromonal sociality in the Hymenoptera. In Birch MC (ed.), *Pheromones*. Amsterdam: North-Holland Publishing Company, pp. 222–49.

Blum MS, Brand JM. (1972). Social insect pheromones: Their chemistry and function. *Am Zool* 12:553–76.

Blum MS, Crewe RM, Kerr WE, Keith LH, Garrison AW, Walker MM. (1970). Citral in stingless bees: Isolation and functions in trail laying and robbing. *J Insect Physiol* 16:1637–48.

Boogert NJ, Hofstede FE, Aguilar Monge I. (2006). The use of food source scent marks by the stingless bee *Trigona corvina* (Hymenoptera: Apidae): The importance of the depositor's identity. *Apidologie* 37:366–75.

Breed MD, Stocker EM, Baumgartner LK, Vargas SA. (2002). Time-place learning and the ecology of recruitment in a stingless bee, *Trigona amalthea* (Hymenoptera, Apidae). *Apidologie* 33:251–58.

Contrera FAL, Nieh JC. (2007). Effect of forager-deposited odors on the intra-patch accuracy of recruitment of the stingless bees *Melipona panamica* and *Partamona peckolti* (Apidae, Meliponini). *Apidologie* 38:584–94.

Costa MA, Del Lama MA, Melo GAR, Sheppard WS. (2003). Molecular phylogeny of the stingless bees (Apidae, Apinae, Meliponini) inferred from mitochondrial 16s rDNA sequences. *Apidologie* 34:73–84.

Cruz-Landim C da. (1967). Estudo comparativo de algumas glândulas das abelhas (Hymenoptera, Apoidea) e respectivas implicações evolutivas. *Arqu Zool Estado São Paulo* 15:177–290.

Cruz-Landim C da, Camargo IJB de. (1970). Light and electron microscope studies of the mandibular gland of *Lestrimelitta limao* (Hym., Meliponidae). *Rev Bras Biol* 30:5–12.

Cruz-Landim C da, Ferreira A. (1968). Mandibular gland development and communication in field bees of *Trigona (Scaptotrigona) postica* (Hymenoptera: Apidae). *J Kansas Entomol Soc* 41:474–81.

Cruz-Landim C da, Puga FR. (1967). Presença de substâncias lipídicas nas glândulas do sistema salivar de *Trigona* (Hym., Apoidea). *Pap Avul Zool* 20:65–74.

Cruz-Landim C da, Moraes RLMS de, Salles HC, Reginato RD. (1998). Note on glands present in meliponinae (Hymenoptera, Apidae) bees legs. *Rev Bras Zool* 15:159–65.

Cruz-López L, Aguilar S, Malo EA, Rincón M, Guzman M, Rojas JC. (2007). Electroantennogram and behavioral responses of workers of the stingless bee *Oxytrigona mediorufa* to mandibular gland volatiles. *Entomol Exp Appl* 123:43–47.

Dahl F. (1885). Die Fussdrüsen der Insekten. *Arch Mikroskop Anat* 25:236–62.

Dambacher J. (2006). Nest specificity in the trail pheromone of a stingless bee, *Trigona corvina* (Apidae, Meliponini). Diploma thesis, University of Ulm, Germany.

Dambacher J, Jarau S, Twele R, Aguilar I, Francke W, Ayasse M. (2007). Nest specific information in the trail pheromone of a stingless bee, *Trigona corvina* (Hymenoptera, Apidae, Meliponini). In *Proceedings of the 23rd ISCE Ann Meeting*, Jena, Germany, p. 248.

Eltz T. (2006). Tracing pollinator footprints on natural flowers. *J Chem Ecol* 32:907–15.

Engels E, Engels W, Schröder W, Francke W. (1987). Intranidal worker reactions to volatile compounds identified from cephalic secretions in the stingless bee, *Scaptotrigona postica* (Hymenoptera, Meliponinae). *J Chem Ecol* 13:371–86.

Esch H, Esch I, Kerr WE. (1965). Sound: An element common to communication of stingless bees and to dances of the honey bee. *Science* 149:320–21.

Federle W, Brainerd EL, McMahon TA, Hölldobler B. (2001). Biomechanics of the movable pretarsal adhesive organ in ants and bees. *Proc Natl Acad Sci USA* 98:6215–20.

Francke W, Lübke G, Schröder W, Reckziegel A, Imperatriz-Fonseca V, Kleinert A, Engels E, Hartfelder K, Radtke R, Engels W. (2000). Identification of oxygen containing volatiles in cephalic secretions of workers of Brazilian stingless bees. *J Braz Chem Soc* 11:562–71.

Francke W, Schröder W, Engels E, Engels W. (1983). Variation in cephalic volatile substances in relation to worker age and behavior in the stingless bee, *Scaptotrigona postica*. *Z Naturforsch* 38c:1066–68.

Free JB. (1987). *Pheromones of Social Bees*. Ithaca, NY: Comstock Publishing Associates of Cornell University Press.

Freitas RIP de, Raw A. (1996). The perception of time by foraging workers of stingless bees (Hymenoptera: Apidae, Meliponini). In *Proceedings of the 6th IBRA Conference on Apiculture in Tropical Climates*, San José, Costa Rica, pp. 101–4.

von Frisch K. (1965). *Tanzsprache und Orientierung der Bienen* [The Dance Language and Orientation of Bees]. Berlin: Springer Verlag.

Gawleta N, Zimmermann Y, Eltz T. (2005). Repellent foraging scent recognition across bee families. *Apidologie* 36:325–30.

Gilbert F, Azmeh S, Barnard C, Behnke J, Collins SA, Hurst J, Shuker D, and the Behavioural Ecology Field Course. (2001). Individually recognizable scent marks on flowers made by a solitary bee. *Anim Behav* 61:217–29.

Goulson D, Chapman JW, Hughes WOH. (2001). Discrimination of unrewarding flowers by bees; direct detection of rewards and use of repellent scent marks. *J Insect Behav* 14:669–78.

Haynes KF, Birch MC. (1985). The role of other pheromones, allomones and kairomones in the behavioral responses on insects. In Kerkut GA, Gilbert LI (eds.), *Comprehensive Insect Physiology, Biochemistry, and Pharmacology*. Vol. 9. *Behaviour*. Oxford: Pergamon Press, pp. 225–55.

Hebling NJ, Kerr WE, Kerr FS. (1964). Divisão de trabalho entre operárias de *Trigona* (*Scaptotrigona*) *xanthotricha* Moure. *Pap Avuls Dep Zool* 16:115–27.

Hemmeter K. (2008). Wegpheromone und Futterrekrutierung bei der stachellosen Biene *Scaptotrigona pectoralis* (Apidae, Meliponini). Diploma thesis, University of Ulm, Germany.

Howse PE, Stevens IDR, Jones OT. (1998). *Insect Pheromones and Their Use in Pest Management*. London: Chapman & Hall.

Hrncir M, Barth FG, Tautz J. (2006). Vibratory and airborne-sound signals in bee communication (Hymenoptera). In Drosopoulos S, Claridge MF (eds.), *Insect Sounds and Communication: Physiology, Behaviour, and Evolution*. Boca Raton, FL: CRC Press, Taylor & Francis Group, pp. 421–36.

Hrncir M, Jarau S, Zucchi R, Barth FG. (2000). Recruitment behavior in stingless bees, *Melipona scutellaris* and *M. quadrifasciata*. II. Possible mechanisms of communication. *Apidologie* 31:93–113.

Hrncir M, Jarau S, Zucchi R, Barth FG. (2004). On the origin and properties of scent marks deposited at the food source by a stingless bee, *Melipona seminigra*. *Apidologie* 35:3–13.

von Ihering H. (1904). Biologie der stachellosen Honigbienen Brasiliens. *Zool Jahrb Abt Syst Geogr Biol Tiere* 19:179–287, tables 10–22.

Jarau S, Hrncir M, Ayasse M, Schulz C, Francke W, Zucchi R, Barth FG. (2004b). A stingless bee (*Melipona seminigra*) marks food sources with a pheromone from its claw retractor tendons. *J Chem Ecol* 30:793–804.

Jarau S, Hrncir M, Schmidt VM, Zucchi R, Barth FG. (2003). Effectiveness of recruitment behavior in stingless bees (Apidae, Meliponini). *Insect Soc* 50:365–74.

Jarau S, Hrncir M, Zucchi R, Barth FG. (2004a). A stingless bee uses labial gland secretions for scent trail communication (*Trigona recursa* Smith 1863). *J Comp Physiol A* 190:233–39.

Jarau S, Hrncir M, Zucchi R, Barth FG. (2005). Morphology and structure of the tarsal glands of the stingless bee *Melipona seminigra*. *Naturwissenschaften* 92:147–50.

Jarau S, Schulz CM, Hrncir M, Francke W, Zucchi R, Barth FG, Ayasse M. (2006). Hexyl decanoate, the first trail pheromone compound identified in a stingless bee, *Trigona recursa*. *J Chem Ecol* 32:1555–64.

Johnson LK. (1983). Foraging strategies and the structure of stingless bee communication in Costa Rica. In Jaisson P (ed.), *Social Insects in the Tropics*. Paris: Université Paris-Nord, pp. 31–58.

Johnson LK. (1987). Communication of food source location by the stingless bee *Trigona fulviventris*. In Eder J, Rembold H (eds.), *Chemistry and Biology of Social Insects*. München: Verlag J. Peperny, pp. 698–99.

Johnson LK, Haynes LW, Carlson MA, Fortnum HA, Gorgas DL. (1985). Alarm substances of the stingless bee, *Trigona silvestriana*. *J Chem Ecol* 11:409–16.

Kerr WE. (1960). Evolution of communication in bees and its role in speciation. *Evolution* 14:386–87.

Kerr WE. (1969). Some aspects of the evolution of social bees (Apidae). *Evol Biol* 3:119–75.

Kerr WE. (1973). Sun compass orientation in the stingless bee *Trigona (Trigona) spinipes* (Fabricius, 1793) (Apidae). *An Acad Bras Ciên* 45:301–8.

Kerr WE. (1994). Communication among *Melipona* workers (Hymenoptera: Apidae). *J Insect Behav* 7:123–28.

Kerr WE, Blum M, Fales HM. (1981). Communication of food source between workers of *Trigona (Trigona) spinipes*. *Rev Brasil Biol* 41:619–23.

Kerr WE, Cruz C da C. (1961). Funções diferentes tomadas pela glândula mandibular na evolução das abelhas em geral e em "*Trigona (Oxytrigona) tataira*" em especial. *Rev Bras Biol* 21:1–16.

Kerr WE, Ferreira A, Mattos NS de. (1963). Communication among stingless bees—Additional data (Hymenoptera: Apidae). *J NY Entomol Soc* 71:80–90.

Kerr WE, Rocha R. (1988). Communicação em *Melipona rufiventris* e *Melipona compressipes*. *Ciên Cult* 40:1200–2.

Lensky Y, Cassier P, Finkel A, Delorme-Joulie C, Levinsohn M. (1985). The fine structure of the tarsal glands of the honeybee *Apis mellifera* L. (Hymenoptera). *Cell Tissue Res* 240:153–58.

Lindauer M. (1956). Über die Verständigung bei indischen Bienen. *Z vergl Physiol* 38:521–57.

Lindauer M, Kerr WE. (1958). Die gegenseitige Verständigung bei den stachellosen Bienen. *Z vergl Physiol* 41:405–34.

Lindauer M, Kerr WE. (1960). Communication between the workers of stingless bees. *Bee World* 41:29–41, 65–71.

McCabe SI, Hartfelder K, Santana WC, Farina W. (2007). Odor discrimination in classical conditioning of proboscis extension in two stingless bee species in comparison to Africanized honeybees. *J Comp Physiol A* 193:1089–99.

Menzel R. (1999). Memory dynamics in the honeybee. *J Comp Physiol A* 185:323–340.

Michener CD. (1974). *The Social Behavior of the Bees: A Comparative Study*. Cambridge, MA: Harvard University Press.

Michener CD. (2000). *The Bees of the World*. Baltimore: Johns Hopkins University Press.

Morgan ED. (1990). Insect trail pheromones: A perspective of progress. In McCaffery AR, Wilson ID (eds.), *Chromatography and Isolation of Insect Hormones and Pheromones*. New York: Plenum Press, pp. 259–70.

Müller F. (1874). The habits of various insects. *Nature* 10:102–3.

Nieh JC. (1998). The role of a scent beacon in the communication of food location by the stingless bee, *Melipona panamica*. *Behav Ecol Sociobiol* 43:47–58.

Nieh JC. (2004). Recruitment communication in stingless bees (Hymenoptera, Apidae, Meliponini). *Apidologie* 35:159–82.

Nieh JC, Barreto LS, Contrera FAL, Imperatriz-Fonseca VL. (2004a). Olfactory eavesdropping by a competitively foraging stingless bee, *Trigona spinipes*. *Proc R Soc Lond B* 271:1633–40.

Nieh JC, Contrera FAL, Nogueira-Neto P. (2003a). Pulsed mass recruitment by a stingless bee, *Trigona hyalinata*. *Proc R Soc Lond B* 270:2191–96.

Nieh JC, Contrera FAL, Yoon RR, Barreto LS, Imperatriz-Fonseca VL. (2004b). Polarized short odor-trail recruitment communication by a stingless bee, *Trigona spinipes*. *Behav Ecol Sociobiol* 56:435–48.

Nieh JC, Ramírez S, Nogueira-Neto P. (2003b). Multi-source odor-marking of food by a stingless bee, *Melipona mandacaia*. *Behav Ecol Sociobiol* 54:578–86.

Nieh JC, Tautz J, Spaethe J, Bartareau T. (1999/2000). The communication of food location by a primitive stingless bee, *Trigona carbonaria*. *Zoology* 102:238–46.

Nogueira-Neto P. (1997). *Vida e Criação de Abelhas Indígenas Sem Ferrão*. São Paulo: Editora Nogueirapis.

Noll FB. (1997). Foraging behavior on carcasses in the necrophagic bee *Trigona hypogea* (Hymenoptera: Apidae). *J Insect Behav* 10:463–67.

Pompeu MS, Silveira FA. (2005). Reaction of *Melipona rufiventris* Lepeletier to citral and against an attack by the cleptobiotic bee *Lestrimelitta limao* (Smith) (Hymenoptera: Apidae: Meliponina). *Braz J Biol* 65:189–91.

Portugal-Araújo V de. (1958). A contribution to the bionomics of *Lestrimelitta cubiceps* (Hymenoptera, Apidae). *J Kans Entomol Soc* 31:203–11.

Pouvreau A. (1991). Morphology and histology of tarsal glands in bumble bees of the genera *Bombus*, *Pyrobombus*, and *Megabombus*. *Can J Zool* 69:866–72.

Quezada-Euán JJG, González-Acereto JA. (2002). Notes on the nest habits and host range of cleptobiotic *Lestrimelitta niitkib* (Ayala 1999) (Hymenoptera: Meliponini) from the Yucatan Peninsula, Mexico. *Acta Zool Mex* 86:245–49.

Rinderer TE, Blum MS, Fales HM, Bian Z, Jones TH, Buco SM, Lancaster VA, Danka RG, Howard DF. (1988). Nest plundering allomones of the fire bee *Trigona (Oxytrigona) mellicolor*. *J Chem Ecol* 14:495–501.

Rodrigues AV, Roselino AC, Hrncir M. (2008). Stingless bees (*Melipona scutellaris*) learn to associate their "footprints" with rewarding food sources. In *Anais VIII° Encontro sobre Abelhas*, Ribeirão Preto, SP, Brazil, p. 546.

Roselino AC, Hrncir M, Zucchi R. (2007). Stingless bees (*Melipona scutellaris*) learn to associate their "footprints" with depleted food sources. In *Proceedings of the Annual Meeting of the French Section of the IUSSI*, Toulouse, France, p. 75.

Roubik DW. (1983). Nest and colony characteristics of stingless bees from Panamá (Hymenoptera: Apidae). *J Kansas Entomol Soc* 56:327–55.

Roubik DW. (1989). *Ecology and Natural History of Tropical Bees*. Cambridge, MA: Cambridge University Press.

Roubik DW, Smith BH, Carlson RG. (1987). Formic acid in caustic cephalic secretions of stingless bee, *Oxytrigona* (Hymenoptera: Apidae). *J Chem Ecol* 13:1079–86.

Sakagami SF, Laroca S. (1963). Additional observations on the habits of the cleptobiotic stingless bees, the genus *Lestrimelitta* Friese (Hymenoptera, Apoidea). *J Fac Sci Hokkaido Univ* 15:319–39.

Sakagami SF, Roubik DW, Zucchi R. (1993). Ethology of the robber stingless bee, *Lestrimelitta limao* (Hymenoptera: Apidae). *Sociobiology* 21:237–77.

Saleh N, Chittka L. (2006). The importance of experience in the interpretation of conspecific chemical signals. *Behav Ecol Sociobiol* 61:215–20.

Saleh N, Ohashi K, Thomson JD, Chittka L. (2006). Facultative use of the repellent scent mark in foraging bumblebees: Complex versus simple flowers. *Anim Behav* 71:847–54.

Saleh N, Scott AG, Bryning GP, Chittka L. (2007). Distinguishing signals and cues: Bumblebees use general footprints to generate adaptive behaviour at flowers and nest. *Arthropod-Plant Interact* 1:119–27.

Sánchez D, Kraus FB, Jesús Hernández M de, Vandame R. (2007). Experience, but not distance, influences the recruitment precision in the stingless bee *Scaptotrigona mexicana*. *Naturwissenschaften* 94:567–73.

Sánchez D, Nieh JC, Hénaut Y, Cruz L, Vandame R. (2004). High precision during food recruitment of experienced (reactivated) foragers in the stingless bee *Scaptotrigona mexicana* (Apidae, Meliponini). *Naturwissenschaften* 91:346–49.

Sánchez D, Nieh JC, Vandame R. (2008). Experience-based interpretation of visual and chemical information at food sources in the stingless bee *Scaptotrigona mexicana*. *Anim Behav* 76:407–14.

Schmidt VM, Schorkopf DLP, Hrncir M, Zucchi R, Barth FG. (2006). Collective foraging in a stingless bee: Dependence on food profitability and sequence of discovery. *Anim Behav* 72:1309–17.

Schmidt VM, Zucchi R, Barth FG. (2003). A stingless bee marks the feeding site in addition to the scent path (*Scaptotrigona* aff. *depilis*). *Apidologie* 34:237–48.

Schmidt VM, Zucchi R, Barth FG. (2005). Scent marks left by *Nannotrigona testaceicornis* at the feeding site: Cues rather than signals. *Apidologie* 36:285–91.

Schmitt U, Lübke G, Francke W. (1991). Tarsal secretion marks food sources in bumblebees (Hymenoptera: Apidae). *Chemoecology* 2:35–40.

Schorkopf DLP, Jarau S, Francke W, Twele R, Zucchi R, Hrncir M, Schmidt VM, Ayasse M, Barth FG. (2007). Spitting out information: *Trigona* bees deposit saliva to signal resource locations. *Proc R Soc B* 274: 895–98.

Schorkopf DLP, Zucchi R, Barth FG. (2004). Memory of feeding time. In *Proceedings of 8th International Conference on Tropical Bees*. In *Anais VI° Encontro sobre Abelhas*, Ribeirão Preto, SP, Brazil, p. 643.

Schulz S. (2001). Selectivity in chemical communication systems of arthropods. In Barth FG, Schmid A (eds.), *Ecology of Sensing*. Berlin: Springer Verlag, pp. 237–52.

Schwarz HF. (1948). Stingless bees (Meliponidae) of the western hemisphere. *Bull Am Mus Nat Hist* 90:i–xvii, 1–546.

Seeley TD. (1989). Social foraging in honey bees: How nectar foragers assess their colony's nutritional status. *Behav Ecol Sociobiol* 24:181–99.

Shorey HH. (1973). Behavioral responses to insect pheromones. *Annu Rev Entomol* 18:349–80.

Simpson J, Riedel IBM. (1964). Discharge and manipulation of labial gland secretion by workers of *Apis mellifera* (L.) (Hymenoptera: Apidae). *Proc R Entomol Soc Lond A* 39:76–82.

Slaa EJ. (2003). Foraging ecology of stingless bees: From individual behaviour to community ecology. Doctoral thesis, Utrecht University, The Netherlands.

Slaa EJ, Cevaal A, Sommeijer MJ. (1998). Floral constancy in *Trigona* stingless bees foraging on artificial flower patches: A comparative study. *J Apicult Res* 37:191–98.

Slaa EJ, Tack AJM, Sommeijer MJ. (2003). The effect of intrinsic and extrinsic factors on flower constancy in stingless bees. *Apidologie* 34:457–68.

Snodgrass RE. (1956). *Anatomy of the Honey Bee.* Ithaca, NY: Comstock Publishing Associates.

Souza BA, Alves RMO, Carvalho CAL. (2007). Diagnóstico da arquitetura de ninho de *Oxytrigona tataira* (Smith, 1863) (Hymenoptera: Meliponinae). *Biota Neotrop* 7:83–86.

Traniello JFA, Robson SK. (1995). Trail and territorial communication in social insects. In Cardé RT, Bell WJ (eds.), *Chemical Ecology of Insects 2*. New York: Chapman & Hall, pp. 241–86.

Villa JD, Weiss MR. (1990). Observations on the use of visual and olfactory cues by *Trigona* spp. foragers. *Apidologie* 21:541–45.

Weaver N, Weaver EC, Clarke ET. (1975). Reactions of five species of stingless bees to some volatile chemicals and to other species of bees. *J Insect Physiol* 21:479–94.

Wille A. (1961). Las abejas jicotes de Costa Rica. *Rev Univ Costa Rica* 22:1–30.

Wille A. (1983). Biology of the stingless bees. *Annu Rev Entomol* 28:41–64.

Wilms J, Eltz T. (2008). Foraging scent marks of bumblebees: Footprint cues rather than signals. *Naturwissenschaften* 95:149–53.

Wilson EO. (1971). *The Insect Societies*. Cambridge, MA: Belknap Press of Harvard University Press.

Wilson EO. (1990). Success and dominance in ecosystems: The case of the social insects. In Kinne O (ed.), *Excellence in Ecology*. Book 2. Oldendorf/Luhe: Ecology Institute.

Witjes S, Eltz T. (2007). Influence of scent deposits on flower choice: Experiments in an artificial flower array with bumblebees. *Apidologie* 38:12–18.

Wittmann D. (1985). Aerial defense of the nest by workers of the stingless bee *Trigona (Tetragonisca) angustula* (Latreille) (Hymenoptera: Apidae). *Behav Ecol Sociobiol* 16:111–14.

Wittmann D, Radtke R, Zeil J, Lübke G, Francke W. (1990). Robber bees (*Lestrimelitta limao*) and their host: Chemical and visual cues in nest defense by *Trigona (Tetragonisca) angustula* (Apidae: Meliponinae). *J Chem Ecol* 16:631–41.

Wyatt TD. (2003). *Pheromones and Animal Behaviour. Communication by Smell and Taste*. Cambridge, MA: Cambridge University Press.

Yokoi T, Fujisaki K. (2007). Repellent scent-marking behaviour of the sweat bee *Halictus (Seladonia) aerarius* during flower foraging. *Apidologie* 38:474–81.

Yokoi T, Goulson D, Fujisaki K. (2007). The use of heterospecific scent marks by the sweat bee *Halictus aerarius*. *Naturwissenschaften* 94:1021–24.

13 The Use of Scent Marks by Foraging Bumble Bees

Dave Goulson

CONTENTS

INTRODUCTION

It has long been observed that bumble bees (*Bombus* spp.) can distinguish between more and less rewarding flowers of the same plant species without actually sampling the reward available. Typically, the bee hovers briefly next to a flower with its antennae extended and nearly touching the corolla, and then either proceeds to land and attempt to feed, or instead rejects the flower without landing and moves on (Figure 13.1). In some circumstances the bee may be directly assessing the reward level, or perhaps examining correlates of reward, such as flower size and symmetry (Brink and deWet 1980; Stanton and Preston 1988; Cresswell and Galen 1991; Møller 1995; Møller and Eriksson 1995). However, there is now strong evidence that perhaps the most important cues used by bees to decide whether to probe or reject a flower are chemical clues left by bees on previous visits (Cameron 1981; Marden 1984; Kato 1988; Schmitt and Bertsch 1990; Goulson et al. 1998; Stout et al. 1998).

Bumble bees, honey bees, sweat bees (Halictidae), carpenter bees (*Xylocopa* sp.), stingless bees (*Trigona* sp.), and flower bees (*Anthophora* sp.) have all been found to leave repellent marks on flowers that they visit, and conspecifics are able to use these to discriminate between visited and unvisited flowers (Núñez 1967; Frankie and Vinson 1977; Cameron 1981; Wetherwax 1986; Gilbert et al. 2001; Giurfa and Núñez 1992; Giurfa 1993; Giurfa et al. 1994; Goulson et al. 1998, 2001; Stout et al. 1998; Williams 1998; Gawleta et al. 2005; Yokoi and Fujisaki 2007; Yokoi et al. 2007). This behavior presumably increases foraging efficiency by reducing the time spent landing on and handling flowers that have recently been emptied by another bee, and thus contain little or no reward (Kato 1988; Schmitt and Bertsch 1990; Goulson et al. 1998).

SCENT MARK CHEMISTRY

In bumble bees the scent marks appear to consist of a chemical cue found on the tarsi, and probably are produced at least in part by the tarsal glands (Schmitt et al. 1991). The tarsi of queen, worker, and male bumble bees all contain a substantial secretory gland described in detail by Pouvreau (1991). However, these compounds also occur in considerable quantities elsewhere on the body,

FIGURE 13.1 *Bombus hortorum* inspecting a flower of *Echium vulgare*. Note the extended antennae, which almost touch the flower corolla. See color insert following page 142. (Photograph by Ben Darvill.)

and may be the result of a blend of secretions from several different glands (Goulson et al. 2000). For *Bombus terrestris* the components of both tarsal glands and the deposited scent marks have been identified and are similar (Schmitt 1990; Schmitt et al. 1991). Tarsal glands produce primarily straight-chain alkanes and alkenes of between twenty-one and twenty-nine carbon atoms, with compounds with odd numbers of carbons predominating. The alkenes are thought to be mostly (Z)-9 and (Z)-11 configurations (Schmitt 1990; Schmitt et al. 1991). These compounds are common cuticular hydrocarbons found in a broad range of insects, and they probably have the primary function of reducing water loss (Lockey 1980; Blum 1981, 1987).

Although broadly similar compounds are found on the cuticles of diverse insects, there are notable differences between species in the precise composition, even among congeners. Comparisons of the hydrocarbons on the tarsi of three sympatric European *Bombus* species, *B. lapidarius, B. pascuorum*, and *B. terrestris*, have revealed clear differences (Goulson et al. 2000). For example, while tricosane was found in significant quantities in all three species, tricosene was only found in abundance in *B. lapidarius*. Pentacosenes were major constituents of the extracts of *B. lapidarius* and *B. pascuorum*, but were virtually absent in *B. terrestris* (Table 13.1). Differences between species have previously been described in the composition of labial gland secretions of male bumble bees (where they presumably help in mate recognition) (Bergström et al. 1981) and also in Dufour's gland secretions of bumble bees (Tengö et al. 1991). Oldham et al. (1994) analyzed cuticular hydrocarbons in *B. lapidarius, B. pascuorum*, and *B. terrestris*, and also compared *B. terrestris terrestris* from mainland Europe with the UK race, *B. terrestris audax*. Although they did not examine tarsal glands, they concluded that the mix of cuticular hydrocarbons was constant across different body parts, but that the species and the two *B. terrestris* subspecies differed in the relative quantities of different compounds. The composition of tarsal extracts described by Goulson et al. (2000) closely follows that for cuticular hydrocarbons found over the rest of the body (Oldham et al. 1994). It seems

TABLE 13.1
Amounts of Each Compound Present in Tarsal Extracts (ng/tarsus ± SE)
of Three Bumble Bee Species, Based on Four Replicate Samples per Species

Compound	MW	*B. terrestris*	*B. pascuorum*	*B. lapidarius*
Heneicosane	296	12.5 ± 2.41	–	+
Tricosenes	322	9.38 ± 6.63	5.90 ± 0.95	70.5 ± 15.1
Tricosane	324	110 ± 13.4	99.3 ± 1.21	94.8 ± 8.63
Methyl-tricosane	324	–	–	+
Tetracosenes	336	–	12.5 ± 6.03	+
Tetracosane	338	–	+	+
Pentacosenes	350	+	174 ± 12.3	155 ± 11.5
Pentacosane	352	114 ± 17.9	106 ± 5.98	170 ± 6.06
Heptacosenes	378	–	64.5 ± 13.2	+
Heptacosane	380	174.5 ± 28.6	35.5 ± 5.60	+
Nonacosenes	406	102.9 ± 26.1	+	+
Total		514 ± 68.7	491 ± 30.5	490 ± 32.4

Source: After Goulson et al. 2000. With permission.

Note: MW, molecular weight; +, trace; –, not detected. Samples were prepared by cutting the tarsi and approximately half of the tibia from five individuals of one species and combining them in 0.5 ml of pentane. The samples were analyzed with a VG-Analytical 70-250SE mass spectrometer coupled to a Hewlett Packard 5790 gas chromatograph. The column was a BP1 of dimension 25 m × 0.33 mm with a film thickness of 0.25 μm, and the carrier gas was helium. Temperature programming was as follows: 60°C for 3 min; heating 20°C/min; 300°C for 10 min; 280°C for 12 min. Nonadecane was used as an internal standard to quantify the amounts of compounds present.

likely that, during grooming, these compounds are distributed across the surface of the body and limbs. During foraging, many parts of the bumble bee body may come into contact with the corolla, depending upon the shape of the flower, not just the tarsi. Thus, it seems probable that scent marks are not exclusively placed by the feet.

In bumble bees, despite differences in the composition of scent mark deposited by different species, it appears that marks deposited by other species are readily detected. For example, interspecific tests between *B. terrestris*, *B. hortorum*, *B. pascuorum*, and *B. pratorum* reveal that each is repelled by scent marks deposited by the other species (Goulson et al. 1998; Stout et al. 1998). Also, tarsal extracts obtained by putting *B. terrestris* legs in organic solvents, when artificially applied to flowers, mimic the repellency of natural scent marks, and induce repellency in a range of *Bombus* species (Stout et al. 1998; Goulson et al. 2000). Even applications of a range of pure synthetic chemical constituents of scent marks (rather than the mixtures that naturally occur) produce more or less the same repellent response. It seems that *Bombus* species exhibit a generalized response to flowers that are contaminated with any of the common hydrocarbons found on the cuticles of bumble bees, be they conspecifics or heterospecifics. This makes sense, for many flower species are commonly visited by a range of *Bombus* species with overlapping resource use (Goulson et al. 2005). The advantage to be gained from detecting empty flowers would be small if only those flowers visited by conspecifics could be detected. Since these compounds are common to most insects, not just *Bombus* sp., it seems likely that bumble bees may be able to detect and reject flowers that have been visited by other insects. Recent studies suggest that *Bombus* species are able to detect scent marks deposited by honey bees (Figure 13.2) and by the solitary bee *Anthidium manicatum*, and vice versa

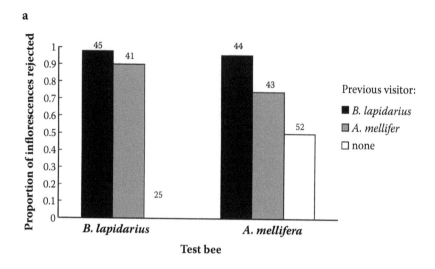

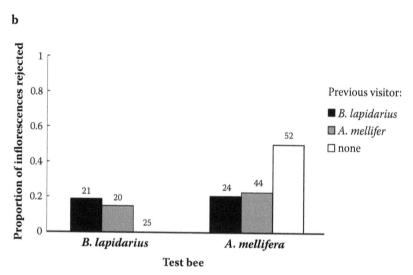

FIGURE 13.2 The proportion of flowers rejected by *Bombus lapidarius* and *Apis mellifera* workers. (a) Less than 3 min after the first visitor; (b) 24 h after the first visitor. The frequency of rejection of flowers that had no previous visitors is also shown. Numbers above the bars represent sample sizes. Responses are recorded as acceptance if bees landed and probed for nectar, or as rejection if the bees approached the flower, but departed without landing or feeding. (After Stout and Goulson 2001. With permission.)

(Stout and Goulson 2001; Gawleta et al. 2005; but see Williams 1998 for conflicting evidence). Bumble bees also seem able to avoid flowers previously visited by hover flies (Diptera, Syrphidae) (Reader et al. 2005). The generality of these observations requires further investigation.

HOW BUMBLE BEES MAKE USE OF SCENT MARKS

The repellent effect of scent marks wanes over time. Thus, when visiting *Symphytum officinale*, foraging *B. terrestris* rejected nearly all flowers that had been visited in the previous 3 min, but by 40 min the rejection response had disappeared (Figure 13.3a) (Stout et al. 1998). This broadly

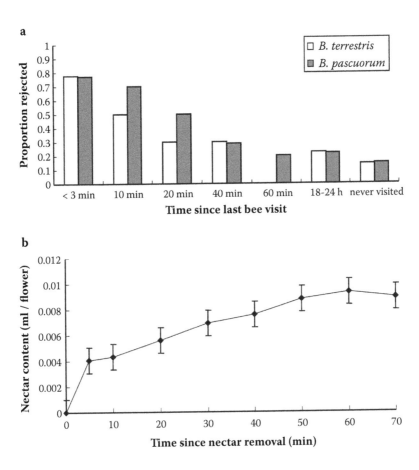

FIGURE 13.3 (a) The proportion of flowers of *Symphytum officinale* rejected by foraging *Bombus terrestris* and *B. pascuorum*, according to the time that had elapsed since the previous visit to the flower. (b) The rate of accumulation of nectar in flowers of *S. officinale* after they have been emptied. (After Stout et al. 1998. With permission.)

matches the rate of accumulation of nectar in *Symphytum officinale*; 40–60 min after being emptied, flowers have refilled (Figure 13.3b) (Stout et al. 1998). However, different flower species vary greatly in the rate at which they secrete nectar, so a fixed repellent response of 40 min duration would not be appropriate for all flower species. For flowers that replenished nectar more rapidly than *S. officinale*, this would result in bees rejecting many flowers that were full of nectar, and conversely, if the secretion rate were slower, many of the acceptable flowers would contain little nectar. Also, if visitation rates are high or flowers are scarce, we would predict that bees should be less choosy (i.e., have a lower threshold for acceptance of a flower), and hence be more likely to accept flowers that were visited quite recently.

It seems probable that bees learn to use an appropriate concentration of scent mark as the threshold for rejection depending on the circumstances (Stout et al. 1998; Stout and Goulson 2002). Given that most individual bees are flower constant (i.e., they tend to visit the same flower species over and over again), they have the opportunity to learn an appropriate threshold concentration of scent mark for their preferred flower species. It is known that bumble bees do sample available floral rewards and modify their behavior accordingly (Dukas and Real 1993a, 1993b). If bumble bees can gain information on the time that has elapsed since a scent mark was deposited from its strength (as suggested by Schmitt et al. 1991 and Stout et al. 1998), then it would be possible for them to learn what concentration of scent corresponds to an appropriate threshold for acceptance of a flower. There is evidence that this does indeed occur. Williams (1998) found that repellency was of very short duration (about 2 min) when *Bombus* species were foraging on *Borago officinalis*, which has an unusually high rate of nectar secretion. Conversely, Stout and Goulson (2002) found that repellent scent marks deposited by *B. lapidarius* on *Lotus corniculatus* flowers lasted for 24 h; *L. corniculatus* has a low nectar secretion rate and was extremely abundant at the study site, so that bees could afford to select only the most rewarding flowers.

Until very recently it was assumed that the mechanism by which rejection responses waned over time was evaporation of the scent mark, or at least evaporation of the more volatile components (Goulson 2003). A mark that evaporated at a predictable rate would enable a bee to estimate the time since the mark was deposited from its strength. However, recent work by Eltz (2006) draws this assumption into question. He demonstrates that scent marks left behind by foraging bumble bees do not wane to any appreciable extent within 2 h of deposition, and that a flower may contain detectable traces of scent marks left behind by the sequence of insects that have visited it over a number of days. In fact, the main hydrocarbons involved in scent marks are large and have low volatility, so this result is not surprising. However, this finding raises questions as to how bees are able to judge the time since a scent mark was deposited over timescales of less than 1 h, which they are clearly able to do. The most likely explanation at present is that the hydrocarbons are slowly adsorbed into the hydrophobic surface of the corolla, so that although still detectable by solvent extraction and gas chromatography, they are not available for detection by the bee in the headspace volatiles surrounding the flower. Although plausible, this explanation remains to be tested. It is also not consistent with the findings of Saleh et al. (2006) that scent marks on complex artificial plastic flowers remain repellent to bumble bees for longer than marks on simple artificial flowers. Presumably in this circumstance the scent marks cannot readily be adsorbed into the plastic surface; if they are not evaporating either, how can bees judge their age? This is a fascinating aspect of scent marking that deserves further investigation.

There is a notable anomaly in studies of scent marking in bumble bees that requires an explanation. The first studies to describe deposition of scent marks on flowers by foraging bumble bees found that the marks were attractant rather than repellent, and were used to mark rewarding flowers (Schmitt and Bertsch 1990; Schmitt et al. 1991). Similar results have been found in honey bees (Ferguson and Free 1979) and stingless bees (see Chapter 12). All subsequent studies of scent marking in bumble bees have only found repellent effects, whether using natural marks, tarsal extracts, or synthetic compounds (Goulson et al. 1998; Stout et al. 1998; Williams 1998). It has previously been suggested that fresh scent marks might be initially repellent, but that as they evaporate (or are adsorbed into the flower) they may become attractants (Stout et al. 1998). However, attempts to test this hypothesis suggest that this is not so. When applying dilution series of tarsal extracts to flowers, even when the lowest concentrations contained less than one molecule per flower, there was no evidence for an attractant response at any concentration (Goulson et al. 2000).

An alternative possibility is that the more volatile or easily adsorbed components produce repellency, and the less volatile ones produce attraction. However, when a range of pure synthetic compounds present in natural extracts were bioassayed, all induced repellency (Goulson et al. 2000). Thus, this seems unlikely. It is possible that the changing composition of a scent mark over time, as some compounds are adsorbed more quickly than others, could result in attractive marks. However, bumble bees tend to reject flowers of *S. officinale* for about 40 min following a visit, but flowers visited 1, 4, or 24 h previously have acceptance rates equal to flowers that have never been visited (Stout et al. 1998). At no point were flowers that had previously been visited found to be more attractive than controls. Generally, unvisited (and unmarked) flowers receive very high rates of acceptance, so there was little scope for a scent mark to increase attractiveness of flowers (Goulson et al. 1998, 2000; Stout et al. 1998). Overall, it seems unlikely that attractant marks are in operation when bumble bees forage at natural flowers.

Examination of the experimental design used by Schmitt et al. (1991) suggests another explanation (Goulson 2003). Their study used artificial flowers that were either always rewarding (regardless of whether they had been visited) or never rewarding. In this (unnatural) circumstance bees would inevitably spend longer feeding on the rewarding flowers, so that rewarding flowers would become liberally covered in cuticular hydrocarbons. Given that bees are readily able to learn associations between sensory cues and rewards (reviewed in Menzel and Müller 1996), it is likely that they may have learned to preferentially visit the marked flowers, since these were the rewarding ones. Witjes and Eltz (2007) set out to test whether bumble bees would behave differently when presented with artificial flowers with realistic, small, and slowly replenishing rewards. In this situation *Bombus terrestris* workers were repelled by scent marks, behaving much as they do on real flowers. Similarly, Saleh and Chittka (2006) found that bumble bees were attracted to scent marks left on artificial flowers when the flowers continuously contained a reward, but were repelled by the scent marks when the flowers were completely emptied by only one visit. Hence it is clear that bumble bees are able to adjust their response to scent marks according to experience and the context in which they are presented.

It is not easy to imagine how short-range attractant marks could operate with real flowers that never provide unlimited rewards. It is possible that attractant marks may be used to indicate plants with unusually high nectar secretion rates; studies to date have not explicitly examined whether differences in reward rates between patches influence how bees interpret scent marks.

THE BIOLOGICAL SIGNIFICANCE OF SCENT MARKS

The value of using scent marks is presumably that it allows the bee to save time that it would otherwise spend in handling flowers that contain little or no reward. The time savings to be gained thus depends upon the handling time of the flower in question. When visiting complex flowers one might predict that bees should be more sensitive to scent marks, whereas when visiting simple flowers where the reward can be swiftly accessed and assessed directly, using scent marks would be of little advantage. Saleh et al. (2006) tested this hypothesis in a laboratory situation using bumble bees foraging on artificial flowers that were either simple or complex. As predicted, bees were more than twice as likely to reject scent-marked flowers that were complex compared to simple flowers.

Since bumble bees do not forage randomly, they rarely encounter inflorescences that they themselves have just visited, so that the evolutionary benefit gained by leaving scent marks is not immediately apparent. Presumably they help in avoiding errors in systematic foraging (Goulson 1999, 2000). In social bees such as bumble bees the depositors of scent marks may also benefit through improving the foraging efficiency of siblings. However, bumble bee colonies are rather small (compared to honey bees), so that the majority of beneficiaries of marks left by bumble bees are often probably not siblings. Genetic studies of patterns of relatedness between foragers have demonstrated that the worker bumble bees foraging on any one patch of flowers contain representatives of numerous nests, and that rather few of them are sisters (Darvill et al. 2004; Knight et al. 2005). Competition between bee species is known to occur in some communities (Inouye 1978; Pyke 1982), and thus scent marking may benefit both siblings and probable competitors. It seems probable that the action of scent marking did not initially evolve as a benefit to the marker or her siblings. Indeed, there is no evidence that repellent scent marks are deliberately deposited. They are comprised of alkanes and alkenes, which commonly occur on the cuticles of diverse insect species (Lockey 1980; Blum 1981, 1987), and which are bound to be left behind in tiny amounts if any part of the body comes into contact with flower parts. Indeed, Saleh et al. (2007) recently demonstrated that the same compounds are left behind on any substrate bumble bees walk on, including, for example, the entrance to their nest. If surfaces that bees have walked over are placed by artificial flowers, the flowers become repellent, strongly indicating that scent marking is passive, i.e., scent marks should be regarded as cues that are accidentally deposited rather than as signals (Wilms and Eltz 2008). Most insects are able to detect and recognize conspecifics by scent, and this ability is highly advanced in social insects that commonly have complex chemical recognition and communication systems. It seems likely that the ability to detect the scent of other insects evolved long before flowers or indeed bees existed. Thus, the only step required for a system of repellent scent marking to evolve is for foragers to learn to associate the scent of another insect on a flower with a low reward. Thus, scent marking should not be regarded as a form of communication, since the information-carrying cue is likely to be accidental.

There is some evidence that scent marks may not just be a means of avoiding landing on empty flowers. *Bombus terrestris* exhibit a stronger avoidance response to flowers visited by the aggressive, territorial solitary wool carder bee *Anthidium manicatum* than they do to flowers visited by conspecifics (Gawleta et al. 2005). Similarly, in *Anthophora plumipes* bees appear to respond to scent marks as territorial markers as well as indicators of low reward, with dominant females responding differently to those of low rank (Gilbert et al. 2001).

CONCLUDING REMARKS

It has only relatively recently become apparent that the use of scent marks by bees when choosing which flowers they are going to visit is not confined to honey bees. As yet we do not know how widespread this phenomenon is. As previously mentioned, scent marking has recently been described in the Neotropical stingless bee *Trigona fulviventris* (Goulson et al. 2001), in the solitary bees *A. manicatum* (Gawleta et al. 2005) and *A. plumipes* (Gilbert et al. 2001), and in the sweat bee *Halictus aerarius* (Yokoi and Fujisaki 2007; Yokoi et al. 2007). Clearly this behavior is not confined to social species. Are all bees able to use scent marks, and are scent marks used in all circumstances? Do naïve bumble bees respond to scent marks, or is it something that they learn from experience? Are they used by flower-visiting insects other than Hymenoptera, such as hoverflies or butterflies? Since the compounds used are widespread, it is likely that interspecific interactions could occur between distantly related taxa, but this has not yet been investigated.

REFERENCES

Bergström G, Svensson BG, Appelgren M, Groth I. (1981). Complexity of bumblebee marking pheromones: Biochemical, ecological and systematical interpretations. In Howse PE, Clément JL (eds.), *Biosystematics of Social Insects*. London: Academic Press, pp. 175–83.
Blum MS. (1981). *Chemical Defenses of Arthropods*. New York: Academic Press.
Blum MS. (1987). Specificity of pheromonal signals: A search for its recognitive bases in terms of a unified chemisociality. In Eder J, Rembold H (eds.), *Chemistry and Biology of Social Insects*. München: Verlag J. Peperny, pp. 401–5.
Brink D, deWet JMJ. (1980). Interpopulation variation in nectar production in *Aconitum columbianum* (Ranunculaceae). *Oecologia* 47:160–63.
Cameron SA. (1981). Chemical signals in bumble bee foraging. *Behav Ecol Sociobiol* 9:257–60.
Cresswell JE, Galen C. (1991). Frequency-dependent selection and adaptive surfaces for floral character combinations—The pollination of *Polemonium viscosum*. *Am Nat* 138:1342–53.
Darvill B, Knight ME, Goulson D. (2004). Use of genetic markers to quantify bumblebee foraging range and nest density. *Oikos* 107:471–78.
Dukas R, Real AL. (1993a). Learning constraints and floral choice behaviour in bumble bees. *Anim Behav* 46:637–44.
Dukas R, Real AL. (1993b). Effects of recent experience on foraging decisions in bumble bees. *Oecologia* 94:244–46.
Eltz T. (2006). Tracing pollinator footprints on natural flowers. *J Chem Ecol* 32:907–15.
Ferguson AW, Free JB. (1979). Production of a forage-marking pheromone by the honeybee. *J Apic Res* 18:128–35.
Frankie GW, Vinson SB. (1977). Scent-marking of passion flowers in Texas by females of *Xylocopa virginica texana* (Hymenoptera: Anthophoridae). *J Kansas Entomol Soc* 50:613–25.

Gawleta N, Zimmermann Y, Eltz T. (2005). Repellent foraging scent recognition across bee families. *Apidologie* 36:325–30.

Gilbert F, Azmeh S, Barnard C, Behnke J, Collins SA, Hurst J, Shuker D, and the Behavioural Ecology Field Course. (2001). Individually recognizable scent marks on flowers made by a solitary bee. *Anim Behav* 61: 217–29.

Giurfa M. (1993). The repellent scent-mark of the honeybee *Apis mellifera ligustica* and its role as communication cue during foraging. *Insect Soc* 40:59–67.

Giurfa M, Núñez JA. (1992). Honeybees mark with scent and reject recently visited flowers. *Oecologia* 89: 113–17.

Giurfa M, Núñez JA, Backhaus W. (1994). Odour and colour information in the foraging choice behavior of the honeybee. *J Comp Physiol A* 175:773–79.

Goulson D. (1999). Foraging strategies of insects for gathering nectar and pollen, and implications for plant ecology and evolution. *Persp Plant Ecol Evol Syst* 2:185–209.

Goulson D. (2000). Why do pollinators visit proportionally fewer flowers in large patches? *Oikos* 91:485–92.

Goulson D. (2003). *Bumblebees. Their Behaviour and Ecology.* Oxford: Oxford University Press.

Goulson D, Chapman JW, Hughes WOH. (2001). Discrimination of unrewarding flowers by bees; direct detection of rewards and use of repellent scent marks. *J Insect Behav* 14:669–78.

Goulson D, Hanley ME, Darvill B, Ellis JS, Knight ME. (2005). Causes of rarity in bumblebees. *Biol Conserv* 122:1–8.

Goulson D, Hawson SA, Stout JC. (1998). Foraging bumblebees avoid flowers already visited by conspecifics or by other bumblebee species. *Anim Behav* 55:199–206.

Goulson D, Stout JC, Langley J, Hughes WOH. (2000). Identity and function of scent marks deposited by foraging bumblebees. *J Chem Ecol* 26:2897–911.

Inouye DW. (1978). Resource partitioning in bumblebees: Experimental studies of foraging behavior. *Ecology* 59:672–78.

Kato M. (1988). Bumblebee visits to *Impatiens* spp.: Pattern and efficiency. *Oecologia* 76:364–70.

Knight ME, Martin AP, Bishop S, Osborne JL, Hale RJ, Sanderson RA, Goulson D. (2005). An interspecific comparison of foraging range and nest density of four bumblebee (*Bombus*) species. *Mol Ecol* 14: 1811–20.

Lockey KH. (1980). Insect cuticular hydrocarbons. *Comp Biochem Physiol* 65B:457–62.

Marden JH. (1984). Intrapopulation variation in nectar secretion in *Impatiens capensis*. *Oecologia* 63:418–22.

Menzel R, Müller U. (1996). Learning and memory in honeybees: From behavior to neural substrates. *Annu Rev Neurosci* 19:379–404.

Møller AP. (1995). Bumblebee preference for symmetrical flowers. *Proc Natl Acad Sci USA* 92:2288–92.

Møller AP, Eriksson M. (1995). Pollinator preference for symmetrical flowers and sexual selection in plants. *Oikos* 73:15–22.

Núñez JA. (1967). Sammelbienen markieren versiegte Futterquellen durch Duft. *Naturwissenschaften* 54: 322–23.

Oldham NJ, Billen J, Morgan ED. (1994). On the similarity of the Dufour gland secretion and the cuticular hydrocarbons of some bumblebees. *Physiol Entomol* 19:115–23.

Pouvreau A. (1991). Morphology and histology of tarsal glands in bumble bees of the genera *Bombus*, *Pyrobombus* and *Megabombus*. *Can J Zool* 69:866–72.

Pyke GH. (1982). Local geographic distributions of bumblebees near Crested Butte, Colorado: Competition and community structure. *Ecology* 63:555–73.

Reader T, MacLeod I, Elliott PT, Robinson OJ, Manica A. (2005). Inter-order interactions between flower-visiting insects: Foraging bees avoid flowers previously visited by hoverflies. *J Insect Behav* 18: 51–57.

Saleh N, Chittka L. (2006). The importance of experience in the interpretation of conspecific chemical signals. *Behav Ecol Sociobiol* 61:215–20.

Saleh N, Ohashi K, Thomson JD, Chittka L. (2006). Facultative use of the repellent scent mark in foraging bumblebees: Complex versus simple flowers. *Anim Behav* 71:847–54.

Saleh N, Scott AG, Bryning GP, Chittka L. (2007). Distinguishing signals and cues: Bumblebees use general footprints to generate adaptive behaviour at flowers and nest. *Arthropod-Plant Interact* 1:119–27.

Schmitt U. (1990). Hydrocarbons in tarsal glands of *Bombus terrestris*. *Experientia* 46:1080–82.

Schmitt U, Bertsch A. (1990). Do foraging bumblebees scent-mark food sources and does it matter? *Oecologia* 82:137–44.

Schmitt U, Lübke G, Francke W. (1991). Tarsal secretion marks food sources in bumblebees (Hymenoptera: Apidae). *Chemoecology* 2:35–40.

Stanton ML, Preston RE. (1988). A qualitative model for evaluating the effects of flower attractiveness on male and female fitness in plants. *Am J Bot* 75:540–44.

Stout JC, Goulson D. (2001). The use of conspecific and interspecific scent marks by foraging bumblebees and honeybees. *Anim Behav* 62:183–89.

Stout JC, Goulson D. (2002). The influence of nectar secretion rates on the responses of bumblebees (*Bombus* spp.) to previously visited flowers. *Behav Ecol Sociobiol* 52:239–46.

Stout JC, Goulson D, Allen JA. (1998). Repellent scent marking of flowers by a guild of foraging bumblebees (*Bombus* spp.). *Behav Ecol Sociobiol* 43:317–26.

Tengö J, Hefetz A, Bertsch A, Schmitt U, Lübke G, Francke W. (1991). Species specificity and complexity of Dufour's gland secretion of bumble bees. *Comp Biochem Physiol* 99B:641–46.

Wetherwax PB. (1986). Why do honeybees reject certain flowers? *Oecologia* 69:567–70.

Wilms J, Eltz T. (2008). Foraging scent marks of bumblebees: Footprint cues rather than pheromone signals. *Naturwissenschaften* 95:149–53.

Williams CS. (1998). The identity of the previous visitor influences flower rejection by nectar-collecting bees. *Anim Behav* 56:673–81.

Witjes S, Eltz T. (2007). Influence of scent deposits on flower choice: Experiments in an artificial flower array with bumblebees. *Apidologie* 38:12–18.

Yokoi T, Fujisaki K. (2007). Repellent scent-marking behaviour of the sweat bee *Halictus* (*Seladonia*) *aerarius* during flower foraging. *Apidologie* 38:474–81.

Yokoi T, Goulson D, Fujisaki K. (2007). The use of heterospecific scent marks by the sweat bee *Halictus aerarius*. *Naturwissenschaften* 94:1021–24.

14 Information Transfer and the Organization of Foraging in Grass- and Leaf-Cutting Ants*

Flavio Roces and Martin Bollazzi

CONTENTS

INTRODUCTION

Even though it is tempting to consider a social insect colony as a unit that collectively decides about, for instance, the selection of a given food source, a colony's foraging response ultimately arises from the decisions made by each individual worker, and from their integration at the colony level. Individual foragers are expected to behave in a way that maximizes food intake at the colony level, because one can assume that the foraging performance of a colony, measured as the delivery rate of food, correlates with colony fitness.

In social insects, most of the resources collected by foraging individuals are not for their own consumption, but are fed to the brood and nestmates, or stored. This complicates the analysis of individual foraging performances and its extrapolation to account for overall colony responses. Put into economic terms, the costs and benefits of the foraging decisions of individual workers are paid for and received by different workers. Moreover, returning foragers deliver two commodities to the colony: food and information. Workers could potentially favor the delivery of one of these commodities at the expense of the other; i.e., workers may spend less time collecting at a food source, thus reducing their own food delivery rate, but allocate more effort to information transfer by promoting recruitment of nestmates, which in turn increases the performance of the entire colony. In one hypothetical extreme, a forager that discovers a high-rewarding food source should quickly return to the colony in order to deliver the appropriate information, without investing time for loading at the food source. At the other extreme, each worker could exploit a food source in order to individually maximize its food delivery rate. When this dual aspect is kept in mind, it becomes clear that the

---

* Dedicated to Professor. J.A. Núñez on the occasion of his eighty-fourth birthday.

optimal policies for delivering each of the commodities might be different, and that the maximization of information transfer is incompatible with a concomitant maximization of food delivery at the level of the individual.

There is ample evidence that foraging social insects (honey bees, wasps, ants) often return to the nest with partial loads (Núñez 1966; Pflumm 1975; Roces 1990a; Roces and Núñez 1993; Josens et al. 1998). For central place foragers, optimal foraging theory predicts the collection of partial loads when animals feed in a patch where intake rate decreases with time (for instance, due to resource depression), or when the net yield of energy increases at a diminishing rate, due to increased foraging costs (food search, transport, etc.). When collecting at *ad libitum* sources, animals should load themselves maximally (Orians and Pearson 1979). Contrary to this expectation, social insect foragers often return with partial loads even when collecting food at *ad libitum* sources, as observed in nectar-feeding ants and leaf-cutting ants. This pattern does not match the expectations of optimality theory. It seems that social insect workers forage in a suboptimal way, at least based on considerations of performance at the level of the individual.

It is unclear to what extent the foraging decisions of social insects are the outcome of a trade-off between these two competing tendencies, i.e., maximizing both loading and information transfer. For instance, foragers may—depending on their individual energy needs—invest more effort into loading than into information transfer. In addition, worker decisions may be influenced by the colony status, i.e., by both the presence of colony stores and the behavioral state of their nestmates. As a consequence, worker decisions do not only influence, but also are influenced by the decisions of other colony members.

Leaf-cutting ants (Attini, genus *Atta* and *Acromyrmex*) appear to be well suited to explore the interplay between individual and social aspects of foraging. Leaf-cutting ants are conspicuous polyphagous herbivores of the Neotropics that cut vegetation into small fragments, which are then transported to the nest as substrate for a symbiotic fungus. The fungus garden represents the sole food source of the developing brood, whereas adult workers obtain a large proportion (more than 90%) of their energy requirements from the plant sap of the harvested material. At the individual level, research has focused on the rules workers use to decide about the size of the cut leaf fragments (Barrer and Cherrett 1972; Wetterer 1990; Roces and Lighton 1995). At the colony level, a number of studies have provided detailed information about plant selection and foraging activity of colonies in the field (Howard 1988, 1991), and different hypotheses addressing the evolution of foraging decisions have been advanced (Burd 1996b; Roces 2002; Burd and Howard 2005b).

Even though Central American leaf-cutting ants have attracted the attention of many researchers since the first descriptions published by Belt (1874), not all leaf-cutting ants live in tropical rainforests, nor are all of them true leaf-cutters. In fact, ca. 70% of the recognized species of leaf-cutting ants are found in the subtropical regions of South America (Fowler et al. 1986), and a number of them predominantly cut blades of grasses or harvest a mixture of both grasses and dicotyledonous plants. The question about the place of origin of leaf-cutting ants (genera *Atta* and *Acromyrmex*), which derived from the more basal fungus-growing ants, is controversial. It has been suggested that they may have originated in the tropics, probably in the Amazon Basin (Kusnezov 1963; Weber 1972). If so, they may have used dicots as fungal substrate, suggesting that the habit of using grass as a fungal substrate is derived. On the other hand, based on the high species richness of both genera, *Atta* and *Acromyrmex*, in the southern subtropical regions of the American continent, it has been argued that they may have originated as grass-cutting ants in the savannas or caatingas (Fowler 1982, 1983). This view is also supported by a much simpler fungal substrate preparation procedure in grass-cutting ants than in leaf-cutting ants. At present, however, this issue remains unsolved (Mayhé-Nunes and Jaffé 1998; Schultz and Brady 2008).

Knowledge about the origins of harvesting behavior in grass- and leaf-cutting ants may help to understand the forces that have driven the evolution of decision making during foraging in these insects. As long as this issue remains unsolved, however, side-by-side comparison of grass- and leaf-cutting ant foraging may be useful in highlighting trade-offs that may underlie the organization of collective foraging, despite the differences in harvesting behavior in these two groups. Most research on leaf-cutting ants has concentrated on species harvesting dicots, particularly those of economic relevance (Lofgren and Vander Meer 1986; Vander Meer et al. 1990; Wirth et al. 2003), so that any comparison with grass-cutting ants may still be incomplete.

Plant selection by leaf-cutting ants is mainly based on both chemical and physical features of the leaves. Colonies are extremely polyphagous, using up to 50–80% of the available plant species in often highly diverse plant communities (Wirth et al. 2003). Nevertheless, foragers show marked preferences for leaves of certain plant species, individual plants, and even leaves within individual plants (Hubbell and Wiemer 1983; Howard 1987, 1990; Nichols-Orians and Schultz 1990; Meyer et al. 2006). In this chapter, we will not treat with the criteria used by workers for plant selection. Instead, we will first compare the behavioral rules used by both leaf- and grass-cutting ant foragers upon discovery of a suitable plant, thereby concentrating on the trade-off between food collection and information transfer, and then briefly comment on the theoretical arguments used so far to understand the foraging processes in leaf-cutting ants.

THE DETERMINATION OF FRAGMENT SIZE: LEAF-CUTTING ANTS

While foraging, scout workers from a leaf-cutting ant colony search for suitable resources and, upon their discovery, decide whether a given resource is worth communicating to other nestmates. The acceptance of a plant is mainly based on chemical and physical features of its leaves. If the source is attractive, workers may decide to return to the nest, laying a chemical trail, or to cut a fragment and carry it back to the nest.

It has repeatedly been observed that polymorphic leaf-cutting ant foragers frequently harvest leaf pieces that correspond in mass to that of the ants' body size (e.g., Lutz 1929). This, in part, results from the geometric method of leaf cutting. Since workers anchor themselves on the leaf edge with their hind legs and pivot around them while cutting arcs out of the leaves, the load size selected may be directly determined by a fixed reach while cutting, dependent on worker body size (Weber 1972). However, there is evidence that not all workers cut fragments of maximal size, suggesting a more flexible mechanism of load size determination in leaf-cutting ants. For instance, foraging distance was shown to affect load size selection in *Acromyrmex lundi* in such a way that ants cut smaller fragments close to the nest (Roces 1990a). In *Atta cephalotes*, however, this relationship was not found (Wetterer 1991b). For an ant of a given size, a negative correlation between leaf area density (leaf mass/leaf area) and the size of the harvested fragment was also found (Cherrett 1972; Rudolph and Loudon 1986; Roces and Hölldobler 1994; Burd 1995).

Do leaf-cutting ant workers use such kind of flexible cutting rules to trade-off loading for information transfer? The extent to which the decision to transfer information about a food discovery influences individual cutting behavior was analyzed in leaf-cutting ants under controlled experimental conditions (Roces 1993; Roces and Núñez 1993). The rationale of the experimental approach was as follows: in independent assays, scout workers of *Acromyrmex lundi* were first exposed to droplets of scented sugar solution of either of two concentrations: 1% or 10% sucrose. Scouts detected these droplets and returned to the colony, leaving a chemical recruiting trail. When the recruited workers arrived, they did not encounter sugar solution, but sheets of sugar-free Parafilm impregnated with the same scent (Figure 14.1). Since all recruited workers found the same standardized material (Parafilm as a pseudoleaf), differences in cutting behavior between workers initially recruited to either a 1% or a 10% sucrose solution must be the result of the information transmitted by the scout worker that initially found the sugar solution. Several behavioral parameters of the recruited workers depended on the information about food quality they had received from a single recruiting scout. Workers recruited by scouts that found 10% sucrose will be referred to as 10% workers, and those recruited by scouts that found 1% sucrose as 1% workers, but it is important to keep in mind that the recruits of both groups encountered a sugar-free pseudoleaf of the same quality at the source. It was observed that 10% workers (1) cut smaller fragments of

FIGURE 14.1 A leaf-cutting ant worker (*Atta sexdens*) cutting a fragment out of a pseudoleaf of Parafilm impregnated with orange scent. See color insert following page 142. (Photograph by H. Heilmann.)

Parafilm, (2) returned to the nest at higher velocities, and (3) displayed more active recruiting behavior (by laying chemical trails) than 1% workers. In another experiment, recruited workers could only collect small pieces of paper of the same size. They still traveled at a faster pace if they had been recruited by 10% scouts, indicating that the differences in speed were not caused by differences in loading. In regard to the amount of collected material, however, the greater velocity did not compensate for the reduction in fragment size: 10% workers, despite their higher velocity, showed a lower delivery rate to the nest than 1% workers (Roces and Núñez 1993). These results contradicted the most obvious functional hypotheses, and led to a vivid discussion (Kacelnik 1993; Clark 1994; Roces 1994a; Ydenberg and Schmid-Hempel 1994). First, it was surprising that workers cut fragments of different sizes, since all found a similar source (Parafilm). Moreover, a single Parafilm leaf represents a nondepleting patch for the individual ant, and animals should load themselves maximally when collecting at *ad libitum* sources. Particularly puzzling was the observation that workers recruited to the more attractive 10% sucrose source cut smaller fragments than workers recruited to the less attractive 1% source. Roces and Núñez (1993) advanced a hypothesis to account for these *a priori* unexpected results, called the information transfer hypothesis, which was based on the original hypothesis developed to account for partial crop filling and seemingly suboptimal foraging behavior in honey bees (Núñez 1982). The information transfer hypothesis states that a newly discovered food source motivates a worker to shorten its loading time and return not completely filled/loaded to the nest, in order to recruit additional nestmates. A worker seems to sacrifice its individual delivery rate in order to return earlier to the colony for further recruitment. Its performance as a food collector is therefore reduced, but the colony as a whole, which represents the level at which natural selection in eusocial insects predominantly takes place, increases its harvesting rate due to the recruited workers that participate in the resource gathering activity. This hypothesis was consistent with the higher travel speed and the reduced leaf fragment size in 10% workers, because by cutting smaller fragments the 10% workers saved cutting time, and by increasing speed they saved travel time. The information transfer hypothesis could also explain the more intense trail-marking behavior observed in 10% workers.

The observed differences in load size determination have a number of consequences on the foraging performance: Regarding the cutting process, cutting length, and therefore cutting time, decreases with decreasing fragment size, so that ants cutting smaller fragments spend less time at the source and run more rapidly on the trail because of their lighter loads. Regarding the effects of fragment size on the transport, travel speed decreases with increasing load mass, so that travel time per round-trip increases (Lighton et al. 1987; Burd 1996a). In spite of their slower speed when carrying larger fragments, workers may achieve higher material transport rates owing to the larger loads they carry. But longer travel times, with the concomitant reduction in round-trip frequency, may negatively affect the probability of information transfer, and therefore the intensity of recruitment (Roces and Núñez 1993; Roces and Hölldobler 1994). Furthermore, longer travel times might result in a longer exposure to predators and parasites (Feener and Moss 1990) or in desiccation—and thus in increased foraging risks.

Why do recruited workers run more quickly when they have been informed about a richer source? The adaptive value of this response might be related to a rapid transmission of information about the newly discovered food source as it allows a faster buildup of workers. At the initial phases of trail development, saving time would be of great importance, in order to monopolize a food source as soon as possible. The evaluation of resource quality by recruited workers and, therefore, the probability of reinforcing the chemical trail partially depend on the information they received; i.e., recruits partially rely on the decisions of the first successful ants whether to amplify the recruitment process. As mentioned above, *Acromyrmex lundi* workers recruited to rich sources cut smaller fragments than those recruited to poorer ones, although both groups found the same standardized pseudoleaves at the source. The recruits' decisions appear, therefore, to be "channeled" by those of the scouts, and it is an open question whether this nondemocratic decision-making system (Jaffé et al. 1985) may be related to both the complexity of assessment of plant quality by leaf-cutting ants, and the need for rapid responses to monopolize newly discovered sources. Finally, it is important to emphasize that the predictions of the information transfer hypothesis refer to the phase of discovery of new resources, when information needs may affect load size selection and may lead to the harvesting of small, suboptimal fragments that fail to maximize individual delivery rates (Roces and Núñez 1993, Roces and Hölldobler 1994). Whether information needs also influence load size determination under routine foraging conditions on well-established trails is unclear (Burd 2000b).

THE DETERMINATION OF FRAGMENT SIZE: GRASS-CUTTING ANTS

Grass-cutting ants display a very distinct cutting behavior. During foraging, workers climb on a grass blade and cut across its width, which results in the selection of a long, more or less rectangular grass fragment (Figure 14.2). The fragment lengths harvested in the field are larger than the maximal reach of workers, so they do not anchor their hind legs at the grass end while cutting. This implies that any mechanism underlying fragment size determination would not be a simple function of body geometry, because the body reach cannot be used as a reference during fragment cutting as in leaf-cutting ants. Therefore, matching between ant and fragment size would not be expected as a result of the geometry of cutting, and in fact, no or only a weak correlation was found between ant size and load size at the cutting site in the grass-cutting ant *Atta vollenweideri* (Röschard and Roces 2003b).

Grass-cutting ant foragers probably use a cutting rule connected with the pronounced pattern of searching along a grass blade before they begin. By walking along a grass blade, especially by walking down from the tip, ants might be able to assess the distance covered, thereby obtaining information about the length of the fragment they are going to cut. Interestingly, ants showed the same behavior, i.e., walking along a grass blade from one end to the other, even if it was lying on the ground, but too large for them to carry away. If ants assess length by walking along the blade, one might *a priori* expect a correlation between ant size and fragment length. This has been reported for *Atta vollenweideri* grass-cutting ants when harvesting the soft grass *Leersia hexandra*, but not

FIGURE 14.2 (a) Grass-cutting ant workers (*Atta vollenweideri*) cutting a grass blade. (b) A foraging trail of grass-cutting ants (*Atta vollenweideri*), Province of Formosa, Argentina. See color insert following page 142. (Photographs by O. Giessler (a) and F. Roces (b).)

for two hard grasses, which suggests little biological relevance of body size on the decision about the fragment length to be cut (Röschard and Roces 2003b). However, if we consider that fragments are carried over considerable distances on foraging trails, which sometimes exceed 100 m, size matching may be important to enhance the transport performance, which refers not only to energetic costs of locomotion (Lighton et al. 1987) but also to time saving (Röschard and Roces 2002). Single foraging trips may require several hours on long trails (Lewis et al. 1974), and since fragments are often observed to be transported sequentially by a number of workers, as will be discussed below, it is conceivable that size matching improves as the fragments are passed on to other workers along the trail. In fact, it has been reported that the size of the grass fragments chosen by foragers prior to transport indeed depended on the distance from the nest, so that size matching between worker and load increased nearer to the nest (Röschard and Roces 2003b). These findings indicate that workers drop or transfer the transported fragments along the trail, thus leading to the formation of transport chains (Röschard and Roces 2003a). Further evidence indicates that large fragments dropped at the source are cut again prior to transport, suggesting that workers make specific decisions to determine the length of the piece to be cut. However, the precise mechanisms involved in the determination of fragment length by grass-cutting ants remain unknown.

FRAGMENT SIZE AND THE ECONOMICS OF LOAD TRANSPORT

As outlined above, both grass- and leaf-cutting ant foragers select fragment sizes using particular rules. Based on standard optimality arguments, foragers are expected to select loads that optimize their performance. Interpretation of the foraging performance of individual workers, and particularly its extension to the colony level, is, however, complicated by the fact that a considerable percentage of leaf-cutting ant workers on an active foraging trail return to the nest without loads (between 13 and 75%) (Hodgson 1955; Cherrett 1972; Lugo et al. 1973; Lewis et al. 1974). While a number of these unladen workers can be involved in trail clearing (Daguerre 1945) or transport of plant sap (Stradling 1978), others seem to be engaged in the reinforcement of the chemical trail or in a combination of recruiting activities and load transport (Jaffé and Howse 1979). In addition, the workers' foraging decisions may differ at different times following the onset of a daily foraging cycle (Roces and Hölldobler 1994; Bollazzi and Roces 2001), or when foraging behavior is scrutinized along established foraging trails (Burd 2000b; Dussutour et al. 2007), where the role of recruitment communication in the regulation of collective foraging is still not well understood.

As indicated above, load mass (and load length for grass-cutting ants) has a large influence on the travel speed of the foragers and, therefore, on their rate of plant tissue delivery (Rudolph and Loudon 1986; Wetterer 1994; Burd 2000a). For this reason, foragers are expected to carry loads that maximize their delivery rate, i.e., the amount of plant tissue carried to the nest per unit of foraging time. It has been suggested that most fragment masses carried by foragers of the leaf-cutting ant *Atta cephalotes* lie within the range that maximizes the delivery rate of plant tissue (Rudolph and Loudon 1986). A closer scrutiny showed, however, that the fragment masses needed to maximize the rate of tissue delivery, or maximize the energetic efficiency of delivery, should be greater than the average fragment masses actually carried by *Atta* leaf-cutting ant foragers (Burd 1996a, 1996b). In other words, the fragment sizes carried by *Atta* foragers during established foraging are well below the sizes that would maximize the delivery rate (Burd 1996a, 1996b). Interestingly, if fragment mass is experimentally increased by adding small weights to the fragments while they are carried, workers indeed attain higher rates of load delivery, despite the detrimental effects of the increased load on absolute walking speed (Burd 2000a).

The question arises why workers do not cut larger fragments, considering that they would be able to carry them and thereby achieve higher delivery rates. Cutting behavior was shown to be indeed flexible, as already discussed earlier in the context of food quality: foragers reduced their reach while cutting smaller fragments of highly attractive leaves. However, extending the reach while cutting, in order to cut larger fragments, may be constrained by morphology: since leaf-cutting ant workers anchor their hind legs at the leaf edge and rotate around their body axis, fragment size is limited by the maximal reach of the ant while cutting (Wetterer 1991a). Observations and experimental manipulations suggest that when cutting tender leaves, the ant's reach is at its maximum. Anchoring of the rear legs at the edge is necessary for fragment size determination, and losing this reference mostly leads to irregularly shaped fragments or to a total cessation of harvesting (Römer 1997). In evolutionary terms, any selection for the harvesting of larger fragments should go along with an increased reach while cutting, i.e., either with an enlargement of the rear legs or with a completely different cutting mechanism. Interestingly, leaf-cutting ants tend to have longer metathoracic legs than grass-cutting ants (Fowler et al. 1986), suggesting that selection has in fact favored longer legs for ants using a geometric mode of leaf cutting. Besides the constraints imposed by leg morphology, selection for larger fragments might have in addition been constrained by both handling requirements and the biomechanics of carriage. Larger fragments might be particularly difficult to maneuver into the carrying position, and because of the larger backwards displacement of their gravity center with increasing size, they may not be easy to carry. In this speculative scenario, the largest fragment sizes harvested by leaf-cutting ants might fail to maximize individual delivery rates because of morphological constraints underlying the cutting mechanisms.

Two different alternative hypotheses have been advanced to account for the selection of leaf fragment sizes that do not maximize leaf delivery rates during established foraging, thus extending the scope of the information transfer hypothesis, which focused on the discovery phase and the initial establishment of a foraging column. The first hypothesis considered the occurrence of interferences among workers over access to cutting sites on the leaf margin (Burd 1996b). By modeling leaf-cutting ant harvesting activity as a queuing process, Burd predicted that delivery rates of plant tissue by a group of foragers may be enhanced when individuals cut small fragments that are suboptimal from an individual perspective. These arguments are based on the assumption that only a limited number of foragers can simultaneously cut on a single leaf, basically determined by the length of its margin, so that queuing delays may occur at high ant densities. In other words, new cutting attempts by arriving foragers may be delayed until an engaged forager finishes its actual cut and returns to the nest (Burd 1996b). To this day, queues at the cutting sites have not been reported in the vast literature on leaf-cutting ants, and field studies have provided little support for predictions derived from this hypothesis (Burd 2000b).

The second alternative hypothesis, recently advanced and explored by Burd and Howard (2005a, 2005b), suggests that in leaf-cutting ants the selection of small fragment sizes, which are below the size that would maximize the forager's delivery rate, was favored during evolution because of the underground processing rates of the fragments that occur after their delivery. Once incorporated into the nest, plant fragments need to be transferred between nest chambers, cleaned, and shredded before the final addition of fungal hyphae. Leaf fragment size was experimentally shown to affect the time needed to perform these sequential processes. Since the time needed for the underground activities would often be larger than the time invested to retrieve the fragments from the foraging site, Burd and Howard (2005a) argued that the maximization of the delivery rate by foragers may not maximize the overall rate of resource acquisition, i.e., the rate that includes both delivery and processing via underground activities. By using a mathematical model that includes both leaf transport and the internal tasks of tissue processing previously measured for a laboratory colony, Burd and Howard (2005b) argued that the leaf fragment sizes that maximize a colony's overall foraging rate correspond well with the fragment sizes naturally harvested by leaf-cutting ants, and that they are below the sizes that would maximize the individual delivery rate by foragers.

The rationale for these arguments is the following: leaf harvesting and underground handling and processing are considered an integrated sequence, a "production line," in which a match between the incorporation and processing rates might be advantageous. In this line of arguments, the underground activities might constrain the behavior of foragers when harvesting outside. It is argued that there would be little advantage at the colony level if foragers maximize their own performance (by cutting larger fragments) but cause bottlenecks and poor performance in tissue processing within the nest. Under such task partitioning, the model predicts that colonies can forage at a maximum rate because foragers working outside the nest perform below the maximal rate for their task (Burd and Howard 2005b); i.e., colony performance is maximized at the expense of the individual performance of foragers. These arguments rest on the assumption that foraging and underground processing represent sequential handling tasks, so that mismatches between their corresponding rates should be disadvantageous in evolutionary terms. However, the observation that a colony's leaf delivery rate may exceed its leaf processing rate could be an artifact resulting from the laboratory settings used (Burd and Howard 2005a). First, fungus chambers in the investigated laboratory colonies were serially connected with pipes, so that fragments transported downstream may have experienced marked delays when passing through the preceding fungus chamber. In *Atta* field colonies, fungus chambers are not connected serially through a single tunnel, but located at the side of a main tunnel, laterally joined to it (e.g., Eidmann 1935; Moreira et al. 2004). Therefore, in natural nests, the transport of fragments to distant chambers is not impaired at all by the processing occurring in the preceding fungus chambers. Second, while laboratory colonies normally accept food around the clock, foraging in field colonies follows a marked daily rhythm, being either diurnal or nocturnal (Lewis et al. 1974). Leaf tissue processing is expected to occur around the clock in field colonies, based on observations of the activities related to fungus maintenance (Herz et al. 2007). During a given foraging period, any additional intake of fragments greater than the actual processing rate would result in a temporary accumulation of fragments inside the nest (Eidmann 1935), which may be recovered and processed later, when no foraging activity takes place. If so, the hypothesis that the demands imposed by the underground processing may have shaped the evolution of fragment size determination by foragers appears to be unlikely.

At present it remains unresolved whether load size selection by foraging leaf-cutting ants may have evolved to optimize the underground processing activities or to maximize the rate of leaf tissue delivery under certain morphological constraints. Comparative studies on both delivery and processing rates in grass-cutting ants may help to answer this question. As indicated above, grass-cutting ants have a much simpler method of fungal substrate preparation than leaf-cutting ants. In leaf-cutters, the collected vegetation is triturated and chewed before it is used as fungal substrate, and the complete tissue processing is time demanding. By contrast, grass-cutters often incorporate whole fragments into the fungus without additional cutting, so that fragments can clearly be recognized in the fungal matrix, or they cut them only in a couple of relatively large pieces (Fowler et al. 1986; Garcia et al. 2005). Vegetation processing in grass-cutters is therefore less time-consuming than in leaf-cutting ants. Because grass-cutting ant foragers do not use their body reach as a reference during fragment size determination, as is the case in leaf-cutting ants, most fragments are longer than the ants' body length, averaging, for instance, 43 mm in *Atta vollenweideri* (Daguerre 1945). As already mentioned above, workers climb on a grass blade and cut across its width, which results in the selection of a long, more or less rectangular grass fragment. Therefore, cutting effort is directly determined by the width of the grass, which usually does not differ very much along the length of a blade except near its tip. Hence, cutting a larger (longer) fragment implies neither a higher cutting effort nor a longer cutting time, provided that grass toughness remains unchanged along the blade. Workers may therefore be able to harvest more material per unit cutting effort by simply cutting longer fragments. If only the energy investment during cutting is considered, the harvesting of very long fragments would be expected. Since fragments need to be transported to the nest, however, fragment length may have substantial effects on maneuverability and speed of transport. During transport, a worker takes a fragment with its mandibles at one end and carries it in a more or less vertical position, usually inclined backward at an angle of 30–70° to the ant's body axis (Figure 14.2b). Because of the marked displacement of the gravitational center of the loaded ant, the transportation of longer fragments reduces the walking speed (Röschard and Roces 2002). Furthermore, workers carrying longer fragments might be more likely to be hindered by obstacles on their way to the nest. Thus, the sizes of the fragments cut by grass-cutting ants under natural conditions may represent the outcome of an evolutionary trade-off between maximizing harvesting rate at the cutting site and minimizing the effects of fragment size on the delivery rate of vegetation. It is speculated that such factors may have set an upper limit to the fragment size (length) selected by grass-cutting ant workers during harvesting (Röschard and Roces 2002), irrespective of the rate of underground processing of plant tissue.

COOPERATIVE LOAD TRANSPORT: PLANT TISSUE DELIVERY AND INFORMATION FLOW

Besides exploring the criteria used by workers during selection of fragment sizes, which may largely influence the economics of load transport, any attempt to understand the organization of collective foraging must consider the occurrence of cooperative vegetation delivery in the form of transport chains. In several leaf-cutting ant species, the leaf fragments are not carried to the nest by the workers that cut them. The harvested fragments may be dropped to the ground or initially transported a short distance, and then dropped or passed on to other nestmates for further transport. Considerations about foraging strategies must, therefore, go beyond the focus on flexible cutting rules, as discussed above, and encompass the adaptive value of sequential cooperation for load carriage. The occurrence of transport chains in leaf-cutting ants was first described for *Atta sexdens rubropilosa* and *Atta cephalotes* (Fowler and Robinson 1979; Hubbell et al. 1980). More recently, transport chains were reported for *Atta colombica*, in which ca. 20% of the harvested fragments are transported to the nest by more than one worker (Anderson and Jadin 2001; Hart and Ratnieks 2001), as well as in a number of *Acromyrmex* species, both grass- and leaf-cutters (Lopes et al. 2003).

The adaptive value of the formation of transport chains was investigated in greater detail in the grass-cutting ant *Atta vollenweideri*. The complete foraging process can be easily analyzed in this ground foraging species. Röschard and Roces (2003a, 2003b) observed that on long foraging trails, more than half of the fragments were transported by chains; i.e., besides the cutter, generally two or three carriers sequentially transported the load. It is an open question which variables motivate workers to drop their fragments, thus leading to the formation of transport chains: (1) Ants might decide to drop fragments that are not sufficiently attractive, thus rejecting them. Since all dropped fragments are retrieved again, however, this appears to be very unlikely. (2) Dropping might occur because of a mismatch between body and fragment size; i.e., either the carrier is too small for the fragment or the fragment is too large to be carried. This seems plausible when the detrimental effects of large loads on transport rates are taken into account (Röschard and Roces 2002). In fact, sequential transport via transport chains in *Atta vollenweideri* leads to a slightly better size matching between the workers and their loads. Size matching is stronger in carriers at the end of the transport chain, indicating that carriers either collect dropped fragments according to their body mass, or cut fragments out of the dropped fragments in sizes corresponding to their body mass, prior to transport. This suggests that the formation of transport chains might in fact improve the delivery rate of the involved fragments. (3) Transport chains may be formed to improve information transfer along the trail, because more workers will be informed about the kind of resource being harvested, either by direct transfers of the collected material or by finding a dropped fragment on the trail. Due to new recruitment, improved information transfer may lead to an increased overall rate of resource transportation. Based on this information transfer hypothesis, the behavioral trait of directly or indirectly transferring fragments may have been selected for because it has positive effects on the information flow among foragers.

TRANSPORT CHAINS IN GRASS-CUTTING ANTS: MAXIMIZATION OF LEAF DELIVERY RATE OR IMPROVED INFORMATION TRANSFER?

Fragment dropping may allow cutters to quickly return to the selected plant for further harvesting. While this argument accounts for the dropping behavior of cutters, the question arises why carriers drop the fragments, too, and then walk back toward the source. Is the sequential transport via a transport chain faster than the transport by single carriers, thus enhancing colony-wide material intake rates? For the sake of simplicity, we would like to call these arguments the economic transport hypothesis. It should be noted, however, that *economic* in this context refers to the maximization of the transportation speed of a leaf fragment, which may result in an increased overall rate of resource transportation at the colony level. Maximization of leaf transportation has been proposed as a foraging criterion for workers of three leaf-cutting ant species (*Atta sexdens*, *A. cephalotes*, and *A. colombica*) that transfer loads or cache fragments on the ground (Fowler and Robinson 1979; Hubbell et al. 1980; Anderson and Jadin 2001). Direct leaf transfer between *Atta colombica* workers, which occurred only in 9% of the transported fragments (79% of the fragments were not transferred, and the remaining 12% were cached), resulted in a higher transportation speed after transfer. But surprisingly, the transferred fragments did not travel more rapidly than those transported to the nest by only one worker. The reason for this was the transfer delays caused by handing the loads over from one ant to another (Anderson and Jadin 2001). In a different study on the same species, however, fragments recovered from a cache were transported back to the nest more slowly than normally foraged leaf fragments (Hart and Ratnieks 2001). The adaptive value of this response remains obscure. In *Atta vollenweideri*, the transport time of fragments carried by a chain of workers was 25% longer (on average 8 min longer) than that of fragments carried by a single worker all the way to the nest. Mostly, this was due to both the waiting time of the dropped fragments and the handling time needed by the foragers that subsequently took them up. Thus, in terms of foraging time and material transport rates, sequential transport by chains was less efficient than transport by single carriers (Röschard and Roces 2003a).

Do transport chains accelerate the transfer of information about the plant species actually harvested, even if they do not increase the transport rate? Several processes may contribute to speed up the transfer of information and to rapidly build up worker numbers at the discovered source. First, fragment dropping after a given distance may allow cutting workers to quickly go back to the harvested plant, making it easier for them to find the source again by following the freshly deposited pheromone trail (Fowler and Robinson 1979; Hubbell et al. 1980). Second, moving along a short trail sector during foraging may enable workers to locally reinforce the pheromone marking better than when walking all the way to the nest. Third, fragments dropped on the trail may act as "signal posts." It has been shown that leaf-cutting ant foragers learn the odors of the harvested resources, and that the foragers' food choices at the patch depend on the material that is currently transported on the trail by their nestmates (Roces 1990b, 1994b; Howard et al. 1996). The odor of the fragments on the trail might function as a stimulus for olfactory learning. Most foragers, even those that continued on their way to the cutting site without loads, were observed to antennate the dropped fragments they encountered on their way. Thus, outgoing foragers may obtain information about the harvested resources both by contacting laden nestmates along the trail and by finding dropped fragments on the trail. This information may stimulate outgoing workers to search for the experienced plant species, thus leading them from the trail to the newly discovered plant.

The predictions of the two competing hypotheses discussed above (economic transport hypothesis and information transfer hypothesis) were tested in a field study (Röschard 2002). Based on the arguments of the economic transport hypothesis, fragment dropping (or transfer) by foragers occurs because of a mismatch between load size and body size at the individual level. Chains are therefore considered a way to improve carrying efficiency at the colony level. Thus, the probability for the occurrence of transport chains would be expected to depend on fragment size and not necessarily on fragment quality, and should be higher for larger fragments that are difficult to carry. Alternatively, the information transfer hypothesis states that the behavioral response of transferring fragments has been selected for because of its positive effect on the information flow. By dropping the load, a worker may be able to return earlier to the foraging site, to more quickly reinforce the chemical trail, and thus to enhance recruitment. In addition, the transferred fragments could themselves act as cues, which contain information about the plant that is worth harvesting. This hypothesis predicts that the formation of transport chains should strongly depend on fragment quality rather than fragment size.

To distinguish between the two hypotheses, standardized paper fragments that differed either in size or in quality (sugar- vs. tannin-rich fragments) were presented to workers of a field colony of *Atta vollenweideri* (Röschard 2002). The occurrence of transport chains was quantified by following individually marked fragments all the way to the nest. The results demonstrated that neither an increase in fragment mass nor an increase in fragment length modifies the probability of the occurrence of transport chains. In addition, the transport by transport chains took longer than the transport by a single carrier. However, the frequency of the occurrence of transport chains increased with increasing fragment quality, independent of fragment size. In addition, high-quality fragments were transferred after shorter distances than less attractive ones, i.e., attractive loads were dropped more frequently and at shorter intervals. All together, these results strongly suggest that a transport chain increases the information flow about the harvested plant rather than the delivery rate of the involved fragment.

CONCLUDING REMARKS

Foraging by grass- and leaf-cutting ants involves finding an attractive source by scout workers, the communication of their discovery to nestmates via chemical and vibrational signals, cutting of leaf fragments by the recruits, and the transport of the fragments back to the nest, where the plant tissue is processed and serves as the substrate for symbiotic fungi. The leaf fragments often are several times heavier than the ants that carry them, and heavy loads reduce the walking speed of loaded foragers. A larger fragment may provide more fungal substrate, but it may be carried to the nest at a slower speed than a smaller one. Therefore, it seems intuitive to think of intermediate fragment sizes as a compromise between increased load mass and decreased carriage velocity. Such arguments refer to the economics of load carriage, and do not take into account the fact that foragers may, in addition, deliver information while running back to the nest. For social insects in general, the analysis of worker behavior in the framework of optimality arguments is particularly complex, because the maximizing agent is the colony as a whole rather than each individual worker. Whether foragers indeed sacrifice their own foraging performance in favor of colony performance is unclear, as theoretical and empirical research so far has not properly addressed these issues. Central place foraging theory has been unable to provide a compelling explanation for the occurrence of partial loading in foraging social insects. The first models, addressing honey bee foraging, have discussed the foragers' decisions in terms of energetic considerations at the individual level, without any reference to information exchange (Schmid-Hempel et al. 1985; Kacelnik et al. 1986). Empirical work on flexible load size determination and information transfer in leaf-cutting ants (Roces and Núñez 1993) called renewed attention to the issue of partial loading in foraging social insects, and although not specifically formulated in a model, the role of information transfer in modulating individual foraging decisions was explicitly acknowledged in later theoretical accounts (Kacelnik 1993; Ydenberg and Schmid-Hempel 1994). Surprisingly, more than 20 years after the development of the first model discussing partial crop filling in honey bees, which was based on central place foraging theory (Schmid-Hempel et al. 1985), recent theoretical work—asking the same question—restricted its focus to the benefits that partial crop loads and the resulting time savings might confer in terms of *collecting* information, but omitted the analysis of the benefits that partial crop loads might provide in terms of *information transfer*, for instance, via the dance behavior (Dornhaus et al. 2006). Despite the bulk of research carried out on the foraging activities of grass- and leaf-cutting ants and the different hypotheses advanced to account for the empirical data, a compelling explanation for the observed selection of fragment sizes, the described flexible cutting rules, and the resulting rates of plant tissue delivery is still lacking. To ultimately understand these topics, both refined theory and further experimental work are required.

ACKNOWLEDGMENTS

We deeply thank Josué A. Núñez for enthusiastic and tireless discussions over the years, Bert Hölldobler for support and exchange of ideas, and Oliver Geißler, Jacqueline Röschard, and Christoph Kleineidam for discussions about information transfer in foraging ants. Our own field work reported here was performed at the wonderful Reserva Ecológica El Bagual (Alparamis SA–Aves Argentinas) in eastern Chaco, Province of Formosa, Argentina. We are very much indebted to the ornithologist Alejandro G. Di Giacomo, his field assistants, and especially the Götz family for providing facilities at the biological station, daily help during our stays, and invaluable logistical support throughout the past 10 years of ant research. Financial support was provided by the DFG, Germany (SFB 554 and SFB 567).

REFERENCES

Anderson C, Jadin JLV. (2001). The adaptive benefit of leaf transfer in *Atta colombica*. *Insect Soc* 48:404–5.
Barrer PM, Cherrett JM. (1972). Some factors affecting the site and pattern of leaf-cutting activity in the ant *Atta cephalotes* L. *J Ent A* 47:15–27.
Belt T. (1874). *The Naturalist in Nicaragua*. London: Edward Bumpus.
Bollazzi M, Roces F. (2001). The control of foraging decisions in leaf-cutting ants: Individual vs. collective aspects. In *Proceedings of the Meeting of European IUSSI Sections*, Berlin, Germany, p. 20.
Burd M. (1995). Variable load size-ant size matching in leaf-cutting ants, *Atta colombica* (Hymenoptera: Formicidae). *J Insect Behav* 8:715–22.
Burd M. (1996a). Foraging performance by *Atta colombica*, a leaf-cutting ant. *Am Nat* 148:597–612.
Burd M. (1996b). Server system and queuing models of leaf harvesting by leaf-cutting ants. *Am Nat* 148:613–29.

Burd M. (2000a). Body size effects on locomotion and load carriage in the highly polymorphic leaf-cutting ants *Atta colombica* and *Atta cephalotes*. *Behav Ecol* 11:125–31.

Burd M. (2000b). Foraging behaviour of *Atta cephalotes* (leaf-cutting ants): An examination of two predictions for load selection. *Anim Behav* 60:781–88.

Burd M, Howard JJ. (2005a). Central-place foraging continues beyond the nest entrance: The underground performance of leaf-cutting ants. *Anim Behav* 70:737–44.

Burd M, Howard JJ. (2005b). Global optimization from suboptimal parts: Foraging sensu lato by leaf-cutting ants. *Behav Ecol Sociobiol* 59:234–42.

Cherrett JM. (1972). Some factors involved in the selection of vegetable substrate by *Atta cephalotes* (L.) (Hymenoptera: Formicidae) in tropical rain forest. *J Anim Ecol* 41:647–60.

Clark CW. (1994). Leaf-cutting ants may be optimal foragers. *TREE* 9:63.

Daguerre JB. (1945). Hormigas del género *Atta* Fabricius de la Argentina (Hymenop. Formicidae). *Rev Soc Entomol Argent* 12:438–60.

Dornhaus A, Collins EJ, Dechaume-Moncharmont FX, Houston AI, Franks NR, McNamara JM. (2006). Paying for information: Partial loads in central place foragers. *Behav Ecol Sociobiol* 61:151–61.

Dussutour A, Beshers S, Deneubourg JL, Fourcassié V. (2007). Crowding increases foraging efficiency in the leaf-cutting ant *Atta colombica*. *Insect Soc* 54:158–65.

Eidmann H. (1935). Zur Kenntnis der Blattschneiderameise *Atta sexdens* L., insbesondere ihrer Ökologie. *Z angew Entomol* 22:185–436.

Feener DH Jr, Moss KAG. (1990). Defense against parasites by hitchhikers in leaf-cutting ants: A quantitative assessment. *Behav Ecol Sociobiol* 26:17–29.

Fowler HG. (1982). Evolution of the foraging behavior of leaf-cutting ants (*Atta* and *Acromyrmex*). In Breed MD, Michener CD, Evans HE (eds.), *The Biology of Social Insects*. Boulder: Westview Press, p. 33.

Fowler HG. (1983). Latitudinal gradients and diversity of the leaf-cutting ants (*Atta* and *Acromyrmex*) (Hymenoptera: Formicidae). *Rev Biol Trop* 31:213–16.

Fowler HG, Forti LC, Pereira-da-Silva V, Saes NB. (1986). Economics of grass-cutting ants. In Lofgren CS, Vander Meer RK (eds.), *Fire Ants and Leaf-Cutting Ants—Biology and Management*. Boulder: Westview Press, pp. 18–35.

Fowler HG, Robinson SW. (1979). Foraging by *Atta sexdens* (Formicidae: Attini): Seasonal patterns, caste and efficiency. *Ecol Entomol* 4:239–47.

Garcia MG, Forti LC, Verza SS, Noronha NC Jr, Nagamoto NS. (2005). Interference of epicuticular wax from leaves of grasses in selection and preparation of substrate for cultivation of symbiont fungus by *Atta capiguara* (Hym. Formicidae). *Sociobiology* 45:937–47.

Hart AG, Ratnieks FLW. (2001). Leaf caching in the leafcutting ant *Atta colombica*: Organizational shifts, task partitioning and making the best of a bad job. *Anim Behav* 62:227–34.

Herz H, Beyschlag W, Hölldobler B. (2007). Assessing herbivory rates of leaf-cutting ant (*Atta colombica*) colonies through short-term refuse deposition counts. *Biotropica* 39:476–81.

Hodgson ES. (1955). An ecological study of the behavior of the leaf-cutting ant *Atta cephalotes*. *Ecology* 36: 293–304.

Howard JJ. (1987). Leafcutting ant diet selection: The role of nutrients, water, and secondary chemistry. *Ecology* 68:503–15.

Howard JJ. (1988). Leafcutting ant diet selection: Relative influence of leaf chemistry and physical features. *Ecology* 69:250–60.

Howard JJ. (1990). Infidelity of leafcutting ants to host plants: Resource heterogeneity or defense induction? *Oecologia* 82:394–401.

Howard JJ. (1991). Resource quality and cost in the foraging of leaf-cutter ants. In Huxley CR, Cutler DF (eds.), *Ant-Plant Interactions*. Oxford: Oxford University Press, pp. 42–50.

Howard JJ, Henneman LM, Cronin G, Fox JA, Hormiga G. (1996). Conditioning of scouts and recruits during foraging by a leaf-cutting ant, *Atta colombica*. *Anim Behav* 52:299–306.

Hubbell SP, Johnson LK, Stanislav E, Wilson B, Fowler H. (1980). Foraging by bucket-brigade in leaf-cutter ants. *Biotropica* 12:210–13.

Hubbell SP, Wiemer DF. (1983). Host plant selection by an Attine ant. In Jaisson P (ed.), *Social Insects in the Tropics*. Paris: Université Paris-Nord, pp. 133–54.

Jaffé K, Howse PE. (1979). The mass recruitment system of the leaf cutting ant, *Atta cephalotes* (L.). *Anim Behav* 27:930–39.

Jaffé K, Villegas G, Colmenares O, Puche H, Zabala NA, Alvarez MI, Navarro JG, Pino E. (1985). Two different decision-making systems in recruitment to food in ant societies. *Behavior* 92:9–21.

Josens RB, Farina WM, Roces F. (1998). Nectar feeding by the ant *Camponotus mus*: Intake rate and crop filling as a function of sucrose concentration. *J Insect Physiol* 44:579–85.

Kacelnik A. (1993). Leaf-cutting ants tease optimal foraging theorists. *TREE* 8:346–48.

Kacelnik A, Houston AI, Schmid-Hempel P. (1986). Central-place foraging in honey bees: The effect of travel time and nectar flow on crop filling. *Behav Ecol Sociobiol* 19:19–24.

Kusnezov N. (1963). Zoogeografía de las hormigas en Sudamérica. *Acta Zool Lilloana* 19:25–186.

Lewis T, Pollard GV, Dibley GC. (1974). Rhythmic foraging in the leaf-cutting ant *Atta cephalotes* (L.) (Formicidae: Attini). *J Anim Ecol* 43:129–41.

Lighton JRB, Bartholomew GA, Feener DH Jr. (1987). Energetics of locomotion and load carriage and a model of the energy cost of foraging in the leaf-cutting ant *Atta colombica* Guer. *Physiol Zool* 60:524–37.

Lofgren CS, Vander Meer RK. (1986). *Fire Ants and Leaf-Cutting Ants— Biology and Management.* Boulder: Westview Press.

Lopes JFS, Forti LC, Camargo RS, Matos CAO, Verza SS. (2003). The effect of trail length on task partitioning in three *Acromyrmex* species (Hymenoptra: Formicidae). *Sociobiology* 42:87–91.

Lugo AE, Farnworth EG, Pool D, Jerez P, Kaufman G. (1973). The impact of the leaf cutter ant *Atta colombica* on the energy flow of a tropical wet forest. *Ecology* 54:1292–301.

Lutz FE. (1929). Observations on leaf-cutting ants. *Am Mus Novit* 388:1–21.

Mayhé-Nunes AJ, Jaffé K. (1998). On the biogeography of Attini (Hymenoptera: Formicidae). *Ecotropicos* 11:45–54.

Meyer ST, Roces F, Wirth R. (2006). Selecting the drought stressed: Effects of plant stress on intraspecific and within-plant herbivory patterns of the leaf-cutting ant *Atta colombica*. *Funct Ecol* 20:973–81.

Moreira AA, Forti LC, Andrade APP de, Boaretto MAC, Lopes JFS. (2004). Nest architecture of *Atta laevigata* (F. Smith, 1858) (Hymenoptera: Formicidae). *Stud Neotrop Fauna Environ* 39:109–16.

Nichols-Orians CM, Schultz JC. (1990). Interactions among leaf toughness, chemistry, and harvesting by attine ants. *Ecol Entomol* 15:311–20.

Núñez JA. (1966). Quantitative Beziehungen zwischen den Eigenschaften von Futterquellen und dem Verhalten von Sammelbienen. *Z vergl Physiol* 53:142–64.

Núñez JA. (1982). Honeybee foraging strategies at a food source in relation to its distance from the hive and the rate of sugar flow. *J Apic Res* 21:139–50.

Orians GH, Pearson NE. (1979). On the theory of central place foraging. In Horn DJ, Stairs GR, Mitchell RD (eds.), *Analysis of Ecological Systems*. Columbus: Ohio State University Press, pp. 155–77.

Pflumm W. (1975). Einfluß verschiedener Zuckerwasser-Konzentrationen und der Anwesenheit von Artgenossen auf das Sammelverhalten von Wespen (Hymenoptera: Vespidae). *Entomol Germanica* 2:7–21.

Roces F. (1990a). Leaf-cutting ants cut fragment sizes in relation to the distance from the nest. *Anim Behav* 40:1181–83.

Roces F. (1990b). Olfactory conditioning during the recruitment process in a leaf-cutting ant. *Oecologia* 83:261–62.

Roces F. (1993). Both evaluation of resource quality and speed of recruited leaf-cutting ants (*Acromyrmex lundi*) depend on their motivational state. *Behav Ecol Sociobiol* 33:183–89.

Roces F. (1994a). Cooperation or individualism: How leaf-cutting ants decide on the size of their loads. *TREE* 9:230.

Roces F. (1994b). Odour learning and decision-making during food collection in the leaf-cutting ant *Acromyrmex lundi*. *Insect Soc* 41:235–39.

Roces F. (2002). Individual complexity and self-organization in foraging by leaf-cutting ants. *Biol Bull* 202:306–13.

Roces F, Hölldobler B. (1994). Leaf density and a trade-off between load-size selection and recruitment behavior in the ant *Atta cephalotes*. *Oecologia* 97:1–8.

Roces F, Lighton JRB. (1995). Larger bites of leaf-cutting ants. *Nature* 373:392–93.

Roces F, Núñez JA. (1993). Information about food quality influences load-size selection in recruited leaf-cutting ants. *Anim Behav* 45:135–43.

Römer D. (1997). Untersuchungen zur Bestimmung der Blattfragmentgröße bei Blattschneiderameisen. Diploma thesis, University of Würzburg, Germany.

Röschard J. (2002). Cutters, carriers and bucket brigades—Foraging decisions in the grass-cutting ant *Atta vollenweideri*. Doctoral thesis, University of Würzburg, Germany.

Röschard J, Roces F. (2002). The effect of load length, width and mass on transport rate in the grass-cutting ant *Atta vollenweideri*. *Oecologia* 131:319–24.

Röschard J, Roces F. (2003a). Cutters, carriers and transport chains: Distance-dependent foraging strategies in the grass-cutting ant *Atta vollenweideri. Insect Soc* 50:237–44.

Röschard J, Roces F. (2003b). Fragment-size determination and size-matching in the grass-cutting ant *Atta vollenweideri* depend on the distance from the nest. *J Trop Ecol* 19:647–53.

Rudolph SG, Loudon C. (1986). Load size selection by foraging leaf-cutter ants (*Atta cephalotes*). *Ecol Entomol* 11:401–10.

Schmid-Hempel P, Kacelnik A, Houston AI. (1985). Honeybees maximize efficiency by not filling their crop. *Behav Ecol Sociobiol* 17:61–66.

Schultz TR, Brady SG. (2008). Major evolutionary transitions in ant agriculture. *Proc Natl Acad Sci USA* 105:5435–40.

Stradling DJ. (1978). The influence of size on foraging in the ant, *Atta cephalotes*, and the effect of some plant defense mechanisms. *J Anim Ecol* 47:173–88.

Vander Meer RK, Jaffé K, Cedeno A. (1990). *Applied Myrmecology—A World Perspective*. Boulder: Westview Press.

Weber NA. (1972). *Gardening Ants—The Attines*. Vol. 92. Philadelphia: Memoirs of the American Philosophical Society.

Wetterer JK. (1990). Load-size determination in the leaf-cutting ant, *Atta cephalotes. Behav Ecol* 1:95–101.

Wetterer JK. (1991a). Allometry and the geometry of leaf-cutting in *Atta cephalotes. Behav Ecol Sociobiol* 29: 347–51.

Wetterer JK. (1991b). Source distance has no effect on load size in the leaf-cutting ant, *Atta cephalotes. Psyche* 98:355–59.

Wetterer JK. (1994). Forager polymorphism, size-matching, and load delivery in the leaf-cutting ant, *Atta cephalotes. Ecol Entomol* 19:57–64.

Wirth R, Herz H, Ryel RJ, Beyschlag W, Hölldobler B. (2003). *Herbivory of Leaf-Cutting Ants—A Case Study on Atta colombica in the Tropical Rainforest of Panama*. Berlin: Springer Verlag.

Ydenberg R, Schmid-Hempel P. (1994). Modelling social insect foraging. *TREE* 9:491–93.

Part III

Modeling Social Insect Foraging

15 An Evolutionary Simulation of the Origin of Pheromone Communication

Yoshiyuki Nakamichi and Takaya Arita

CONTENTS

INTRODUCTION

Colonies of eusocial insects, such as ants and bees, demonstrate a variety of complex and adaptive behaviors, even though the individuals of a colony follow a relatively simple and limited set of rules (see Chapters 2 and 14). Due to the increasing interest in developing artificial systems inspired by social insects, the last decade has seen an explosion of research in the scientific areas referred to as swarm intelligence or collective intelligence (Bonabeau et al. 1999).

In ants, the communication by means of pheromones is crucial for the social organization of a colony, since it mediates various different work processes such as, for instance, the exploitation of food sources (Hölldobler and Wilson 1990; Deneubourg et al. 1990; Bonabeau et al. 1999; see also Chapter 2). The fundamental question of how such complex pheromone communication systems have evolved, is of interest not only to biologists but also to other fields of research, such as linguistics (Arita 2002) and practical engineering (ant colony optimization [ACO]; Dorigo 1992; Bonabeau et al. 2000; swarm robotics, Payton et al. 2001).

As a first step toward an understanding of the origin of pheromone communication, we will focus on its adaptability in a population of agents in the present chapter. In fact, there are two computational approaches to this issue. The first one is to assign in advance both the meaning of the pheromone and the behavioral rules to the agents. The second one is to let the agents establish communication autonomously through learning or evolution. The latter approach is especially effective in agent populations that have little or no *a priori* knowledge about how to solve a particular problem. But can pheromone communication really evolve without *a priori* knowledge? In fact, pheromone communication will not be possible at all until the following two rules autonomously emerge in accordance with each other: a rule that assigns particular conditions in which the agents deposit a pheromone, and a rule that determines the reactions of the agents when detecting the pheromone.

Kawamura and co-workers conducted evolutionary simulations in which pheromone communication indeed evolved when two colonies of artificial ants competed for food (Kawamura and Ohuchi 2000; Kawamura et al. 2001). Encouraged by these studies, we addressed the following questions: Is it possible for populations of artificial ants to solve more complex problems through pheromone communication that they autonomously establish by means of learning or evolution? And if it is, how do the agents attach a meaning to one particular pheromone in settings in which they are able to deposit several different kinds of pheromones? This chapter reports on the current state of our research based on computer simulations that use a simple model for the evolution of pheromone communication, in which the ant agents' neural networks evolve as the result of their foraging activity.

THE ANT FORAGING MODEL

BACKGROUND AND DESIGN INTENT

Two models of artificial ant agents, which search for food in a grid environment, served as the basis for our experimental design. In AntFarm (Collins and Jefferson 1992), ant agents left their nest with the purpose to return with food. In this model, the foragers were composed of neural nets that obtained the following inputs: the availability of food in the surroundings, the presence or absence of the nests, the direction of the nest, the amount of pheromones present, whether or not the forager was carrying food, and random values to diversify the ants' behavior. According to the output of the neural net, the agents moved around, picked up or put down food, and deposited pheromones. The genetic algorithm, which evolved the weights of the neural net, used the number of food items brought back to the nest as an evaluation function. In this model, the artificial ant agents evolved to gather more food, yet coordinated behaviors based on pheromone communication did not emerge.

In Ants War (Kawamura and Ohuchi 2000; Kawamura et al. 2001), by contrast, adaptive pheromone communication emerged. This model simulated the competition for food sources among ant colonies. Ants War involved two types of artificial ant agents (two different colonies), which searched for food in a grid environment and competed with each other in regard to the number of gathered food items. Here, the foragers were constructed as neural nets with the following inputs: the presence of food and nestmates (agents of the same type) in the surroundings and pheromone gradients. According to the output values of the neural net, the agents moved around, waited, searched for food, and deposited pheromones. As in AntFarm, a genetic algorithm evolved the weights of the neural network based on the amount of gathered food. In Ants War, however, cooperative behavior in the form of assembling nestmates at food sources by means of pheromone communication emerged.

What are the possible reasons that adaptive pheromone communication emerged in Ants War but not in AntFarm? (1) The agents in AntFarm only detected pheromone levels, whereas the agents in Ants War detected pheromone gradients, which carry directional information. This capacity of the foragers enabled them to combine pheromone information with their traveling behavior. (2) In AntFarm, the agents had to search for food and the nests, whereas in Ants War their only task was to search for food. This simplification of the problem for the agents, and stimulating the combination of pheromones and food-searching behavior, gave rise to cooperative activity in Ants War. In fact, two types of pheromones emerged in Ants War, attractive and repellent pheromones, which both facilitated the food-searching activity of the foragers.

Adaptive pheromone communication likely emerges when an increase in problem complexity multiplies the potential behaviors of agents that need to be facilitated. Yet, which types of pheromone communication will emerge if the number of pheromones is increased?

Building on these findings and on the open questions, we designed the following model of ant foraging behavior in order to investigate what kinds of adaptive pheromone communication arise when the foragers can use more than one pheromone in a setting in which they are confronted with a complex task (Nakamichi and Arita 2004, 2005, 2006). Thus, our model differed from Ants War in both the problem setting and the number of pheromones used by the agents.

The problem setting was actually the same as that in AntFarm, involving the search for food and its transport back to the nest. Consequently, the behaviors that the foragers had to carry out efficiently were "getting to the food source" and "returning to the nest." The design of the agents allowed them to detect pheromone gradients in the same manner as in Ants War. The agents' behavior was determined by a neural net, whose weights and threshold values were expressed as "genes," and which evolved in the framework of a genetic algorithm. Furthermore, the foragers were capable of detecting and secreting two different pheromone types. Thus, in line with the increased complexity of the problem compared to Ants War, the emergence of a pheromone communication, in which each type of pheromone serves a different purpose, was possible. Up to this day, maximally two different pheromones, for which the function had been assigned in advance, have been considered in computer experiments. So far, no model has considered the evolution of communication when multiple types of pheromones are available to the ant agents. However, in order to get a more profound insight into the potential origin and evolution of pheromone communication, it proves necessary to study the potential emergence of communication in complex task settings in which foragers are able to use multiple types of pheromones.

ENVIRONMENT

Our model was based on multiagent modeling and, similar to Ants War, was inspired by foraging ants in nature. The agents moved around an $X \times Y$ grid environment with several food resources, a nest, and pheromone markings (Figure 15.1). Each colony consisted of N_a foragers with the aim to look for food and carry as many food items as possible back to their nest. The food resources were located on grids, and more than one food item could exist on the same spot. The nest was defined as a certain area of the grid environment. As soon as the foragers delivered food into the nest, these food items were stored. In other words, they disappeared from the environment.

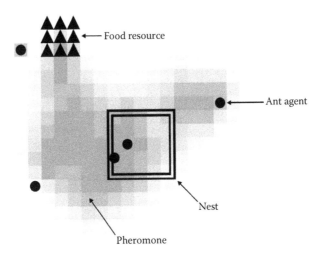

FIGURE 15.1 The grid environment of the ant foraging model used in the present study, containing several food resources, a nest, pheromone markings, and ant agents. (From Nakamichi and Arita 2006. With permission.)

PHEROMONE

The ant agents were able to secrete several types of pheromones, which did not affect each other, at the same point in the environment. In the following, the type of pheromone is identified by the subscript v ($v = 1, 2, ...$). Each forager was capable of depositing pheromones, which subsequently gradually evaporated and diffused. The ant agents could detect only diffusing pheromones. In the following, deposited pheromone and diffusing pheromone are represented by $T_v(x,y)$ and $P_v(x,y)$, respectively, where $T_v(x,y)$ is the intensity of deposited pheromone at position (x,y) and $P_v(x,y)$ is the intensity of diffusing pheromone at position (x,y). The diffusion process was defined by a partial differential equation as follows:

$$T_v^*(x,y) = (1 - \gamma_{eva}) T_v(x,y) + \Delta T_v^k(x,y)$$ (15.1)

$$\Delta T_v^k(x,y) = \begin{cases} Q_p & \text{if } k\text{-th ant agent on the grid } (x,y) \text{ deposits the pheromone} \\ 0 & \text{otherwise} \end{cases}$$ (15.2)

$$P_v^*(x,y) = P_v(x,y) + \gamma_{dif}(P_v(x-1,y) + P_v(x,y+1) + P_v(x,y-1) + P_v(x+1,y) - 5P_v(x,y)) + \gamma_{eva}T_v(x,y)$$ (15.3)

where the parameters γ_{eva} and γ_{dif} are the evaporation rate and the diffusion rate of the pheromone per unit time, respectively. The superscript asterisk denotes the intensity of pheromone at time $t+1$, and Q_p is the intensity of the pheromone deposited by an ant agent.

ANT AGENT

All ant agents of a colony were homogenous and their behavior followed the same decision mechanism. The agents carried out the following sequence of actions (Figure 15.2):

1. *Sense the environment and the proper state*: The ant agent sensed whether a food item existed on the grid, whether the grid was part of the nest, and it recognized whether it was carrying food or not.
2. *Deposit a pheromone*: Ant agents deposited pheromone in accordance with the output of their neural network. Each ant agent used N_v types of pheromone. We set $N_v = 0, 1,$ or 2 in the experiments.
3. *Pick up or put down a food item*: When the ant was at the food source and carried no food, it picked up one food item. When the ant agent was in the nest and carried a food item, it put the food down.
4. *Select and carry out one action*: If the ant agent had not been able to execute the previous action (3), it selected and carried out one of the following five options: "wait (do nothing)," "move forward," "turn backward," "turn left," or "turn right." If the forager selected an action that was impossible to carry out, it waited instead. The action was stochastically selected according to the output of each agent's neural network.

All ant agents had the same simple two-layer neural network with $3 + 4N_v$ sensory input neurons and $5 + 2N_v$ output neurons. The output value I_i ($i = 1, ..., 3 + 4N_v$) of the i-th sensory input neuron of an ant agent was defined as follows:

$$I_1 = \begin{cases} 1 & \text{if food is present} \\ 0 & \text{otherwise} \end{cases}$$ (15.4)

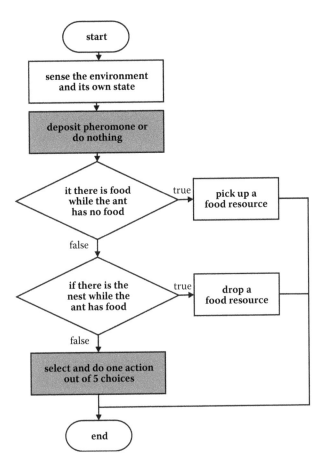

FIGURE 15.2 Behavior of ant agents. At each time unit, the foragers carried out the same sequence of behavioral actions, as shown in the flowchart. The action in the gray squares was stochastically selected according to the output of each agent's neural network. The five choices referred to in the lower gray square were "wait," "move forward," "turn backward," "turn left," or "turn right." (From Nakamichi and Arita 2006. With permission.)

$$I_2 = \begin{cases} 1 & \text{if the ant is on the nest} \\ 0 & \text{otherwise} \end{cases} \tag{15.5}$$

$$I_3 = \begin{cases} 1 & \text{if the ant has food} \\ 0 & \text{otherwise} \end{cases} \tag{15.6}$$

$$I_{4+4(v-1)\dots7+4(v-1)} = \begin{cases} 1 & \text{if } P_v^s \text{ is satisfied} \\ 0 & \text{otherwise} \end{cases} \tag{15.7}$$

where the position of an ant agent is assumed to be (x,y), and the adjacent position is (x',y') $\in \{(x-1,y),(x+1,y),(x,y-1),(x,y+1)\}$. $P_v^s(x',y')$ is the probability of firing a pheromone-sensitive neuron. This stochastic firing function, which was used to obtain discrete values for $I_{4+4(v-1)\dots7+4(v-1)}$, depended on the pheromone gradient and was defined as follows:

$$P_v^s(x',y') = \frac{1}{1+\exp\left(-\frac{P_v(x',y')-P_v^\#(x,y)}{T}\right)} \tag{15.8}$$

$$P_v^\#(x,y) = \frac{P_v(x-1,y)+P_v(x+1,y)+P_v(x,y-1)+P_v(x,y+1)}{4} \tag{15.9}$$

The output value O_j ($j = 1, ..., 5+2v$) of the j-th output neuron was calculated using the following regular equations:

$$O_j = f\left(\sum_{i=1}^{3+4N_v} \omega_{ij} I_i - \theta_j\right) \tag{15.10}$$

$$f(x) = \frac{1}{1+\exp(-x)} \tag{15.11}$$

where ω_{ij} and θ_j are the weight and threshold of the neural network.

The output values O_{1-5} corresponded to "wait," "move forward," "turn backward," "turn left," and "turn right," respectively. The output values $O_{6+2(v-1)}$ and $O_{6+2(v-1)+1}$ corresponded to the actions "deposit a pheromone v" and "do not deposit a pheromone v," respectively.

EVOLUTION

A genetic algorithm (GA) was applied to evolve the neural networks of the ant agents (Figure 15.3). All ant agents of a colony were homogenous, as described in the previous subsection. Therefore, an ant colony could be expressed as one genotype, which was shared by all ant agents in this colony. A set of weights and thresholds describing a neural network was directly encoded onto the "genes" of an ant's genotype. In the system there were N genotypes, the genes of which were initially set to random values of [–0.5, 0.5]. A fitness function was defined to be 1 plus the number of food items that had been stored in the nest. A new generation of N genotypes was generated by roulette wheel selection and mutation. The mutation operator was defined as the addition of a random real number of [–0.5, 0.5] to the current gene value with the mutation probability μ. These processes were repeated until the final generation g_{max}, which was predefined for each trial, was reached.

```
Initialize GA
For g = 1 to g_max do
        For n = 1 to N do
                Initialize ant foraging
                For t = 1 to t_max do
                        For k = 1 to N_a do
                                Ant agent k acts according to Figure 15.2

        For n = 1 to N do
                Select one genotype by roulette selection
                Mutate the genotype and add it to pool of next generation
End
```

FIGURE 15.3 Genetic algorithm (GA), in pseudocode, used to evolve the neural network of the ant agents. (From Nakamichi and Arita 2006. With permission.)

COMPUTER EXPERIMENTS

SETTING

The nest was located in the center of the grid environment of X × Y size and occupied an area of 5 × 5 grid cells. Seventy-two food items were placed on a randomly selected 3 × 3 area at time $t = 0$. Each grid in this food area contained eight food items. Fifteen trials were conducted for each number of pheromones N_v ($N_v = 0, 1, 2$) used by the agents. Details of the parameter settings for both the ant foraging and the genetic algorithm (GA) are given in Table 15.1.

NUMBER OF PHEROMONES USED

To answer the question of whether ants in our model evolve to gather more food, we counted the average number of stored food items (score) gathered by foragers between the 15,001st and 25,000th generation (Figure 15.4). Those trials with a score between 6 and 12 food items (low-score trials) were labeled as class a, while those with scores of 14 to 22 food items (high-score trials) were labeled as class b. In all trials where no pheromone was present ($N_v = 0$), the ant agents stored six to eight food items (class a). By contrast, when foragers used pheromones ($N_v \geq 1$), at least in some trials the agents collected between fourteen and twenty-two food items (class b). The trial with the highest score (20–22 food items) was observed when the ant agents used only one type of pheromone ($N_v = 1$) (Figure 15.4).

As can be seen in Figure 15.5, the average number of stored foods steeply increased to between seven and ten items within the first three thousand generations in all studied cases ($N_v = 0, 1, 2$). Thereafter, transitions differed depending on the number of pheromones (N_v) used by the ant agents. In the case of $N_v = 0$, the number of stored foods remained at a level of around seven items (Figure 15.5a). In the case of $N_v \geq 1$, we found two different types of transitions, which indicates that there are at least two evolutionary pathways. In one type (cases 1a and 2a; low-score trials), the average number of stored foods remained in a range between seven and ten items, which is similar to the case of $N_v = 0$. In the other type (cases 1b and 2b; high-score trials), however, the average number of stored foods increased beyond ten items. In case 1b, the moving average of the stored food oscillated between fifteen and twenty-five items, whereas it gradually increased beyond seven in case 2b (Figure 15.5b and c).

The results of our experiments can be summarized as follows: Ant agents that used pheromones ($N_v \geq 1$) were clearly more efficient in collecting and storing food than those without pheromone use ($N_v = 0$). Surprisingly, however, the efficiency of the ants did not increase with the number of available pheromones. In particular, case 1b (one pheromone used) outperformed all the other cases

TABLE 15.1
Parameter Settings of Ant Foraging and the Genetic Algorithm (GA)

			Ant Foraging						GA	
X	Y	N_a	t_{max}	Q_p	γ_{eva}	γ_{dif}	T	N	g_{max}	μ
50	50	1,000	1,000	1	0.1	0.1	0.002	20	20,000	0.05

Source: From Nakamichi and Arita 2006. With permission.

Note: X, grid width; Y, grid height; N_a, number of ant agents; t_{max}, number of unit time in one generation; Q_p, intensity of the pheromone deposited; γ_{eva}, evaporation rate of the pheromone per unit time; γ_{dif}, diffusion rate of the pheromone per unit time; T, sensitivity of pheromone sensors; N, number of phenotypes (population size); g_{max}, number of generations in a trial; μ, mutation rate per gene.

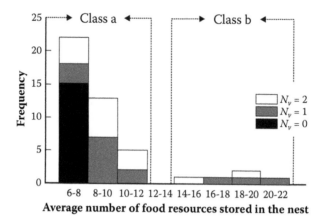

FIGURE 15.4 Histogram of the average number of stored food items. The results were classified into low-score trials (class a) and high-score trials (class b). All trials in which agents did not use pheromones ($N_y = 0$; black bars) belonged to class a, whereas some trials in which the agents used one (gray bars) or two (white bars) pheromones ($N_y \geq 1$) belonged to class b. (From Nakamichi and Arita 2006. With permission.)

in terms of food items stored (Figure 15.5). This finding points to the difficulty in establishing the meaning of a pheromone when ants use multiple types of pheromone (see "Effectiveness of Evolved Pheromone Communication" section below).

The analysis of the neural network revealed that in case 1b the ant agents behaved as follows. Generally, the ant agents moved forward with a probability of about 90%, and turned right with a probability of about 10%. Yet, when they encountered a decreasing pheromone gradient, the ant agents turned right with a probability of about 40%. Furthermore, ant agents without food hardly secreted pheromone, whereas those that carried food items secreted pheromone with a probability of 40%. In the nest, the ant agents secreted pheromone with a probability of 20%, independently of whether they carried food or not. It must be noted that the pheromone and food/nest information were not strictly tied. The probability of secreting a pheromone, therefore, was similar when agents carried food and when they were in the nest. Consequently, the pheromone moderately indicated the presence of both food and nest.

EFFECTIVENESS OF EVOLVED PHEROMONE COMMUNICATION

In order to investigate the effect of pheromone on the ants' foraging efficiency, the pheromone sensors of ant agents of the last generation ($g = 20,000$) were inhibited, so that they perceived a constant pheromone value. Thereafter, these pheromone-inhibited ant agents performed one thousand foraging tasks. The comparison of the average number of food items stored by normal and by pheromone-inhibited ant agents revealed that, in all cases of $N_y \geq 1$, normal ant agents collected more food than pheromone-inhibited foragers (Table 15.2). This finding clearly demonstrates the emergence of a pheromone communication that increased the efficiency of foraging in normal colonies. It should be noted here that the difference between normal and pheromone-inhibited ant agents was larger in case 2b than in 1b (Table 15.2). From this, it can be concluded that the neural networks had evolved to be more adapted to processing pheromones when a higher number of pheromones was available to the ant agents. However, normal ant agents in case 2b (two pheromones) had a lower score than those in 1b (one pheromone) (Table 15.2). This indicates that ant colonies using multiple types of pheromone do not necessarily perform better in regard to the number of food items stored in the nest than those using only one type.

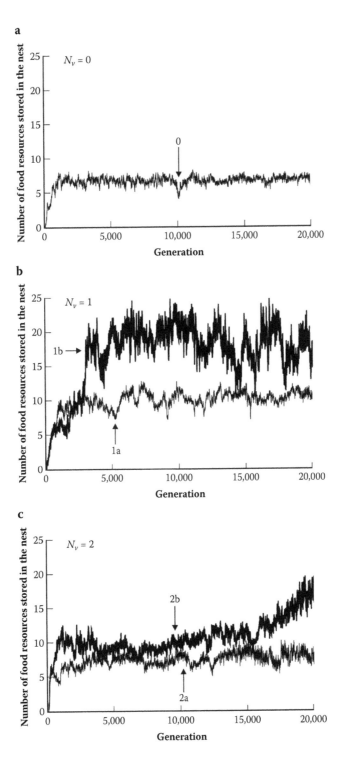

FIGURE 15.5 Moving average of food resources stored in the nest when agents used (a) no pheromones ($N_v = 0$), (b) one type of pheromone ($N_v = 1$), or (c) two types of pheromone ($N_v = 2$). Note the differences between low-score trials (1a, 2a) and high-score trials (1b, 2b) in cases with pheromone use. (From Nakamichi and Arita 2006. With permission.)

TABLE 15.2
Average Number of Food Items Stored by "Normal" Ants and Pheromone-Inhibited Ants

N_v	Class	Normal Ants	Pheromone-Inhibited Ants
0	a	8.8	(Same as normal)
1	a	18.9	12.3
1	b	28.0	15.6
2	a	14.4	13.8
2	b	24.7	5.7

Source: From Nakamichi and Arita 2006. With permission.

Note: N_v, number of types of pheromone use.

COMPARISON WITH HUMAN-DESIGNED PHEROMONE COMMUNICATION

In order to evaluate the effectiveness of evolved pheromone communication, the results of the experiments we described and discussed above were compared with the results from experiments using human-designed pheromone communication (Figure 15.6). In these experiments, the ant agents used the following two types of pheromone: (1) a food pheromone and (2) a nest pheromone. Similar settings are, in fact, commonly applied in pheromone communication models (e.g., Gambhir et al. 2004). In our experiments, the food pheromone, which was deposited by ant agents on encountering a food resource, indicated the presence of food resources. If ant agents without food detected this pheromone, they moved along its gradient toward higher pheromone concentrations. The nest pheromone indicated the location of the nest. If ant agents were in the nest, they deposited this pheromone. If ant agents that carried food items detected this pheromone, they moved along its gradient toward higher pheromone concentrations. We expected that strictly tying different pheromones to either the food source or the nest would improve the food collecting efficiency of the ant agents.

We used four different settings—resulting from the combination of two pheromone-depositing rules and two movement rules—for our human-designed ant agents. Food pheromone was deposited either when an ant agent encountered a location that contained food (food present) or when it was carrying food (carrying food). In regard to traveling and changing direction, one rule was to "move forward, face backward, face left, face right" (with moving direction correlating with facing direction; nondirectional movement), while the other excluded "facing backward" (directional movement). With each of these four types of human-designed ant agents (food present + nondirectional; food present + directional; carrying food + nondirectional; carrying food + directional), we conducted one thousand foraging tasks. The average number of food resources stored in the nest by these foragers is shown in Table 15.3.

Surprisingly, human-designed ant agents had less foraging success—in terms of food items stored in the nest—than the ant agents with an evolved pheromone communication ($N_v \geq 1$) (compare Tables 15.2 and 15.3). This finding indicates that evolved pheromone communication is more adaptive than human-designed pheromone communication, and shows that an effective pheromone communication, which cannot simply be imitated by human design, emerged through evolution in the computer experiments. The details of evolved pheromone communication are expected to be clarified in future studies by closely investigating the relation between sensory inputs and output actions of the neural networks.

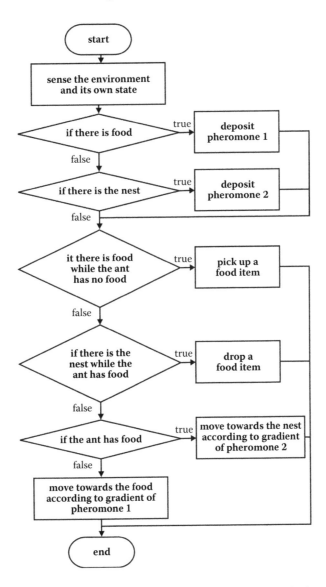

FIGURE 15.6 Behavior of human-designed ant agents. In these experiments, the foragers moved between a food source and the nest, using a food pheromone (pheromone 1) and a nest pheromone (pheromone 2). Every time unit, they carried out the same sequence of behavioral actions, itemized in the flowchart. (From Nakamichi and Arita 2006. With permission.)

SUPERIORITY OF EVOLVED PHEROMONE COMMUNICATION

The results of the second set of experiments demonstrated that, concerning the foraging success, evolved pheromone communication is more efficient than human-designed pheromone communication. But why, actually, was the human-designed pheromone communication outperformed by the evolved pheromone communication? To answer this question, we closely analyzed the behavior of the foragers and the spatial distribution of the deposited pheromone during the experiments. In the case of the human-designed ant agents, the ideal state for an efficient food collection was an equal amount of both pheromone types—the food pheromone and the nest pheromone—in the environment (Figure 15.7a). Yet, we frequently observed ill-balanced states (Figure 15.7b and c), which resulted in

TABLE 15.3
Average Number of Stored Food Items in Human-Designed Ants Following Different Pheromone-Depositing Rules (Food Present/Carrying Food) and Different Movement Rules (Nondirectional/Directional) (See Text for Details)

	Food Present	Carrying Food
Nondirectional movement	12.4	9.3
Directional movement	7.7	11.4

Source: From Nakamichi and Arita 2006. With permission.

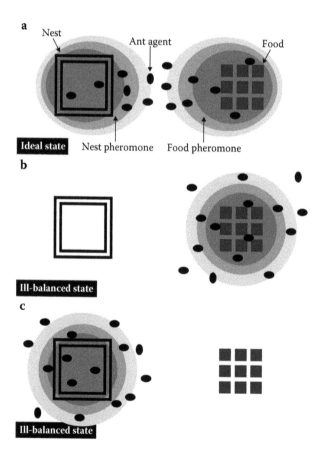

FIGURE 15.7 Ideal and ill-balanced states in the case of human-designed ant agents, using a food pheromone and a nest pheromone. In the ideal state (a), equally distributed amounts of both pheromone types resulted in a coordinated foraging behavior of the agents. When the pheromones were unevenly distributed (ill-balanced states), the foragers crowded either at the food source (b) or the nest (c). (From Nakamichi and Arita 2006. With permission.)

a congregation of the foragers either around the food sources or in the nest. These ill-balanced states had not been designated by our settings. Their unintended—and unexpected—emergence points to the difficulty in controlling the behavior of pheromone-communicating ant agents, which was caused by the dynamic properties—diffusion and evaporation—of the pheromones with time.

In contrast to the human-designed ant agents, foragers rarely crowded together at the food sources or in the nest in the case of evolved pheromone communication. This is particularly true in the most efficient case, 1b (Figure 15.5b), in which ant agents secreted a single type of pheromone with high probability both inside the nest and when carrying food. In consequence, the pheromone moderately indicated the presence of both the food sites and the nest to searching foragers. A detailed analysis of the evolved network showed that the ant agents had evolved a probabilistic tendency to move either forward or toward locations with a high pheromone concentration. In addition, they showed a weak tendency to turn either right or left, depending on the respective trial, especially when heading down the pheromone concentration gradient. In consequence of these right or left turns, the ant agents tended to move around both the food sites and the nest, sometimes in circles (Nakamichi and Arita 2006). In fact, the foragers did not behave optimally because they had no possibility of distinguishing between food sites and the nest by using only one type of pheromone. Yet, it is significant to note that ill-balanced states, which could frequently be observed in the case of human-designed ant agents, only rarely occurred in ant colonies with evolved pheromone communication.

Taking the best performing case (case 1b; Figure 15.5b) as an example, the advantageous features of the evolved pheromone communication over the human-designed model can be summarized as follows: A single type of pheromone had evolved to moderately indicate the presence of both the food sites and the nest. This, apparently, prevented any excessive accumulation of ants at either the food sites or the nest, as was sometimes the case in the human-designed model in consequence of an imbalanced distribution of pheromones. Therefore, in evolved pheromone communication, both the diversity of pheromone distribution and the diversity of the ant agents' collective behavior were maintained at appropriate levels for the colonies to accomplish a robust and efficient food collection task.

CONCLUDING REMARKS

In this chapter, we introduced a model for the evolution of pheromone communication in ants, in which the neural networks of ant agents evolved in accordance with their foraging success. Our computer experiments revealed the following features of this evolved pheromone communication: (1) Evolved pheromone communication can be adaptive, but it is not necessarily true that the efficiency of the pheromone communication increases with the number of available pheromone types. (2) Evolved pheromone communication outperformed human-designed pheromone communication in terms of foraging success of the ant agents, which suggests that an effective pheromone communication, which cannot simply be imitated by human design, emerged through evolution. This, eventually, points to the importance of an evolutionary mechanism for the establishment of an effective pheromone communication with its dynamic properties of diffusion and evaporation of pheromones over time.

We are currently conducting further experiments to investigate how—and what kind of—effective pheromone communication could emerge in the above presented model. In fact, our model can be extended in several directions. One is to consider more complex task settings, in which diverse behaviors and various forms of communication could emerge. Another challenge is the application of our model in the field of swarm robotics.

REFERENCES

Arita T. (2002). *Artificial Life: A Constructive Approach to the Origin/Evolution of Life, Society, and Language.* Tokyo: Igaku-Shuppan [in Japanese].

Bonabeau E, Dorigo M, Theraulaz G. (1999). *Swarm Intelligence: From Natural to Artificial Systems.* Oxford: Oxford University Press.

Bonabeau E, Dorigo M, Theraulaz G. (2000). Inspiration for optimization from social insect behaviour. *Nature* 406:39–42.

Collins RJ, Jefferson DR. (1992). AntFarm: Towards simulated evolution. In Langton CG, Taylor C, Farmer JD, Rasmussen S (eds.), *Artificial Life II*. Redwood City, CA: Addison-Wesley, pp. 579–601.

Deneubourg JL, Aron S, Goss S, Pasteels JM. (1990). The self organizing exploratory patterns of the Argentine ant. *J Insect Behav* 3:159–68.

Dorigo M. (1992). Optimization, learning and natural algorithms. PhD thesis, Polytechnic Institute of Milan, Italy [in Italian].

Gambhir M, Guerin S, Kauffman S, Kunkle D. (2004). Steps toward a possible theory of organization. Paper presented at Proceedings of ICCS-2004, Boston. Accessible at http://necsi.org/events/iccs/2004 proceedings.html.

Hölldobler B, Wilson EO. (1990). *The Ants*. Cambridge, MA: Belknap Press of Harvard University Press.

Kawamura H, Ohuchi A. (2000). Evolutionary emergence of collective intelligence with artificial pheromone communication. In *Proceedings of IECON-2000*, Nagoya, Japan, pp. 2831–36.

Kawamura H, Yamamoto M, Ohuchi A. (2001). Investigation of evolutionary pheromone communication based on external measurement and emergence of swarm intelligence. *Trans Soc Instrum Control Eng* 37:455–64.

Nakamichi Y, Arita T. (2004). An evolutionary simulation of the origin of pheromone communication. In *Proceedings of SEAL-2004*, Busan, Korea, pp. 1–6.

Nakamichi Y, Arita T. (2005). Effectiveness of emerged pheromone communication in an ant foraging model. In *Proceedings of the 10th ISAROB*, Oita, Japan, pp. 324–27.

Nakamichi Y, Arita T. (2006). An evolutionary simulation of the origin of pheromone communication. *Trans Inform Process Soc Jpn* 47:78–88 [in Japanese].

Payton D, Daily M, Estowski R, Howard M, Lee C. (2001). Pheromone robotics. *Auton Robot* 11:319–24.

Mathematical and Neural Network Models of Medium-Range Navigation during Social Insect Foraging

Allen Cheung

CONTENTS

INTRODUCTION

Foraging is a complex task, particularly in a social animal. This chapter will focus on one critical aspect of social insect foraging, namely, navigation, without reiterating all the important aspects related to social insects' communication and foraging behaviors. I will briefly provide an overview of some important, but distinct, ways in which medium-range navigation in foraging insects have been modeled, with a particular focus on the effects of social interactions.

It is clear from the preceding chapters that a great deal of behavioral data related to social insect foraging has been gathered. Furthermore, there is a complex interplay among the self, the society, and the resources, in generating behavior. In order for foraging strategies and information transfer systems to be successful in generating useful foraging behavior, there must be a robust navigation system at the foundation.

Just how do insects navigate? At the time of writing of this chapter, no complete, published model of insect navigation—either top-down or bottom-up, either in canonical form or neural network based—is available. Numerous hurdles exist. One problem has been the technical difficulties of obtaining *in vivo* cellular-level data during medium- to long-range navigational tasks. As such, neither the connectivity nor functionality of the neurocircuitry of medium-range navigation (see below) has been directly measured. Another significant challenge to modeling lies in not knowing whether the various subsystems of navigation can be considered separate modules, or whether the observed behaviors are necessarily the consequence of the complex interactions of all the subsystems. Third, it is now increasingly accepted that the behavior of, say, a colony of honey bees should probably be considered as one superorganism rather than the sum of many individuals. Does that mean that the navigational system of the colony is some sort of supersystem?

Even though a complete answer is not currently available, many partial solutions still exist. This chapter will briefly describe some of the most important groups of published models in an attempt to weave together large patches of what may eventually become a complete solution. It should be noted at the outset that a complete model will not be offered here, but some ideas for future research will be briefly discussed. It should also be noted that most of the mathematics and detailed neural network descriptions have been omitted in order to maintain perspective on the larger picture, and perhaps also to appeal to a wider readership.

Figure 16.1 shows a schematic representation of one way of considering the complex relationship and interactions between the various aspects of social insect foraging and the task of navigation.

FIGURE 16.1 Schematic representation of information flow during social insect foraging. In this scheme, navigation (medium range) directly controls locomotion. In turn, higher-level commands and strategies direct the navigation system. Due to the strong interdependence between all components—direct or indirect—this representation serves as a reminder that, in general, mathematical and neural network models of insect navigation always make assumptions and simplifications of reality.

Even such a simplified scheme shows clearly that navigation cannot be considered in isolation but instead must be in context. The corollary of that argument is that foraging strategies, communication, and so forth, are not independent of navigational properties and limitations. Hence, when one considers a model relevant to medium-range navigation during social insect foraging, it is imperative to be mindful of the assumptions, simplifications, and focus of the model.

MEDIUM-RANGE NAVIGATION

Locomotion is a defining characteristic of nearly all animals. For the current discussion, navigation will be defined as the process of controlling nonrandom locomotion. This definition includes searching strategies, path integration (dead reckoning), map-based navigation (e.g., pilotage), and grid-based navigation (e.g., view-based homing).

For most insects, long-range migration is not the type of movement used for day-to-day foraging, and it will not be considered here. At the other extreme, there are moment-to-moment locomotory mechanisms, which are fundamental to keep an insect moving. Such mechanisms are generally very robust, and tend to be predictable under given local environmental cues. Hence, in the context of foraging, their direct influence on behavior ought to be small. Medium-range navigation, therefore, would include journeys of typically a few meters to several hundred meters. Even with these restrictions, however, giving a detailed description of all existing models in this chapter is impossible due to space and time limitations. Instead, the major groups of models will be sampled to give an overall impression of the current state of progress.

SEARCHING FOR A RANDOM LOCATION

The first group of models considered here includes top-down mathematical descriptions of searching behavior. The ecological context in connection with foraging is to find a resource without prior knowledge of its location. A common example would be an insect searching for a new source of food.

The problem is fundamentally one of searching for sparsely located targets. The critical issue is the sparseness of resource. Otherwise, any kind of movement will likely result in success, with little cost in terms of time, energy, or risk exposure to the insect. When the cost of searching becomes significant, then, at least in theory, there ought to be significant natural selective pressures to optimize searching. The following four distinct search strategies will be considered here: (1) random walk, (2) idiothetic directed walk, (3) allothetic directed walk, and (4) Lévy walk (actually a Cauchy walk) (Figure 16.2).

It should be noted that in this context, the term *walk* does not necessarily imply terrestrial locomotion. It is a mathematical term denoting some form of continuous physical movement, and implies that finite time is taken between "steps." In comparison, a *flight*, as in Lévy flight, describes a series of jumps that may or may not be instantaneous. Indeed, a plot of a Lévy flight is sometimes called Lévy dust since only the sparsely distributed endpoints of each step are shown (Mandelbrot 1983).

Strategies 1, 2, and 4 have historical significance and mathematical relevance. They share the property that with an increasing number of steps, the probability of searching any given position approaches unity. This recurrence property holds for one- and two- but not three-dimensional space. Strategy 3 was included as a control to illustrate the importance of recurrence in a searching strategy. Some of the most important properties, as well as the expected radial displacements of the different search strategies, are summarized in Tables 16.1 and 16.2.

Random Walk

The random walk (RW) was described and characterized several decades before biologists began to recognize its potential use (Pearson 1905, 1906; Rayleigh 1905, 1919). Skellam (1951) was one of

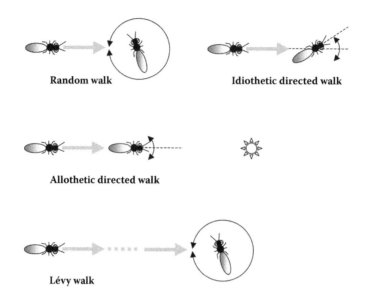

FIGURE 16.2 Schematic illustration of four searching strategies. During a random walk, an insect may turn in any direction at each step. In contrast, during an idiothetic directed walk it will make a small turn relative to its current body axis following each step, irrespective of its actual compass heading. During an allothetic directed walk, an insect would continually reorientate its heading relative to some external directional reference, e.g., the azimuth of the sun. During a Lévy walk, an insect may turn in any direction following a cluster of steps. The key difference in a random walk is the potentially large number of steps in the cluster, during which an insect does not change direction.

the first to show how animal dispersion may be modeled as a diffusion-like process, amenable to be described by the mathematics of a random walk.

The fundamental unit of a random walk consists of a step plus a totally random turn (Figure 16.2). Since each turn is totally random, there is no correlation whatsoever between the net displacement

TABLE 16.1
Comparison of Searching Strategies

Strategy	Individual Property	Population Property	Steering Mechanism
Random walk (RW)	Special case of PRW	Special case of CRW	Radially symmetric heading at each step
Idiothetic directed walk (IDW)	Persistent random walk (PRW) is a special case of IDW	Correlated random walk (CRW) is a special case of IDW	Restricted heading distribution, centered on egocentrically forward axis
Allothetic directed walk (ADW)	Goal directed: searching a narrow path rather than an area	Radial spoke pattern: missing sectors between spokes	Restricted heading distribution, centered on external direction
Lévy walk (LW)	Significant probability of long step clusters; biologically equivalent to many actual steps	Severe truncation of LW's length distribution results in a RW	Radially symmetric heading at each step cluster; presumably ADW during cluster; truncated heavy-tailed distribution for step cluster length

Note: CRW, correlated random walk; PRW, persistent random walk.

TABLE 16.2
Expected Radial Displacement of Different Search Strategies

Searching Strategy	Expected Radial Displacement, $\langle R \rangle$
Random walk	$\dfrac{\sqrt{n\pi}}{2}$
Idiothetic directed walk	$\lim\limits_{n \to \infty} \langle R \rangle = \dfrac{2}{\pi\sigma^2} \displaystyle\int_0^\infty z(\mu_X + z) e^{\frac{-z^2}{2\sigma^2}} EllipticE\left(\dfrac{\pi}{2}\middle|\dfrac{4z\mu_X}{(\mu_X + z)^2}\right) dz$

This result is an approximation first published in Cheung et al. (2007b).

$$\mu_X = \langle\cos\Delta\rangle \dfrac{1 - \langle\cos\Delta\rangle^n}{1 - \langle\cos\Delta\rangle}$$

σ^2 is the variance of the asymptotic bivariate normal distribution.

Allothetic directed walk	$n\langle\cos\Delta\rangle$

This is, in fact, the expected displacement along the intended heading direction, but closely approximates the expected radial displacement, especially as the number of steps increases.

Lévy walk	∞

Note: Δ is the turn angle from one step to the next. *EllipticE* denotes the incomplete elliptic integral of the second kind.

of successive steps. Hence, from basic statistical theory, the variance must increase linearly with the number of steps, while the expected radial displacement increases proportionally to $\sqrt{n}$ (n = number of steps) (Table 16.2). Since the probability distribution is radially symmetric (bivariate normal; see also Figure 16.3a), the actual dispersal of organisms is attributable purely to the increasing variance of each individual's position.

While the asymptotic solutions of the random walk are perfectly applicable to diffusion-like processes, they neglect the mechanistic properties of actual animal navigation. At timescales pertinent to an animal's locomotion, movement is discrete, and movement dynamics may be complex. The growth and spread of a newly established population of insects over many generations in a static environment may indeed behave like diffusion, but that would surely be the exception rather than the norm. How can insect movements be modeled individually, over timescales relevant to foraging?

IDIOTHETIC DIRECTED WALK

The first significant advance from the treatment of insect movements as random walk was the recognition that the heading distribution was not totally random at each step (Kitching 1971; Karieva and Shigesada 1983; Bovet and Benhamou 1988; Byers 2001). Macroscopically, the distribution of foraging insects within a population may appear to be diffusive, yet microscopically there is structure to the movement paths. In fact, the heading (allocentric direction of movement) of an animal at one step influences its heading at the subsequent step. In other words, the headings of successive steps are correlated. Cheung et al. (2007a) quantified the first and second positional moments of just such a walk, which, from a mechanistic point of view, is equivalent to an animal attempting to move in a straight line using only its idiothetic senses—hence the term *idiothetic directed walk* (IDW). Unlike the RW, the dispersal mechanism of the IDW is twofold (Figures 16.2 and 16.3b; Table 16.2). Initially, being a *directed walk*, the expected position of individual animals moves away from the starting point (e.g., home) along the direction of interest. Thus, a population of insects undergoing IDWs will begin dispersing in a ring-like

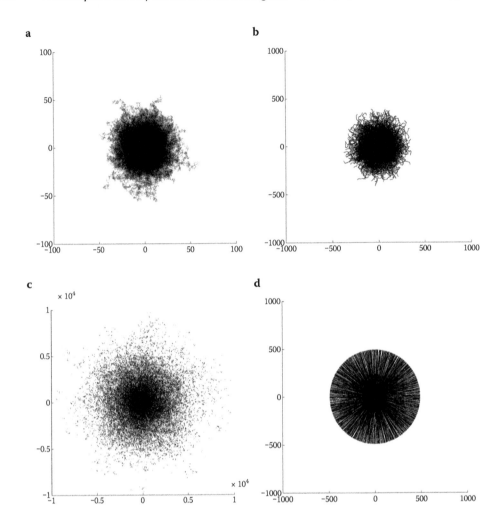

FIGURE 16.3 Computer simulation of ten thousand agents carrying out independent searching over five hundred steps. Shown are the endpoints of each agent following each step. All step sizes are 1 unit except in panel c (see text for details). (a) The result of agents searching by individually carrying out random walks. (b) The results of agents searching using correlated random walks with Gaussian errors of 0.2 rad standard deviation between steps. (c) The result of searching using truncated Lévy walks with a scale constant $\gamma = 1$, and truncated to a maximum length of 1,000 units, thus equating to a mean step length of approximately 49.5 units. (d) The results of searching using allothetic directed walks, equally spread in all directions, with Gaussian errors of 0.2 rad standard deviation between steps.

fashion radially outward. Later, as the heading errors accumulate, the positional variances approach a steady rate of increase, while the expected displacement asymptotes to a maximum value (Cheung et al. 2007a). The positional distribution will evolve toward a bivariate normal distribution. For the special case where locomotion may be described as a simple elementary step, the IDW is equivalent to a *persistent random walk* (PRW) (Wu et al. 2000; Bracher 2004), while a homogeneous radially distributed population of agents undergoing a simple IDW will behave as a *correlated random walk* (CRW) (Karieva and Shigesada 1983; Bovet and Benhamou 1988; Byers 2001).

An insect population undergoing IDWs will spread outward much faster than insects that move at random (compare Figure 16.3a with 16.3b, noting the change in scale). The asymptotic rate of spread is positively related to the correlation of heading between successive steps. With perfect positive correlation, the path is a straight line and a population of insects will simply travel radially outward—like a wheel spoke pattern (similar to Figure 16.3d). With complete independence (implying zero correlation) between steps, the IDW reduces to a RW.

It should be emphasized that there is a subtle but extremely important distinction between the use of an IDW as a searching strategy and the degeneration of an intended straight path into an IDW due to the lack of a compass (Cheung et al. 2007a). In the first case, the turns at each step are made purposefully and the resulting headings are measured directly using a compass. In the second case, the current heading is inferred/estimated from accumulated internal measures of rotation. The significance of this distinction will become more evident in the later section on path integration.

ALLOTHETIC DIRECTED WALK

It is now well understood that the expected maximum displacements of IDWs are limited (Cheung et al. 2007a). This is due to the fact that only an internal measure of "forward" is used at each step as the reference direction. Furthermore, it is known that using an external directional reference (i.e., a compass) to maintain heading avoids this limit. In addition, it also dramatically reduces positional uncertainty. Given this information, it is pertinent to ask whether a population of animals carrying out search via *allothetic directed walks* (ADWs; Figure 16.2) would have an advantage. Clearly, a population of individuals would have a larger expected radial displacement for a given number of steps (see Table 16.2). Indeed, it is the expected gain in position for each individual rather than its positional uncertainty that generates the population spread. However, does optimizing the rate of radial spread minimize the time to find sparsely located food sources?

There is one critical deficiency in this form of searching, which is the lack of recurrence. Even though the expected position continues to shift with searching, there is a tendency to move in more or less a straight line radially outward from the starting point. This means any location not on a line of search would be missed altogether, and even a source close to a line of search may be missed (by chance) the first time and never be approached again. *A priori*, it is assumed that the location of a resource is unknown. This type of searching may lead to vast distances covered with no guarantee that even nearby resources will be found.

LÉVY WALK

Viswanathan et al. (1999) used analytical arguments as well as computer simulations to find the optimal step size distribution in searching for sparse targets. The authors assumed that the turns are random (like a RW) but the distances between turns follow power-law distributions. The motivation for the latter assumption was based on experimentally observed distributions of intervals between turns across a range of species (Levandowsky et al. 1988; Cole 1995; Schuster and Levandowsky 1996; Viswanathan et al. 1996).

General power-law distributions have the property

$$\lim_{l \to \infty} pdf(l) \sim l^{-(1+\alpha)} \tag{16.1}$$

where *pdf* denotes the probability density function; $0 < \alpha < 2$, which leads to an infinite variance; and *l* denotes the length of movement before a turn is made. Once α reaches 2, the distribution tail is no longer heavy and the variance becomes finite. Furthermore, all distributions with power-law tails are scale-free or scale-invariant since

$$pdf(kl) = k^{-(1+\alpha)} l^{-(1+\alpha)} \propto l^{-(1+\alpha)} \tag{16.2}$$

where *k* is a proportionality constant. In other words, stretching or compressing the distribution (through multiplication of the length by a proportionality constant) does not alter its intrinsic shape (hence the transformed *pdf* is still proportional to the original *pdf* in Equation 16.1).

An important class of heavy-tailed distributions is the *Lévy skew alpha-stable distribution* (Lévy SαS distribution). A complete review of its properties is well beyond the scope of this chapter, but one very important characteristic is worth noting. In essence, the linear combination of randomly drawn values from a Lévy SαS distribution will follow the same distribution. For example, if undersampled data show that one measured interval or step during a journey followed a Cauchy distribution (a special form of Lévy SαS), it is quite possible that it is made up of many smaller steps, each also following a Cauchy distribution in length.

For completeness, the probability density of a Lévy SαS distribution may be written as the Fourier transform of its characteristic function:

$$f(l;\alpha,\beta,\gamma,\mu) = \frac{1}{2\pi}\int_{-\infty}^{+\infty}\varphi(t)e^{-itl}dt \tag{16.3}$$

where α is the exponent mentioned previously, γ is a scale parameter that stretches the width of the distribution, β is the skew parameter, μ is the shift, $\varphi(t) = e^{it\mu - |\gamma t|^\alpha(1 - i\beta \operatorname{sgn}(t)\phi)}$ is the characteristic function, and $sgn(t)$ is the sign of t (a real number), while

$$\phi = \begin{cases} -\dfrac{2}{\pi}\ln|t| & \alpha = 1 \\ \tan\left(\dfrac{\alpha\pi}{2}\right) & otherwise \end{cases} \tag{16.4}$$

The most important result obtained by Viswanathan et al. (1999) was that the optimal value of α for finding sparse targets was 1 (refer to Table 16.3), with the implication that all previous work assuming that the search should follow Gaussian, Rayleigh, Poisson, or other classical distributions

---

TABLE 16.3
Analytically Expressible Lévy Skew Alpha-Stable Distributions

α	Mean	Variance	Probability Density Functions	Special Name(s)
.5	∞	∞	$f(l) = \sqrt{\dfrac{\gamma}{2\pi}}\dfrac{e^{\frac{-\gamma}{2(l-\mu)}}}{(l-\mu)^{3/2}}\ l \in \{\mathbb{R} > \mu\}$	Lévy (distinct from Lévy SαS distribution)
1	Undefined	∞	$f(l) = \dfrac{1}{\pi}\dfrac{\gamma}{\gamma^2+(l-\mu)^2}\ l \in \{\mathbb{R}\}$	Cauchy/Lorentz
			Obeys standard central limit theorem	
2	μ	σ^2	$f(l) = \dfrac{1}{\sigma\sqrt{2\pi}}e^{\frac{-(l-\mu)^2}{2\sigma^2}}\ l \in \{\mathbb{R}\}$	Gaussian/normal
			$2\gamma^2 = \sigma^2$	

Note: Not all heavy-tailed distributions are stable. Brcich et al. (2005) proposed a statistical test for the stability of a heavy-tailed distribution. However, all alpha-stable distributions obey the generalized central limit theorem (CLT): Sums of independent identically distributed random variables from an alpha-stable distribution tend toward the same alpha-stable distribution. Generally, there is no closed-form expression for the density and cumulative distribution functions. Exceptions are the Gaussian, Cauchy, and Lévy distributions. These may be considered to be special cases of the Lévy skew alpha-stable distribution. Note μ is the shift parameter and γ the scale parameter. The Gaussian and Cauchy distributions are defined by setting the skew parameter $\beta = 0$ (unskewed), whereas the usual form of the Lévy distribution as shown here assumes $\beta = 1$ (maximum positive skewness).

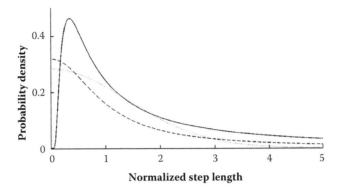

FIGURE 16.4 Comparison of the Gaussian (dotted line), Cauchy (dashed line), and Lévy (solid line) distributions. Shift parameter $\mu = 0$; scale parameter $\gamma = 1$. Note that the Gaussian distribution depicted here is equivalently denoted as $N(0,2)$, and that the Gaussian and Cauchy distributions are symmetrical about the shift parameter μ (the negative portions are not shown).

in step intervals would be suboptimal. Figure 16.4 graphically illustrates the difference between a Gaussian distribution and two heavy-tailed distributions (a Cauchy and a Lévy distribution). Notice that although all three functions are defined (have finite thickness) throughout the domain of the positive real axis, the heavy-tailed distributions are relatively thicker far from the origin.

A *random walk* whose step lengths are distributed according to a Lévy skew alpha-stable distribution may be loosely called a Lévy walk (Figure 16.2). However, it should be noted that the only density function defined in closed form for $\alpha = 1$ is the Cauchy or Lorentz distribution, *not* a Lévy distribution (see Table 16.3). Nonetheless, without adding further confusion to the developing field, the term *Lévy walk* will be adopted for now. The advantage of Lévy walks is that the probability of returning to a previously visited site is smaller than it is for random walk behavior (Shlesinger and Klafter 1985). However, this is valid only for sparse targets defined as $\lambda \gg r_v$, where λ is the mean free path between targets, and r_v is the (constant) distance from which a target can be detected. Search efficiency was defined as the inverse of the average path length taken to find a target (in contrast to the average path length taken per Lévy jump). This treatment ignores the challenges of moving large distances in a straight line and the costs in terms of time and neural resources an individual needs to return home (compare Figure 16.3c with the other panels of Figure 16.3 to see the differences in potential areas from which homing must occur following different search strategies), the benefits from information exchange with other individuals, and the overlap in searched area by itself or other foragers.

Searching for Home

After finding a resource, an effective navigation system should be able to direct a foraging animal's path homeward, more or less directly. In principle, only near the termination of the foraging task will an insect require the second phase of searching to begin—the search for home. Hence, the search for food and the search for home correspond to two distinct and noncontiguous phases of foraging. Surprisingly, however, recent evidence suggests that the optimal strategy of searching for home (a unique, stable target) is identical to searching for one of an unknown number of sparsely located targets (Reynolds et al. 2007).

These recent findings are astonishing since, with no starting information, sparsely located targets could be anywhere in the environment. Thus, the *a priori* probability distribution for the goal of navigation should be considered as flat. In contrast, when returning home, an insect is not searching randomly, but rather has information about the path traveled, e.g., through path integration or landmarks. Thus, the *a priori* distribution of home is not flat. Indeed, Cheung et al. (2007a) showed that cumulative sensorimotor noise will result in a positional uncertainty that is distributed according to a bivariate normal distribution. Wehner and Srinivasan (1981) found that when displaced desert ants searched for their nest, the searching distribution roughly followed a normal distribution, i.e., the distribution was not heavy-tailed. The reciprocal argument presented by these authors was that the optimal searching algorithm should follow a similarly shaped distribution.

Thus there appears to be a contradiction between the recent theoretical results (Reynolds et al. 2007) with past experimental and theoretical evidence (Wehner and Srinivasan 1981; Cheung et al. 2007a). By using harmonic radars, Reynolds et al. (2007) reported evidence that honey bees use scale-free, looping search paths to find their hive if displaced far from familiar landmarks. Furthermore, they reported unpublished work that apparently shows that the optimal strategy in searching for a single target consists of looping search paths whose loop lengths are distributed according to an inverse-square law ($\alpha = 1$), much like what has been observed in desert ants (e.g., Wehner and Srinivasan 1981). According to Reynolds et al. (2007), there is an unpublished theory that shows that if the initial search progresses without yield, then eventually, a freely roaming Lévy walk should be adopted. It is unclear, however, whether the analysis for optimality carried out by these authors incorporates the *a priori* probability distribution of home at the beginning of the search. Nonetheless, in view of this and related work, it is important to reanalyze the existing data in order to determine whether the distributions reported in the past may have been generated by looping Lévy walks, which represented the first part of an optimal biphasic searching strategy.

FINDING HOME

In the previous section, I discussed some of the most important mathematical concepts underlying searching theory. The focus was on searching for an unknown resource, like a scout insect might do when searching for food. The analytical similarity to the situation of searching for home following an unexpected/unmeasured displacement was also briefly discussed. The second group of models considered here is a mix of top-down and bottom-up descriptions of navigating homeward. This is a compulsory task for every foraging animal irrespective of whether it performs scouting functions. One might consider this to be even more important than searching for food.

PI MODELS

Path integration (PI), *dead reckoning*, and *vector navigation* are terms used synonymously in the animal navigation literature. Loosely defined, they all refer to the process whereby the path taken from some place *A* to another place *B* is measured from moment to moment, and somehow, the information is combined so that the location of *B* relative to *A* (or vice versa) may be known. In animals, the questions of what exactly is measured from one moment to the next and how the information is combined, as well as the question about the basis for spatial representation, are still subjects of much experimental and theoretical research. Among insects, the ability to carry out PI is best exemplified by species of the desert ant genus *Cataglyphis* (e.g., Müller and Wehner 1988), but there is evidence in other species as well (e.g., Australian desert ant, *Melophorus bagoti*, Narendra 2007a, 2007b; honey bee, *Apis mellifera*, von Frisch 1967; Chittka et al. 1995; Collett et al. 1996; Dyer et al. 2002; a subsocial shield bug, *Parastrachia japonensis*, Hironaka et al. 2007) (Table 16.4).

Although it is difficult to make sense of this maze of existing PI models that are based largely on experimental data from studies with insects (see Table 16.4), some general comments can be made. First, the spectrum of models includes different points of reference, namely, egocentric and allocentric, as well as Cartesian (including Cartesian-like) and polar representations of space. For the current discussion, a Cartesian-like representation is one where the axes may not be at right

TABLE 16.4
Compilation of Published Mathematical and Neural Network Models of Path Integration That Are Based Completely or Mostly on Insect Experimental Data

Author(s)	Brief Description
Jander (1957)	Time-weighted angular mean: $\Phi_n = \sum_{j=1}^{n} t_i \theta_i \Big/ \sum_{j=1}^{n} t_i$
Mittelstaedt and Mittelstaedt (1980, 1982)	Bicomponent model (allocentric Cartesian): $$X_n = \sum_{j=1}^{n} L_i \cos\theta_i \qquad Y_n = \sum_{j=1}^{n} L_i \sin\theta_i$$
Müller and Wehner (1988)	Empirically derived model (allocentric polar): $$\Phi_{n+1} = \Phi_n + L_{n+1} \frac{\kappa(180 - \delta_n)(180 + \delta_n)\delta_n}{R_n}$$ $$R_{n+1} = R_n + L_{n+1}(1 - \delta_n/90)$$ $$\kappa = 4.009 \times 10^{-5} \deg^{-2}$$
Benhamou et al. (1990)	Egocentric Cartesian: $$X'_{n+1} = R_n \cos(\Phi'_n - \Delta_n) - L_{n+1} \qquad Y'_{n+1} = R_n \sin(\Phi'_n - \Delta_n)$$
Gallistel (1990); corrected by Benhamou and Séguinot (1995)	Continuous (egocentric polar): $$\frac{d\Phi'}{dt} = -\frac{d\theta}{dt} + \frac{ds}{dt}\frac{\sin\Phi'}{R}$$ $$\frac{dR}{dt} = -\frac{ds}{dt}\cos\Phi'$$ where $\frac{ds}{dt}$ is the instantaneous speed
Hartmann and Wehner (1995)	Linear and circular neural chains + approximate goniometric functions; based on Müller and Wehner (1988), i.e., on allocentric polar representation
Wittmann and Schwegler (1995)	Phasor (spatial domain) representation of positional vector
Kim and Hallam (2000)	General neural ring (like a circular neural chain) for storage of vectorial information—pseudocode calculation of home vector
Vickerstaff and Di Paolo (2005)	Artificially evolved path integration neural network; "leaky" equivalent of bicomponent model
Merkle et al. (2006)	Continuous (egocentric Cartesian): $$\frac{dX}{dt} = -v + \omega Y$$ $$\frac{dY}{dt} = -\omega X$$ where v is the forward speed and ω the turning rate
Haferlach et al. (2007)	Artificially evolved path integration neural network; heading inputs from simulated polarization-sensitive neurons, and vectors represented in neural ring structure

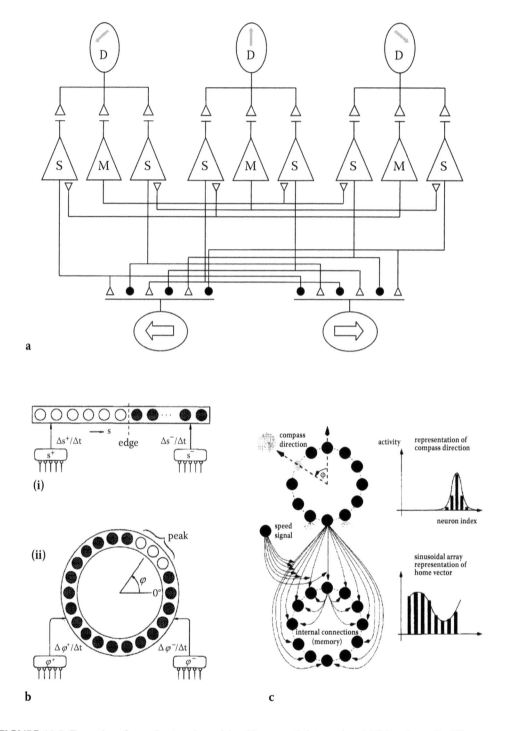

FIGURE 16.5 Examples of neural network models of insect path integration. (a) Direction cells (D) measure headings (perhaps like polarization-sensitive neurons found in many insects do), while memory cells (M) integrate the firing histories of the direction cells and project them to sigmoid filters (S) along with the current heading signals (D) to direct steering (left/right arrows). In this model, only three directional cells were used due to the considerable time consumption of the genetic algorithm. (Redrawn from data in Haferlach et al. 2007.) (b) The neural network model of path integration consisted of two major components: (i) a neural chain and (ii) a neural ring, to represent the polar coordinates of the insect's current position relative to its starting point.

angles and there may be different numbers of axes, e.g., a neural ring (Figure 16.5a and c) where each neuron (at least three; for example, see Figure 16.5a) represents a distinct but fixed allocentric direction. Second, among the explicit neural network models, there is a distinct lack of map-based structures or neural grids, or indeed anything that may resemble a "cognitive map" (coined by Tolman in 1948). This perhaps reflects the relative lack of evidence that insects possess such maps (but see Gould 1986; Gould and Towne 1987; Mizunami et al. 1998). In contrast, mammalian PI models have been based mostly on map-like structures (O'Keefe and Nadel 1978; Samsonovich and McNaughton 1997; McNaughton et al. 2006; Burgess et al. 2007).

It is now clear that for path integration to be effective beyond just a few steps, very specific types of sensory information must be available (Cheung et al. 2007a). The first is a measure of distance. The involved apparatus is termed an odometer, and the respective information that is used could arise from such diverse sources as proprioception (Wittlinger et al. 2007) or optic flow (Esch and Burns 1996; Srinivasan et al. 1996). The second vital piece of information is a measure of direction, specifically the current heading, in absolute terms. It is important to distinguish between a relative or inferred heading and an absolute heading. In principle, an animal might be able to keep track of rotations via proprioceptive, vestibular, or visual information. However, due to unavoidable imperfections in its sensors, noise will accumulate. It can be proven mathematically that cumulative errors in rotational measurements will lead to disastrous consequences, whereas cumulative errors in odometry generally do not (Cheung et al. 2007a). A simplified, intuitive explanation for this result is as follows: angular errors accumulate in a way that spreads the positional uncertainty in angular space. Therefore, the area of positional uncertainty rapidly increases (in fact accelerates) as the navigating insect moves farther and farther from home. In contrast, the linear errors increase positional uncertainty in a more or less linear fashion, and would only be significant if they were much larger than their angular counterparts. Therefore, a successful PI system capable of carrying out medium-range navigation must in fact have direct access to—and make direct use of—allocentric heading information. Theoretically valid examples include a geomagnetic compass, the use of the corrected solar azimuth, or a distant landmark. Therefore, a valid PI model must receive information from some sort of compass *and* an odometer.

Some examples of published PI neural network models are shown in Figure 16.5. The models of Haferlach et al. (2007) and Wittmann and Schwegler (1995) (Figure 16.5a and c, respectively) assumed direct compass and odometric input into the PI system, which is theoretically valid. In contrast, Hartmann and Wehner (1995) described two versions of the neural ring portion of their model (Figure 16.5b, panel ii), where the first measured directions by integrating rotations, while the second used external compass cues. Since the first lacks information from an external compass, we now understand that this version of the Hartmann and Wehner model is theoretically invalid beyond a few steps.

FIGURE 16.5 (CONTINUED) The filled circles represent quiescent neurons, while open circles represent active neurons. Thus, in the neural ring, the activity peak represents the angular component of the insect's position vector, while the edge position represents the linear distance from the starting point of the journey. $\Delta s^{+/-}$ denotes linear displacements, whereas $\Delta \varphi^{+/-}$ denotes angular displacements. (After Hartmann and Wehner 1995. With permission.) (c) The neural network model consisted of two neural rings: a compass layer and an integration layer. Each neuron in the compass layer was assumed to have a sinusoidal tuning curve whose period was exactly 2π radians. The simulated neuronal activity peak would occur at the position on the ring corresponding to an allocentric directional cue, e.g., solar azimuth. At each step along the journey, the insect would add the compass neurons' sinusoidal activity pattern to the integration layer. Due to the properties of sine functions, the integration layer's activity pattern is always sinusoidal. In this way, the amplitude and phase of the resultant sinusoidal activity pattern directly represented the distance and direction of home, respectively. (After Wittmann and Schwegler 1995. With permission.)

VIEW-BASED HOMING

In the past decade, several biomimetic approaches, using robots that performed navigational tasks, have tried to provide an insight into biological navigation behavior (reviewed by Franz and Mallot 2000). A range of view-based homing models has been proposed to explain the way an insect may navigate in visually rich environments. Note that the models are algorithmic in nature and do not necessarily have direct correspondence to insect neuroanatomy or neurophysiology. The underlying principle for most of these models assumes that a scene as viewed from the goal location may be used to determine the course to take according to the currently perceived scene. However, the exact procedures for determining the course vary (Franz and Mallot 2000). For example, Hong et al. (1992) extracted a normalized image intensity profile at the horizon (Figure 16.6a) and landmarks based on the normalized intensity contrast and the local frequency of occurrence. During homing, the angular offset between current and target (home) landmarks was found by the navigating robot, and it moved to reduce the set of angular discrepancies. Lehrer and Bianco (2000) used a different algorithm, which first involved finding image segments that strongly differed from their neighbors, and stored these as landmark templates. As the robot retreated from the starting point, it followed a turn-back-and-look (TBL) path (shown in Figure 16.6b), testing the reliability of potential landmarks. The TBL step was inspired by bees. During homing the robot is steered in a direction that reduces the discrepancy between the relative positions of the landmarks and their sizes (between the starting and current positions). The two previous models were implemented in indoor environments. Zeil et al. (2003) not only took the problem outdoors, but also found a simplified procedure for homing (Figure 16.6c and d). The essence was simply to move to reduce the mismatch between the current view and the target image (stored as a full panorama).

Dale and Collett (2001) used artificial evolutionary algorithms to generate neural networks that controlled virtual "animats" endowed with various simulated sensors and actuators (see Figure 16.7 for an overall anatomical plan; see Figure 16.8 for an overview of the genetic code and simulated evolution). Their main selection criterion for evolution was the precision with which animats reached a visual target (Figure 16.8, lower right panel). Using a simplified two-dimensional virtual world, the authors showed that the animats with greater degrees of freedom of movement performed better. Interestingly, however, the optimal evolved strategy seemed to be one where the target landmark is kept at a fixed retinotopic position. This lends support for the notion that turn-back-and-look (TBL; see Figure 16.6b) and view-based navigation are in fact efficient, perhaps even optimal, strategies under certain circumstances.

The optimal wasp-like animats of Dale and Collett (2001) showed four phases (P1 to P4) in their navigational strategy:

P1: Maintain retinotopicity of landmark, being free to change heading to do so.
P2: Optional phase—used if the animat ended up inside the landmark (hence devoid of landmark information); yaw until current heading activated a sideways motor (coupled to a compass sensor), moving the animat sideways out of landmark.
P3: Pivot around landmark by maintaining retinotopicity of landmark, until certain heading is reached.
P4: Maintain heading direction (compass) and retinotopicity of landmark.

Clearly, this model is based on a gross oversimplification of reality, because it used two dimensions rather than three, simple idealized landmarks, absolute compasses, simplified biomechanics, and thriving evolutionary intermediates. Nonetheless, it seems at least plausible that a number of important messages may be pertinent to insects because of the apparent robustness of the outcomes to starting conditions. First, optimized strategies were multiphasic. Second, the optimally evolved subserving neural networks were anatomically modularized. As Dale and Collett (2001) suggested, this may confer a high degree of adaptability to an otherwise complex system. Third, simple

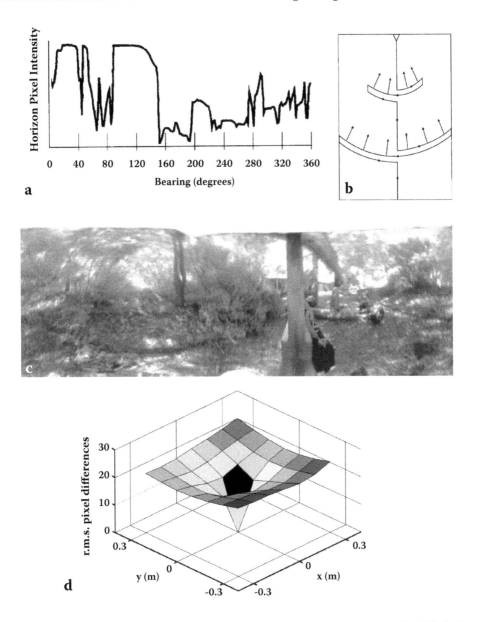

FIGURE 16.6 (a) Location signature expressed as a one-dimensional pixel intensity profile of the horizon at some location. Regions of high local contrast (landmarks) were extracted and compared via a correlation procedure to the home location signature. The most likely home direction could then be calculated. (After Hong et al. 1992. With permission.) (b) The TBL (turn-back-and-look) phase of biomimetic robot navigation. TBL was first reported by Lehrer (1991) for bees following departure from a novel reward site. Two-dimensional image segments were picked using correlation techniques to be used as landmarks for navigation. (After Lehrer and Bianco 2000. With permission.) (c) Panoramic two-dimensional image of a real outdoor environment, which was used for direct pixel-per-pixel comparison in order to calculate the overall image mismatch. (d) The RMS (root mean squared) pixel difference between the current and reference image as a function of the viewpoint displacement from the reference position. It is clear that any reliable form of gradient descent algorithm would allow a navigating agent to reach the starting position (x = 0, y = 0). (c, d: After Zeil et al. 2003. With permission.)

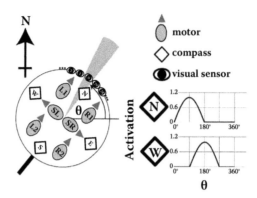

FIGURE 16.7 The animat anatomy: thirty visual sensors output the amount of visual landmark occlusion, four wide-field compass neurons measure current heading (N, E, S, W), and L and R actuators generate a forward thrust at rostral (L1, R1) and caudal (L2, R2) positions. For ant-like animats, lateral movement had to be preceded by differential thrust, which generated torque and hence yaw ($\Delta\theta$). For wasp-like animats, there was the added option of sideways thrust to the left (SL) or right (SR) side, which could translate the animat independently of any other changes in body orientation (θ). The lower right panels correspond to the angular tuning (sensitivity) curves for the north (N) and west (W) compass neurons.

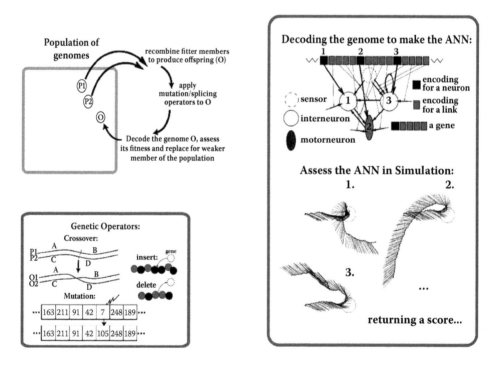

FIGURE 16.8 The animat genome: the neural network for the animat described in Figure 16.7 was encoded in a simulated genome (upper right panel) consisting of designations for neuron type, connectivity, synaptic weights, activation threshold, and time constant used in the recurrent artificial neural network's (ANN) state update equation. The progeny of a population was generated by random pairing of parents, weighted in favor of those with relative navigational success (upper left panel). Genetic reassortment and mutations also occur, including crossover, point mutations, deletions, and insertions (lower left panel). Typically, forty to one hundred thousand generations were required to reach optimal navigational success.

navigational strategies such as using local landmarks as beacons (early part of the journey) or image matching (late part of the journey) to complete the homing process may, in fact, be optimal in more ways than their computational simplicity.

One point to note is the fact that vectorial (honey bees), trail (ants), and other information can (almost certainly) be communicated between insects as a component of foraging. However, it seems unlikely that details about a visual scene may be communicated between social foraging insects. As such, if this is, in fact, a ubiquitous insect navigational strategy, there is still the question of its significance in social, as opposed to solitary, foragers.

DYNAMICAL MAPS

Even though there is at best only weak evidence that insects have neural maps (Gould 1986; Mizunami et al. 1998), one interesting dynamical neural map model of honey bee foraging shall briefly be discussed here. Inoue et al. (2001) presented a computational model consisting of a sequence route map, a cognitive spatial map, and a matching recognition map. The task was to relate landmarks to food sources, and to relate spatial locations to landmarks such that a learned path can be recalled and navigated through a sequence of food sources. The sequence route map was a dynamical attractor neural network, which received sensory input from the current location, and gave the landmark representation of the next location as output. Thus, the cognitive spatial map allowed the physical location of the next landmark to be recalled, and the coincident inputs from current and goal place cells determined the next place cell to be active. The matching recognition map took landmark representation as input, and remained in a limit cycle attractor state, switching between food source representation and landmark representation. On receiving direct food source sensory input, it switched to a point attractor in the food source representation (recognition). Following foraging at one location, a uniform transient firing input (intention) triggered the sequence route map to output the next landmark, thereby starting the entire process again. This is one of few neural network models in the literature that captures the essence of insect foraging at multiple locations during one trip in an orderly, memorized sequence. It shows in principle how a neural network may be used to represent and recall the necessary information for a foraging animal, such as a honey bee, to find a sequence of goals at different locations.

FORAGING AS A GROUP

Stored vectors and snapshots allow an individual insect to revisit a resource of its own accord. Previous sections have reviewed a range of models, which were either based on or inspired by insect biology, that are seemingly capable of performing various aspects of this task. However, as alluded to earlier, social insects present an added level of complexity in that there is the *potential* for useful navigational/foraging information to be gained via interactions with others.

MEASURING SUCCESS

Optimal foraging theory refers to a family of theories developed since the 1960s (Emlen 1966; MacArthur and Pianka 1966; see review by Pyke et al. 1977) aimed at describing the foraging behavior of animals. The theories assume that the fitness of an animal to forage has been optimized through biological selective forces. The currency (Schoener 1971) of fitness has often been assumed to be energy gained through foraging. Over the years, the family of theories expanded to incorporate other biologically relevant parameters, including memory (Pyke 1978), risk-taking strategies (Caraco 1980; Stephens 1981; Stephens and Krebs 1986), and navigational constraints (Cain 1985). Fundamentally, the foraging success of a colony would be measured by the net gain of energy.

Pyke (1978) studied the optimality of foraging of bumble bees. He found experimentally that there was a strong correlation, with zero bias, between a forager's heading on approaching an inflorescence and its heading when leaving the patch. Hence, the optimal searching strategy may be well approximated by an idiothetic directed walk (IDW)* and not a random walk (RW). Pyke (1978) further examined the effect of being able to quantify and remember the total energy obtained at the recent inflorescence. Assuming that low energy yield is due to recent foraging, experiencing a low yield at an inflorescence indicates an increased probability of foraging in a plot with already visited and nectar-depleted flowers. Pyke (1978) found analytically that the optimal solution is to make the correlation between movements (of an IDW) inversely proportional to the energy yield. Hence, by increasing the straightness of its flight trajectory, the optimal insect tends to leave overused areas. By contrast, by visiting nearby resources and changing directions more frequently (or making larger turns at the same frequency), insects stay longer near fresh, nectar-rich inflorescences. This idea can be seen clearly in Figure 16.3a (low correlation) and b (high correlation—note the change in scale reflecting the change in area covered by the different search algorithms).

Cain (1985) developed a computational model of herbivorous insect foraging with a range of variables, including the density and arrangement of food plants. It is interesting to note that the variables considered included navigational parameters like movement lengths, turn angle distributions, effective food detection distance, and foraging mortality rates (an important natural selective force favoring efficient navigational strategies/systems), and so on. The underlying search strategy was an IDW. The author concluded that the effectiveness of search, and hence of foraging, depends heavily on all navigational parameter values, which in turn are responsive to environmental conditions. In essence, a biologically realistic navigation system is likely to be highly dynamic, and adapts to environmental conditions. The converse of this argument is that oversimplification of the navigational system may oversimplify the foraging model as a whole, leading to conclusions that are difficult to interpret.

Following from optimal foraging theory, due consideration must be given to the relative costs and benefits of awaiting information, as opposed to searching based on currently available information. Clearly, the cost-benefit ratio of both strategies depends on how readily one's own navigational system will find a new source of food, or be able to return to an old one. The optimal strategy would also depend on the time before the next scout returns, and how reliable/precise the information provided by it is, as well as on a host of environmental factors affecting resource dynamics. For example, suppose the available information tells an insect there is a low-quality food source far away. The forager may expect a low return on energy if it leaves immediately. However, whether the low return is optimal depends on what future information brings. If a scout returns shortly with information about a high-quality food source nearby, the cost of waiting would likely be outweighed by the benefits. A variety of models that explain how different sociobiological variables affect foraging success at the population level will be briefly discussed below.

DIVISION OF LABOR

For the good of the colony, a social insect forager has to perform one of a number of tasks. Actively searching for a fresh resource, revisiting known resources, following trails/dances, and waiting for information provided by other individuals are all tasks that need to be partitioned among the foraging population to maximize the colony's energy gain. One could think of this as a meta-control of navigation. In other words, even if we could develop biologically realistic models of path integration, searching, view-based homing, and so on, the true behavioral output of a social foraging insect depends critically on when, or in what context, each system is used. For example, when should a

---

* The argument assumes that the probability of finding the goal in any area, if it is actually there, is proportional to the time spent there. To maximize the *a priori* probability of finding the goal at the next step, the location corresponding to the peak of the *a priori* target distribution should be searched. As searching continues without finding the target, the *a priori* target distribution should be adjusted to reject regions where the target was not found—and therefore are increasingly less likely to contain the target. Eventually, the searching distribution would match the original target distribution.

forager activate a search program, as opposed to return to a previously visited site, or as opposed to staying home? One way of looking at this is to consider a social insect colony as a whole, and consider what factors and strategies may be beneficial or harmful at the colony level. Beekman et al. (2007) recently modeled the division of labor in a honey bee foraging workforce. Their model used colony-based differential equations to describe the dynamics of five subpopulations of worker bees: unemployed foragers, recruits, scouts, exploiting foragers, and dancers. While this type of model may be able to describe pertinent aspects of subpopulation dynamics, some significant biological functions may be implicitly assumed to occur, including navigation and communication. It is a challenge to incorporate details of insect navigation into such a description.

INFORMATION QUALITY

Consider the subset of foragers who are active and receive information from others. There is no doubt that communication, such as the waggle dance of the honey bees (see Chapters 6 and 7), the vibrations produced by stingless bee foragers (see Chapter 11), or trail laying in ants (see Chapters 2 and 6) and stingless bees (see Chapter 12), can affect both individual and collective decision making. In Chapter 15 Yoshiyuki Nakamichi and Takaya Arita look at agent-based evolutionary models of foraging ants and the effects of pheromone communication. They present a novel and interesting model that shows that, in the context of foraging, scent trail communication may indeed evolve in order to increase a colony's food intake. Perhaps navigation systems may be another product of coevolution with communication systems in foraging insects.

GENETIC DIVERSITY

Cox and Myerscough (2003) presented a flexible model of foraging by a honey bee colony, which accounted for the effects of individual behavior as determinants of foraging success. The parameters considered per foraging cycle included (1) the visitation rate to a resource, (2) the probability of dancing, (3) the duration of dancing, and (4) the probability of abandonment at a resource. First, the model showed that the individual's probability of abandoning a resource affects the colony's nectar intake nonlinearly. Second, and perhaps more interestingly, for a given number of colonies, and a given spread of any foraging parameter value, it was more advantageous for the entire population (of all colonies) if the variability in parameters was predominantly within each colony, rather than between colonies. In other words, it was better for each colony to have high and low values (of any foraging parameter) simultaneously than for some to have low values and others to have high values. The implication is that high variability in a colony's behavioral parameters, originating from genetic variability, may be advantageous. However, genetic heterogeneity that arises from polyandry, i.e., from different patrilines, decreases the average relatedness of colony members. According to this theory, the increased success of foraging is sufficient to offset the reduced genetic relatedness in preserving eusociality. Furthermore, heterogeneity in genotype, and hence phenotype, may well be sufficient to spontaneously generate division of labor.

Cox and Myerscough (2003) extended the work of Camazine and Sneyd (1991) by explicitly allowing for resource dynamics in their model, thereby increasing its biological plausibility. However, one weakness of both models is the assumption that all sites are found with equal ease. In setting up differential equations, it should be theoretically possible to incorporate a variability term to account for the range of sites at which a foraging colony may attempt to gather resources. Perhaps a more challenging problem is to quantitatively model the genotypic/phenotypic *covariability* of navigational systems and communication characteristics (e.g., honey bee dance threshold), since navigational machinery must clearly be under genetic control as well.

A related problem is the quantification of profitability versus foraging costs. Much of the success—or failure—of foraging strategies depends critically on the accuracy, reliability, and

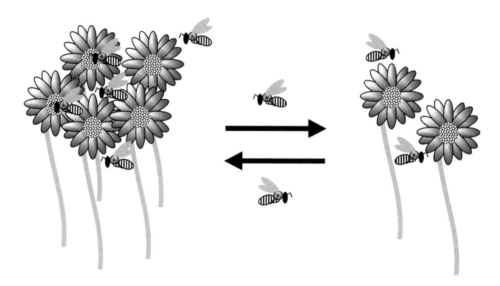

FIGURE 16.9 Schematic representation of the ideal free distribution (IFD). Given two unequal patches of resource (flowers), foragers should distribute themselves dynamically between the two patches in such a way that the number of foragers at each patch is proportional to the available resource.

efficiency of the navigation system. However, like similar models, Cox and Myerscough (2003) implicitly assumed that navigation success is constant and essentially errorless.

RESOURCE QUALITY: TOP-DOWN IFD MODELS

In an *ideal free distribution* (IFD; Fretwell and Lucas 1970) the proportion of animals in a region matches the proportion of resource in that area (Figure 16.9). Such an equilibrated distribution could occur if all animals are equally competitive and movement bears no cost to an animal. Thence, each animal could, in principle, explore the entire habitat and detect the region of maximum resource–to–population density ratio—and move there. If this process were carried out repeatedly, then the IFD would be reached (or at least closely approached). Obstacles to reaching an IFD include dynamic resources, unequal competitiveness, and lack of information due to large distances between resources or to difficulty in analyzing the environment.

Fretwell and Lucas (1970) proposed the IFD to explain the distribution of foragers in an environment with heterogeneously distributed resources. *Ideal* implies perfect knowledge of the environment, while *free* implies zero risk or cost in movement. However, empirical studies found that undermatching often occurs. In other words, regions with more resources tend to be underutilized, while regions with less resources tend to be overutilized (Abrahams 1986; Kennedy and Gray 1993). Therefore, two major types of modified IFDs spawned. One group of theories focused on relaxing the assumption of equal competitors (Sutherland 1983; Parker and Sutherland 1986; Houston and McNamara 1988; Hugie and Grand 1998; Ruxton and Humphries 1999). Unequal competitors implied unequal competitive advantage at one particular site, which in turn implied that the rate of movement between resource areas became species specific. As the models became more sophisticated, it became common practice to use computer simulations to model the evolution of the system toward equilibria states. Starting conditions, such as the number of resource patches, the number of phenotypes of competitors, the assumed perfect/imperfect information, and infinite/finite populations, were often somewhat arbitrary theoretical assumptions/approximations, so the results were difficult to interpret.

Houston and McNamara (1988) used statistical mechanics to calculate the probability of various equilibrium states. In this way, they were able to show that the number of individuals of a particular species on the best resource patch increased with competitive ability, and vice versa. Consequently, relatively few, but more competitive, individuals were at the best patches, while more, but less competitive, individuals were at the poor patches. This led inevitably to undermatching, consistent with observations (Kennedy and Gray, 1993). Yates and Broom (2005) developed a stochastic model of an individual's movements, and found analytical solutions for the equilibrium states. This way, the final steady-state distribution of foragers could be predicted analytically, and rare events, in principle, could be predicted with some confidence.*

Despite the apparent success at explaining certain observations, these models have issues with biological plausibility. Details of the kinetics, such as navigation between resource patches, were largely ignored. According to the differential equations, movements in opposite directions between two resource patches occur completely independently of each other, implying that there is no communication and no memory of previous visits. Furthermore, the rate parameters were defined by the increase in resource intake rate, which would be conferred onto an animal if it were to move. Without specific communication from scouts, this seems highly unlikely since it implies that all individuals have an *a priori* quantitative knowledge of all dynamic patch resources!

The second group of modified IFD theories relaxed the assumption of perfect knowledge (Abrahams 1986; Gray and Kennedy 1994; Spencer et al. 1995, 1996; Ranta et al. 1999). However, many of these approaches still assumed that resource patches were immediately and precisely recognized. A variation of this approach assumed that foragers are uninformed and must acquire information about resource distribution through learning (reviewed by Tregenza 1995; see also Ollason 1987; Ollason and Yearsley 2001). Learning rules themselves suffer from instability problems (Beauchamp 2000). Overall, this latter group of theories and models gives more plausible explanations of the mechanism of how an individual animal is to behave, particularly when moving between resource patches, in order to account for the observed population behaviors.

However, unequal competitors and perceptual limits are likely to coexist in nature, which suggests a further modification of IFD theory to incorporate both factors to further increase its biological plausibility. Koops and Abrahams (2003) did this and combined biological competitiveness and imperfect information with memory in one coherent model. Not surprisingly, competitiveness and information state both affected the probability of a foraging insect's being at a resource patch of a given quality, as well as the rate of leaving that patch. An interesting result was that under the model assumptions, deviations from the IFD were more likely a result of perceptual limits (imperfect information) than variability in competitiveness. While that was an advance as far as understanding population foraging dynamics is concerned, there is still one very important gap in the overall picture. Like Fretwell and Lucas (1970), Koops and Abrahams (2003) still assumed that travel time and costs are negligible (see also Chapter 5).

Higginson and Barnard (2004) found that foraging decisions in honey bees are influenced by the accumulation of wing damage with age. In particular, there appeared to be a tendency for the older animal to accept lower-quality resources (fewer flowers per inflorescence). Higginson and Gilbert (2004) developed an analytical model that predicted the optimal threshold number of flowers needed to accept an inflorescence, based on the accumulated wing damage, and hence on age, of honey bee foragers. They found that the cost of foraging (wing damage) over time altered the theoretically optimal navigational decision, so that wing-damaged foragers should forage at a low-quality inflorescence with greater probability than their wing-intact counterparts.

Trivialization of the process of navigation should be treated with caution. Even though a range of models may be able to capture certain population characteristics, particularly at equilibrium conditions,

---

* Remember that a walk is a mathematical description that does not make any assumption about the actual type of locomotion—terrestrial, aerial, or aquatic.

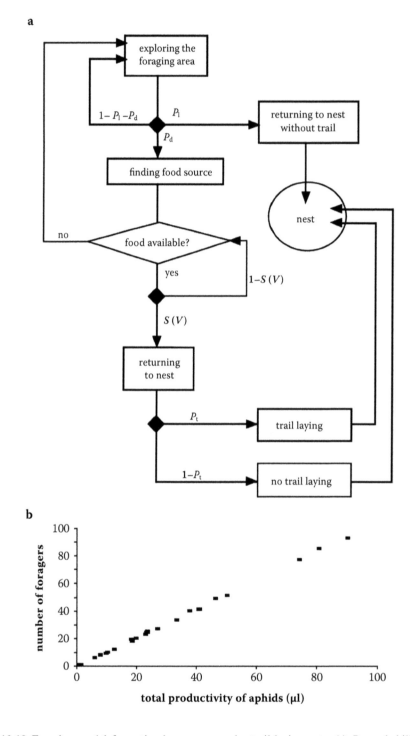

FIGURE 16.10 Foraging model for optimal resource use by trail-laying ants. (a) P_d, probability of food (honeydew) discovery; P_l, probability of leaving the area; $S(V)$, probability of stopping food ingestion; P_t, probability of laying a trail. (b) Theoretical results showing the relationship between the number of foragers recruited to a resource and its quality (proportional to the total volume of honeydew produced by aphids). (After Mailleux et al. 2003. With permission.)

there are issues about the necessary biological mechanisms that would lead to those equilibria. For example, many top-down models assume that a biological system, for example, foragers of a social insect colony, behaves to give the optimal solution to a problem like obtaining energy.

RESOURCE QUALITY: BOTTOM-UP IFD MODELS

Mailleux et al. (2003) developed a model to explain the optimal use of resources by trail-laying ants (Figure 16.10). In their model, the scout ant followed a simple, mechanistically plausible rule to decide in an all-or-none fashion whether to lay a trail. The critical input was whether an ant reached a threshold volume of honeydew intake within a brief exploratory period. It was demonstrated that this simple rule, repeated across the colony of foragers, allowed the recruitment of a proportionate number of ants to the source of a given quality, much like the IFD (see Figure 16.10b). This represents an example where communication allows the relevant information to be distributed across the colony to optimize population foraging, but individuals do not necessarily have all the information. Some interesting features are worth noting in this model. First, at the agent level, events as well as the agent's choices are assumed to be stochastic (represented as black diamonds in Figure 16.10). For example, the probability of food discovery (P_d), the probability of leaving the area (P_l), the probability of stopping food ingestion ($S(V)$), and the probability of laying a trail (P_t) are all random events with a certain probability value. Second, the individual ant does not have information about the rest of the colony.

Dechaume-Moncharmont et al. (2005) considered the importance of time constraints and the transience of food availability (Figure 16.11). Time constraint was twofold. First, they explicitly modeled the duration of food availability. Second, there was a probabilistic term describing the rate of finding food either with or without recruitment information, hence implicitly affecting the time taken to find food. The authors were able to quantify the optimal ratio of proactive foragers (independent searchers) to reactive foragers (recruitment-guided searchers) with a range of time constraints. The most important finding was that across a wide range of time constraint parameters, the optimal strategy was 100% proactive foraging. This result highlights the interdependence of communication, navigation, and resource dynamics during social insect foraging. It is interesting to note that Dechaume-Moncharmont et al. (2005) did not explicitly model navigation mechanisms, which may make foraging dynamics even more unpredictable.

AGENT-BASED MODELING

In the preceding section, bottom-up models, which were able to capture important aspects of insect foraging behavior, were described. The common goal of these models was to capture—with as much biological fidelity as possible—the key determinants of foraging behavior at the individual and colony level.

In response to the need for formal analytical descriptions of social animals, including insects, a number of new, sophisticated agent-based models have been developed. According to Wooldridge and Jennings (1995), collaborative agents should display three main characteristics: autonomy, learning, and cooperation. Autonomy refers to the principle that agents can operate on their own, that is that they have individual states and goals. Learning allows reaction and adaptation to the environment. Cooperation is specific to social animals.

Camazine et al. (2001) considered self-organization in a system to be a slow process in which patterns at a global level emerge from numerous interactions among the lower-level components. Perhaps even more significantly, and in contrast to top-down approaches, they believed that the rules governing interactions among the system's components depend only on local information, without any reference to the global pattern. Clearly, there is a focus on mechanistic plausibility with the assumption that sufficient biological fidelity will lead to the emergence of the actually observed behaviors.

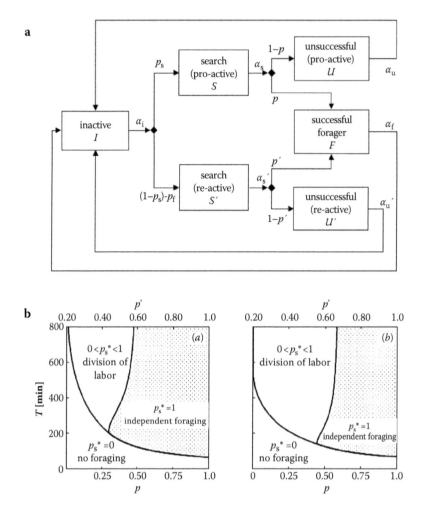

FIGURE 16.11 Foraging model considering the importance of time constraints and the transience of food availability. (a) Flow diagram of the behavioral decision points of each worker. All α values, rate constants; p, P(finding food|scout) [P denotes probability]; p', P(finding food|recruit); p_s, P(scout)/P(scout + recruit); p_f, P(successful recruitment). I, S, S', U, U', and F represent the relative population of workers in the various states. Navigational dynamics are incorporated in the variables p and p'. Foraging as described by the flow diagram continues for a total time T, and the colony's net energy gain is calculated for that period. Thus, the duration of food availability T does not directly affect the individual's behavior. (b) Phase diagram of Dechaume-Moncharmont et al. (2005), which shows empirically determined optimal values of p_s (denoted as p_s^*) for a range of values of p (probability of finding source) and T (duration of source). It more realistically assumes that p and p' are dependent—although the relationship is hypothetical. Notable results include: (1) food sources have to reach a critical level of duration of availability before any foraging is beneficial; (2) waiting for scout information is generally more optimal when probability of finding food is low (finding the source is difficult); (3) if the cost of waiting for information is sufficiently great, workers should become scouts to maximize the colony's energy intake. The only difference between the models used to generate the two phase diagrams shown here lies in the function used to describe the recruitment rate. The default recruitment function (corresponding to the left phase diagram) assumes that the rate of successful recruitment is directly proportional to the number of successful foragers. The alternative recruitment function assumes that large numbers of recruiters will interfere negatively with the recruitment process, so the recruitment rate saturates with the number of successful foragers. (After Dechaume-Moncharmont et al. 2005. With permission.)

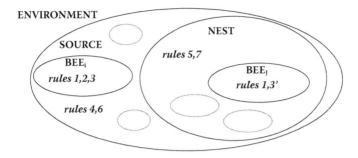

FIGURE 16.12 A simplified membrane model system of a honey bee colony carrying out collective foraging. Rule 1, navigation; rule 2, nectar uploading at source; rule 3, communication of load, position, and identity; rule 4, pass nectar from environment to hive + waggle dance; rule 5, exchange of nectar, then either stay or begin next trip; rule 6, restart searching for nectar; rule 7, attend a dance (with imperfect communication). (After Gheorghe et al. 2001. With permission.)

Membrane Systems

Membrane system models are a construct that is best represented as a hierarchical nested diagram (Venn diagram), as often used in set theory. There are no intersections, and the entire structure is encased by a superset (Figure 16.12). Each set or subset represents an entity such as a bee, the hive, or the food source. It is described as a membrane model because the subsets (membrane structures) may include behaviors like membrane dissolution or exocytosis/endocytosis of single entities (Păun, 2000). Rules of varying priority may be associated to each set or subset, and applied in parallel throughout the model. Membrane system models are attractive for modeling animal colonies because of the inherent parallelism of the description, but the true power of the method may be lost by simplifying each animal to a single membrane unit. Perhaps each animal should better be modeled as a superset of membrane organelles to capture complex behaviors. One critical issue is the synchronous output of this modeling system. A complex description of navigation, for example, does not lend itself to be modeled as an event occurring over a single discrete time step. Perhaps more complex descriptions of the individual animal can circumvent this problem, but that is yet to be seen.

Communicating Stream X-Machines

A communicating stream X-machine (cSXM) represents an advanced version of the finite state machine (FSM) in that there is memory attached to the machine, and transitions are functions of both inputs and memory values. Formally, a cSXM consists of eight variables and may be written as

$$cSXM = \begin{pmatrix} Input, Output, States, Memory, MemoryTransition, \\ StateTransition, InitialState, InititalMemory \end{pmatrix}$$

In other words, cSXM models are agent-based states and rules that describe individual animals, such as a single worker honey bee, or some other entity, such as the nest, a food source, or even the entire habitat. In the example shown in Figure 16.13, a number of important principles of cSXM models are illustrated. First, there are a number of cSXMs in the system, which may include animals, e.g., bees in different foraging states, as well as any relevant inanimate entities/objects, e.g., the environment, nest, food source. Second, each cSXM can have practically unlimited levels of complexity inside (see the cSXM of Figure 16.13 labeled as the "foraging bee"). Third, there is memory, as stated earlier, which is vital to tasks like navigation. Finally, there are intrinsic communication

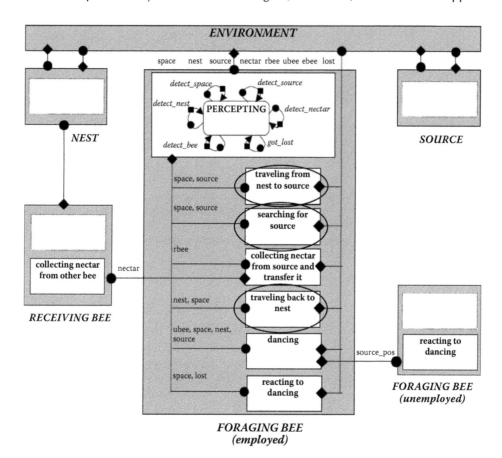

FIGURE 16.13 Communicating stream X-machine (cSXM) models of three individual bees (an unemployed bee = ubee; an employed bee = ebee; and a receiving bee = rbee), the environment, the nest, and a food source. The black-filled circles represent input ports, while the black-filled boxes represent output ports. The ports represent channels explicitly built into the model for individual components to receive/transmit information from/to other components. The components that directly involve medium-range navigation systems are circled. Note that cSXMs are more advanced and flexible variants of finite state machines (FSMs). The main differences are that cSXMs have a memory and their transitions can be described by very complex functions of inputs and memory values (as opposed to direct probabilistic mapping of input to output states). (After Gheorghe et al. 2001. With permission.)

ports that allow the exchange of information, which could as easily represent insect-environment interactions, or indeed insect-insect communication.

An example was reported by Jackson et al. (2004), who modeled Pharaoh's ants as cSXMs in an attempt to reproduce the patterns of formation of pheromone trails between nest and food. Each ant (agent) is a cSXM, but the model ant itself is made up of X-machine components. The authors were able to systematically vary navigation strategies, such as the number of steps to take before actively turning, the magnitude of the turn angle, and in principle, many other parameters, to determine the effect on the trail structure. Although preliminary in nature, the results appear promising in having been able to predict the emergence of pheromone trail networks based on relatively simple rules at the individual level. Perhaps more importantly, it is clear from a modeling perspective that the cSXM construct allows for ease of modification of the model as new data become available. As Jackson et al. (2004) suggested, this method is well suited to incremental developments in modeling in tandem with empirical research.

FUTURE DIRECTIONS

Our understanding of insect navigation is currently incomplete. That is surely one of the main reasons that many models of social insect foraging greatly simplify the details of biological navigation systems. There is perhaps also a naïve hope that somehow the details of navigation may be implicitly but adequately captured by higher-level descriptions of the dynamics of foraging. As we have seen, in models that have incorporated navigation as a core feature, the outcomes do tend to depend on the navigation parameters. Similarly, there exist detailed models of navigation, e.g., path integration (PI) models, which describe a part of the navigation system in isolation. Considering the concepts encapsulated in Figure 16.1 once more, it is evident that the modeling efforts to date have been focused on portions of the overall picture.

Fundamental principles and limits of insect navigation systems are continually being unraveled and understood. It seems at least plausible that some of the missing pieces in understanding social insect foraging lie in the lack of understanding of the way in which navigation affects—and is affected by—foraging strategies and social communication.

At a neural network level, developments in navigation theory are forthcoming. For example, there is continuing progress in the understanding of PI systems. Recently, analytical results have been developed that show that representational systems, such as allocentric Cartesian or egocentric polar, are qualitatively different in terms of tolerance to noise (Cheung 2007; Cheung and Vickerstaff, in preparation). Insights of this type will hopefully guide further development of PI and other navigation models.

At an agent level, interesting avenues might include the application of evolutionary algorithms to study the simultaneous interactions of neural networks subserving navigation, individual-based foraging strategies, and social communication, respectively, with the view of maximizing colony fitness. However, given the complexity of the interactions, this will likely require considerable time, patience, and more experimental data.

Modern biology is developing a growing acceptance and understanding of terms like *emergent behavior* and *collective intelligence*. As such, agent-based formalisms like membrane systems, X-machines, or their variants may well become the staple modeling tool for future workers in the field. With their flexibility, entire social communities, including complex environments, may be modeled. As parallel computing becomes the norm in a modeling laboratory, it is to be expected that emergent behaviors will be routinely predicted and described in a bottom-up manner.

What about top-down approaches? There is certainly scope for continuing development in their own right, but also as a powerful adjunct to the bottom-up approaches. Indeed, it is already common practice to include sets of rules in a model formalism that are based on empirical observations or pure assumptions in describing one portion of the whole system. At the same time, another part of the system may be described by a much more detailed, anatomically and physiologically correct model. Such hybrid models may not appeal to all, but due to their flexibility, they are likely to become more prevalent, especially during the initial stages of modeling complex systems like insect colonies.

Overall, medium-range navigation in socially foraging insects presents an interesting and challenging set of problems to the modeling community, with many new and exciting avenues to explore.

REFERENCES

Abrahams MV. (1986). Patch choice under perceptual constraints: A cause for departures from an ideal free distribution. *Behav Ecol Sociobiol* 19:409–15.

Beauchamp G. (2000). Learning rules for social foragers: Implications for the producer-scrounger game and ideal free distribution theory. *J Theor Biol* 207:21–35.

Beekman M, Gilchrist AL, Duncan M, Sumpter DJT. (2007). What makes a honeybee scout? *Behav Ecol Sociobiol* 61:985–95.

Benhamou S, Sauvé JP, Bovet P. (1990). Spatial memory in large scale movements: Efficiency and limitations of the egocentric coding process. *J Theor Biol* 145:1–12.

Benhamou S, Séguinot V. (1995). How to find one's way in the labyrinth of path integration models. *J Theor Biol* 174:463–66.

Bovet P, Benhamou S. (1988). Spatial analysis of animals' movements using a correlated random walk model. *J Theor Biol* 131:419–33.

Bracher C. (2004). Eigenfunction approach to the persistent random walk in two dimensions. *Physica A* 331:448–66.

Brcich RF, Iskander DR, Zoubir AM. (2005). The stability test for symmetric alpha-stable distributions. *IEEE T Signal Process* 53:977–86.

Burgess N, Barry C, O'Keefe J. (2007). An oscillatory interference model of grid cell firing. *Hippocampus* 17: 801–12.

Byers J. (2001). Correlated random walk equations of animal dispersal resolved by simulation. *Ecology* 82:1680–90.

Cain M. (1985). Random search by herbivorous insects: A simulation model. *Ecology* 66:876–88.

Camazine S, Deneubourg J, Franks NR, Sneyd J, Theraulaz G, Bonabeau E. (2001). *Self-Organisation in Biological Systems*. Princeton, NJ: Princeton University Press.

Camazine S, Sneyd J. (1991). A model of collective nectar source selection by honey bees: Self-organization through simple rules. *J Theor Biol* 149:547–71.

Caraco T. (1980). On foraging time allocation in a stochastic environment. *Ecology* 61:119–28.

Cheung A. (2007). Theory and neural network nodels of insect navigation. PhD thesis, Australian National University, Canberra.

Cheung A, Zhang SW, Stricker C, Srinivasan MV. (2007a). Animal navigation: The difficulty of moving in a straight line. *Biol Cybern* 97:47–61.

Cheung A, Zhang SW, Stricker C, Srinivasan MV. (2007b). Animal navigation: Pitfalls and remedies. In *Proceedings of the 63rd Annual Meeting of ION*, Cambridge, MA, pp. 270–79.

Chittka L, Kunze J, Shipman C, Buchmann SL. (1995). The significance of landmarks for path integration of homing honey bee foragers. *Naturwissenschaften* 82:341–42.

Cole BJ. (1995). Fractal time in animal behaviour: The movement activity of *Drosophila*. *Anim Behav* 50: 1317–24.

Collett TS, Baron J, Sellen K. (1996). On the encoding of movement vectors by honeybees. Are distance and direction represented independently? *J Comp Physiol A* 179:395–406.

Cox MD, Myerscough MR. (2003). A flexible model of foraging by a honey bee colony: The effects of individual behavior on foraging success. *J Theor Biol* 223:179–97.

Dale K, Collett TS. (2001). Using artificial evolution and selection to model insect navigation. *Curr Biol* 11: 1305–16.

Dechaume-Moncharmont FX, Dornhaus A, Houston AI, McNamara JM, Collins EJ, Franks NR. (2005). The hidden cost of information in collective foraging. *Proc R Soc Lond B* 272:1689–95.

Dyer FC, Gill M, Sharbowski J. (2002). Motivation and vector navigation in honey bees. *Naturwissenschaften* 89:262–64.

Emlen JM. (1966). The role of time and energy in food preference. *Am Nat* 100:611–17.

Esch HE, Burns JE. (1996). Distance estimation by foraging honeybees. *J Exp Biol* 199:155–62.

Franz MO, Mallot HA. (2000). Biomimetic robot navigation. *Rob Autonom Syst* 30:133–53.

Fretwell SD, Lucas HL. (1970). On territorial behavior and other factors influencing habitat distribution in birds. I. Theoretical development. *Acta Biotheor* 19:16–36.

von Frisch K. (1967). *The Dance Language and Orientation of Bees*. Cambridge, MA: Harvard University Press.

Gallistel CR. (1990). *The Organization of Learning*. Cambridge, MA: Bradford Books/MIT Press.

Gheorghe M, Holcombe M, Kefalas P. (2001). Computational models of collective foraging. *Biosystems* 61: 133–41.

Gould JL. (1986). The locale map of honeybees: Do insects have cognitive maps? *Science* 232:861–63.

Gould JL, Towne WF. (1987). Evolution of the dance language. *Am Nat* 130:317–38.

Gray RD, Kennedy M. (1994). Perceptual constraints on optimal foraging: A reason for departure from the ideal free distribution. *Anim Behav* 47:469–71.

Haferlach T, Wessnitzer J, Mangan M, Webb B. (2007). Evolving a neural model of insect path integration. *Anim Behav* 15:273–87.

Hartmann G, Wehner R. (1995). The ant's path integration system: A neural architecture. *Biol Cybernet* 73: 483–97.

Higginson AD, Barnard CJ. (2004). Accumulating wing damage affects foraging decisions in honeybees (*Apis mellifera* L.). *Ecol Entomol* 29:52–59.

Higginson AD, Gilbert F. (2004). Paying for nectar with wingbeats: A new model of honeybee foraging. *Proc R Soc Lond B* 271:2595–603.

Hironaka M, Tojo S, Nomakuchi S, Filippi L, Hariyama T. (2007). Round-the-clock homing behavior of a subsocial shield bug, *Parastrachia japonensis* (Heteroptera: Parastrachiidae), using path integration. *Zoolog Sci* 24:535–41.

Hong J, Tan X, Pinette B, Weiss R, Riseman M. (1992). Image-based homing. *IEEE Contr Syst* 12:38–45.

Houston AI, McNamara JM. (1988). The ideal free distribution when competitive abilities differ: An approach based on statistical mechanics. *Anim Behav* 36:166–74.

Hugie DM, Grand TC. (1998). Movement between patches, unequal competitors and the ideal free distribution. *Evol Ecol* 12:1–19.

Inoue S, Kashimori Y, Yang Y, Liu H, Hoshino O, Kambara T. (2001). Neural processes of self-organization of control system for foraging trips of honeybees. *Neurocomputing* 38–40:1325–34.

Jackson AL, Humphries S, Ruxton GD. (2004). Resolving the departures of observed results from the ideal free distribution with simple random movements. *J Anim Ecol* 73:612–22.

Jackson D, Holcombe M, Ratnieks F. (2004). Coupled computational simulation and empirical research into the foraging system of Pharaoh's ant (*Monomorium pharaonis*). *Biosystems* 76:101–12.

Jander R. (1957). Die optische Richtungsorientierung der Roten Waldameise (*Formica rufa*). *Z vergl Physiol* 40:162–238.

Kareiva PM, Shigesada N. (1983). Analyzing insect movement as a correlated random walk. *Oecologia* 56: 234–38.

Kennedy M, Gray RD. (1993). Can ecological theory predict the distribution of foraging animals? A critical analysis of experiments on the ideal free distribution. *Oikos* 68:158–66.

Kim DE, Hallam JCT. (2000). Neural network approach to path integration for homing navigation. In *Proceedings of the 6th International Conference on Simulation of Adaptive Behavior*. Cambridge, MA: MIT Press, pp. 228–35.

Kitching R. (1971). A simple simulation model of dispersal of animals among units of discrete habitats. *Oecologia* 7:95–116.

Koops MA, Abrahams MV. (2003). Integrating the roles of information and competitive ability on the spatial distribution of social foragers. *Am Nat* 161:586–600.

Lehrer M. (1991). Bees which turn back and look. *Naturwissenschaften* 78:274–76.

Lehrer M, Bianco G. (2000). The turn-back-and-look behaviour: Bee versus robot. *Biol Cybernet* 83:211–29.

Levandowsky M, Klafter J, White BS. (1988). Swimming behavior and chemosensory responses in the protistan microzooplankton as a function of the hydrodynamic regime. *Bull Mar Sci* 43:758–63.

MacArthur RH, Pianka ER. (1966). On optimal use of a patchy environment. *Am Nat* 100:603–9.

Mailleux AC, Deneubourg JL, Detrain C. (2003). Regulation of ants' foraging to resource productivity. *Proc R Soc Lond B* 270:1609–16.

Mandelbrot B. (1983). *The Fractal Geometry of Nature*. New York: Freeman & Company.

McNaughton BL, Battaglia FP, Jensen O, Moser EI, Moser MB. (2006). Path integration and the neural basis of the 'cognitive map.' *Nature Rev Neurosci* 7:663–78.

Merkle T, Rost M, Alt W. (2006). Egocentric path integration models and their application to desert arthropods. *J Theor Biol* 240:385–99.

Mittelstaedt ML, Mittelstaedt H. (1980). Homing by path integration in a mammal. *Naturwissenschaften* 67: 566–67.

Mittelstaedt H, Mittelstaedt ML. (1982). Homing by path integration. In Papi F, Wallraff HG (eds.), *Avian navigation*. Berlin: Springer Verlag, pp. 290–97.

Mizunami M, Weibrecht JM, Strausfeld NJ. (1998). Mushroom bodies of the cockroach: Their participation in place memory. *J Comp Neurol* 402:520–37.

Müller M, Wehner R. (1988). Path integration in desert ants, *Cataglyphis fortis*. *Proc Natl Acad Sci USA* 85: 5287–90.

Narendra A. (2007a). Homing strategies of the Australian desert ant *Melophorus bagoti*. I. Proportional path-integration takes the ant half-way home. *J Exp Biol* 210:1798–803.

Narendra A. (2007b). Homing strategies of the Australian desert ant *Melophorus bagoti*. II. Interaction of the path integrator with visual cue information. *J Exp Biol* 210:1804–12.

O'Keefe J, Nadel L. (1978). *The Hippocampus as a Cognitive Map*. Oxford: Clarendon.

Ollason JG. (1987). Learning to forage in a regenerating patchy environment: Can it fail to be optimal? *Theor Pop Biol* 31:13–32.

Ollason JG, Yearsley JM. (2001). The approximately ideal, more or less free distribution. *Theor Pop Biol* 59: 87–105.

Parker GA, Sutherland WJ. (1986). Ideal free distributions when individuals differ in competitive ability: Phenotype-limited ideal free models. *Anim Behav* 34:1222–42.

Păun G. (2000). Computing with membranes. *J Comp Syst Sci* 61:108–43.

Pearson K. (1905). The problem of the random walk. *Nature* 72:294.

Pearson K. (1906). Mathematical contributions to the theory of evolution. XV. Drapers' Company Research Memoirs. Reproduced in Pearson ES (ed.), *Karl Pearson's Early Statistical Papers* (1948). Cambridge, MA: Cambridge University Press.

Pyke GH. (1978). Optimal foraging: Movement patterns of bumblebees between inflorescences. *Theor Pop Biol* 13:72–98.

Pyke GH, Pulliam HR, Charnov EL. (1977). Optimal foraging: A selective review of theory and tests. *Quart Rev Biol* 2:137–54.

Ranta E, Lundberg P, Kaitala V. (1999). Resource matching with limited knowledge. *Oikos* 86:383–85.

Rayleigh JWS. (1905). The problem of the random walk. *Nature* 72:318.

Rayleigh JWS. (1919). On the problem of random vibrations and random flights in one, two and three dimensions. *Philos Mag* 37:321–47.

Reynolds AM, Smith A, Menzel R, Greggers U, Reynolds DR, Riley JR. (2007). Displaced honey bees perform optimal scale-free search flights. *Ecology* 88:1955–61.

Ruxton GD, Humphries S. (1999). Multiple ideal free distributions of unequal competitors. *Evol Ecol Res* 1: 635–40.

Samsonovich A, McNaughton BL. (1997). Path integration and cognitive mapping in a continuous attractor model. *J Neurosci* 17:5900–20.

Schoener TW. (1971). Theory of feeding strategies. *Annu Rev Ecol Syst* 2:369–404.

Schuster FL, Levandowsky M. (1996). Chemosensory responses of *Acanthamoeba castellani*: Visual analysis of random movement and responses to chemical signals. *J Eukar Microbiol* 43:150–58.

Shlesinger MF, Klafter J. (1985). Comment on "Accelerated diffusion in Josephson junctions and related chaotic systems." *Phys Rev Lett* 54:2551.

Skellam JG. (1951). Random dispersal in theoretical populations. *Biometrika* 38:196–218.

Spencer HGM, Kennedy M, Gray RD. (1995). Patch choice with competitive asymmetries and perceptual limits: The importance of history. *Anim Behav* 50:497–508.

Spencer HGM, Kennedy M, Gray RD. (1996). Perceptual constraints on optimal foraging: The effects of variation among foragers. *Evol Ecol* 10:331–39.

Srinivasan MV, Zhang SW, Lehrer M, Collett TS. (1996). Honeybee navigation en route to the goal: Visual flight control and odometry. *J Exp Biol* 199:155–216.

Stephens DW. (1981). The logic of risk-sensitive foraging preferences. *Anim Behav* 29:628–29.

Stephens DW, Krebs JR. (1986). *Foraging Theory*. Princeton, NJ: Princeton University Press.

Sutherland WJ. (1983). Aggregation and the ideal free distribution. *J Anim Ecol* 52:821–28.

Tolman EC. (1948). Cognitive maps in rats and men. *Psychol Rev* 55:189.

Tregenza T. (1995). Building on the ideal free distribution. *Adv Ecol Res* 26:253–307.

Vickerstaff RJ, Di Paolo EA. (2005). Evolving neural models of path integration. *J Exp Biol* 208:3349–66.

Viswanathan GM, Afanasyev V, Buldyrev SV, Murphy EJ, Prince PA, Stanley HE. (1996). Lévy flight search patterns of wandering albatrosses. *Nature* 381:413–15.

Viswanathan GM, Buldyrev SV, Havlin S, da Luz MGE, Raposo EP, Stanley HE. (1999). Optimizing the success of random searches. *Nature* 401:911–14.

Wehner R, Srinivasan MV. (1981). Searching behaviour of desert ants, genus *Cataglyphis* (Formicidae, Hymenoptera). *J Comp Physiol A* 142:315–38.

Wittlinger M, Wehner R, Wolf H. (2007). The desert ant odometer: A stride integrator that accounts for stride length and walking speed. *J Exp Biol* 210:198–207.

Wittmann T, Schwegler H. (1995). Path integration—A neural model. *Biol Cybernet* 73:569–75.

Wooldridge M, Jennings NR. (1995). Intelligent agents: Theory and practice. *Knowl Eng Rev* 10:115–52.

Wu HI, Li BL, Springer TA, Neill WH. (2000). Modelling animal movement as a persistent random walk in two dimensions: Expected magnitude of net displacement. *Ecol Model* 132:115–24.

Yates GE, Broom M. (2005). A stochastic model of the distribution of unequal competitors between resource patches. *J Theor Biol* 237:227–37.

Zeil J, Hofmann MI, Chahl JS. (2003). Catchment areas of panoramic snapshots in outdoor scenes. *J Opt Soc Am* 20:450–69.

Social Insects and the Exploitation of Food Sources
Concluding Thoughts

Stefan Jarau and Michael Hrncir

> *"...think of an insect colony as a diffuse organism, weighing anywhere from less than a gram to as much as a kilogram and possessing from about a hundred to a million or more tiny mouths. It is an animal that forages ameba-like over fixed territories..."*
>
> **Edward O. Wilson, 1971**

In social insects, foraging individuals gather food that predominantly supports the rearing of the brood and satisfies the energetic demands of nonforaging individuals. Thus, the foragers' activities serve the benefit of the colony rather than their personal profit. Food collectors sometimes even "sacrifice" their individual performance and gather food at suboptimal rates (Núñez 1982; Roces and Núñez 1993), in exchange for a faster return to the colony and the activation of additional foragers (Roces and Núñez 1993; see also Chapter 14). At the individual level, therefore, social insect foraging does not necessarily fit into "simple" concepts, theories, and models of "optimal foraging" (Pyke 1984; Stephens and Krebs 1986) or "social foraging" (Giraldeau and Caraco 2000), which have been developed for vertebrates (or vertebrate groups) that optimize their food-collecting activity in accordance with their individual needs. Yet, the optimality of food exploitation in social insects has to be considered at the level at which natural selection takes place—that is, the entire colony.

In analogy to higher organisms that are composed of numerous single cells, colonies of eusocial insects have frequently been considered "superorganisms" (Wilson 1971; Seeley 1989; Moritz and Fuchs 1998), in which the workers adopt the role of the somatic cells. In contrast to multicellular organisms, however, colonies of eusocial insects have no (or only a weak) central control that regulates and coordinates the behavior of their subunits. The decision of workers to start, continue, or stop a specific task, e.g., foraging, is based on local stimuli, such as the interactions with nestmates or cues from the environment (Seeley 1989; Biesmeijer et al. 1998; see also Chapter 2). Thus, because of the lack of a central control and the individual decision-making by foragers, it is quite astonishing that social insect societies are able to coordinate their foraging forces for an optimized exploitation of food sources at the colony level.

The amazing coordination among the individuals is a result of self-organization processes through interactions, feedback loops, and local information transfer among workers. The principle of self-organization was originally introduced in physical and chemical systems in order to explain the emergence of macroscopic patterns (e.g., rippled sand dunes) from interactions at the microscopic level (e.g., between sand grains). This concept has been adapted to describe how complex collective behaviors in social insect colonies arise from interactions between individuals that follow rather simple behavioral rules* (Bonabeau et al. 1997). In their book, Scott Camazine and his co-authors

---

* It should be kept in mind, however, that the principle of self-organization largely ignores the complexity and variability that occurs at the individual level in social insects, and allows them to adapt to changing environments (see also Chapter 6).

give a fascinating and detailed overview of self-organization in biological systems (Camazine et al. 2003), including the emergence of foraging patterns in eusocial insects. Using nectar source selection by honey bees, trail formation in ants, and swarm raids of army ants as examples, the authors demonstrate that colonies are able to efficiently solve particular foraging problems in spite of the fact that the single individuals do not possess a global knowledge of the "problem" at hand.

An important prerequisite for the emergence of self-organized patterns is the interaction among the subunits of a system—in regard to our book's topic, the workers of a social insect colony. The two basic modes of such interactions are positive and negative feedback loops (Camazine et al. 2003), in which the activity of an individual is affected by the performance of others. Robin Moritz and Stefan Fuchs (1998) defined these feedbacks as follows: "Workers are engaged in a specific activity and by doing so, modify the local conditions to exceed the stimulus response thresholds of other workers (positive feedback). Also the opposite can be possible: the activity of a worker reduces the stimulus below the response threshold which then stops being involved in that task" (Moritz and Fuchs 1998, p. 12). Hence, positive feedback between subunits, in general, promotes changes within a system, whereas negative feedback helps to stabilize certain processes (Camazine et al. 2003). Such feedbacks might be particularly adaptive for the emergence of colony-level foraging patterns when single workers have different thresholds for a particular task. A threshold variability among the individuals of a colony has, for example, been demonstrated to be related to caste differences (e.g., the different response thresholds for recruitment displays found in the different castes of *Pheidole pallidula* ants; Detrain and Pasteels 1991), or to a genetic variability among individuals (e.g., foraging thresholds in pollen- or nectar-collecting honey bee workers; Fewell and Page 2000; Pankiw and Page 2000).

Probably the best-studied example concerning the importance of feedback loops for the emergence of self-organized patterns from the social insect world is the dance communication of the honey bee (*Apis mellifera*). Here, food-collecting workers inform unemployed foragers about the quality of a discovered food source through the liveliness of their waggle dance (duration and rate of waggle runs performed in the hive; Seeley et al. 2000; see also Chapter 6). This information from the dance, as well as gustatory information received through trophallactic interactions with an active forager (see Chapter 10), influences the decision of unemployed foragers whether to initiate foraging or to stay in the hive. In case the intensity of the information exceeds the foraging threshold of the bees, they will start collecting at the advertised food source. Thus, bit by bit, the colony's foraging force will increase at this food source through an accumulation of attractive signals (positive feedback). Yet, if too many foragers have been activated, the nectar influx might exceed the food processing capacity of a colony. In this case, in order to maintain the food exploitation at an efficient level, the foraging activity should be downregulated, or alternatively, nectar processing should be enhanced. Here, information about the colony's nectar processing status is provided to the foragers through the time they need to unload their crop (Seeley 1995). In case nectar foragers, which return from profitable food sources, experience long search times to find food receivers, they are stimulated to perform tremble dances. On the one hand, these dances activate additional nectar receivers and, consequently, increase the food processing capacity of the colony (positive feedback). On the other hand, the tremble dances have an inhibitory effect on waggle dances (negative feedback) so that no further foragers are activated and, consequently, the nectar-collecting rate is stabilized (Seeley 1995). As can be seen from this example, the emergence of feedback loops is directly related to the capacity of the individuals to process relevant information obtained both from other individuals and from the local environment. Thus, the management of information at the individual level constitutes the core of self-organized processes at the colony level.

Several models exist that attempt to link the pattern of task performance at the individual and colony levels in order to explain division of labor. These models are based on both external and internal factors, such as ontogeny, genetic components, motivation, experience, environmental factors, and probably most important, the flow of information between the individuals of a colony (for a comprehensive review see Beshers and Fewell 2001). It is apparent that the organization of foraging in honey bee hives is largely based on *direct* social cues (interaction rate between individuals), such as encounters between nestmates and the concomitant direct exchange of chemical cues, which have also been demonstrated to be a major channel of information transfer in ant colonies (Gordon 1999; see also Chapter 2). However, as Claire Detrain and Jean-Louis Deneubourg illustrate in Chapter 2, foraging efficiency in ants is also considerably enhanced by the detection of *indirect* social cues, which can be by-products of the ants' activity or excreted compounds left behind. The use of indirect cues to estimate the activity or density of nestmates and foreign workers allows an asynchronous transfer of information. Basically, this means that such cues can contain the recent history of frequentation of an area. Furthermore, this information is available immediately, without requiring much sampling effort. Ants that use such indirect cues and accordingly tune their foraging and recruitment activities can increase their food collection efficiency at both the individual and colony level.

When looking at the adaptiveness of foraging strategies one must also consider sensory biases and learning abilities of the involved individuals. As demonstrated by Nigel Raine and Lars Chittka in Chapter 1, bumble bee workers from eight species (and from eight subspecies within *Bombus terrestris*) show a strong innate color preference for violet-blue, which therefore seems to be a phylogenetically ancient trait. Since violet and blue flowers also contain high nectar rewards in a variety of habitats, this innate preference likely is advantageous in terms of foraging success of bees (an innate violet and blue preference has also been demonstrated for *Apis mellifera*). Indeed, *B. terrestris* colonies with workers that showed a stronger preference for violet over blue flowers in a laboratory assay also harvested more nectar per unit of time in a natural habitat where violet flowers on average contained twice as much sugar reward than blue flowers (see Chapter 1 for details). Likewise, there was a positive correlation between the learning speed of the workers from different bumble bee colonies in the lab and their foraging performance (nectar intake per unit of time) under field conditions, indicating that learning speed is closely associated with colony fitness (Chapter 1).

Certainly, learning (or more generally spoken, experience) strongly influences the foraging decisions of insects. In Chapter 7, Ellouise Leadbeater and Lars Chittka show that bumble bee workers, which have used the holes cut into broad bean flowers by previous nectar robbers, thus acting as secondary robbers, are more likely to cut a hole at the right place (the base of a flower's corolla) when subsequently visiting intact flowers than robbing-inexperienced bees. How well this transmission of social information—the use of holes cut by others—may guide the decision-making of social insect foragers is illustrated by the authors' observation that broad bean plants grown in the open may stay intact for weeks, but then, within hours, every flower can be robbed. Workers of certain ant species can learn the route to a food source by following experienced leaders (Chapter 7), and flower-visiting solitary and social bees learn to interpret scent marks left by previous visitors in accordance with the reward level of a food source—scent-marked flowers, which are depleted after a single visit, are avoided, whereas the odors left behind at *ad libitum* food sources attract the bees (see Chapters 9, 12, and 13 for honey bees, stingless bees, and bumble bees, respectively). The adaptive value of using scent marks to reject recently visited flowers is that the bees save time and energy that would otherwise be spent with landing and handling a depleted resource. Instead, they quickly fly on until rewarding flowers are encountered, thereby increasing their foraging efficiency. Scent marks left at *ad libitum* sugar solution feeders, on the other hand, are a reliable cue that indicates food, and bees learn to associate them with a reward, just as they do with other sensory cues (Menzel and Müller 1996). The duration of the repellent effect of the scent marks encountered on flowers may reflect their nectar secretion rate, and it seems possible that bees learn the appropriate concentration of scent mark as the threshold for rejection of a particular flower species (Chapter 13). It is tempting to speculate that a mechanism similar to ants' ability to use area markings might be at work: from the model presented by Claire Detrain and Jean-Louis Deneubourg in Chapter 2 we learn that ants may be able to estimate the original amount of area markings as well as the time elapsed since their deposition when such markings consist of at least two compounds with different decay rates (which they certainly do, in both ants and bees). In such a case, the ratio between the compounds will change over time, and the age of a marking can be deduced by the relative concentrations of its single compounds. Within the perception thresholds for the single compounds and the compounds' ratio, respectively, an insect can potentially use this information. However, one problem with this idea in terms of scent marks left at flowers by visitors is that their original compositions already vary considerably at the time of deposition, depending on the species the depositor belongs to.

Due to the overlap of foraging ranges of neighboring social insect colonies with similar food requirements, both inter- and intraspecific competition for resources can be high. Not surprisingly, it is apparently widespread (although little studied) that foragers use information about the presence of other individuals—nestmates and non-nestmates alike—as an indication for food. As illustrated by Judith Slaa and William Hughes in Chapter 8, there are different ways such information can influence the foraging decisions of insects. The logic behind this is simple: where individuals are present, food may be as well. Thus, a food-searching worker can make use of visual or olfactory cues passively provided by nestmates or workers of foreign colonies, or even of other species, when deciding where to forage. Such local enhancement might facilitate the formation of feeding groups (if attraction is to nestmates) or simply the finding of food. Its use, however, will only be adaptive if a food source is large enough to support several foragers, or if the attracted individual can drive away the original exploiter. But where an insect feeds (or recently has fed), a resource might as well be depleted. In such cases, foragers are repelled by cues that indicate the presence of others, as in the example of repellent scent marks used by the flower-visiting bees mentioned above. Likewise, foragers may also avoid food sources occupied by more aggressive heterospecific foragers (see Chapters 3 and 8). Thus, local inhibition can lead to the avoidance of depleted resources and may help to evade direct competition. Several studies have demonstrated that workers of a particular wasp species avoid occupied feeders in areas where they co-occur with larger, more aggressive species, but are attracted to collecting foragers at sites where the dominant competitors are absent. As pointed out by Robert Jeanne and Benjamin Taylor in Chapter 3, this strongly suggests that the respective responses to other workers at food sources are learned, and depend on the local species composition. Here, the parallels to what is known about the resource-dependent interpretation of food-marking odors by bees (see Chapters 12 and 13) are noteworthy.

Where communication occurs in a network environment, there is also potential for parasitism on *signals* (as opposed to cues) released by individuals in order to recruit their nestmates. Non-nestmate "eavesdroppers" can capitalize on this information by following it to the indicated food source. Since the recruitment signals of others are a reliable signpost pointing toward the location of food, following them may be a beneficial strategy to find resources. So far, eavesdropping in social insects has been observed only in scent trail recruiting ants and a stingless bee, which likely is due to the nature of the longer-lasting chemical signals involved.

Another source of useful information is natural food odors, which are learned by (bee) workers and can play an eminent role in their foraging decisions. Our knowledge about the underlying mechanisms has advanced enormously during the past years (for honey bees see Chapters 9 and 10, for stingless bees, see Chapter 12; for bumble bees, See Dornhaus and Chittka 1999, 2004). Experienced foragers associate, and orient toward, the scents of rewarding flower species. In Chapter 9, Judith Reinhard and Mandyam Srinivasan show for honey bees that scent-associated memories of food sources can become quite complex, including not only a food source's nectar reward, but also its shape and color, its location in space, and even the navigational route to reach it. These scent-food associative memories can be recalled by the presentation of the respective odor in the hive, which quickly reactivates experienced foragers to revisit known food locations, thereby enhancing the foraging efficiency of the colony.

In Chapter 10, Walter Farina and Christoph Grüter emphasize the importance of trophallactic interactions within honey bee hives as a means of social information sharing. Recruited bees associatively learn food odors during trophallaxes with nectar-distributing foragers, and later arrive at flowers with the same odor in the field. Interestingly, the frequency of short trophallactic contacts that likely do not provide much nectar, but information about the collected food, increases with increasing food quality. This corroborates the idea that information about food quality is transmitted via trophallaxis. Since food receivers also offer the food to other bees (second-order receivers), the incoming nectar is rapidly distributed among hive bees of all ages. Consequently, potential cues present in the collected food could rapidly be spread within the colony and serve as "global information" about the characteristics of currently available food sources. Yet, trophallaxis offers information not solely to the workers inside the nest, but also provides foragers with chemosensory information about resources collected by others. Based on such information, a forager may reevaluate its own food source and decide whether to continue exploiting it, abandon foraging, or switch to another, more rewarding food source.

Similar to the rapid propagation of information about food characteristics by means of trophallaxis in honey bee hives, the transport chains observed in leaf- and grass-cutting ants, through which a cut fragment is transported into the nest sequentially by a number of workers, may serve to transfer information about food quality and odor. As argued by Flavio Roces and Martin Bollazzi in Chapter 14, this could be of particular importance during the early phase of recruitment following the discovery of a food source. Dropped fragments along the trail may, for instance, act as cues that enable outgoing foragers to learn the odor of the harvested plant by contacting them, and then quickly locate it at the forage site. Thus, at the colony level, the increased rate of social information transfer may offset the suboptimal food delivery at the individual level. Analogous to these transport chains in leaf- and grass-cutting ants, foragers of *Macrotermes subhyalinus* termites also transfer a piece of food to other workers while returning to the nest, and then rapidly run back to the food source (Traniello and Leuthold 2000). While this behavior might speed up the food collection process, the possibility of information transfer via food cues during these interactions might well be worth consideration.

Of course, the foraging efficiency of social insect colonies depends not only on the use of information that emanates from cues. A large variety of true signals, which are deliberately emitted by an individual to communicate with its nestmates, have evolved among the social insects, e.g., trail pheromones in ants (Hölldobler and Wilson 1990; see also Chapters 2 and 6), stingless bees (see Chapter 12), and termites (e.g., Pasteels and Bordereau 1998; Traniello and Leuthold 2000); Nasonov pheromone released at food sources by honey bee workers; or their dance communication inside the nest (von Frisch 1965, 1967; Seeley 1995; see also Chapters 6 and 9). The social wasps, however, might be an exception here. As illustrated in Chapter 3 by Robert Jeanne and Benjamin Taylor, cues seem to be the primary information used to make foraging decisions in this group of insects. To date, only a single wasp species has been described to use true signals (an attractant pheromone) during the recruitment of nestmates (see Chapter 3 for details).

The sheer fact that pheromone communication in connection with food collection evolved in such a vast number of social insects makes it quite plain that it must have an adaptive value. However, testing this assumption in nature is difficult. One possibility could be the application of a phylogenetic approach, i.e., studying related species with and without pheromone communication under the same conditions (as was done in studies on innate color preferences of bumble bees; see Chapter 1). This, however, might be critical, since several biological characteristics may vary between species apart from their mode of recruitment communication. Here, the modeling approach can provide valuable insights. In Chapter 14, Yoshiyuki Nakamichi and Takaya Arita present an agent-based evolutionary model, in which the agents ("ant foragers") had to collect food items in a grid environment and store them in their nest. The agents could either deposit pheromone(s) or not, and their behavior was determined by neural networks, which evolved in the framework of a genetic algorithm and in accordance with their foraging activities. The outcome of this model not only affirmed that adaptive pheromone communication indeed evolves, but also demonstrated that agents with pheromone-inhibited sensors collected fewer food items than agents that could make use of pheromone information. Thus, the evolution of pheromone communication indeed increases foraging efficiency.

Whether social information use is adaptive in nature depends on the actual situation and environmental conditions under which the foragers act. In Chapter 6, Madeleine Beekman and Audrey Dussutour demonstrate that the range of flexibility of social insects to react to changing conditions (e.g., availability of rich food sources) largely depends on the recruitment mechanism they apply, which, in turn, likely is tuned to the environment in which they evolved. Experiments with honey bees, for instance, have revealed that recruitment through dances indeed leads to an increased foraging efficiency (nectar intake rate) when floral resources are clustered (tropical habitat or Californian winter), but not in landscapes with dispersed flowering plants (Central Europe and Spain, or Californian summer and autumn) (see Chapters 1 and 7). The stingless bees comprise many different species, most of which require similar resources. Competition for food, therefore, might be particularly high, and recruiting bees should rather aim at quickly mobilizing the available foraging force than spending much time with communication at pinpointing a particular food patch. This may be a good reason why a precise, but time-consuming, referential communication mechanism inside the nest, comparable to the dances of honey bees, has not evolved in this group of social insects. As Michael Hrncir shows in Chapter 11, stingless bee foragers indeed quickly activate the available foraging force by jostling as well as by means of vibrations and sounds emitted inside the nest, when returning from rich food sources. Once activated, the recruits' decision on where to collect is largely influenced by field-based cues and signals, which are predominantly of a chemical nature (see Chapter 12).

The colony foraging patterns that emerge from the decisions of individual workers and from their interactions with each other are also influenced by the environment itself. In all ecosystems, the food availability is subject to changes in the course of the year. In addition to food abundance, abiotic factors such as temperature and relative humidity can vary on a daily and seasonal basis and, consequently, influence the foraging activity of social insects. In Chapter 4, Flávia Medeiros and Paulo Oliveira present a case study on *Pachycondyla striata* ants from a semideciduous forest habitat in southeastern Brazil, where summers are rainy and hot and winters dry and cooler. Their study shows that the ants adjust their foraging behavior to the different seasons, with higher activities and larger home ranges during the rainy season. The increased foraging activity corresponds to a threefold increase in the abundance of potential litter-dwelling prey, mainly arthropods, which might well explain the observed differences. Changing foraging areas of single *Pogonomyrmex* harvester ant colonies were also revealed during a detailed long-term study in the Arizona dessert (Gordon 1999). In addition to seasonal variations, the age, and thus the developmental stage of a colony, significantly affected the behavior of *Pogonomyrmex* workers, specifically in relation to territorial conflicts and the return rate to places where foreign ants were encountered (Gordon 1999). In analogy to the findings in ants, several species of termites also change their foraging patterns according to the season. Some species increase the collection of leaf litter with rainfall, whereas others show their peak foraging activity during the dry season (Bignell and Eggleton 2000). Without repeating all details here, these examples clearly show how foraging decisions of social insect workers, as well as the resulting foraging patterns at the colony level, are influenced by biotic (e.g., competition, natural enemies, availability of prey, age of a colony) and abiotic (e.g., temperature, precipitation, relative humidity) factors alike. One certainly has to keep this in mind when evaluating the foraging behavior of social insect workers and their activities at the colony level, especially in studies that are carried out during relatively short periods of time.

Since eusocial insects are central place foragers, the foraging range is a fundamental aspect of their ecology. Whereas it is relatively easy to investigate how far ants walk when searching and collecting food (Gordon 1999; Chapter 4), insects that forage on the wing cause a bit of a problem, because it proves difficult to follow their flight path. Not surprisingly, therefore, our knowledge about foraging ranges in bees is poor, the main exception being the honey bees, which encode the distance of visited food sources in the length of the waggle run performed by a returned forager inside the nest, where it can be measured by an observer (e.g., von Frisch 1965, 1967; Seeley 1995; Oldroyd and Wongsiri 2006). In Chapter 5, Dave Goulson and Juliet Osborne review what we know about foraging ranges in bumble bees. This problem was approached by a variety of methods, which all have their drawbacks. Nevertheless, all available studies indicate that bumble bees do not necessarily forage at the food plants closest to the nest. Goulson and Osborne argue that the most likely explanation for the low worker densities near their nest could be that flight to and from food patches is relatively cheap compared to the energy spent within a patch, and that intracolony competition would be quite high if many workers stayed close to the nest (see Chapter 5). This assumption seems to be supported by the evidence that the foraging ranges of different bumble bee species are positively correlated with their respective colony sizes (adult worker populations). In this regard, it would be very interesting to study the third group of eusocial bees in the tribe Apini, the stingless bees. Since this bee group shows a great variability in colony and worker sizes among the several hundred known species, they would be well suited to test the assumption brought forth for bumble bees by Goulson and Osborne. Unfortunately, the few data available on flight distances (rather than foraging distances) of stingless bees do not yet allow coming up with a meaningful interpretation of the distribution of foragers in relation to their colony characteristics.

One critical aspect that certainly has to be considered when thinking about foraging ranges of insects is how they navigate, which, in turn, may influence not only their ability to find a resource without prior knowledge of its location or to return to a known feeding site, but also their ability to find their way back home after food uptake. Thus, in order to fully understand how an insect manages to carry out its day-to-day foraging activities, knowledge about the underlying medium-range navigation mechanisms is required. Apart from the empirical studies carried out on a few species, different mathematical and neural network models have been developed to address this question, many of which are reviewed by Allen Cheung in Chapter 16. As argued by the author, models dealing with navigation systems often treat them in isolation (i.e., not in a foraging context), whereas most of the available models of social insect foraging greatly simplify the details of biological navigation systems (which likely is due to our incomplete understanding of insect navigation). But still, from the models presented in Chapter 16 we can learn, for instance, how the competitive ability and information state of a foraging insect can affect the probability of its presence at a resource patch, how a forager's age might influence its threshold to accept a resource of lower quality (e.g., in honey bees due to higher flight costs caused by accumulating wing damage), or how it may find its way back home according to the navigation strategy applied (e.g., path integration, view-based homing, map-based navigation). It is clear that much of the success of a foraging strategy depends critically on the accuracy, reliability, and efficiency of the underlying navigation system. In social insects, foragers can also potentially gain useful navigational (foraging) information via interactions with nestmates.

To come to a close, we think that the most significant recent advance in our understanding of the mechanisms underlying the foraging decisions and recruitment behavior of social insect workers, and ultimately foraging patterns of social insect colonies, is the recognition that, aside from true signals, learning and the context-dependent interpretation and use of cues play a central role. It is also clear that foraging patterns at the colony level are not fixed features, but underlie a strong influence of environmental as well as current colony conditions, and that the workers' ability to search the nest's surroundings for resources and to successfully return with food depends on their navigation mechanism(s). We may still be far from providing a complete picture of how all these mechanisms work together to produce the collective, self-organized foraging processes observed in social insect societies. Yet, we expect that the combination of empirical ecological and behavioral studies with theoretical models will prove to be a fruitful approach to get us further toward a profound understanding of how social insects exploit food sources.

REFERENCES

Beshers SN, Fewell JH. (2001). Models of division of labor in social insects. *Annu Rev Entomol* 46:413–40.
Biesmeijer JC, van Nieuwstadt MGL, Lukács S, Sommeijer MJ. (1998). The role of internal and external information in foraging decisions of *Melipona* workers (Hymenoptera: Meliponinae). *Behav Ecol Sociobiol* 42:107–16.

Bignell DE, Eggleton P. (2000). Termites in ecosystems. In Abe T, Bignell DE, Higashi M (eds.), *Termites: Evolution, Sociality, Symbioses, Ecology*. Dordrecht, The Netherlands: Kluwer Academic Publishers, pp. 363–87.

Bonabeau E, Theraulaz G, Deneubourg JL, Aron S, Camazine S. (1997). Self-organization in social insects. *Trends Ecol Evol* 12:188–93.

Camazine S, Deneubourg JL, Franks NR, Sneyd J, Theraulaz G, Bonabeau E. (2003). *Self-Organization in Biological Systems*. Princeton, NJ: Princeton University Press.

Detrain C, Pasteels JM. (1991). Caste differences in behavioural thresholds as a basis for polyethism during food recruitment in the ant, *Pheidole pallidula* (Nyl.) (Hymenoptera: Myrmicinae). *J Insect Behav* 4:157–76.

Dornhaus A, Chittka L. (1999). Evolutionary origins of bee dances. *Nature* 401:38.

Dornhaus A, Chittka L. (2004). Information flow and regulation of foraging activity in bumble bees (*Bombus* spp.). *Apidologie* 35:183–92.

Fewell JH, Page RE. (2000). Colony-level selection effects on individual and colony foraging task performance in honeybees, *Apis mellifera* L. *Behav Ecol Sociobiol* 48:173–81.

von Frisch K. (1965). *Tanzsprache und Orientierung der Bienen*. Berlin: Springer.

von Frisch K. (1967, second printing 1993). *The Dance Language and Orientation of Bees*. Cambridge, MA: Harvard University Press.

Giraldeau LA, Caraco T. (2000). *Social Foraging Theory*. Princeton, NJ: Princeton University Press.

Gordon DM. (1999). *Ants at Work: How an Insect Society Is Organized*. New York: The Free Press.

Hölldobler B, Wilson EO. (1990). *The Ants*. Cambridge, MA: Belknap Press of Harvard University Press.

Menzel R, Müller U. (1996). Learning and memory in honey bees: From behavior to neural substrates. *Annu Rev Neurosci* 19:379–404.

Moritz RFA, Fuchs S. (1998). Organization of honey bee colonies: Characteristics and consequences of a superorganism concept. *Apidologie* 29:7–21.

Núñez JA. (1982). Honeybee foraging strategies at a food source in relation to its distance from the hive and the rate of sugar flow. *J Apic Res* 21:139–50.

Oldroyd BP, Wongsiri S. (2006). *Asian Honey Bees: Biology, Conservation, and Human Interactions*. Cambridge, MA: Harvard University Press.

Pankiw T, Page RE. (2000). Response thresholds to sucrose predict foraging division of labor in honeybees. *Behav Ecol Sociobiol* 47:265–67.

Pasteels JM, Bordereau C. (1998). Releaser pheromones in termites. In Vander Meer RK, Breed MD, Espelie KE, Winston ML (eds.), *Pheromone Communication in Social Insects*. Boulder: Westview Press, pp. 193–215.

Pyke GH. (1984). Optimal foraging theory: A critical review. *Annu Rev Ecol Syst* 15:523–75.

Roces F, Núñez JA. (1993). Information about food quality influences load-size selection in recruited leaf-cutting ants. *Anim Behav* 45:135–43.

Seeley TD. (1989). The honey bee colony as superorganism. *Am Sci* 77:546–53.

Seeley TD. (1995). *The Wisdom of the Hive—The Social Physiology of Honey Bee Colonies*. Cambridge, MA: Harvard University Press.

Seeley TD, Mikheyev AS, Pagano GP. (2000) Dancing bee tunes both duration and rate of waggle-run production in relation to nectar-source profitability. *J Comp Physiol A* 186:813–19.

Stephens DW, Krebs JR. (1986). *Foraging Theory*. Princeton, NJ: Princeton University Press.

Traniello JFA, Leuthold RH. (2000). Behavior and ecology of foraging in termites. In Abe T, Bignell DE, Higashi M (eds.), *Termites: Evolution, Sociality, Symbioses, Ecology*. Dordrecht, The Netherlands: Kluwer Academic Publishers, pp. 141–68.

Wilson EO. (1971). *The Insect Societies*. Cambridge, MA: Belknap Press of Harvard University Press.

Index

9 780367 385613